Cavitation and Associate

T0256401

Dmitry A. Biryukov

Thermophysics Dept.
Moscow Power Engineering Institute
Moscow, Russian Federation

Denis N. Gerasimov

Thermophysics Dept.
Moscow Power Engineering Institute
Moscow, Russian Federation

Eugeny I. Yurin

Thermophysics Dept.
Moscow Power Engineering Institute
Moscow, Russian Federation

CRC Press
Taylor & Francis Group
Boca Raton   London   New York

CRC Press is an imprint of the
Taylor & Francis Group, an **informa** business

A SCIENCE PUBLISHERS BOOK

Cover credit: Cover illustrations reproduced by kind courtesy of the authors.

First edition published 2022
by CRC Press
6000 Broken Sound Parkway NW, Suite 300, Boca Raton, FL 33487-2742

and by CRC Press
2 Park Square, Milton Park, Abingdon, Oxon, OX14 4RN

ISBN: 978-0-367-42528-9 (hbk)
ISBN: 978-1-032-24384-9 (pbk)
ISBN: 978-0-367-85349-5 (ebk)

DOI: 10.1201/9780367853495

Typeset in Times New Roman
by Radiant Productions

Preface

Almost every physics book begins from a statement similar to 'this phenomenon is a common thing in nature'. Cavitation is a slightly different matter: actually, this is rather a technological derivative than a natural phenomenon. Of course, it is possible to observe cavitation in natural intense flows; however, the appearance of cavitation can be expected rather in technological devices, e.g., in pumps.

Thus, the statement that cavitation has artificial nature as a rule is not too ambiguous. It is a child of technology, and we may say that it is a dangerous child. While scientists try to investigate cavitation, engineers try to avoid it—almost everywhere, except for some special cases also discussed in this book. Mainly, this book is aimed at the first group: this is a story about the nature of cavitation, not about the technical applications, exactly, technical misapplications of cavitation.

As for the physical description of cavitation, this is quite a usual story: the overall characteristics of this phenomenon are well-understood; however, the description of almost every detail meets serious problems. Not to venture too far from the technical side of things: we know that cavitation contains studies on void cavern collapses; thus, it is clear that these collapses lead to damages of solid surfaces nearby. That's right—oh, that's right! But can we describe the exact mechanisms of the destruction? From this point onward, we have divarication: many ways to destroy a solid surface with collapsing voids can be proposed. For instance, there are theories stating that enormously high temperatures can be reached in a collapsing bubble; if so, one may see the traces of thermal destruction of the surface. Does this phenomenon take place? We will discuss it in this book. The physics of cavitation is a much more complicated subject than a brief analysis of the most distinct sides of it.

From your student days, you probably remember that in general physical courses cavitation is much more frequently mentioned than discussed or studied systematically. Short questions met simple answers, but here simplicity does not seem to be a perfect thing. 'Why we see a two-phase flow here, without any heating?—Oh, that's because of cavitation.' 'What was the cause of the damage to that screw?—Cavitation, obviously.' 'What happened to that turbine?—A sudden cavitation...' Then, what is cavitation? Many books were written to answer this question, and many books will be written in the future.

Bubbles can grow in a boiling liquid due to heating, but they may do the same during cavitation because of a drop in pressure. The physics of this process seems to be similar to boiling only at first glance, and the theoretical description of cavitation is quite different indeed. Cavitation occurs in special conditions: at low or even initially negative pressures, at very short time scales and/or on very small spatial scales. Studying cavitation, it is useful to remember about physical limitations of well-known physical laws; we can easily oversee the boundary of their corrections. The simplest example: at the initial stage of cavitation, the size of a cavitating bubble may be less than (or comparable to) a mean-free path of a molecule of gas; thus, any macroscopic description of the gas phase is ruined.

The next interesting feature of cavitation: it is a multi-faceted phenomenon. It shows an expert its different sides and any of its projections looks pretty complete. Moreover, any of these sides may take away the expert's whole life, because of the complexity of every such projection. However, this does not mean that this projection gives a complete set of information about cavitation as a whole.

We discuss this matter and its consequences later, but initially here we have to answer the initial questions that must arise from every reader.

Cavitation is an old phenomenon, and many books were written about it to date. So, the first question that must be answered: what is the point of writing a new book devoted to cavitation? Isn't there enough yet?

Our book is devoted to special problems concerning cavitation, it combines classical elements of the science of cavitation and also some special topics that are related to cavitation, such as electrization, light emission, etc. Actually, we think that special topics are the main ones in this book; however, they cannot be suspended in a vacuum. Ad exemplum: in order to consider electrization and light emission from a bubble, we have to deliberate the basic problems of bubble dynamics during acoustic cavitation. Generally, to bundle the themes altogether, we need a consistent uniform explanation of all the problems concerning cavitation, from the very beginning. We cannot refer to other books all the time; this way is inconvenient for readers and confusing for authors.

The next important question is: for whom is this book written?

This book is aimed at all researches, technicians and students who are interested in cavitation. This is not a textbook in a strict meaning of this term: it does not contain special exercises, homework assignments, etc. However, we believe that the systematic description presented in this book makes it useful for everyone who possesses the basic knowledge of physics or engineering. Any university course on general physics is enough to understand the material of this book; for sure, we include appendixes with the basic information from thermo- and hydrodynamics.

As we described above, cavitation is a complex problem, and, with time, some approaches, some definite points of view on this science were formed under the influence of particular 'projections' mentioned previously. These distinct approaches led to special forms of the 'cavitation science' that can be studied almost independently.

The first view on cavitation that can be easily recognized in scientific literature is based on a very serious practical meaning of cavitation. Technical applications demand certain answers—or, at least, firmly proved recommendations—for technological devices like pumps and turbines. The exact physical theory of such systems is very complicated, and it cannot provide definitive answers for the problems; thus, experiments and empiric recommendations play a significant role for such technological objects.

The second set of problems concerning cavitation is a dyad 'acoustic cavitation—dynamics of a cavitating bubble'. This 'projection' is rather a part of pure physics—experimental, as well as numerical and theoretical. In numerical calculations and theoretical studies, this class of problems necessarily includes the Rayleigh equation as the key object of investigations. In experiments, this is a wide area for research of single-bubble cavitation in an acoustic field.

We should also highlight another—slightly forgotten—view on cavitation. Up to the end of the XX century, every physical problem was analyzed with the whole apparatus of theoretical physics. This way—the way of exact solutions for approximations of fundamental equations—led to tremendous, complicated formulas for physical parameters. Sometimes these formulas provide good agreement with experiments, sometimes they do not. However, in any case these 'strict' approaches may clarify the qualitative physical picture of the phenomena, if the reader can break through the mathematical thorns—like velocity potentials, for example. For common reasons, today such approaches are not very popular in physics (except for areas of physics that consist of approximately 100% mathematics, such as the string theory); despite this fact, we suppose that the benefits of this approach outweigh the difficulties. Especially, this approach is useful to understand the hydrodynamic instabilities on the liquid–vapor interface.

If we take a careful look at many books about cavitation, we may easily discover the types of 'projections' of cavitation the book is devoted to. In our book, we try to hold all the elements of this mosaic together, adding new elements that seem to be even more impressive than the 'pure', classical cavitation.

Thus, in agreement with all the previous—and, possibly, superfluous—explanations, we can briefly describe the structure of this book.

The first two chapters are devoted to the everyday, technological, practical applications of cavitation. In the first chapter, we depict the problem with simple or even domestic traits. Knocking the bottom of a bottle is a popular exercise on the internet or even in movies (see the underrated pulp-noire 'Gutshot Straight' as an example), and this is a good illustration for the power of cavitation, especially with the most impressive trick: destroying a bottle by removing the cork. The second chapter contains the basic (and not only the basic) 'technological approach' for cavitation. Beyond many other reasons, this material is useful for the qualitative, physical understanding of the problem.

Chapter 3 is devoted to the (surprisingly) rarely observed question in books about cavitation—the phenomenon of negative pressures. The fact that intermolecular forces may provide stretching tenses in media deserve special attention. Of course, such terms as 'liquid rupture' and 'breakaway tension'—for a fluid!—must be well-understood as the basic physical principles of cavitation. Moreover, physical processes during the rupture may also shed light on other problems associated with cavitation.

The fourth chapter is a tribute to the classical, pure physics, including its theoretical part. We should not forget about physical theories even in the middle of a pile of engineering problems. Hydrodynamics may say much more about a cavitating flow than could be imagined at first glance. Possibly, this chapter is the most difficult and requires some deeper knowledge in mathematics. However, we believe that it is impossible to describe the dynamics of a cavitating cavern without discussions of instabilities of the interphase boundary.

Chapter 5 describes the very important side effect of cavitation—the hydraulic shock. This phenomenon can play a big role for a cavitating liquid near a solid surface; however, this process can also be considered separately. This chapter adds specific information about the interaction between a cavitating fluid and a solid surface and, what is even more important, about the physical property of a liquid in this process.

Chapter 6 describes experiments on the acoustically induced cavitation. Probably, this is the purest appearance of cavitation, which is very convenient for investigation. On a common complicated background of cavitation (described in the first two chapters), this subject looks like an island of calm. The mystical disturbance on this island—sonoluminescence—will be considered in the last chapter.

In chapter 7, we turn our attention to the popular aspect of modern applied physics—bubble dynamics. Despite the fact that there exists whole books devoted to this problem, we depict here the nuances that are usually missed: boundary conditions on the bubble wall; in truth, for these problems the final answer strictly (and sometimes evidently) depends on the statement of conditions. The interaction of two oscillating bubbles is considered there too; this effect can be easily observed during acoustic cavitation described in Chapter 6.

Chapter 8 is an additional part again. Spontaneous electrization of liquid is an intriguing physical process, which can be observed in various conditions. Triboelectricity and triboluminescence as scientific disciplines are far from their final state, despite the fact that electrization caused friction was discussed even by the ancient Greeks: this is one of the oldest scientific problems. Today we know much more about the electric properties of bubbles that can sometimes turn into sources of non-equilibrium plasma.

The last chapter contains the critical analysis of physical theories of sonoluminescence—the long-build problem in physics. Different explanations of sonoluminescence are based on various physical principles; their fundamentals are given in eight previous chapters. We discuss these theories as well as their experimental roots; actually, these procedures were provided in special books, but we do it in another manner. As for the experimental fundament of sonoluminescence, this is probably the weakest part of this scientific problem. There may be many theories which may explain the light emission from a fluid under a mechanical impact; the question is—which exact mechanism is responsible for the given glow. This question leads to the counter question: what are the properties of the 'given glow'? Interpretations of experimental spectra are ambiguous, so the theories on this effect are still 'hanging in the air'. Observing the theories of sonoluminescence, the reader may come to the conclusion that any theory cannot describe this phenomenon correctly; in this case, we have to find another one. In short, sonoluminescence is the most intriguing point concerning cavitation, and all other aspects must be studied in preparation for it.

All chapters are consequential, and we think that they can be read as a tutorial for cavitation and associated problems. We start from the common observation of the investigated subject, move from simple questions to the complex ones and finish on an undiscovered problem; this way satisfies the classical structure of teaching a university course.

However, we may propose a different way to read our book for those who are interested only in some separate topics of cavitation. They may divide the whole text into 'sub-books', each of these sub-books considers some special traditional sub-themes. In this manner, we may consider Chapters 1–2, 4–5 and 6–7 as separate parts of the common problem of cavitation: Chapters 1&2—for the overall description and technological applications, Chapters 4&5—for physical hydrodynamics, Chapter 6&7—for bubble dynamics during acoustic cavitation. Chapters 3, 8 and 9 are special items that correspond to the 'associated phenomena' part of the title. The last chapter is sort of above all other matters: it covers the most marvelous and mystical manifestation of cavitation. We suppose that even a student may read all sub-books with minimal inquiries into other material. However, to understand the special chapters, all other material must be studied beforehand. Of course, we think that all the sides of cavitation should be explored.

Finally, we must emphasize that cavitation is an open area for physicists. We feel that a long way is needed to obtain precise physical descriptions for all the aspects of this complex phenomenon.

Dmitry A. Biryukov
Denis N. Gerasimov
Eugeny I. Yurin

Contents

CHAPTER 1

Morphology of Cavitation

1.1 Cavitation vs. boiling

1.1.1 Inside a vessel

We believe that everybody knows what the process of boiling is. Everyone has seen bubbles inside a pot, and, surely, we all know that boiling is the process of formation of vapor bubbles in a vessel due to heating. Exactly, the whole cycle includes the creation of a bubble on a solid heated surface, the bubble's detachment from that surface, and its floating toward a liquid surface.

Thus, boiling is a phase transition—the process in which liquid turns into vapor. The verb in this phrase is very important: in a successful case, i.e., when the amount of heat is sufficient, all the mass of the given substance transforms from the liquid phase into the vapor one. In an intermediate case, when we have a partial dose of the required heat at our disposal, the final result will be a mixture of liquid and vapor.

Boiling is a part of our life. If this book was dedicated to boiling, this chapter would be made out of enormous parts, and the reader would be submerged into stories about boiling in natural conditions, boiling of cryogenic liquids, boiling in technological devices and power equipment; being tired of this information, the reader would come to the kitchen to make a cup of tea where, to his consternation, he would realize that when he put down a teapot on a heater (a gas burner or something), boiling takes place again…

But this book is not about boiling. This is a story about the poor unhappy sister of boiling–cavitation.

Usually, cavitation is explained in the same manner as boiling, but without mentioning heating. This phenomenon can be explained as the formation of bubbles causing a drop in pressure—for example, in the modification of a pot presented in Fig. 1.1.1. In such a 'modified' pot we may—if we are sufficiently strong—pull up the lid which is hermetically connected to it, by decreasing the pressure inside. As a result, we will see the process of bubble formation, approximately, like in a regular pot during heating. By the way, actually, you can conduct this experiment without a special pot in domestic conditions by using a syringe and obtain the same result (see Fig. 1.1.2 and the next section).

We can say directly that the process in the regular pot is called boiling, while the phenomenon in the modified pot is called cavitation. Explaining things by their origin, one may formulate that there are two ways to summon bubbles in a fluid: by heating (boiling), or by decompression (cavitation). From this point of view, the difference is clear. On the other hand, explaining things by their consequences, one may muss up these processes: indeed, all the components of the problem are the same—a liquid, and bubbles inside it, and, moreover, the increase of the amount of the gaseous phase with time. Are boiling and cavitation absolutely different in their nature, or they have many similarities too? Indeed, this is a very interesting matter, and the answer to this question is less important than the process of obtaining it.

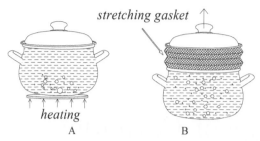

Fig. 1.1.1. Two devices for creating the gaseous phase: a pot and a modified pot. The modified pot represents a hermetically closed construction, where the lid is connected to the vessel by a stretching gasket (similar to bellows). In the first case, we observe bubbles that are formed on the bottom of the heated pot; then bubbles float to the liquid surface. In the second case, when we lift up the lid of the pot, bubbles are formed everywhere throughout the volume.

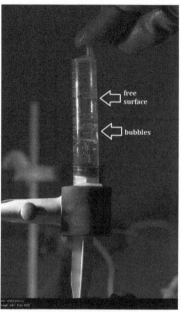

Fig. 1.1.2. A domestic way to observe cavitation. Take a syringe, fill it with water, close the nozzle, pull out its piston, and watch the two-phase perturbations inside it. See details in the next section.

We will consider this problem below, but before we go further, we may answer an eventual naive question: can we boil a liquid in the modified pan? Actually, that depends on the meaning of the term 'to boil' used in the sentence. In the domestic language, 'to boil' means to 'warm up' a liquid to a temperature high enough for this liquid (or at least a part of it) to turn into vapor. In this sense—no, we cannot boil a liquid with the second device presented in Fig. 1.1.1: water will not be heated in this way. Temperature will stay the same; exact, it will even slightly decrease during the process (see below).

However, the scientific terminology defines boiling without mentioning heating—from the thermodynamic point of view, boiling is a liquid–vapor phase transition. Full stop. If so, to figure out similarities and differences between boiling and cavitation, we need to consider other thermodynamic notions first. It is possible that some part of liquid has been boiled during cavitation—let us figure it out.

1.1.2 Inside a thermodynamic diagram

Boiling is a physical process which is studied worthily; it is examined for a long time with a great energy (energy of scientists, we mean, not the total power output of their setups). In this way, during

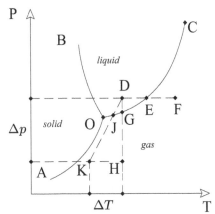

Fig. 1.1.3. The state diagram of a substance.

the treatment of boiling, many results were obtained; first of them, of course, was brought from thermodynamics, which explains many experimental results. Let us follow this way, picking up cavitation on the side.

We consider a thermodynamic diagram—the graph where we plot points corresponding to the pressure of substance p and its temperature T. Each point corresponds to some stable thermodynamic phase—the aggregate state of this substance: solid, liquid or gas[1]—except for the lines which separate the phases (saturation curves): on these, we have two corresponding phases simultaneously, in equilibrium; see Fig. 1.1.3. There are three such saturation curves on this diagram related to three possible types of phase transition: the melting curve OB (corresponding to the solid–liquid transition), the boiling curve OC (liquid–gas), and the sublimation curve OA (solid–gas). In other words, when we read 'gas' on some area on the diagram, this means that the stable aggregate state corresponding to that area is gas; for a short time, another thermodynamic phase may exist at such external parameters—this is important to understand the following matters.

Note that the boiling curve ends at a certain point—this is the so-called critical point C; at parameters higher than the parameters of the critical point, phase transition is impossible since there are no separate phases; in that region, the substance is the so-called 'supercritical fluid', which is neither a liquid nor a gas.

All three phases may coexist at a single point O—the triple point.

Along any saturation curve, the Clapeyron–Clausius equation can be written as:

$$\frac{dp}{dT} = \frac{r_{12}}{T(v_2 - v_1)}, \tag{1.1.1}$$

where r_{12} is the latent heat (the difference of enthalpies of two phases, see Appendix A) of the phase transition $1 \rightarrow 2$, and v_1 and v_2 are the specific volumes (per a unit of mass) of phases. For instance, for boiling $q_{12} > 0$ at $v_2 > v_1$ (the specific volume of vapor is higher than of liquid); therefore, the curve OC has a positive slope.

In accordance with the dependence (1.1.1), for example, one may follow from point E on the phase diagram to point G and further to point J, etc. We are interested in other ways, however.

Let us assume that we consider the phase diagram in Fig. 1.1.3 for water, and choose the initial point D with domestic parameters $p = 10^5$ Pa, $T = 25°C$. The usual boiling process (in a teapot, for

---

[1]   Indeed, this is a point of permanent battle: how to 'correctly' term the gaseous phase of the substance–vapor or gas? Usually, the term 'vapor' is used for the gaseous phase of a substance, the aggregate state of which in normal conditions is liquid, and 'gas' for a substance that is normally a gas. We will use these terms synonymously, except for the special topics, where we will distinct the gaseous phase appearing from the evaporation of a liquid (vapor) and the background, extraneous one (gas).

example) corresponds to heating at a constant pressure, since the external (atmospheric) pressure can be considered as a fixed value. Thus, the heating process—the heat conduction to the system—in such circumstances corresponds to

- ✓ moving from point *D* to *E*—the heating of liquid water;
- ✓ the phase transition at point *E*, where liquid transforms into vapor; all the heat that was conducted to the system is spent on the liquid–vapor conversion, so the temperature remains constant during the whole process of the phase transformation; this process exactly is termed as boiling;
- ✓ moving from point *E* to *F*—the heating of vapor.

Thus, under boiling, as we understand it domestically, we usually mean the process *D–E–F*. Strictly, only the process at point *E* corresponds to phase transition, but when we ask our spouse to 'boil the teapot' and forget to turn off the heating in time, we exactly follow the whole line *D–E–F*.

Now consider the abrupt pressure drop, as it was mentioned in the previous section—a drop so intense that the final point belongs to the stable gaseous phase, below the boiling curve. Note especially that the final aggregate state of this sudden process is not the stable vapor, but the unstable liquid. What can thermodynamics say about this process, and where exactly the final point must be set on that diagram?

Philosophically speaking, the usual 'zero problem' of thermodynamics is to determine the process which takes place in the given system. A piston moves in a cylinder of a power machine; thermodynamics may help us to calculate the work of this process, but first, we have to choose the kind of process that takes place in that cylinder: adiabatic, isothermal, or something else. The same situation takes place in any case: before thermodynamics helps us, we must help it first—we must know exactly, what is the process we are dealing with. The most famous example: when Isaac Newton considered the propagation of sound in a medium, he supposed that this process is isothermal. Alas, with time, it turned out that sound propagates adiabatically—this fact was pointed out by Pierre-Simon Laplace.

Thereby, we have to establish the physical nature of the pressure drop on our own.

The first presumption, of course, is that the temperature during this process is constant:

$$T = const. \tag{1.1.2}$$

With the condition (1.1.2), the pressure drop follows the line *D–G–H* on the phase diagram.

At first glance, the persistence of temperature looks logical: whether one may point out the reason of the temperature change or not? Unfortunately, yes. Temperature remains constant when the heat exchange has time to equalize it in the system. In other words, one may expect (1.1.2) for a slow process. On the contrary, as we described the considered process above, here we actually deal with an abrupt pressure drop. This rather means adiabatic conditions, when the heat exchange does not happen at all. For many, adiabatic conditions mean constant entropy directly

$$S = const. \tag{1.1.3}$$

For example, for a perfect gas it follows from (1.1.3) that

$$p^{1-\gamma} T^{\gamma} = const, \tag{1.1.4}$$

where γ is the adiabatic exponent—the ratio of isobaric heat capacity to the isochoric one. The equation (1.1.4) is the so-called Poisson equation (see Appendix A); we emphasize that this expression is correct only for a perfect gas; it does not fit a real gas or, all the more so, a liquid, that is, the process of a sudden pressure drop in a liquid does not obey equation 1.1.4. Nevertheless, the process of a sudden decrease in pressure in accordance with (1.1.3) must cause some temperature decrease and, consequently, can be depicted generally on the phase diagram as some curve *D–J–K*, which does not correspond to (1.1.4) (we repeat it again).

Does the consideration above mean that the correct way to the pressure drop corresponds to (1.1.3)? Once again, not necessarily. The condition (1.1.3) reflects a reversible adiabatic process: this matter can be easily forgotten, but indeed the correlation (1.1.3) and an adiabatic process are not synonymous. For the very fast change in pressure, the process runs in another, irreversible way. In a limiting case, if the piston is pulled away from a syringe incredibly fast, then the change in pressure in a syringe is similar to the adiabatic expansion in a vacuum, when the system does no expansion work. This process (the Joule process) obeys the condition for the internal energy of a substance in the form

$$U = const. \tag{1.1.5}$$

For a perfect gas, the equation (1.1.5) means exactly (1.1.2), because its internal energy depends only on temperature (this statement is the Joule law). For a liquid, the situation differs insignificantly: for example, for water at the considered parameters $P = 10^5$ Pa, $T = 25°C$), the specific internal energy is $u = 104.819$ kJ/kg; during the pressure drop in condition $u = const$, we get on the saturation curve *OC*, the pressure $p_s = 3.16959$ kPa and the corresponding temperature $T = 24.9982°C$; all the data is taken from (NIST). Seriously, without scholastic reasoning, this fact means that the condition of constant temperature (1.1.2) suits not only a very slow process, but also a very fast process (an abrupt pressure drop) too. Thus, one may see that the circle is closing from some point of view: conditions (1.1.2) and (1.1.5) mean the same.

On the other hand, it is difficult to provide the exact conditions for (1.1.5) in practice: it is hard to reach velocities high enough to assume (1.1.5). Thus, in a general case, we have some intermediate adiabatic process—neither isentropic, nor the process at a constant internal energy. Following this way, we may conclude that after a sudden pressure decrease by Δp, temperature may decrease by some value ΔT, but this value is expected to be comparatively small. From here, we will now consider the path *D–J–K* as the representation of a common cavitation process, corresponding to some irreversible adiabatic process. Despite all the nuances, the final point—the point *K* in our common case—belongs to a stable vapor phase. Thermodynamics taught us that the liquid phase there is unstable, and the story must end when the entire condensed phase turns into the vapor one. In other words, sooner or later the system at the parameters corresponding to point *K* must represent pure vapor.

The key matter, however, is how to maintain parameters corresponding to point *K*. Practically, returning to the modified pot or to a syringe, we may note that after we pull the pot's lid or the syringe's piston, we do not regulate the pressure anymore. We only establish the initial volume at which the initial pressure in the system corresponds to point *K*. After that, the system will tend to return to the stable (equilibrium) state,[2] but it should not be expected that in such conditions the liquid will completely transform into vapor: any extra amount of gas evaporated from the liquid will increase the pressure in that closed volume, returning the pressure value back to the saturation curve *OC*. For instance, let the pressure after a drop be 170 Pa—much lower than the saturation pressure $p_s = 3.1659$ kPa at 25°C; see above. To increase the pressure by 3 kPa, that is, to return to the saturation curve, the additional vapor mass must be

$$M = \frac{\mu p_s V}{RT}, \tag{1.1.6}$$

(this is the rewritten Clapeyron equation, of course). For water, $\mu = 18$ g/mole; thereby, for example, considering a syringe, for the free volume in a syringe $V \sim 5$ cm^3, we have the mass $M \approx 10^{-7}$ kg, or

---

[2]   From some point of view, it can be considered as the manifestation of the Le Chatelier principle: deviation from equilibrium causes a process that tends to return the system to equilibrium.

the corresponding liquid volume of 0.1 mm³. In simple words, when 100 cubic micrometers of water will evaporate inside a syringe, the whole process of phase transition will stop, since the equilibrium state will be established—we will return from point *K* on the phase diagram to point *J*. Note that evaporation is accompanied by the decrease in temperature: this is one additional reason to get lower temperature after cavitation.

The practical output, measured in the additional vapor mass obtained by the pressure drop, is approximately the same for the 'modified pot' from Fig. 1.1.1. Actually, this 'modified pot' can boil up a negligible amount of liquid, and this construction is almost useless for such purpose. The pressure decrease may cause some weak phase transition, but not even close to the one obtained with the plain, ordinary heat transfer to the system, with boiling in its 'domestic' sense.

But we are interested in the process inside a liquid under a very low pressure, not because of the weak phase transition caused by it. The complicated phenomenon that takes place after an abrupt pressure drop—cavitation—is accompanied by many concomitant processes of their intrinsic value, and the set of those processes is impressively wide. Possibly, the last meaning of cavitation is the phase transition from liquid to vapor.

Thus, answering the question from the previous section: indeed, cavitation and boiling have many common features, especially if we consider theoretical conditions of a constant low pressure in the system. In practice, if we define cavitation as the complex phenomenon taking place in a system during and after a sudden pressure decrease, only a single small part of cavitation concerns boiling, i.e., the phase transition directly.

Finally, we state that even a consideration of physical mechanisms shows that boiling and cavitation are related, but nothing more. These processes are not identical; moreover, one may conclude that boiling (phase transition) is only a part of cavitation.

Therefore, we may return to the simple explanation of the difference between cavitation and boiling: contrary to boiling, cavitation does not have the thermal nature, but rather a mechanical one. This matter manifests itself in every detail, including the simplest features.

1.1.3 The dynamics of phase transition

Let us consider a liquid at a low pressure, so low that its value is lower than the saturation pressure at the corresponding temperature. Above, we stated that from the point of view of thermodynamics, this phase is unstable—it tends to transform into vapor. Here we discuss this process more narrowly.

The first type of the liquid–vapor phase transition is evaporation: the detachment of molecules from the liquid surface. This process is caused by the simple fact that at any temperature, some molecules have sufficient energy to overcome the potential energy that holds them at the surface— molecules are distributed on their velocities, and some of them are much faster than the average particle. Thus, evaporation takes place at any temperature, not only at the boiling temperature— we can make sure of this by simply leaving a glass of water on a table for some time. Actually, evaporation is always accompanied by condensation—the process of attachment of molecules from the vapor phase to the liquid.

At the boiling temperature, when a liquid is warmed enough that it can transform into vapor by all of its volume, we have another type of phase transition—vaporization. Inside the unstable liquid phase, the nuclei of vapor arise everywhere—molecules separate from each other not only on the free surface, but also within the bulk of the liquid. This process of homogeneous nucleation leads to the global effervescence of the liquid. On the molecular level, small caverns appear in the liquid during this process, then these caverns grow since this process is thermodynamically advantageous.

Now we realize that the phase transition can be initiated not only by heating, but also by the low pressure in the medium (see above). Consequently, certain dynamics of bubbles appearing in a liquid will depend on the specific conditions that rule the phase transition.

The first reason for bubble growth is the high temperature of the liquid, which corresponds to boiling, in the ordinary sense of this term. Due to evaporation from the liquid surface, the vapor mass inside a bubble increases, while the liquid mass around it decreases. At first glance, the velocity of the bubble's growth can be calculated as for the single-phase Stefan problem:

$$v = \frac{q}{h_{LG}\rho},$$

(1.1.7)

where q is the heat flux on the liquid surface, h_{LG} is the latent heat of vaporization, ρ is the density of the liquid. This equation means that the liquid turns into vapor, and no other processes take place: the liquid simply 'vanishes' due to transition into vapor, and the cavity grows at the rate of this 'vanishing'.

We may estimate the velocity of the interface according to (1.1.7); even for the heat flux $q \sim 10^7$ W/m², for $h_{LG} \sim 10^6$ J/kg, $q \sim 10^3$ kg/m³ we have $v \sim 1$ cm/s. For lower heat fluxes, the rate of growth will be lower, but anyway, it is obvious that this mechanism cannot explain the rate of bubble growth during cavitation.

On the other hand, this physical description implies that the pressure inside the vapor phase increases enormously: the rate of mass growth $\dot{M}$ can be expressed through the mass flux on the bubble wall $\dot{m} = q/h_{LG}$ as $\dot{M} = \dot{m}S$, where S is the surface area of the bubble wall; thus, if we assume that the surface area and the volume change slowly, which is reasonable because of estimations made above, then

$$\dot{p} = \frac{\dot{m}}{\mu}\frac{SRT}{V} = \frac{3qRT}{\mu r h_{LG}};$$

(1.1.8)

the last equation is correct for a spherical bubble. Thus, for a bubble of radius r = 1 mm, we have for the rate of pressure increase for water at $T = 373$ K the enormous value $\dot{p} \approx 2 \cdot 10^9$ Pa/s. Of course, for smaller heat fluxes the result will be smaller, but the general conclusion is the same: the pressure inside a bubble grows very rapidly.

Thus, we see that even the simplest approach which considers the 'thermal scheme' of bubble growth actually demands to take into account the mechanical forces that push on the bubble wall from within. We have to discuss the second reason for cavern growth—the high pressure inside the bubble in comparison with the outer pressure. The 'thermal scheme' considered above for boiling means that the inner pressure is large while the pressure outside is normal. In an opposite case, for cavitation, we have a different pressure distribution: the 'normal pressure' inside a bubble and the low outer pressure.

In any case, the pressure difference Δp drives the process of bubble expansion. The rate of bubble growth can be determined by the balance of forces at the cavern's boundary. Approximately (more precise equations will be obtained in the following chapters, especially in Chapter 7), we may assume that the work that was done by the bubble was spent on the kinetic energy of the surrounding liquid:

$$\Delta p \Delta V \sim mv^2/2,$$

(1.1.9)

where ΔV is the volume change of the bubble and m is the corresponding liquid mass in that volume. Thus, we have the estimation for the velocity of the liquid at the bubble's wall—that is, the velocity of the bubble's wall itself as

$$v \sim \sqrt{\Delta p/\rho},$$

(1.1.10)

where ρ is the density of the liquid; we omit constants since this expression is only a rough estimation by an order of magnitude. Thus, even if the pressure difference is only 10^3 Pa, then the velocity of the bubble's wall is ~1 m/s. This is a noticeable growth rate, which causes many dramatical issues of cavitation.

Now we consider the following point: what is a bubble inside a liquid, i.e., what is its physical nature? We can imagine a bubble in a boiling liquid, filled with pure vapor, but does a liquid contain a dispersed gaseous phase at any temperature? Generally, the answer is positive. Not necessarily, but a liquid may contain air bubbles. To say precisely, it takes a lot of effort to get rid of air bubbles. To exist in a stationary case in normal conditions, these bubbles must be filled with a gas at approximately atmospheric pressure; actually, the surface tension may also play a role for very small caverns, but we may not consider this correction for bubbles that can be seen with the naked eye. Then, except for air, such bubbles must also contain vapor—thermodynamics demands the phase equilibrium inside bubbles too. The pressure of vapor is equal to the saturation pressure—the pressure on the curve *OC* in Fig. 1.1.3.

Thus, some air–vapor bubbles can be found in a liquid. Besides them, pure vapor caverns can be formed in a liquid, when some molecules break bounds with their environment: being detached, these molecules 'push away' the liquid, freeing the space for the new ones. When temperature is low and pressure is high, they have no chance to succeed—such a nucleus cannot survive, and caverns collapse as soon as they appear. But the situation can be changed during cavitation, when the outer pressure drops below the saturation value. In such conditions, any formation of a vapor nucleus inside it is thermodynamically favorable now, then 'pure vapor' bubbles can be formed and grow.

1.1.4 What pressure is required for cavitation?

The process of abrupt pressure drop, which leads to the partial phase transition in a liquid, is cavitation itself. From the point of view of thermodynamics, this part of the phenomenon is the process of the liquid-vapor phase transition anyway, i.e., it can be considered as boiling. Actually, sometimes this description can be discussed in the literature devoted to cavitation (Pirsall 1972), but it is not a good way to mix such different processes and call cavitation 'boiling'. This terminology is not applied more or less commonly, because, indeed, the weak phase transition during the pressure drop is not the main matter that one can observe in such conditions.

A much more important consequence of the pressure drop is the development of bubbles in the liquid, as can be seen in Fig. 1.1.2. Usually, exactly that process is called cavitation—the growth of a bubble in a liquid caused by decompression. In this sense, we are not interested in the phase transition at all; probably, some readers who knew something about cavitation before were surprised by the fact that any phase transition takes place during cavitation too, without heating. For many, cavitation is a purely hydrodynamical phenomenon—the formation of giant bubbles inside a liquid under low-pressure conditions. These huge bubbles are the main object of the cavitation science, so to say.

Now it is time to turn from thermodynamics and various thermal effects to hydrodynamics and processes of expansion and collapse of a bubble. Let us explore the fundamental matter—how low must the outer pressure be after decompression? Is it necessary to lower the pressure below the saturation value in order to observe the formation of 'giant bubbles', or we can stop somewhere in the 'liquid area' on the phase diagram, that is, above the *OC* line?

The last question looks reasonable if we take air bubbles into account. Indeed, there exist bubbles filled with air at atmospheric pressure. When the outer pressure drops to values $\sim 10^{3-4}$ Pa, slightly above the saturation curve,[3] we may expect that the inner pressure will inflate these air bubbles up to huge sizes; therefore, the sub-saturation pressure is not necessary.

Unfortunately, the potential of air bubbles to grow is very restricted. First of all, today it is difficult to say, how strongly the air mass inside a bubble can be changed during cavitation: generally, air may infiltrate from the liquid to the bubble or can be dissolved in the condensed phase. Usually, it is assumed that the air mass in a bubble is constant; then, we will consider the problem

---

[3]   The saturation pressure of 10^4 Pa for water corresponds to the temperature of 46°C.

with this approach. If so, then, in conditions of constant temperature, we may conclude that pressure and volume are related to each other as

$$pV = const. \tag{1.1.11}$$

Thus, if pressure drops even by two orders, down to 10^3 Pa, then the bubble radius will be increased only by 4.6 times; at the external pressure at 10^4 Pa, the bubble size can only be doubled. At the new size, the inner pressure will be balanced out by the external one, and further expansion would stop. Since usually the largest air bubbles in a liquid do not exceed 1 mm in diameter, we have no chance to observe giant bubbles when the outer pressure drops to values above the saturation pressure.

Thereby, we need the outer pressure below the saturation value. For this case, the lower value of the external pressure plays a role, i.e., we hope to obtain a larger volume from the same equation $V \sim 1/p$ with lower p. The physical nature of the expansion is different, when the outer pressure is lower than the saturation pressure p_s.

As it was mentioned above, the vapor pressure inside a bubble is always p_s—thermodynamics monitors comply with its laws carefully.[4] The pressure inside a cavern is sustained by the conditions of thermodynamic equilibrium: it remains a constant value p_s during all the expansion phase because the processes of evaporation will sustain the pressure at the same level despite the size of the bubble. Thus, even if the pressure difference between the pressure in a bubble and in the surrounding liquid is low, but the outer pressure is lower than the saturation one: $p_{out} < p_s$, then the bubble containing vapor may grow infinitely in such conditions, regardless of the size of the cavern. Independently from the bubble radius, the pressure difference that pushes the bubble remains constant, and it will move bubble's walls further and further.

To increase the meaning of undersaturated pressure, we may repeat that in such conditions new bubbles arise in the bulk of the liquid—the phase transition also helps the formation of new bubbles.

1.1.5 Final comparison between boiling and cavitation

Ergo, boiling corresponds to the horizontal path on the phase diagram in $p - T$ coordinates, while cavitation is an almost vertical descent in that landscape.

In both ways, we reach the point (different points for boiling and cavitation, obviously), the parameters of which correspond to the stable vapor phase, not the liquid one.

For boiling, normally,[5] phase transition takes a lot of time. A large amount of liquid can be transformed into the vapor phase, which depends on the disposable heat. One way says that boiling and phase transition are processes connected directly; moreover, usually the 'liquid–vapor' phase transition is termed as boiling.

On the contrary, the 'liquid–vapor' phase transition is never called cavitation; we never heard the term 'cavitation' in that sense, at least. Utmost, sometimes cavitation can be mentioned as the process of bubble formation, which does not imply phase transition in that context.

The explanation for such terminology is quite clear: for cavitation, the phase transition is an accompanying phenomenon. When the pressure drops below the value on the saturation curve, some—very small—amount of liquid turns into vapor, but this is not a crucial part of the cavitation phenomenon. In normal conditions,[6] in pipes, vessels, etc., the outer pressure will be increased

---

[4]   However, this simple rule can be violated for a very fast process.

[5]   Of course, we may imagine conditions when the temperature of the liquid decreases with time during the boiling process: for instance, it can be boiling on a hot body immersed into a cooling fluid. With time, when the temperature of that body drops enough, the boiling process will stop. Above, we considered more frequent boiling conditions—at a constant heat flux.

[6]   Once again, we refer to 'normal' conditions, because one may propose another scheme: cavitation in a vacuum; for instance, directly in space, where zero value of the outer pressure is maintained by the whole universe, so to say.

when a comparatively small amount of liquid passes into the gas phase. This enhanced pressure will stop the phase transition, but it will not stop the cavitation process as it is. Bubble dynamics and all other processes initiated by the pressure drop, that is, cavitation in our definition, will continue for some time.

Here we come to the main difference between boiling and cavitation.

Boiling (at least with its usual scientific description) is considered as a stationary process. Partially, of course, here some scientific paradigms manifest themselves,[7] but really, when we look at a boiling teapot, we do not want to overcomplicate the problem and consider non-stationary features of the process. We can be satisfied with the time-averaged picture.

Cavitation is a fundamentally different case. This is an unstable, non-stationary phenomenon from the very beginning. Pressure drops, bubbles expand, liquid evaporates, pressure increases, etc. The stationary state comes only at the end of the cavitation itself.

Generally, non-stationarity promises many difficulties and a lot of interesting things. Cavitation delivers on all these promises in abundance; moreover, it offers one of the most mysterious phenomena in modern physics. To know what it is, you may go straight to Chapter 9, but we carry on the explanation of cavitation in series. The next subject is the observation of cavitation in domestic conditions.

1.2 Cavitation in domestic conditions

1.2.1 Cavitation in a syringe

This experiment was briefly mentioned in the first chapter. Here we describe this simple example in details.

Observing cavitation in a syringe is a very simple trick that can be easily reproduced in domestic conditions. We need a syringe of a medium size and water, not necessary the distilled one; tap water suits too.

Our aim is to obtain low pressure, which can be easily achieved with a syringe: actually, this device was designed exactly for this operation; to draw a fluid into a syringe, we create low pressure in its volume by pulling its piston out. One nuance here: for our present goals, we do not need to retract any fluid; on the contrary, we want the free space in the syringe to be formed after the piston motion.

To be exact, we must fill the syringe partially—approximately, up to half way. Then we have to close the nozzle of the syringe somehow. The simplest device for this operation is a finger (see Fig. 1.2.1), but a glove is needed—under the pressure difference, the skin on the finger can be damaged.

Then, we can possibly meet the usual problem that is well-known both in the scientific world and in your everyday life: the absence of the third hand. We need to hold (firmly) the syringe, to seal (reliably) its opening, pull out the piston with some force, and—last but not the least—watch the process inside the syringe; for the last purpose, our own hand must not obstruct the picture. For a second, the pressure difference we are dealing with is about 1 atm; thus, if the surface area of the piston is ~ 1 cm^2, the total force that must be applied is ~ 10 N; this is roughly equal to lifting a weight of ~ 1 kg. Note that usually the friction of a piston inside a syringe can multiply this value by several times (that depends on the type of syringe).

If we believe in our hands, we may execute the whole procedure manually; to fix the process on a camera, it is probably more convenient to fasten the syringe on a tripod, like in Fig. 1.1.2.

Thus, we are ready to provide, probably, our first experiments in cavitation science. As we discussed in the previous section, our goal is to create a low-pressure zone over the free surface of a

---

[7] Scientists and engineers prefer to consider stationary patterns instead of the dynamical picture of the phenomenon. Sometimes, it is justified, but sometimes not, indeed.

liquid. With a closed syringe, this is a comparatively easy task: moving a piston, we create a rarified zone over the liquid. One may say that vacuum appears above the liquid when the piston of a syringe leaves the liquid volume inside of it, moving away.

However, it is not exactly the case: vacuum is created, but not for long. Once the 'vacuum' arises above the liquid (the desired low pressure is created above the liquid surface), the evaporation process leads to the appearance of steam in that free volume above the liquid immediately, i.e., the 'vacuum' stops being a vacuum in the strict sense of this word.[8] Evidently, evaporation increases the gas pressure above the liquid, reducing the pressure difference between the value in the liquid (we mean, in bubbles inside it) and above it.

Thus, we come to a conclusion that the result must depend on the velocity at which the piston is removed from the syringe: the faster the motion of a piston, the lower the pressure that can be achieved above the liquid, since in this case evaporation 'has no time' to create a sufficient pressure in the free volume.

Fig. 1.2.1. Gloves suit well for these experiments: they help to close the hole of the syringe firmly, save your hands, and look stylish.

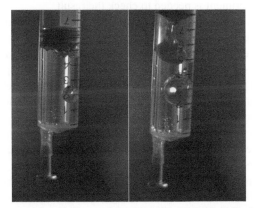

Fig. 1.2.2. Cavitation inside a syringe turned upside down at fast motion of the piston. Initially, there were no distinguishable bubbles in the volume. Captured at two different moments of time; notice the huge size of the bubble that has been formed. One may also notice the alternative way to close the syringe's nozzle in domestic conditions.

---

[8] By the way, the term vacuum has no common definition. Vacuum is a low-pressured gas, generally. In special sciences, some thresholds are established for various sorts of vacuum, but many scientists did not even hear about this scale.

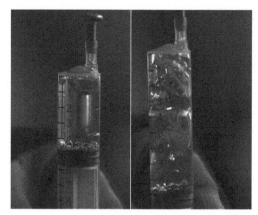

Fig. 1.2.3. Cavitation in an upward-facing syringe. Bubbles form from both ends of the liquid column—both above and below. Two different instants of time of the same process.

In Fig. 1.2.2, we can see the result of 'fast' pulling: a huge bubble is formed in the liquid—this is the evidence of a great pressure difference that was achieved in the syringe.

Note that the orientation of a syringe does not play a crucial role from this point of view. In Fig. 1.2.3, we present the results for a syringe oriented upward: generally, we see the same picture, but with more paddling. Actually, one may expect some differences from many points of view:

- ✓ restriction for the velocity of motion of the interface—the liquid must fall, following the piston; this fact limits the velocity of motion of the upper surface;

- ✓ creation of the second interface below the liquid, immediately on the piston; while the piston moves down, it creates a second decompression zone, at the 'bottom' of the liquid mass;

- ✓ the Rayleigh–Taylor instability on the upper surface adds some nuances to the process, but here we are in no mood to consider such complicated matters (see Chapter 4 for such difficult problems).

Together, these circumstances may lead to a more complicated, more perturbed cavitation pattern. On the other hand, we probably cannot attain a sufficient decompression in both zones here: at the bottom, the liquid tends to overtake the piston, preventing the appearance of a sufficiently low-pressure zone; at the upper surface, the liquid moves too slowly to create a significant decompression. An experiment is needed to check the result.

Finally, we have the following results: when the piston is moved abruptly (as fast as we can), we observe a developed cavitation with large cavitating caverns.

Now let us consider the opposite case—when the piston moves slowly in the syringe. For this case, we may hope to see another situation: due to evaporation, the pressure above the liquid will aspire to the saturation pressure. As a result, the pressure difference between the gas in a bubble and in the surrounding liquid will not be significant, so the rate of bubble expansion is expected to be very restricted, so to say.

Then, we come to the experiments. First, take a look at the initial system (see Fig. 1.2.4): two relatively big bubbles can be spotted there. During the cavitation process, we may expect that these bubbles will become the nuclei for cavitating caverns.

Then we pull the piston away from the syringe. The process is shown in Fig. 1.2.5. In general, this pattern is a good illustration for the reasoning made above: indeed, bubbles grow smoothly and slowly. When the piston is put forward to the utmost, bubbles are much smaller than the bubble shown in Fig. 1.2.2.

Moreover, this slow operation can be reversed (see previous section concerning the reversibility of the process). Beginning with the stage corresponding to the maximum expansion, we will smoothly return the piston to the initial position; this process is shown in Fig. 1.2.6.

Fig. 1.2.4. A syringe filled with water; some air bubbles can be seen even with the naked eye (marked by arrows).

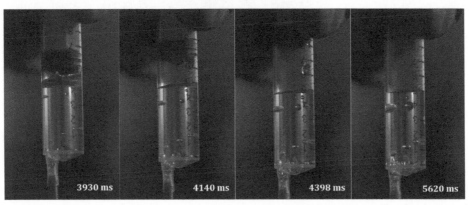

Fig. 1.2.5. The 'soft cavitation mode'. The initial stage was presented in Fig. 1.2.4. Then, pulling the piston gently, we observe only the growth of air bubbles that were in the liquid originally. Their growth is restricted: compare their size with the bubbles shown in Fig. 1.2.2.

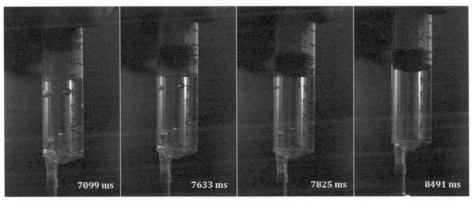

Fig. 1.2.6. The reverse process to the one presented in Fig. 1.2.5. As one can see, we return to the pattern presented in Fig. 1.2.4.

Thus, we can reverse this process almost completely: the only difference is the noticeable displacement of the second (lower) bubble compared to its initial position.

The opposite orientation does not change the situation significantly; see Fig. 1.2.7 for the direct process and Fig. 1.2.8 for the reverse one.

Then, we must consider the obtained results more thoroughly, with formulas and reasoning.

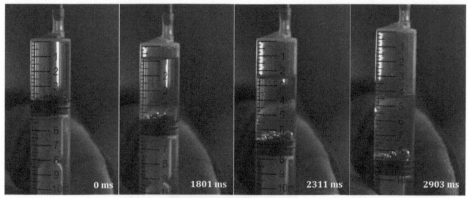

Fig. 1.2.7. Soft cavitation in an upward-facing syringe. On the first frame, we see the initial bubbles on the piston; with time, or with the displacement of the piston, to be more precise, these bubbles grow.

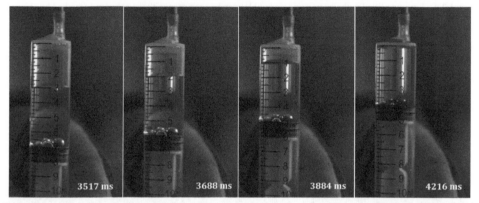

Fig. 1.2.8. To reverse the process shown in Fig. 1.2.7, we return to the state when small air bubbles are settled on the piston.

The first question that must be discussed concerning the matter of establishing the pressure difference in a bubble and the surrounding medium is: why does the pressure created by the same evaporation process differ in two zones—inside vapor bubbles and above the liquid? Indeed, as we know, the saturation pressure depends only on temperature, which is the same throughout the whole system. Thus, one may expect the pressure p_s inside a bubble and the pressure $p_s + \rho_{gh}$ in the liquid around it: here we take into account the additional hydrostatic pressure from the liquid column of height $h \sim 1$ cm, which gives a conspicuous correction; by the bye, this additional term gives ~ 100 Pa, while the saturation pressure for water at 20°C is 2337 Pa.

The answer to this question concerns the matter of the process rate. When we move the piston away, we create a rarified zone between the piston and the liquid. This rarified zone is refilled by a vapor of that liquid, but, actually, it is intuitively clear that the larger the volume of this zone, the greater the time needed to create the same pressure in corresponding objects. To provide some quantitative estimations, let us consider the creation of the same pressure in a bubble of radius r_b and in a cylindrical volume of radius r_c and height h_c (see Fig. 1.2.9).

The pressure in the corresponding volume is defined by the Clapeyron equation

$$p = \frac{M}{\mu} \frac{RT}{V}. \tag{1.2.1}$$

In such an equation, for a bubble we must insert $V_b = 4\pi r_b^3/3$, for a cylinder $V_c = \pi r_c^2 h_c$. The mass originates from the evaporation process, the rate m of which (the mass flux—the mass evaporated per unit of time from a unit of the surface area) is the same for a bubble and a cylinder; correspondingly, the mass evaporated into a bubble with the surface area S_b during time τ_b is

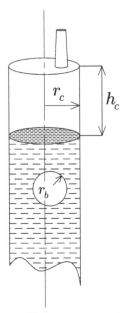

Fig. 1.2.9. Two volumes with the same evaporation rate on their surfaces.

$$M_b = \dot{m}S_b\tau_b = 4\pi \dot{m}r_b^2\tau_b, \tag{1.2.2}$$

while for the cylindrical volume

$$M_c = \dot{m}S_c\tau_c = \pi \dot{m}r_c^2\tau_c. \tag{1.2.3}$$

Finally, we must have equal pressures $p_b = p_c$ for both objects, that is, from (1.2.1) with corresponding volumes and masses from (1.2.3), we get

$$\frac{\tau_b}{\tau_c} = \frac{r_b}{3h_c}. \tag{1.2.4}$$

That is, if we have a bubble of radius ~ 1 mm under the volume of height ~ 3 cm, then the time ratio (1.2.4) is about $\sim 10^{-2}$: due to evaporation, the equilibrium inside a bubble will be established 100 times faster than in the gaseous phase above the liquid.

A possible question: if the time difference is so great that $\tau_c = 100\tau_b$, why doesn't the vapor mass inside the bubble increase after the time τ_b? Why cannot the bubble receive additional mass of vapor during the last time $99\tau_b$? If so, the pressure inside the bubble can be increased more. Of course, this is not impossible. The equilibrium pressure corresponds to the case when the flux of evaporation is balanced out by the flux of condensation: the latter increases with the vapor pressure. Thus, once the saturation pressure is reached, it cannot be increased further. Consequently, after the abrupt volume change (when the piston moves fast) the vapor bubble has approximately τ_c time to grow, until the external pressure increases to balance out the inner one.

The next question is more interesting. Why does the bubble grow at all when we pull the piston slowly? Following the simple explanation of the phenomenon, once the piston has been moved away from the liquid, the saturation pressure has been established above the liquid. Under this pressure difference, the bubble must grow up to a certain size, at which the total pressure is sufficiently low. The further motion of the piston does not lead to the lower pressure above the liquid, it remains at the value of the saturation pressure anyway; thus, the pressure difference between the bubble and the liquid remains the same too, and no growth can be observed.

However, strictly, when we move the piston from the syringe, the pressure above the liquid is always lower that the saturation pressure. Evaporation only follows the volume change; the real consequences are:

- ✓ we pull the piston—the volume changes;
- ✓ the pressure above the liquid decreases temporally;
- ✓ evaporation from the liquid surface supplies the additional vapor mass to provide the saturation pressure, when the further net evaporation does not take place.

Thus, evaporation always only overtakes the decompression above a liquid. The slower the motion of a piston, the more successful this overtake attempt is—the pressure above the liquid is closer to the saturation one.

For the opposite process, when the piston is pushed into the syringe, we have a completely opposite situation: the volume shrink causes the increase in pressure, then the condensation process tries to diminish the pressure there, and so on.

1.2.2 Cavitation in a bottle

Important notice: please do not try to reproduce the experiments described below without protective measures—these exercises may hurt you. Possibly, it is even better to believe our words.

The experiment that we discuss here is a famous trick; for example, it was shown in the pulp movie 'Gutshot Straight'. Jack made a bet with Duffy that he can break a bottle with his hand. He filled the bottle halfway, hit the open neck of the bottle with his palm, and the bottom of the bottle broke off.

How is this possible? Is this real or is this a cinematographic trick-effect?

It may look surprising in our time when films are more like cartoons and 'everything is not what it seems',[9] but this is a real trick which can be reproduced in domestic conditions, with a lot of precautionary measures, which are necessary. Indeed, we warn you to not repeat this stunt, at least directly in the way that was shown in the movie. It may be even more surprising, by the way, that this juggle is a matter of scientific works (Daily et al. 2014, Ganiev and Ilgamov 2018, Pam et al. 2017, Yukisada et al. 2018), since this is a good (and cheap) example of the cavitation impact and hydraulic shock.

Thus, the aim is to break a bottle. This is a barbaric goal, so it can be reached only by the corresponding methods. Here is our recipe: take a glass bottle, gloves, some protective eyewear and a mallet. Yes, a mallet, because even a simple blow with the palm of your hand to the bottle neck is painful.

Pour a little water into the bottle, about a half of it, take the bottle by the neck in one hand so that the sealing surface of the neck is not covered by anything, and the mallet in the other. Now strike the sealing surface of the bottle neck with the mallet sharply. If all went well, the bottle will remain intact, but without the bottom: the bottom of the bottle will be broken off by cavitation; see Fig. 1.2.10.

The variations of this trick involve breaking a bottle by hitting it against a surface or by sharply pulling out the cork, all depending on the dexterity of one's hands. The most impressive stunt is to put a finger in a bottle full of water and pull it out sharply with the same result—a broken bottle. We can honestly admit that we failed this version of the experiment. Weak fingers, or something else. Thus, we concentrate on the classic version of this experiment.

To understand the physical processes that led to such an impressive result, Fig. 1.2.10 is insufficient: we do not see there any detail that may clarify the physics of this destruction. One of

---

[9]   Confucius.

Fig. 1.2.10. After the strike on the neck, the bottle breaks: its bottom falls off.

the first things that may come to mind is that the bottle was broken simply by the mallet's blow, that is, the water in the bottle does not play a role in this process. It is impossible, however, to break an empty bottle in the same manner, i.e., with the separation of its bottom (breaking a bottle with a mallet into shards is a primitive task, of course). By the way, we tried these experiments too, despite the evident negative results: for scientific purity. Yes, empty bottles do not break.

Then, look at the bottom closer; see Fig. 1.2.11.

We see that the blow on the bottle's bottom is caused by the fluid mass falling on the glass surface. Cavitating bubbles collapse, and the liquid column descends upon the bottom of the bottle.

The question about the origin of those bubbles is simple; the complete cycle (of another experiment) is shown in Fig. 1.2.12. Of course, bubbles arise due to cavitation: when the bottle 'moves away from under the water' after the hit, the pressure at the bottom decreases, and bubbles arise there.

The role of collapsing bubbles in the damage, however, is not absolutely clear. The main component of the impact is the strike of the water mass on the bottom; bubbles that are inevitably located at the surface may play a role in the process, accelerating the liquid mass moving toward the surface, but definitely the simple fluid blow is the prime component.

Thus, as a whole, the picture looks as follows. On impact, the bottle begins to move downward, but the water does not move with the bottle, it remains at the same position where it was before the impact. Because of this, the pressure at the bottom of the bottle reduces, which leads to cavitation there—to the formation of bubbles, to be exact, of the bubble cloud. However, the decompression cannot be unlimited, so at a certain moment the water also begins to move down and tries to overtake the bottom of the bottle and take its usual position. At this very moment, the growth of bubbles stops and the process of their collapse begins. The downward motion of fluid is accompanied by the pressure increase, so the bubbles collapse rapidly. The final strike of the liquid on the bottom causes the bottle's breakdown.

Thus, the physical process that one may observe in Fig. 1.2.10 and 1.2.11 can be named, generally, as hydraulic shock. Below, we devote the whole chapter to this phenomenon, which is named exactly so.

Here, at the final point of this destructive engagement, we also discuss a matter that is mentioned sometimes: about the impossibility of destruction of carbonated water. For experiments, we used Georgian mineral water Borjomi in its original bottle (see Fig. 1.2.13). As we see, the result is the same: the broken bottle.

Finally, we pay attention to the uninvited guest here—the phenomenon which will be mentioned many times in the following chapters. This is a very important matter which explains some details

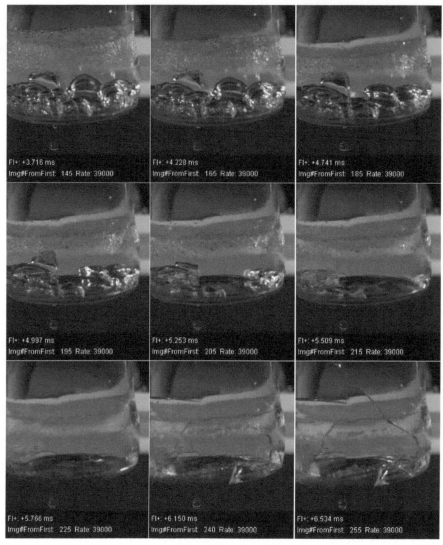

Fig. 1.2.11. The process of bubble collapse at the bottom. The technical information is presented on frames (39000 frames per second; number of frames, etc.). We see that bubbles collapse near the bottom; then, in a pure liquid, without bubbles, the destruction takes place.

of the cavitation impact on the surface (see Fig. 1.2.14). When a bubble collapses at the boundary of the fluid, where the flow cannot exist in a spherically-symmetric pattern, this bubble loses the sphericity of its shape. Some kind of jet is formed, directed away from the fluid to the wall. In Fig. 1.2.14, we may see not the best example of such a jet (this jet is too fat), but this is a good occasional instance of the common phenomenon which belongs, generally, to some other matters of cavitation: the 'point' effects of cavitation on the surface.

At this point, we finish the consideration of bottle cavitation. Many bottles were harmed during the experiments.

1.2.3 Fishing

This funny example belongs rather to outdoor experiments, not to domestic ones. But anyway, this is a nice practical application of cavitation.

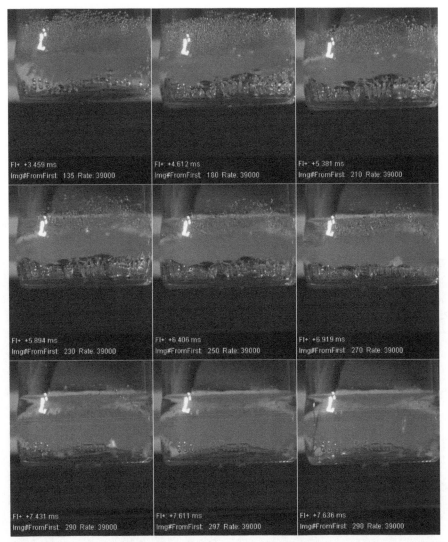

Fig. 1.2.12. Another experiment with the same result. Recorded with a high-speed camera; once again, we leave all the technical information on frames: 39000 frames per second, frame number, etc. On first two frames, we see how a bubble is growing near the 'cavity foam' at the bottom of the bottle. Then the bubble begins to collapse, but it loses the spherical symmetry during this process. First three frames: the formation of the cavitation cloud near the bottom; then, when the pressure there becomes too low, the liquid mass aspires to the glass surface. At this stage (next three frames), that cavitation cloud collapses. Again, as in Fig. 1.2.11, we may see the single-phase fluid (the eighth frame) before the first crack appears. Note that the last two frames here are consequential: the time interval between them is 25 microseconds. To the point: the small 'pimples' that can be seen on the bottle on all the frames are the details of the bottle's relief.

To catch a fish—to hook it—you need a bait on a hook. The simplest kinds of bait are worms, flies, pieces of food, etc. But aside from these 'natural' baits, there exists a class of artificial baits—plastic (or wooden) models of fishes (Fig. 1.2.15). These jerkbaits must attract predatory fish which, in theory, must confuse these plastic fishes with a real prey, becoming a prey for you.

A problem is that predators are clever beasts, so the portrait likeness of the bait to a fish may be insufficient for a wary carnivore. So, we need to add some 'life' to the jerkbait.

While breathing, fishes blow bubbles. Scientists know this fact through long-term observations; predatory fish know this feature because of their deep intuition. Thus, to imitate a living fish, we want to produce bubbles around a jerkbait.

Fig. 1.2.13. Breaking a bottle of Borjomi (mineral sparkling water). Before the impact, we shook the bottle to demonstrate the gaseous bubbles there (first frame); after the blow, the process goes as usual.

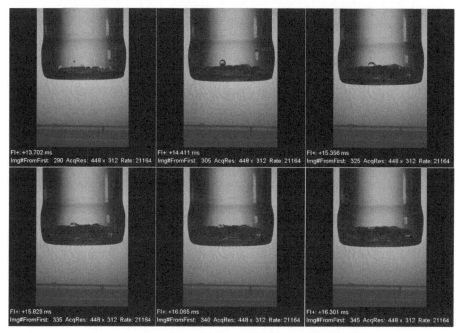

Fig. 1.2.14. The collapse of a spherical bubble in vicinity of the wall.

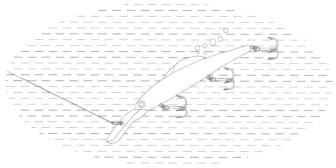

Fig. 1.2.15. This is not a real fish, this is a jerkbait.

The cheapest way to make bubbles from nowhere is cavitation. In this case, we are dealing with another sort of this phenomenon in comparison with cavitation in closed vessels, like we saw in a syringe or a bottle. The external cavitation takes place when a body of irregular shape[10] moves in a fluid; the separation flow in this case may create bubbles which, from some point of view, look like the bubbles emitted by the fish itself.

Thus, by moving that specially constructed jerkbait sharply, we obtain a bubble cloud around the plastic fish. In many cases, the body of evidence—the formal resemblance, the motion itself, the bubbles—assure the predatory fish that it looks at a real, fleshy fish.
With all the consequences.

1.3 Cavitation in technological devices

1.3.1 The origin of cavitation

It will be better for everyone if this matter will be considered without domestic experiments, as in the previous section. Despite the fact that someone may probably find a pump in their shed and try to establish its cavitation limit experimentally, we believe that this experiment will not be of a great scientific meaning. One broken pump more, one broken pump less—for the engineering history this matter is negligible, but that pump may have an importance as a family item or something else. For the subject of this section, it is better to set aside the hands-on experiments and trust the theory.

For the considered matter, the theory says that the main reason behind cavitation—decrease in pressure below the saturation pressure p_s—that was discussed in the previous section, is obviously the same. The cause of cavitation is always the pressure drop, but during the experiments with a syringe the external cause—the macroscopic reason, so to say—was the pulling of the syringe's piston, i.e., the volume increase. Here, two things are important for the present consideration: (a) we cause cavitation artificially, (b) the pressure has been dropped outside the fluid.

Usually, in technological devices, cavitation originates in a slightly different way. First, it arises beyond our will (except for cases considered below, when we summon cavitation for some purposes). Then, as a rule, the pressure drop leading to cavitation occurs not because of the pressure drop somewhere out of the fluid, but due to the decrease in pressure inside this very fluid. In other words, the example of cavitation in technological devices is not necessarily a case of a sudden depressurization on a spaceship or something of that sort: cavitation is much closer than one may imagine.

If we must seek a person who is in charge of cavitation, it must be Daniel Bernoulli (and, probably, his father Johann Bernoulli). According to the Bernoulli law, in two points along the streamline, the following relation must be fulfilled:

$$p_1 + \rho g h_1 + \frac{\rho u_1^2}{2} = p_2 + \rho g h_2 + \frac{\rho u_2^2}{2},$$ (1.3.1)

where p is pressure, ρ is the density of the fluid, g is the acceleration of gravity, h is the height, u is velocity; all these quantities are defined at two different points.

Thus, Bernoulli predicts that if a fluid accelerates, i.e., its velocity increases, or lifts higher, then the pressure inside it decreases. As for the height, this factor usually is negligible, but velocity plays a crucial role in the cavitation process. If u_2 significantly exceeds u_1, then $p_2 < p_1$ and the fluid at point *2* is closer to cavitation than at point *1*.

This very mechanism is responsible for cavitation in most practical cases. When we want an object to move in a fluid at high velocity, we have to be ready for cavitation on this object. It can be, for example, a propeller screw: at high rate of its rotation, cavitation appears on it. A close relative to this object is a hydrofoil—the underwater wing which creates a lifting force; in a particular case, this

---

[10] Actually, the developed cavitation may appear at hydrofoils too.

device also creates a special kind of cavitation—supercavitation, the phenomenon of a giant cavern on a fast-moving body in a fluid. This kind of cavitation is used—yes, cavitation is sometimes not avoided but used—for special ships; see Section 1.5.

In pipes, the scheme described above works directly: in the narrow section of a pipe, velocity increases to maintain the same mass discharge as in the wide section. Therefore, in accordance with (1.3.1), pressure decreases there, and cavitation may appear. Another kind of cavitation takes place in channels with abrupt deviations in the shape of their cross-sections: abrupt turns, sudden constrictions, etc.

Pumps are special objects in context of cavitation. To work, a pump must suck the fluid in. Thus, the low pressure (at the pump entrance) is an inherent part of the whole pump scheme. Consequently, the danger of cavitation is an indefeasible satellite of pumps. Fortunately, special methods to calculate the input parameters of the pump are developed. Thereby, cavitation is a quite avoidable event for a pump system, if all the engineering was properly taken into account.

1.3.2 The process of cavitation

Cavitation is an unstable process, any attempt of its stationarity description meets serious difficulties. The main feature of this process is not the presence of the gaseous phase inside a liquid one, but the dynamics of this gaseous phase.

The central object of the cavitation process is a bubble. During cavitation, such bubbles appear from every small nucleus that was presented in the fluid; they grow, merge, and so on. At the initial stage, cavitation means the increase of the gaseous phase in the fluid. However, the bubble collapse is an even more significant stage for technological application: usually, the main danger of cavitation is erosion caused by the collapsing bubbles. Thus, not only the pressure decrease, but also the following pressure increase can manifest in a fluid in dangerous cases.[11] Due to the non-stationary character of cavitation, both processes are possible, and the collapsing phase of bubble dynamics can be observed during cavitation too, despite the fact that cavitation is initiated by a drop in pressure.

Thereby, the main feature of the cavitation process is its non-stationarity. This property of cavitation is reflected in many ways in the description of cavitation (see, for instance, Chapter 4: attempts to construct a stationary flow pattern for supercavitation); below we consider simple examples.

Above, we mentioned the Le Chatelier principle, according to which some mechanisms tend to return the system to its previous state. For cavitation caused by a sudden external pressure drop, the first three such mechanisms are:

✓ evaporation which tries to establish the equilibrium vapor pressure out of the fluid and, therefore, to prevent the formation of cavitation from the outside; see the illustration in the previous section;

✓ the expansion of the bubble, which can cause the decrease in pressure if the amount of gas is constant there (this scheme does not work for a vapor bubble since the pressure inside it remains constant at the saturation value—the mass of vapor increases due to evaporation with the bubble increase); this way tends to prevent cavitation from the inside;

✓ the global change in external conditions caused by bubble dynamics: not only external fields— of pressure and velocity—influence the bubble, but also the bubble affects the pressure and velocity field; this is the most complicated, nonlinear way.

However, the very fact that some mechanism cause acts in the opposite direction does not mean that the process will be stopped once the initial state will be achieved again. To stop at the initial

---

[11] Supercavitation indeed uses the presence of the quasi-stable low-pressure in the cavitation zone for such type of flow; see below.

point, some dissipation is needed, while a liquid–gas system has very restricted abilities in this sense.

Let us consider the second item from the list above—the expansion of an air bubble in a liquid.[12] During the expansion phase, the pressure inside it drops; at some moment, the pressure inside the bubble becomes equal to the external one. Does this fact mean that the bubble will be stopped immediately when this state will be reached? Of course not. The bubble will continue its motion on inertia, as we can realize based upon our elementary physics intuition. Some calculations following from the solution of mathematical equations will be presented in Chapter 7, but the answer is clear right now: there is no stationary case anymore, and such a bubble will oscillate.

Now we consider the first item—an attempt of the system to return to its initial condition after an external pressure drop caused by the evaporation processes. As we saw from the experiments with a syringe, this method works for such calm, stationary external conditions. If, otherwise, we have a fluid flow, then the free volume out of a liquid phase will be changing too; these variations will cause a change in pressure: the evaporation-condensation process will follow the variation of volume change, but, as we discussed in the previous section, they are always number two. Thus, in a flow, it is questionable to observe the full stabilization of a cavitating flow with this mechanism.

The third item emphasizes that the hydrodynamical problem is a self-consistent task. For instance, for cavitation initiated by the velocity increase, i.e., in accordance with the Bernoulli equation[13] (1.3.1), we see that for constant pressure the velocity distribution must also be stationary. But, in its turn, the fluid velocity must obey some restrictions on the boundaries—both on the liquid–solid interface and on the liquid–vapor ones (see details in Chapter 4). In their turn, the boundaries of the last type move: the position of the liquid–vapor interface, generally, changes because of the fluid flow itself. The only exception is when, for some reason, this interface is steady—a bubble attached to the solid surface somehow. For this particular case, there are, actually, two problems: how will the dynamics of that bubble be stopped at some position (see above), and is this pattern fundamentally possible, i.e., can we describe such a cavern consistently (see Chapter 4)?

Hence, both the distribution pressure and the velocity configuration also depend on the movement of the cavitating caverns, the formation and dynamics of which depend on pressure, etc. When cavitation incepts in a stable flow, this flow ceases to be stable. Of course, for such a flow, some averaged parameters can be calculated too, but at microscales—which are important for the cavitation effect—we have to deal with a non-stationary pattern.

Thus, we may reasonably assume that once a two-phase medium appears, it will try to stay there. In a general case, the common Le Chatelier principle does not predict the disappearance of the gaseous phase or even the elimination of external conditions that lead to cavitation. Moreover, the development of the gaseous phase during cavitation may summon cavitation effects of another level: if cavitating bubbles merge into a giant cavern, the following dynamics of such a two-phase fluid differs significantly both from the dynamics of the initial single-phase fluid and from the behavior of a liquid filled with small bubbles.

Possibly, the reasonings considered in this section look too abstract. That was the point: the analysis of separate details of cavitation processes will be given below in corresponding sections (bubble dynamics, microjet formation, hydraulic shock, etc.), but here we synergize the description into a full, possibly somewhat philosophical, picture.

1.3.3 Results of cavitation

The main consequence of cavitation is the damage of the surface at which cavitation takes place. To be exact, this does not necessarily mean an abrupt destruction of this object—as a rule, this is a gradual erosion of the surface. Of course, sometimes the pulsation of pressure may cause a direct

---

[12] That is, we assume the air mass is constant during the process.

[13] Generally, in non-stationary case, another equation must be written instead of (1.3.1): the Lagrange–Cauchy relation, which differs from (1.3.1) by the additional term—the time derivative of the velocity potential.

macroscale effect called the total destruction, but this is a rare case. Usually, cavitation works as a herd of little worms which gnaw on metal surfaces at enviable rate. The solid surface that suffered from cavitation for a long time can be used—after rescaling—in war movies to demonstrate the devastation after air strikes.

From the scientific point of view, this process is called erosion—the mass carryover. The origin of erosion is mainly the bubbles collapsing near the solid surface. When shrinking in vicinity of the surface, a bubble 'conducts' the liquid mass onto the solid surface; the consequence of the fluid strike on the surface was illustrated in the previous section with examples of broken bottles. In that very section, the main participant of that process was shown: the initially spherical bubble which was deformed under the influence of pressure; the center of that bubble is the 'conductor' of liquid: this central part—the jet—hits the surface at high velocity.

Cavitation erosion is one of the many types of erosion; its closest relative is the erosion of turbines in damp air, caused by water droplets striking the surface. In both cases, we have to deal with strikes of water masses[14] at high velocity onto the solid surface: relative hardness is important for the erosion process, and this parameter is the same for these cases. On the other hand, cavitation erosion, of course, has its own special features, like the special form of dependence of the erosion rate on time (see Section 2.4 for details).

Erosion is the most ordinary effect of cavitation, but the macroscale effects of cavitation are important too. Bubbles can merge with each other, and in conditions where the low-pressure zone is large, giant gaseous caverns may appear. These caverns significantly change the flow structure, initiating macroscale deviations in pressure. In simple words, these deviations provoke pressure pulsations, which affect every element of the hydraulic scheme. In the case of a long-time effect, even weak-magnitude pulsations may cause the destruction of some elements; strong pulsations may damage the setup instantly.

Summing up, we may conclude that cavitation is the natural enemy of technical appliances. This is why engineers try to avoid cavitation or minimize it as well as possible. Chapter 2 will be specially devoted to the technical aspects of cavitation, so, we finish our introduction to the subject here.

1.4 The destructive force of cavitation

1.4.1 Types of destruction

As it was mentioned in the previous section, cavitation may exist in various types that can be, approximately, divided in two groups according to the spatial scale of the process: cavitation with separate bubbles (bubble cavitation) or cavitation when huge gaseous caverns are formed in a liquid (developed cavitation). In the first case, cavitation has a small spatial scale: it corresponds to the size of a single bubble. In the second case, the spatial scale of cavitation is macroscopic—it corresponds to the size of the elements of apparatus: wing, propeller screw, etc.

Correspondingly, the effects of cavitation can be divided in two types: destruction of the whole element of a setup or microlevel injuries, which cavitation brings to the surface it comes in contacts with.

The macroscale appearance of cavitation will be discussed in Section 1.4.5—how a cavitation vortex influences the work of a turbine. Microtraumas from cavitation on surfaces are more frequent phenomena, so we discuss them in the following sections right away. Moreover, the consequences of those microdamages contribute to the macrodamage too.

---

[14] In both cases, these 'water masses' are very small: as for the water droplet from wet air, as for microjet from the collapsing bubble.

1.4.2 Surface damage

As it was stated repeatedly above, for engineers, the most important feature of cavitation is the damage provoked by this process. Above, we already have seen one manifestation of the cavitation impact—the breaking of a bottle, caused by a complicated hydrodynamical mechanism, the main part of which is hydraulic shock; it will be discussed in details in Chapter 5. This very arrangement is responsible to the destruction of pipelines or their elements, especially valves, etc.

But, except for the large-scale effects, cavitation may also have an effect at short spatial scales—not as a part of the complex impact, but solely, damaging a surface by local blows. Therefore, this phenomenon leads to the destruction of units working in a liquid for a long time.

Let us take a look at a new surface (Fig. 1.4.1) and on the surface after approximately 10 hours of intense acoustic cavitation on it (Fig. 1.4.2); see Chapter 6 for details on acoustic cavitation.

In common, we may conclude that the outcome of cavitation is considerable. Of course, that surface (actually, this is the surface of an ultrasound waveguide—the special appliance that is placed in a liquid to produce acoustic vibrations) was made specially to suffer from cavitation, but it should also be taken into account that the result presented in Fig. 1.4.2 is obtained only after a couple of hours. The propeller screws that work for months and years get much more serious damage.

On smaller scales, we may see many potholes and large deep hollows on the surface (Fig. 1.4.3). At a closer look, one may also spot that the traces of cavitation actually represent not only a set of dots, but rather some dashes, i.e., dots are connected to one another. The reason behind such connections is that any defect, any microhole on a surface facilitates cavitation in that location: this means that cavitation produces more suitable conditions for cavitation where it has already took place. In Chapter 5, we will discuss the negative consequences of a positive 'cavern–cavitation' relationship; in Fig. 1.4.4, we may see some interesting facts illustrating this matter.

Fig. 1.4.1. A new metal surface. It was not polished well: regular scratches are the traces of the technical processing of rough polishing.

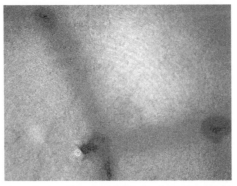

Fig. 1.4.2. The surface after acoustic cavitation. Compare it with Fig. 1.4.1; it may be hard to believe that such a simple phenomenon as cavitation is able to produce such a complex picture.

Fig. 1.4.3. A closer look at the right-hand part of Fig. 1.4.2. This surface is full of small potholes—consequences of microimpacts caused by cavitation.

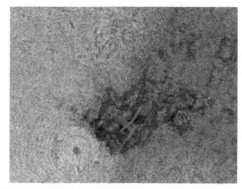

Fig. 1.4.4. A detailed picture of the bottom-left side of the image from Fig. 1.4.4. Here we see a hollow and one more interesting thing: the site that was spared by cavitation.

The large hole presented in Fig. 1.4.4 has no principal difference in comparison with the one shown in Fig. 1.4.3. A more intriguing circumstance: we see that some part of the surface—of a round shape, at the bottom left of Fig. 1.4.4—has suffered much less damage from cavitation than its neighboring sites. Really, we see that cavitation dug a deep hole in vicinity of this 'island of stability', but did not touch it directly.

To understand why this location remained unaffected during cavitation, it is useful to be familiar with the origin of that round site. The nature of this circle lies in the technological process of fabrication of that waveguide.

This waveguide—of a cylindrical form—was made on a lathe in a way approximately presented in Fig. 1.4.5. A blank—a piece of metal stock—was clamped between the live and the dead centers, so that circle is the trace of that center. During the manufacturing process, the future waveguide

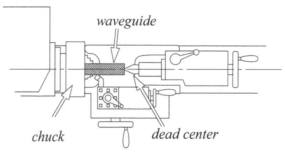

Fig. 1.4.5. The manufacturing of the part presented in previous photos.

was polished naturally at that round site. The result can be distinguished in Fig. 1.4.4: indeed, the structure of the surface looks more 'tidy' on that round trace of a center.

In simple words,the positive causal relationship also means that the absence of the considered matter causes the absence of this matter subsequently. Cavitation has difficulties on a well-polished surface.

1.4.3 The general mechanism of the impact of cavitation on a surface

Cavitation is the formation of bubbles, their growth and collapse. To be exact, the collapse phase was initially presumed as the main reason behind the cavitation damage: from the early works by Stanley S. Cook, this mechanism was considered as the main destructive process.

The collapse of a bubble means, the fast motion of the liquid that surrounds that bubble. When the cavitation process is intense enough—we mean, when the velocity of the bubble wall (i.e., of the liquid on that wall) is high, then one may expect that this moving liquid may affect any obstacle it meets on its path, for example—the solid surface at which cavitation takes place.

The initial problem, the first unanswered question, was to figure out the exact scheme—how can a collapsing bubble transfer the kinetic energy of a fluid to the wall? Actually, some answers are presented in figures from Section 1.2 (especially Fig. 1.2.11), where the bubble collapse in vicinity of a surface is shown. One momentous detail: that bubble was initially formed at this very surface. What we may say for the bubbles that were born in the bulk of a fluid, then approached the solid surface and collapsed close to it—do these bubbles play a role in the cavitation erosion?

Generally, yes. This matter will be thoroughly considered in Chapter 5; here we only depict the ways for a cavitation cavern to cause erosion on a solid surface. First, a bubble oscillating in a cavitating fluid causes pressure waves that may affect the surface directly. Then, the initially spherical bubble in the direct vicinity of the surface may collapse losing its sphericity, as it is shown in Fig. 1.2.13. One may see that this process can be described as follows: a spherical bubble becomes unstable, changes its shape, and the liquid from the central upper part of the bubble streams down toward the surface. This jet—a microjet—plays a crucial role for understanding cavitation erosion.

It should be emphasized that the bubble collapse presented on that figure is far from the ultimate case. That jet (which does not even look like a jet) is thick and, as it can be shown (see Chapter 5), is slow—its velocity is too low; from those snapshots, its velocity can be estimated as several meters per second. Thus, the damage from this jet would be comparatively weak, if it even reaches the surface.

Thus, really, cavitation may produce impacts on a solid surface in vicinity of which the cavitation takes place.

1.4.4 Erosion

Small-scale fluid impacts on the surface cause erosion. The erosion process consists of corrosion of the material—the gradual detachment of pieces from the surface. Moreover, such a detachment (cracking) can be performed even if the stress in the material is lower than the initial ultimate strength.

A periodic collapse of a bubble leads to periodic vibrations of pressure near the surface, that is, to periodic forces acting on the surface. These forces, in their turn, lead to deformations. If the deformation goes beyond the limit, then one may expect the destruction of the given material.

As for the limits, it is worth mentioning here that the limiting parameters of a material are not constant. The well-known Bauschinger effect manifests that for all polycrystal materials (i.e., for almost all the materials that can be met in practice), the plastic deformation of a certain sign leads to worse elastic properties for the deformation of the opposite sign. This effect is usually observed under an external alternating loading; for the surface bombarded by cavitation impacts, we may also presume this effect. Indeed, the same site of the surface gets the compressive stress at one moment

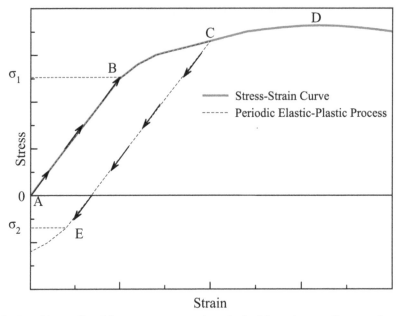

Fig. 1.4.6. The Bauschinger effect. Linear parts correspond to elastic deformations, nonlinear sections of the graph corresponds to the plastic deformations. Initial tensile stress causes plastic strain; further, this material shows reduced elastic properties for compression.

of time (when the fluid impacts it immediately), and the tensile stress at another instant (when the neighboring part of the surface is compressed when the fluid strikes it).

For explanation, let us consider a stress-strain curve (see Fig. 1.4.6). It can be considered either as a dependence of stress in a material during the strain, or the strain caused by the influence of stress. The last one is suitable for our purpose, since erosion is caused by stress, finally.

Point A corresponds to zero stress and strain—the material is not under compression. The line from point A to point B represents the Hooke's law—the deformation is elastic. After point B (yield point), the plastic behavior begins. Until point D, the stress increases with increasing deformation, after—the opposite. It means that with stress variation after the D point (ultimate strength), the fracture of material is inevitable(in our case an elementary erosion event has occurred).

Now let us see what happens to the surface of a material undergoing cyclic elastic-plastic deformation (arrows in Fig. 1.4.6). Consider an element from the beginning of extension: elastic deformation (A to B) occurs first; then goes the plastic deformation (B to C)—the stress is increased. At point C we begin to decrease the stress, but since plastic deformation has taken place, the material does not return to its previous position—shrinking is going in accordance to the Hooke's law (C to E). However, plastic deformation not only caused material displacement, but also decreased the yield stress: $|\sigma_2| < |\sigma_1|$. The subsequent stage of compression with plastic deformation may lead to an even greater yield tensile stress decrease, and so on.

Thus, under the influence of cavitation stresses, the elastic properties of the surface material get worse. As a result, a series of such inelastic deformations sooner or later leads to the rupture of a small part of the surface. In a single word, fatigue occurs—material gets tired of cyclic exposure.

1.4.5 The Sayano-Shushenskaya accident

The large-scale effect of cavitation was observed during the tragedy that took place at the Sayano-Shushenskaya hydro power plant (HPP). The explosion on this plant was astonishing from many points of view; we still remember the question from our colleague when we listened to the radio together radio that day: 'What is there to explode?' Indeed, a HPP has no nuclear reactor or something

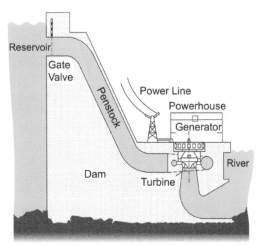

Fig. 1.4.7. The general scheme of the Sayano-Shushenskaya HPP.

else that may blow up: this is, so to say, a huge water mill, where water rotates turbines. Now let us try to figure out, what may explode there.

This accident happened on August 17, 2009 at 8:13 AM. Below, we focus only on the technical part of the accident, and all the details can be found in the accident investigation report by the Federal Environmental, Industrial and Nuclear Supervision Service of Russia (Rostechnadzor 2009, Leonov et al. 2016)

The cross-section of the HPP is given in Figure 1.4.7.

This is a simplified HPP scheme that represents the general principle: power is produced by a generator that is powered by a turbine; the turbine is rotated by the water flow caused by the level difference (e.g., head) between the reservoir and the river level. The Sayano-Shushenskaya HPP is built on the Yenisei river. The construction began in 1963, then in 1978 the first power was produced, and the final construction was finished in 2000. The plant had 10 units (10 turbines, generators, etc.), with a rated power of 640 MW each, so the total HPP power was 6.4 GW. Also, the HPP had a bypass which was used, for example, when the reservoir level was below the gate.

On August 17, 2009, there were 9 working units (Unit No. 6 was shut down for repairs). Unit No. 2 (hereafter, the unit) was under load and also had a priority in power regulation. At 8:12 AM, the automatic power control system reduced the unit's power. At 8:13 AM, the turbine cover studs broke. Due to the influence of water pressure, the unit was thrown upwards. Water began to flow into the powerhouse through the vacated turbine shaft. As a result, the powerhouse was partially destroyed and flooded, and the power generation of the whole HPP was stopped with a blackout. Obviously, the problem with the unit No.2 led to this accident, so we will consider it further.

First, it should be noted that the turbine has a working life of 30 years, and when the accident happened, the turbine worked approximately its full term: 29 years and 10 months. Also, cavitation erosion is a predictable phenomenon for a working turbine. It was fixed twice during overhauls. In 2000, cavitation fractures of the blades (up to 12 mm deep) and cracks were detected. During the repair in 2000, cracks were rectified; cavitation destruction of the blades was also eliminated in 2005.

Concerning the unit operation just before the accident, the last regime points are given in Fig. 1.4.8.

In the figure, the pressure head and power values are indicated, as well as four areas; usually, such diagrams are considered for a constant pressure head. The areas in Fig. 1.4.8 are determined according to the results of a full-scale experiment performed by the manufacturer. Area III is recommended for permanent operation: this area is characterized by the best energy conversion efficiency and minimum pressure pulsations; the rated power corresponds to this mode. Although

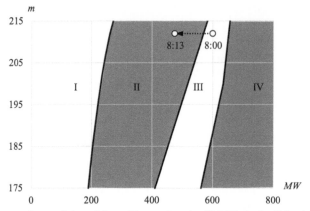

Fig. 1.4.8. The performance characteristics of the turbine used at the SS HPP: hydraulichead (in meters)[15] vs. power (in megawatts). Areas correspond to operating parameters: I—operation is allowed; II—operation is not recommended; III—operation is recommended; IV—operation is prohibited.

Area I is characterized by low energy conversion coefficient, the operation in this mode is allowed. Areas III and IV are separated by a power limiting line, and operation in Area IV is prohibited. Between areas I and III, there is a zone characterized by heavy pressure pulsations, so operation in Area II is not recommended, but obviously it is necessary for the transition from Area I to Area III.

As shown in the figure, the accident happened exactly during operation in Area II. Moreover, in 2009, the unit was in this area more than 200 times. During the last 8 hours of its life, the unit crossed Area II six times, as it had priority in power regulation.

13 minutes before the accident, the radial vibration amplitude (mean value) of the turbine bearing increased from 600 to 840 μm (design value is 160 μm), see Fig. 1.4.9. It should be noted that vibration had exceeded design limits more than a month before the accident. However, the control system was not put into operation (despite being installed)—that is why the unit was not stopped for diagnostics.

Extreme vibration growth caused by regular transitions through Area II led to fatigue failures of the attachment points. Next, we will discuss what causes pressure pulsations in Area II.

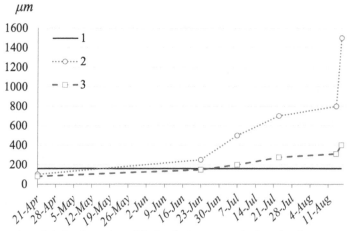

Fig. 1.4.9. Data from the radial vibration sensor of the bearing: pressure pulsations (in micrometers) vs. date. 1—design maximum; 2—actual maximum; 3—mean value.

---

[15] Sometimes it is convenient to use units for pressure other than pascals; here the pressure head in meters of water column is used.

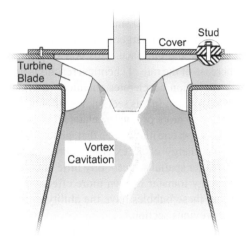

Fig. 1.4.10. Vortex cavitation in a turbine.

At the HPP, turbines of Francis type are used, see Fig. 1.4.10. The water flow is forwarded to the runner (rotor) by the wicket gate. The runner is located in a spiral case and has blades that are mounted between the hub and the crown. At first, the flow moves in radial direction and then in axial. In this type of turbine, a cavitation vortex is formed: it looks like a bubble rope stretching from the runner downstream. Since the flow swirls, this "rope" fluctuates and the vortex core precesses. This precession (extremely intense in Area II) leads to pressure pulsations (pressure waves) of the flow in the turbine. To sum up, it is the vortex cavitation that lead to strong pressure pulsations destroying the cover studs.

Thus, we may conclude and answer the question given above—what can explode where there is nothing to explode. From the common meaning, explosion is an abrupt pressure jump (and it does not matter what was the reason of this jump—TNT, a nuclear reactor or cavitation). The results of this jump are the same anyway.

As for the future of the Sayano-Shushenskaya HPP, it has been producing electricity again since 2011, and in 2014 Unit No. 2 was put into operation. Of course, all units are new—even the ones that were not damaged expended their service life and were replaced. Also, flow (vortex cavitation) stabilizing ribs were installed at the turbine outlet to decrease pressure pulsations in Area II.

1.5 Where can cavitation be used?

1.5.1 Useful features of cavitation

The main point of this section is to demonstrate that cavitation is not the enemy of mankind, as is often presented. It is rather some unique physical process, the abilities of which can be put to good use. The complexity of cavitation, its simple appearance, its destructive potential and non-stationary character have also the reverse—useful—side. Let us take a closer look at this side.

Cavitating bubbles are not the only huge bubbles, the diameter of which is comparable to the size of the vessel, like that giant bubble in a syringe. On the contrary, most bubbles are very tiny caverns, which cannot be observed without special equipment. Thus, cavitation is the process that can provide heterogeneity at very short spatial scales.

The last matter deserves a special consideration. Let us assume that we have one cubic centimeter of liquid. In this volume, one 'huge' bubble may appear—of 1 mm in diameter. The second choice: a thousand smaller bubbles arise in this volume; their total volume corresponds to the volume of a one-millimeter bubble. The question is: how different is the total interface surface in these cases?

The answer is that the volume of a bubble of 1 mm in diameter is 0.52 mm^3; its surface area is π mm^2, obviously. The diameter of one of 1000 small bubbles is 0.1mm; the total interface that is formed by these bubbles is 10 π mm^2—much larger than for a single large bubble.

Thus, first, cavitation 'develops' the interface. This is a very important fact for the so-called heterogeneous catalysis: the increase of the rate of chemical reactions in 2D-geometry in comparison with the full 3D-case. Indeed, cavitation accelerates chemical reactions; for ultrasonic cavitation, this scientific area is called sonochemistry. In this section, we will not discuss this area of application of cavitation; here we pay attention to more primal, mechanical effects.

From the mechanical point of view, one may notice that, generalizing a consideration, cavitation takes the large-scale macroscopic energy of an external source (a pump, an ultrasound acoustic field) and transfers it to much smaller spatial scales—tiny little bubbles. Continuing this reasoning, one may also notice that this energy transfer is even more effective than turbulence, especially if we take into account the fact that these bubbles have the ability to hand over that energy to solid surfaces; we confirmed it in the previous section.

The second fact that can be taken into account is that the total energy that can be invested in cavitation can be very high, almost as high as we want, if we are talking about acoustic cavitation caused by ultrasonic emitters.

In sum, we have a quite attractive mechanism, which may receive substantial portion of energy on a large spatial scale and dissipate it on microscales. It is especially convenient, if we want to affect solid surfaces, but this is optional; see below about emulsion preparation with cavitation mills.

Almost all the examples listed below in this section illustrate this common principle of cavitation use. One important exception concerns the case, when we turn cavitation against itself; see Section 1.5.5.

1.5.2 Cleaning

In Section 1.4, we saw that cavitation can damage surfaces, and this is the dark side of cavitation. However, this side can be easily turned into the light one: cavitation can be used for cleaning surfaces. It cleans a wide range of items, from truck parts to dental implants and jewelry.

A bath is used as a device for cavitation cleaning. Usually, it is made of stainless steel, with sources of mechanical vibration attached to the bottom (see Fig. 1.5.1). The container of the bath is filled with a cleaning solution or water, then the object to be cleaned is immersed in the bath, trying not to touch the walls or the bottom of the bath with this object; usually, special nets or hangers are used for this.

After the part is in the solution, the sources of mechanical vibrations are turned on. Typically, such sources are piezoceramic emitters[16] with a frequency of 22 kHz and higher, depending on the model. Sometimes magneto strictors are used, despite their bulkiness in comparison with piezoceramic

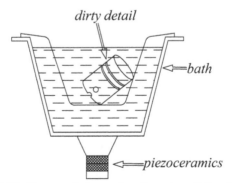

Fig. 1.5.1. The general scheme of an ultrasonic cavitation cleaner.

---

[16] For details about acoustic cavitation and the sources ultrasonic irradiation, see Chapter 6.

emitters; they are more powerful than piezo transducers—this is a significant advantage for the considered problem. The emitters are attached to the body of the bath with glue or sometimes they are screwed or welded in place. In some baths, only one source is used; in others,several devices of the same frequency are applied; there are setups in which ultrasound sources of different frequencies are used. By operating with different frequencies, it is possible to achieve a better cleaning effect on the part being cleaned. Sometimes such baths are equipped with a heater.

The basic principle of operation of this machine is clear: an ultrasonic emitter creates cavitation which damages not the surface of the body surface itself, but destructs dirt and sediments on that surface. To be exact, the work of the source of mechanical vibrations leads to the appearance of compression and tension waves in the tank which, in their turn, cause the appearance of cavitating bubbles. The collapse of these bubbles on the surface of the part being cleaned leads to the mechanical destruction of contaminants. Thus, the added detergents and solution heaters only contribute to the cleaning of the surface, the main work is taken over by cavitation.

Of course, the positive effect of such a machine can be achieved only if the processing time is not too long; otherwise,there is a chance to clean the surface along with the removal of parts of that very surface.

We may also mention the ultrasonic washing machine here—a small device which generates ultrasound waves. This machine is placed in a bath with dirty clothes, where it produces cavitation which, in its turn, provides some cleaning effect. Such appliances were popular 10–20 years ago, but due to their restricted power, the washing output was not very impressive.

1.5.3 Cavitation mills

Cavitation mill is first and foremost a mill. Its function is to disintegrate the substance placed in this device. From this point of view, cavitation mill belongs to a large family, the head of which is, undoubtedly, the old reliable windmill.

In comparison to a windmill, however, a cavitation mill has one great advantage: it is possible to obtain very small size of dispersed fractions with its help. We can reach the micrometer sizes of grains or even less.

The scheme of a cavitation mill is presented in Fig. 1.5.2. A fluid with the dispersed phase that must be fractioned is placed into the cavitation mill, the main part of which is a disc with a developed surface. When rotating, this disc, due to its developed surface, generates intense cavitation. Small

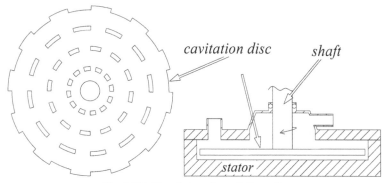

Fig. 1.5.2. The scheme of a cavitation mill.

bubbles do what is beyond the power of solid millstones: disintegrate particles into very small pieces.

Note that a cavitation mill can be used not only to crush solid particles, but also to prepare emulsions: liquids with microscopic droplets of another liquid.

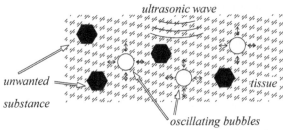

Fig. 1.5.3. Irradiation with ultrasound. Cavitating bubbles appear in the tissue and help ultrasound to eliminate targets—
some undesirable objects that are located in the tissue.

1.5.4 Ultrasonic medicine and cosmetology

Probably, any invention must be tried in two fields: in medicine and in production of military weapons. Here we consider the first option.

Medicine procedures use a very wide range of implements, including irradiation of various physical nature: most famous of them are X-rays, or course, but here we are interested in radiation from a different source—irradiation by ultrasound.

The ultrasonic irradiation means the impact of sound waves (of frequency greater than ~ 20 kHz) from the source on the investigated object, which is tightly connected to that source (see Fig. 1.5.3). This tight connection can be realized in various ways: the source of ultrasound can be located directly in vicinity of a tissue, or a patient can be placed in a water-filled bath, and the acoustic wave is focusing on the processed body organ; in the last case, we use the fact that water has approximately the same acoustic impedance as the human body (see Chapter 6 for details concerning the propagation of sound waves at the boundary of two media), because, as the first approximation, all of us are essentially walking water masses. This is a well-known undeniable fact.

As a particular case of acoustic influence, the ultrasound effect is the impact of pressure waves with a magnitude of ~ 10^7 Pa or may be even higher. These pressure waves may cause different direct actions which, probably, are familiar to you: for instance, the ultrasonic cleaning of teeth where, generally, the interaction of ultrasound and a fluid (a special solution) also provides some cavitation effects.

Here we are interested in a more dramatic interaction of ultrasonic waves and tissues. If the target medium is a liquid-contained substance (like our flesh), then the consequences of such an impact may include cavitation as an important intermediate in the process. The formation of cavitating bubbles may increase the effect of ultrasound on the tissue due to features listed in Section 1.5.1, but, actually, sometimes the role of cavitation looks questionable (Brennen 2015).

We consider two variants of the scheme presented in Fig. 1.5.3: for medical purposes and for the cosmetology ones. In both cases, we want to destruct some objects contained in the given tissue with the help of destruction potential of cavitation. The main idea is the same as in two previous sections: cavitating bubbles create mechanical impacts on targets, accumulating there the energy of the applied ultrasonic field.

The medical effect of cavitation can be illustrated with lithotripsy—the therapy that uses ultrasound to eliminate, for instance, kidney stones. Of course, ultrasonic irradiation cannot disintegrate those stones completely into separate atoms, but it can grind them up to pieces small enough to be ejected through natural biological processes. Thereby, here we see some sort of a cavitation mill again, so to say.

In cosmetology, a similar effect is used for liposuction (removal of adipose tissue) as a part of a common procedure, or as an independent manipulation (that is, without surgical operation). In the considered case, the targets are not stones, but the adipose cells. To destroy them, a very short spatial scale of influence is needed, and cavitation provides impacts on such microscales.

Of course, cavitation may also cause a negative influence in living tissue; we recommend to review (Brennen 2015) where various aspects of the cavitation effect on biological objects are considered.

1.5.5 Hydrofoil boats

This is not precisely an example of the useful properties of cavitation. This is a slightly more complicated matter; it is rather a story about how to adapt to cavitation.

Today, virtually everybody knows the wing principle (see Fig. 1.5.4).

When such an object as depicted in Fig. 1.5.4—it can be called a wing, or a profile, or a foil—moves in a medium, the pressure difference is created on different sides of it. Because of a special shape of this foil, the low-pressure zone is created on the upper side, while the high-pressure zone is created at the bottom side.[17] This is a well-known fact, and almost a half of hydrodynamics is based around that wing.

The idea represented in Fig. 1.5.4 can be applied not only for bodies moving in air, the same principle can also be used for a motion in a liquid; indeed, from the point of view of hydrodynamics, both gases and liquids are approximately the same media, and one may consider them in common; the corresponding science is called 'fluid dynamics'. Moreover, for the first time, the idea of using the lifting force for ships was implemented around the same time as for airplanes: the airplane constructed by the Wright brothers ('Flyer I') took off in 1903,[18] and the hydrofoil boat by Enrico Forlanini has been successfully tested in 1905.[19]

Thus, hydrofoil boats work on the same principles as aircrafts; possibly, the direct translation of the Russian term 'ship on underwater wings' better explains the fundamentals of operation of these objects. Indeed, when in motion, these ships rise above the water like on wings, because they exactly lean on wings, on underwater wings. Most likely, you already have seen such ships.

One may imagine broad horizons for hydrofoil boats. Moving in such a manner so that the main part of a ship's hull is out of the water, this ship meets low hydrodynamic resistance. Thus, really, one may expect high, very high velocities from such boats.

Partially, this is the case. However, unfortunately, our friend—cavitation—decisively interferes with this idyll. As we mentioned above, the upper part of hydrofoil is under a low pressure; the higher the speed of the boat, the lower the pressure in that region. As we know, a sufficiently low pressure causes cavitation, and vapor bubbles arise on the upper part of a hydrofoil when a boat moves at high velocity. These bubbles provide the whole set of cavitation effects: from erosion to the increase in hydrodynamic resistance, and, consequently, a very high speed cannot be achieved by hydrofoil boats. Now we see that cavitation restricts the work of a whole class of ships, at first glance. There exists a radical way to fix this problem.

'If you cannot beat them, join them'. This English proverb reflects the crux of the considered matter. By the way, again, the Russian equivalent of it can be translated as 'If you cannot prevent it, then lead it'—a slightly different shade. Now we apply this proverb to the problem of cavitation on hydrofoils.

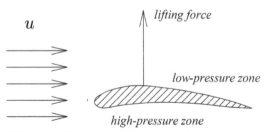

Fig. 1.5.4. The lifting force of wing. In air, this device works more predictably than in water.

---

[17] We consider a general case; indeed, the abnormal pressure—the low one above or the high one below—can be created only at a single side of the wing, while the pressure on the opposite side is 'normal'.

[18] For those who are concerned with this matter: we did not state that Flyer I was the first airplane. For those who are not concerned: there are many versions about 'who was the first' in aircraft construction, about as many as for the invention of radio. We have no personal opinion on this subject.

[19] Again, we do not underestimate the roles of Charles de Lambert, John Isaac Thornycroft and others.

Actually, separate bubbles are not the only form of cavitation. At high velocity, all the bubbles are merged into a single giant cavern on the upper surface of a hydrofoil—this mode is called supercavitation,[20] or developed cavitation. In the case of supercavitation, we lose a significant part of the lifting force—now we can rely only on the lifting force generated by the bottom side of the hydrofoil. But, on the other hand, we even obtain some advantage in the form of reduced drag force and, which is also important, in this mode the cavitation erosion is negligible.

Thus, it is possible to solve this problem 'from the opposite end'. Pre-cavitation mode allows hydrofoil boats to reach high speed, while the supercavitation one lets them get super-high speed. Generally, the considered matter can be related rather to the positive influence of cavitation, although not undeniable.

1.6 Physical problems in cavitation

1.6.1 What does an ordinary human expect from a scientist?

We decided to start a consideration of a fundamental and complicated matter with a much more fundamental and complicated matter. What does society want to get from the scientific community?

Abandoning the philosophical matters, we cut straight to the point. The times when an everyman was admiring pure science are gone. An everyman may possibly listen to a brief about black holes or something similar, but not for long. The taxpayer wants to see what science can bring to our life, and, of course, he or she means something useful, especially useful in everyday life. The reports about creation of a tremendous weapon give the taxpayer mixed feelings; the character of those feelings strongly depends on a certain country, first of all.

Timid attempts of scientists to explain that these expectations must be addressed to engineers, but not to scientists, were refuted long ago: taxpayers are not inclined to pay attention to such a small beer like the difference between a scientist and an engineer.

This battle has ended in the last quarter of the XX century, and the modern generation of scientists accepted the rules of the game. Everybody who remembers Einstein's sentence that 'Science exists for Science's sake, like Art for Art's sake, and does not go in for special pleading' had been eliminated quickly, quietly and decisively. What are the times, such are the customs.

But let us take a look at the problem from another angle. Further development of engineering is unthinkable without the development of the fundament of engineering–science. Today, scientists have to spend more time on practical problems, but this circumstance does not mean that they must retrain as engineers. Moreover, this fact does not even mean that scientists must consider practical problems all the time. In a good balance, it is possible to combine practical science, that is, investigations of the processes that have direct applications today, with the fundamental ones. This method is all the more preferable if we take into account that some matters that look impractical today can become vital tomorrow.

An actual example: this book was written in the time of the incredible pandemic of COVID-19. Quarantine measures were introduced by governments irregularly, some countries fluctuated between different scenarios; some measures were admitted insufficient and ineffective afterwards, and so on. It turned out that we have no proper methods to predict the development of an pandemic— we have no suitable mathematical models which may forecast the spread of a disease depending on key parameters. Even concerning such a basic measure as quarantine, governments had to rely on their own feelings and opinions of those scientists who were sufficiently brave to accept the responsibility. Meanwhile, generally speaking, population evolution[21] is a sufficiently old scientific area, and it is very sad that this area was not ready for a serious test. Possibly, the reason is that this area was always considered as somewhat theoretical, as somewhat almost redundant. Now we are in dire need of such models, but we do not have them. By the way, Nikolay Timofeev–Ressovsky, a

---

[20] This kind of cavitation is discussed in Chapter 4.

[21] Here we mean the virus population, of course.

biologist and specialist in population genetics, predicted more than half a century ago that biology needs proper mathematical models; in his opinion, that matter was one of the actual problems of contemporary (for him) biology.

Returning from such global questions to a particular one, we suppose that a complex scientific description of the problem automatically becomes of practical meaning. This is the line that we follow throughout this whole book which is devoted to cavitation—a phenomenon which has direct use for engineering. It is impossible that a well explained phenomenon would not find practical applications.

The main application of cavitation in technology is specific: strictly speaking, cavitation is damage; even beneficial sides of cavitation use mainly this ability of cavitation. Cavitation distorts any technological object it can reach, that is, any device that works in a fluid. These small bubbles limit the work of giant ships, hydraulic turbines, pipelines and of many other important and, last but not the least, expensive objects. Thus, in many expositions the phenomenon of cavitation is described like a disease, with corresponding attitude to the cavitation science: explore to eliminate.

However, any scientist loves the object of their investigations. Moreover, in comparison with spiders, cockroaches and leeches, cavitation looks quite respectable, and in this book, we concentrate our attention on the scientific problems of cavitation, not on the termination of that phenomenon at all. Below, we will discuss the physical features of cavitation, and we believe that a better understanding of them would help in very utilitarian matters sometime.

1.6.2 Complexity of cavitation

For an engineer, cavitation is a headache. For a scientist, cavitation is a magic box: among the problems related to cavitation, physicists feel themselves as children in a toy store.

Above, we stated that cavitation is a fundamentally non-stationary phenomenon. Thus, from the theoretical point of view, one may leave any attempts to construct a stationary theory of cavitation. This approach is somewhat against our nature, since the main part of the theoretical apparatus of applied physics is based on stationary models. Alas, non-stationarity is the only way; in Chapter 4, we will consider some stationary attempts to describe cavitation (developed cavitation on an obstacle) and what mathematical tricks follow from such attempts.

Another feature of cavitation is the microscale of all the significant processes. Huge bubbles are only the 'top of an iceberg': the most important (we mean the most harmful) effects are caused by small bubbles which are barely visible by eye. Meanwhile, these very tiny bubbles drill holes in surfaces, and these very objects should be investigated, especially if we are really concerned about the application of science.

Combining these two factors—non-stationarity and small scales—we get a very inconvenient object for experimental investigation. This difficulty will be illustrated in Chapter 6, where we present oscillations of a small bubble. To be precise, on a relative scale this bubble is huge—its size is ~ 0.1–1 mm, but as an object of observations, this is a very difficult patient.

On the other hand, one may state something like 'What is the point in watching an oscillating bubble? A spherical bubble alternates its radius; there are no interesting matters here'. This argument is wrong at least in one point. Not in the part of absence of interest—this is, at least, a personal approach. But it is wrong that an oscillating bubble has a spherical shape: observations presented in Chapter 6 show this clearly. During the expansion phase, even a small bubble, the size of which is less than 1 mm, looks like a potato.

Such facts as non-sphericity of bubbles slightly complicates the problem for theoretical investigation, to say the least. The mathematical implementation applied in physics aims at the derivation of clear results, which may help to better understand the physical phenomenon and to obtain some quantitative parameters of the process. As for the last point, recently, computers began to play a similar role in many engineering areas: computational fluid dynamics can provide appropriate results for some problems for which a large set of experimental data exists, that is, the procedure gently called the 'verification of computing code' is possible. But anyway, the first goal—to see a physical picture clearly—cannot be reached without laconic mathematical expressions, and this is a difficult point for theorists. They have a dilemma: either to deal with insolvable equations or to find a solution for equations describing a very simplified problem. What is preferred: to solve the equation for a spherical bubble, knowing that it corresponds rather to a cloud of a common shape, but not to a sphere, or to tinker with a huge mathematical equation, the analytical solution of which is impossible anyway without other simplifications, if not concerning the shape, but for some another feature?

For some purposes, to obtain more common results (to be exact, semi-results), the first way can be used. Solutions in the form of huge series are obtained in these approaches; the truncation of these series may give some recognizable expressions.

Usually, however, theorists prefer the second way—initial simplification,[22] and the poor accuracy of experiments helps them in that difficult choice. Indeed, all believe that verification of a theoretical model can be provided only with an experiment,[23] but, if so, the model may not claim to have a better accuracy than the experimental data, even if the reasons for errors differ in the experiment and in theoretical manipulations. For that example: an experimenter cannot fix the real size of a fast-oscillating bubble with good accuracy, since he has no reliable reference for objects deep in a large vessel (due to the perspective and other similar reasons), while a theorist simply puts the spherical shape of a bubble in calculations. It is quite possible that after a comparison of their results both will be happy.

This way, actually, opens a very wide path for theoretical combinations. If we do not aspire for fine accuracy of the derived formulas, then we may try to use a lot of tricks, and the first of them is the analysis of dimensions. Many expressions can be constructed from the simple reasoning of a sort 'this variable must depend on that and that, and it is easy to see that the second one must be at the power of 2'. We do not abuse this method below, but actually, it helps, especially in situations when other methods failed.

The third side of the complexity of cavitation is the variability of physical processes that can or even must be taken into consideration. Actually, this matter is so wide that it is better to explain every element of the puzzle called 'cavitation' separately.

1.6.3 Bubble dynamics

We have already begun to discuss this point above, so it is logical to continue here. Of course, the dynamics of a bubble plays an invaluable role in cavitation. A bubble is, so to say, the heart of cavitation; thus, continuing this line, we may conclude that this beast has many hearts.

It is also logical to consider a single bubble first and hope to discuss the multi-bubble case later. Here, however, we have two different problems.

The first problem is that bubble dynamic, even of a single bubble, is so complicated that we cannot complete its description properly. This matter concerns some mathematical difficulties that were mentioned above; moreover, we cannot even properly formulate the very physical problem. What are the conditions in that bubble? Does it remain adiabatic, i.e., does it lose or gain heat from the liquid, or it saves its temperature constant? What are the processes at the gas–liquid interface, how much vapor may condensate on that boundary or, otherwise, can be evaporated from there?

---

[22] There also exists a third way: to construct physical models based on the initial physical properties of the process, not on the common mathematical equations. Generally, this a worthy way, but in practice, usually, it gives awful results full of incomplete relations with unknown parameters in need of adjustment to experimental data.

[23] In this book, you will find a couple of examples when the experimental data was interpreted erroneously. A correct experimental data includes a correct theory that provides the propriety of measurement and its interpretation. Thus, a theory cannot be affirmed by an experiment: it can be affirmed only by the combination of an experiment and another theory.

All these matters are discussed in Chapter 7, where some results for various conditions are obtained. But the second problem of the bubble description remains: how do the results for a single bubble relate to bubbles in a cavitation cloud?

Some simple consequence can be derived from the theory of a single bubble and applied to the multi-bubble case. For instance, it turns out that oscillating bubbles interact with each other; this result can be obtained theoretically and corresponds to experiment data (see Chapter 6).

However, one—and the most impressive—effect of cavitation raises a question whether single-bubble oscillations and multi-bubble cavitation are the same phenomena? The statement in question looks strange, but for that effect, called sonoluminescence, everything is strange. We will return to this effect below.

1.6.4 Hydrodynamics of cavitation

This is a continuation of the previous matter, but with specific details. The description of bubble oscillations is based on some simple hydrodynamical equations, but the full picture—the mathematical description of a cavitating flow—is much more complicated.

In Chapter 4, we consider the Navier–Stokes equation and its particular case—equations for velocity potentials—for the cavitation problem. Mainly, we are interested there in developed cavitation—the flow where cavitation represents a giant cavern behind a body.

Instabilities are a substantial part of hydrodynamics, and in Chapter 3 we also discuss two instabilities at the liquid–gas interface: the Kelvin–Helmholtz instability and the Rayleigh–Taylor one. Moreover, these instabilities can be considered together, and this consideration gives amazing results.

1.6.5 Hydraulic shock

This is a phenomenon that has as an intrinsic value as a part of cavitation-related problems. We have already mentioned this effect in previous sections, when the destruction potential of cavitation was discussed.

The simple meaning of this term can be explained with its second name—water hammer. A water mass impacting an obstacle is a hammer indeed, and the pressure of that mass moving at velocity v to an obstacle is

$$p = \rho v c \qquad (1.6.1)$$

where ρ is the density of the fluid and c is the speed of sound there; this equation will be derived in Chapter 5. For water, $c \approx 1500$ m/s; thus, a fluid that moves at velocity of only 1 m/s gives pressure of ~10 atm on the obstacle. Note that the velocity of the fluid in collapsing bubbles was estimated above as several meters per second; therefore, such a jet brings significant energy, but not enormous. On the one hand, this value of pressure does not look critical, since metals may handle much more stress. On the other hand, velocities of ~ 1 m/s are the lower limit of such jets (note that the jet captured on the video was too fat and, therefore, too lazy); normal values are greater by one or two orders; maybe, even by three orders, but this is a matter of discussion, see Chapter 5.

Meanwhile, the main effect of cavitation on a solid surface—cavitation erosion—can be described exactly with this mechanism—by water hammers of microjets.

1.6.6 Negative pressure

This is another 'side' problem of cavitation. Negative pressures are often mentioned together with cavitation, but it is hard to obtain them experimentally for us (it is much easier for plants, but we, unfortunately, are not plants; see Chapter 3). Liquids resist to produce negative pressure inside them, preferring another type of behavior in such conditions—the rupture, which is also called cavitation in our language.

However, this matter—the existence under negative pressure, which can be considered from our point of view as anti-cavitation—is an interesting hypostasis of a liquid. See Chapter 3 for details.

1.6.7 Electrization

This feature opens a set of mysterious sides of cavitation. In some types of cavitating flows—in throttles, for example—the fact of electrization was established experimentally and firmly.

Such electrization, a kind of triboelectricity, causes very special effects in flows of such type. As it often happens, electricity produces plasma wherever it is possible—in this case, inside bubbles. As often happens, plasma produces light, which can be registered and analyzed.

Plasma spectroscopy is a reliable, but delicate instrument to understand the state of a medium. The measured data must be analyzed carefully, otherwise its interpretation can be absolutely wrong. This matter is discussed in chapters 8 and 9: all these questions concerning the difference between the rotational and vibrational temperatures of gas (it is impossible to explain it briefly here, if you need details—go to Chapter 8), the continuum, and so on. But, if successful, we may establish the composition of the light-emitting medium along with some parameters of its state.

All these operations are possible for some types of flow. The reasonable question: does such type of electrization take place in a common case, i.e., during cavitation necessarily? We discuss this matter in Chapter 8 too, but, indeed, there is no certain answer yet. Some theories predict this phenomenon, but, actually, the fundamentals of these theories are questionable, and their goals (to explain light emission during cavitation) may be erroneous. Some experimental data shows something of that sort, but, on the other hand, this data can be interpreted differently. Electrization of a fluid during cavitation is an open matter today, but this fact does not mean that it cannot be discussed. On the contrary, this is a vivid subject, the consideration of which helps to understand not only the possible (or impossible) appearance of electric charges during cavitation, but also the cavitation process itself and, maybe, to take cover from an even more mysterious phenomenon.

1.6.8 Light emission

Light emission from a fluid is usually called sonoluminescence. This is a common term. Taking into account the theory that terminology is required for understanding, one can be satisfied with this term: today, many people have heard this word. However, this term is not exact. 'Sono' means 'sound-related', and, indeed, the first observations of light emission from a fluid were in conditions of acoustic cavitation.

With time, on the other hand, light emission from fluids was observed in quite different conditions: for example, during the flow in the throttle mentioned above. In this case, another term—hydroluminescence—is preferable. The first question, which remains unanswered to date is: are hydroluminescence and sonoluminescence similar phenomenon?

The nature of light emission from a fluid is absolutely mysterious, or, it is better to say, annoyingly mysterious. It seems that science has an answer for much more complicated questions, but a luminous bubble does not surrender,and it is difficult to guess the date when this fortress will fall.

The consideration of that mysterious phenomenon finishes our book with Chapter 9.

1.6.9 Practical problems

Thus, we briefly navigated the reader around our book; around all the chapters except the next one.

First, we will consider engineering problems of cavitation. Usually, the engineering approach means minimum equations and maximum correlations; we will not, however, overuse the relations too. The first aim of that consideration is to give a representation of the practical problems of cavitation.

Engineers cannot wait until scientists propose proper and simple physical models which will allow them to design hydraulic systems with minimal cavitation. Thus, some practical methods were developed: as usual, these methods significantly rely on empirical data. Such an approach allowed people to solve the designing problems, and, first, they are much better than nothing, second, some of them work well, and third, in some problems the engineer's feeling outstrips the scientific methods. Thus, it is useful in every way to begin the investigation of cavitation with the engineering approach.

However, we believe that someday 'cavalry will come to rescue', and scientific approach will also bring a full-scale help into such a practical area as cavitation.

Conclusion

Cavitation is a complex phenomenon which takes place in a fluid when the outer pressure suddenly decreases below the value corresponding to the saturation pressure. At such thermodynamic parameters, the formation of the vapor phase is preferable, and a part of the liquid transforms into vapor; however, usually this is a negligible part. Bubbles—both the initial bubbles that were located in the liquid and the bubbles that have been formed—expand, since the pressure inside them is equal to the saturation pressure (or higher because of the presence of some air inside them) while the outer pressure, in the initial condition, is lower.

Simultaneously, the surface of the fluid evaporates too, so the outer pressure increases, because of the increase of the vapor mass there. In a closed volume, such evaporation establishes the saturation pressure; thereby, the difference in the pressure inside the bubble and out of it vanishes. The bubble expansion tends to stop, but the dynamic picture of cavitation usually prevails, and oscillations of a gaseous cavern can be observed.

Cavitation can be obtained easily in domestic experiments: on the example of one of them, we recommend to explore the morphology of cavitation, but we do not recommend the other one. Cavitation in a syringe is a simple and an almost safe trick. You may orient the syringe differently with respect to gravity, pull the syringe's piston at different rates, and observe various types of cavitation. We are sure that you can get more impressive cavitation patterns than those which are presented on photos in this chapter.

The bottle trick—to break a half-full bottle with a palm of a hand—is dangerous, and not a recommended way to explore the destructive force of cavitation. Indeed, in domestic conditions, this is an almost useless exercise, since you will only see the result—a broken bottle, shards everywhere, and water pouring on your shoes. Without a high speed camera, this experience cannot provide more information. Moreover, in the worst case, you may be hurt with the debris. From the physical point of view, cavitation observed in a bottle is a part of the common hydraulic shock that damages the bottle at its bottom.

Mainly, cavitation interests the engineers because of its potential for destruction. Both the large-scale effect of cavitation, which may damage huge constructions, and cavitation erosion— the microscale process that renders elements of constructions out of order—cause problems that, actually, cannot be solved properly. For some engineers, words 'cavitation' and 'accident' sound synonymous.

On the other hand, we should not consider cavitation as a permanent curse of hydrodynamics. The destructive force of cavitation can be directed to our benefit; the simplest example when destructive forces do useful things is surface cleaning: during this operation, we destroy not the very surface, but the sediments on it. This positive effect can be applied in many fields, including even medicine: the force of cavitation may be used in accordance with our will, and if so, we can get profit from this phenomenon.

Surely, cavitation is a fertile soil for scientists. It provides a lot of interesting riddles, questions and, possibly, new horizons that are still undiscovered (we mean the mysterious light that often accompanies cavitation—the so-called sonoluminescence). As we will see from the following chapters, the 'accompanying phenomena' for sonoluminescence cover an immense range of scientific problems: from the rate of bubble collapse to plasma spectroscopy, from the lifting force of a wing to triboelectricity.

Let us follow this way.

References

Avdeev, A. A. 2016. *Bubble Systems*. Springer Nature. https://doi.org/10.1007/978-3-319-29288-5.

Brennen, C. E. 1995. *Cavitation and Bubble Dynamics*. Oxford University Press: New York.

Brennen, C. E. 2015. Cavitation in medicine. *Interface Focus*, 5: 20150022. https://doi.org/10.1098/rsfs.2015.0022.

d'Agostino, L. and M. V. Salvetti (eds.). 2007. *Fluid Dynamics of Cavitation and Cavitating turbopumps*. SpringerWien, NewYork. https://doi.org/10.1007/978-3-211-76669-9.

d'Agostino, L., and M. V. Salvetti (eds.). 2017. *Cavitation Instabilities and Rotordynamic Effects in Turbopumps and Hydroturbines*. Springer International Publishing. https://doi.org/10.1007/978-3-319-49719-8.

Daily, J., J. Pendlebury, K. Langley, R. Hurd, S. Thomson and T. Truscott. 2014. Catastrophic cracking courtesy of quiescent cavitation. *Physics of Fluids*, 26: 091107-1–2. https://doi.org/10.1063/1.4894073.

Franc, J.-P., and J.-M. Michel (eds.). 2004. *Fundamentals of Cavitation*. Kluwer Academic Publishers: Dodrecht.

Ganiev, R. F., and M. A. Ilgamov. 2018. Liquid-column cavitation under motion of its lower boundary. *Doklady Physics*, 63: 362–5. https://doi.org/10.1134/S102833581809001X.

Kiyama, A., Y. Tagawa, K. Keita Ando and M. Kameda. 2016. Effects of a water hammer and cavitation on jet formation in a test tube. *Journal of Fluid Mechanics*, 7: 224–36. https://doi.org/10.1017/jfm.2015.690.

Leonov, G. A., N. V. Kuznetsov, and E. P. Solivyeva. 2016. Mathematical modeling of vibrations in turbogenerator sets of sayano-shushenskaya hydroelectric power station. *Doklady Physics*, 61: 55–60. https://doi.org/10.1134/S1028335816020105.

Manickam, S., and M. Ashokkumar (eds.). 2014. *Cavitaion. A Novel Energy-Efficient Technique for the Generation of Nanomaterials*. CRC Press.

Minakov, A. V., D. V. Platonov, A. A. Dekterev, A. V. Sentyabov, I. M. Pylev and A. V. Zakharov. 2015. Use of methods of mathematical modeling to analyze low-frequency pressure pulsations in the continuous run of high-head HPP. *Power Technology and Engineering*, 49: 90–97. https://doi.org/10.1007/s10749-015-0580-8.

NIST. Data from NIST Standard Reference Database 69: Chemistry WebBook.

Pan, Z., A. Kiyama, Y. Tagawa, D. J. Daily, S. L. Thomson, R. Hurd and T. T. Truscott. 2017. Cavitation onset caused by acceleration. *PNAS*, 114: 8470–4. https://doi.org/10.1073/pnas.1702502114.

Pflieger, R., S. I. Nikitenko, C. Cairós and R. Mettin. 2019. *Characterization of Cavitation Bubbles and Sonoluminescence*. Springer, Cham.

Pirsall, I. S. 1972. *Cavitation*. Mills and Boon Limited: London.

Platonov, D., A. Minakov and A. Sentyabov. 2019. Numerical investigation of hydroacoustic pressure pulsations due to rotor-stator interaction in the Francis-99 turbine. *Journal of Physics: Conference Series*, 1296: 012009. https://doi.org/10.1088/1742-6596/1296/1/012009.

Federal Environmental, Industrial and Nuclear Supervision Service of Russia (ROSTECHNADZOR). The act of technical investigation into the causes of accident, occurred on august 17, 2009 in the branch of the open joint-stock company "RusHydro" - "Sayano-Shushenskaya GES P. S. Neporozneg". *Technical Report* (in Russian).

Shah, Y. T., A. B. Pandit, and V. S. Moholkar. 1999. *Cavitation Reaction Engineering*. Springer Science. https://doi.org/10.1007/978-1-4615-4787-7.

Tullis, J. P. 1989. *Hydraulics of Pipelines*. John Willey & Sons.

Yasui, K. 2018. *Acoustic Cavitation and Bubble Dynamics*. Springer Nature. https://doi.org/10.1007/978-3-319-68237-2.

Young, F. R. 2005. *Sonoluminescence*. CRC Press.

Yukisada, R., A. Kiyama, X. Zhang and Y. Tagawa. 2018. Enhancement of Focused Liquid Jets by Surface Bubbles. *Langmuir*, 34: 4234–40. https://doi.org/10.1021/acs.langmuir.8b00246.

Zhao, L., Y. Yang, T. Wang, L. Zhou, Y. Li and M. Zhang. 2020. *A Simulation Calculation Method of a Water Hammer with Multipoint Collapsing*, 13: 1103. https://doi.org/10.3390/en13051103.

Zudin, Y. B. 2019. *Non-equilibrium Evaporation and Condensation Processes*. Springer Naure. https://doi.org/10.1007/978-3-030-13815-8.

CHAPTER 2

Cavitation in Engineering

2.1 Cavitation in pipes

2.1.1 Cavitation number: General approach

The initial cause of cavitation is always the same—low pressure in the system. Above, in the first chapter, we reasoned that pressure must be lower than the saturation pressure p_s at the given temperature, which is a value of $\sim 10^3$ (for instance, ~ 2337 Pa at 20°C);[1] for certain estimations, one may expect cavitation exactly at the outer pressure p_s.

There exist many ways to create a low-pressure zone artificially; simple examples were presented in the previous chapter: exercises with a syringe and a bottle. In all those cases, cavitation was summoned after our special efforts, including the health risk during the bottle destruction. Alas, in many cases, cavitation can be caused without our desire—due to the basic law of hydrodynamics. The Bernoulli equation, in its simplest form, states that along the streamline the quantity

$$p + \frac{\rho u^2}{2} \tag{2.1.1}$$

(p – pressure, ρ – density, u – velocity) remains constant, see Appendix B. In a general case, one may take into account the variable density and include gravity; also, here a derivation of the hydrodynamic potential on time may be added. We restrict our consideration to a stationary flow of an incompressible fluid, where gravity does not play a role.

Thus, if pressure and velocity are set at the given point of the flow, for instance at the entrance to a channel, then the pressure at another point is determined by the velocity at that point:

$$p = p_0 + \frac{\rho u_0^2}{2} - \frac{\rho u^2}{2}, \tag{2.1.2}$$

where the parameters determined at the given point are denoted by index '0'.

Then, the velocity in the given cross-section of the flow depends on the mass discharge G in the channel and on the area of this cross section S. For a stationary flow, the mass discharge must be constant; thus,

$$G = \rho u S = const. \tag{2.1.3}$$

Here we consider u as the mean velocity of the flow in the given cross-section: of course, velocity also depends on the coordinate across the channel (of radius of the given point in a round pipe, for example).

---

[1] All the reference data taken from the NIST database.

Thereby, we see the following chain leading to cavitation: area of cross-section decreases → velocity increases according to (2.1.3)→ pressure decreases according to (2.1.2); if the final pressure is equal or below p_s, cavitation takes place.

Let us estimate the parameters that may cause cavitation in a pipe. Let us consider a pipe of a variable cross-section (see Fig. 2.1.1): at the entrance, in the wide section of a pipe at diameter $d_0 = 5$ cm, pressure p_0 is 2 atm (that is, $2 \cdot 10^5$ Pa) and mass discharge of water is 2 kg/s.

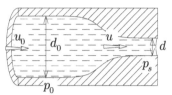

Fig. 2.1.1. A pipe of a variable cross-section. The pressure in the narrow section is equal to the saturation pressure, that is, cavitation begins there.

The question is: what diameter of the narrow section is required for cavitation to appear?

Let us calculate. The velocity of the fluid in the wide section is $u_0 \approx 1$ m/s—a comparatively slow flow, but this is a developed turbulent flow anyway: the Reynolds number Re $= u_0 d_0/v$ (here $v \approx 10^{-6}$ m²/s is the coefficient of the kinematic viscosity of water at room temperature) in that section is $\sim 5 \cdot 10^4$.

Then, velocity in the narrow section is

$$u = \sqrt{u_0^2 + \frac{2}{\rho}(p_0 - p_s)}. \tag{2.1.4}$$

From (2.1.4), we obtain velocity in the narrow section $u \approx 20$ m/s, and the corresponding diameter (from (2.1.3)) is $d \approx 1$ cm. Thus, for the considered parameters, cavitation takes place when the diameter is decreased approximately 5 times.

Usually, in technological devices exactly this chain 'cross-section—velocity—pressure' is responsible for cavitation. As a variant, the drop of pressure can be caused by the local hydraulic resistance due to an abrupt change in the channel configuration.

Now let us try to generalize the problem and introduce a parameter with which we may predict cavitation. Before this procedure, we remind that in this section we consider only horizontal flows, neglecting gravity.

At first, we may be surprised: do we need some additional parameters? Really, can't we calculate the pressure directly at every point of a pipeline to analyze these values by comparing them with the saturation pressure? Actually, this problem is more complicated than one may assume. For a long pipeline, pressure losses are very difficult to calculate with all the peculiarities in it. For this reason, it does not make much sense to represent the equation (2.1.2) through the areas of cross-sections. Moreover, we repeat that now we are looking for a common way to describe the formation of cavitation.

Then, continuing our consideration, this sought-for parameter must contain measurable quantities of the flow, we mean easy measurable parameters. Such parameters are the quantities at the entrance or at the exit; for certainty, we will consider the first case, when the pressure p_0 and the velocity u_0 are known.

We are afraid that the cavitation process may start somewhere inside a pipeline. We do not know the velocity at that hypothetical point, but we know that the pressure there is equal to the saturation pressure. Finally, me may represent our findings for a given streamline as

$$p_0 + \frac{\rho u_0^2}{2} = p_s + ?. \tag{2.1.5}$$

Technically, we know three terms out of four, but that fourth one is a bit of mystery for us; we may calculate it, however, from the Bernoulli equation, i.e., with (2.1.5) directly, but the statement of the problem is different now: what if we know only the averaged parameters of the incoming flow, not for any streamline. Moreover, we may know another velocity, the velocity in the cross-section where $p = p_s$ (see below), but first we discuss the case (2.1.5) exactly. Thus, actually (2.1.5) is everything we know about the problem. The finishing touch: there must be $p_0 > p_s$, otherwise cavitation would start at the entrant cross-section.

Possibly, it should be emphasized that the pressure in the right-hand side of the equation (2.1.5) must not be equal to p_s in the arbitrary pipeline obligatorily: on the contrary, we hope that at any point of the pipe the pressure is higher than p_s. But we consider the most dangerous case, when the saturation pressure was reached at some point.

At first glance, one may assume that simply the larger the left-hand side of (2.1.5), the lesser the chance that the pressure somewhere drops down to p_s. However, it should also be well understood that pressure and velocity are not the equivalent parts of (2.1.5), so to say. If velocity is very high so that $\rho u_0^2/2 \gg p_s$, but $p_0 \sim p_s$ (we mean pressure at the entrance is slightly higher than the saturation pressure), then the risk of cavitation is high anyway: if velocity is increased slightly in some cross-section, then the pressure there would be decreased, summoning cavitation.

Gathering all the reasonings together, we may express the risk of cavitation with the dimensionless cavitation number, which is the ratio[2]

$$Q = \frac{p - p_s}{\rho u^2 / 2},$$ (2.1.6)

where we omit indexes '0' since there are no other parameters there. In theory, $0 < Q < \infty$ (at $Q < 0$ cavitation would begin at the initial point). In practice, its range is much narrower, of course. Operating with the cavitation number, we may determine the 'cavitation reserve' of the flow: if the cavitation number is high, the risk is low, if $Q < 1$ or even $Q \ll 1$, then cavitation will take place almost certainly. For an arbitrary flow, pressure and velocity may be of any value, so the cavitation number in a common case may be of any value; if, however, p and u correspond to cavitation at some point of a flow, then this value of Q is termed as the critical cavitation number: this case exactly corresponds to the case when the pressure at some point is equal to the saturation point.

The representation of the cavitation number in a form (2.1.6) is quite universal: it can be used both for a flow in pipe and for an external flow (see following sections), but indeed there are many forms of the cavitation number in literature. Thus, certain values of p and u, that is, certain sections where these parameters are determined, should be clearly pointed out.

For instance, in some problems (see Section 2.1.3 below), the fluid outflow from a large reservoir to a pipe can be considered. In this case, the velocity in the reservoir (i.e., at the entrance to the pipe) is approximately zero, while the velocity at the exit of the pipe is sufficiently high and can be measured. In this case, the expression (2.1.6) is used with p—pressure at the entrance (in the reservoir), and u—velocity of the fluid at the pipe's exit. For the example considered above (Fig. 2.1.1, where $d_0 \gg d$) we have, with $u_0 = 0$, the corresponding critical cavitation number, in designations of (2.1.2) which can be written as

$$Q_{cr} = \frac{p_0 - p_s}{\rho u^2 / 2} = 1.$$ (2.1.7)

Actually, this is the preferable case: critical cavitation number is equal to unity. But for any physical problem, it is not possible to formulate the cavitation number in such a successful way.

More frequently, another form of the cavitation number is used for pipes. Repeating all the demands—measurable parameters, physical sense of the result—one may represent the cavitation number with the upstream pressure p_u and the downstream pressure p_d:

---

[2]   There exists a relative to this number – the Euler number which can be written exactly so.

$$Q = \frac{p_u - p_s}{p_u - p_d}. \tag{2.1.8}$$

Again, in a general case, definitions (2.1.6) and (2.1.8) are not equivalent, so it is important to realize which representation is used in the given case.

The goal of using Q follows from its meaning. For comparatively non-complex problems, for certain simple geometries one may try to introduce the cavitation number to obtain a definite value of it, which corresponds to the beginning of cavitation.

For more complicated problems, for the given flow in a certain complex system, experimentally the critical cavitation number—value of Q at which cavitation starts—can be determined. On the other hand, all this approach seems not the only way to determine the beginning of cavitation.

2.1.2 The pressure loss in pipes

As it was mentioned at the beginning, finally we are interested in pressure at the given cross-section. Cavitation starts when pressure drops sufficiently.

The pressure drop in a pipeline can be calculated directly, if we know the exact construction of the given pipeline. Using the engineering approach, this procedure can be realized with the help of friction factors.

The Darcy–Weisbach equation gives a correlation of the pressure loss Δp in the section of a pipe of length L and diameter D with the mean velocity u as

$$\Delta p = \xi \frac{\rho u^2}{2} = \lambda \frac{L}{D} \frac{\rho u^2}{2}, \tag{2.1.9}$$

where ξ is the friction factor,[3] which can be represented with a Darcy friction factor λ. Both these coefficients depend on the character of the flow. It can be obtained analytically that for a laminar flow, $\lambda = 64/\mathrm{Re}$; for a turbulent flow, the empirical relation gives $\lambda \sim \mathrm{Re}^{-1/4}$.

This approach, initially empirical, can be spread on a more complicated case than a flow in a straight pipe. Pressure loss takes places at every non-uniform section of a pipeline: at turns, at abrupt expansions or contractions, etc. The corresponding loss on each such element can be represented in the manner of (2.1.9):

$$\Delta p_i = \xi_i \frac{\rho u_i^2}{2}. \tag{2.1.10}$$

Coefficients ξ_i can be determined empirically for any imaginable element of a pipeline; these parameters are collected in references, for instance, in (Idelchik 1986).

This way, we may obtain the total pressure loss in a pipeline; for an incompressible fluid, we get

$$\Delta p_\Sigma = \sum_i \xi_i \frac{\rho u_i^2}{2}, \tag{2.1.11}$$

which can be represented with the mass discharges G_i in every element as

$$\Delta p_\Sigma = \sum_i \xi_i \frac{G_i^2}{2 \rho S_i^2}. \tag{2.1.12}$$

Note that discharge is not necessary constant in every element: there may be different values if a pipe branches out.

---

[3] For an external flow, this value is called the drag coefficient, so sometimes in pipes the same terminology is used.

For instance, let us consider a turn of round pipe by δ degrees (0 < δ < 180); see Fig. 2.1.2.

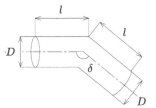

Fig. 2.1.2. The turn of a pipe.

For sufficiently long pipe elbows (if their length l exceeds 10 diameters of the pipe: $l/D > 10$), at high Reynolds numbers (higher than $2 \cdot 10^5$), the friction factor can be represented as

$$\xi = A\xi_m, \tag{2.1.13}$$

where

$$A \approx 0.95 + \frac{33.5}{\delta}, \tag{2.1.14}$$

$$\xi_m = 0.95\sin^2\frac{\delta}{2} + 2.05\sin^4\frac{\delta}{2}. \tag{2.1.15}$$

Another simple scheme is the sudden expansion of a channel. For high Reynolds numbers $Re > 10^4$, the friction factor can be calculated with the Borda–Carnot[4] expression

$$\xi = \left(1 - \frac{F_1}{F_2}\right)^2, \tag{2.1.16}$$

where F_1 is the area of the narrow section, and F_2 is the area of the wide part of the pipe. This expression implies the uniform distribution of velocity at the narrow part of the pipe; otherwise, the pressure loss will be greater and must be calculated with a more complicated equation (Idelchik 1986).

Analogically, for the sudden narrowing of a channel from F_1 to F_2:

$$\xi = \frac{1}{2}\left(1 - \frac{F_2}{F_1}\right). \tag{2.1.17}$$

Thus, using expressions of such sort, one can determine the pressure loss in channels and, therefore, predict a significant pressure drop which may cause cavitation in the pipeline. One nuance that may surprise us: cavitation is already here, in relations (2.1.16) and (2.1.17).

2.1.3 Flow in a throttle

In pipes, there exist a better way to obtain cavitation, which actually leads to pressure losses at a sudden change in the pipe cross-section. Let us take a closer look at the flow at the sudden expansion of a pipe (which gives us the relation (2.1.16)) and the sudden narrowing (which gives (2.1.17)).

For a more convenient consideration, we combine these patterns together and consider the flow in a throttle—the pipe of a special form, where the cross-section initially narrows, and then restores its initial size (see Fig. 2.1.3). Actually, 'thermodynamic throttle' has slightly different proportions with a shorter length of the narrow section, but the throttle shown in Fig. 2.1.3 is more convenient for analysis.

---

[4] One of the authors is Lazare Carnot, the father of famous Sadi Carnot (the author of "Reflections on the Motive Power of Fire").

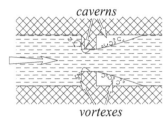

Fig. 2.1.3. The flow pattern in a throttle.

As we can see, flow patterns have really a much more complicated structure than one may assume: one may assume that the fluid is attached to the walls in all three sections, but we see that, actually, it is not. In a general case, even in a pipe, the fluid must not occupy the entire channel section. Instead, it forms complicated flow patterns which consist of jets surrounded by a complex micro-vortex structure and vapor caverns.

These conditions at the entrance to the narrow section are subject to active investigations: the milestone work (Nurick 1976) was published in 1976, but until modern days this problem is treated both experimentally and theoretically (Martynov et al. 2006, Payri et al. 2004). In these researches, so to say, 2/3 of the scheme presented in Fig. 2.1.3 is examined—up to the broadening at the exit of the narrowest part. The critical cavitation number for such a flow was found in (Payri et al. 2013) (in the meaning of (2.1.8)): for the range $Re = 5 \cdot 10^3 \div 10^4$ (the Reynolds number is defined at the exit—in the narrow part), critical cavitation numbers $Q_{cr} = 1 \div 1.2$ were obtained. In these experiments, the cavitation number at which cavitation first arises in the orifice (the incipient cavitation number, in authors' terminology) is significantly higher than the 'regular' cavitation number: the incipient parameter was up to 1.5.

2.1.4 Final thoughts on the cavitation number

The last matter mentioned in the previous subsection is more interesting than it may look at first glance.

The same reasonings can be found in (Brennen 1994) concerning the difference between the cavitation inception number Q_i and the cavitation breakdown parameter Q_b applied to a centrifugal pump.[5] The first one corresponds to the first appearance of cavitation, while the second one describes the pressure head breakdown. These parameters differ significantly from each other (twofold or more), and both of them differ from the 'desinent' cavitation number Q_d, which corresponds to the end of cavitation after the pressure increase.

Cavitation in its initial stage includes phase transition. Due to this very process, initial bubbles appear in the system: in the liquid (homogeneously), or at the solid wall (heterogeneously). This phase transition is sensitive to initial conditions, mainly to impurities in the system (including air microbubbles). In the absence of impurities, the phase transition goes harder, but, on the other hand, once these initial seeds—nuclei—appear, some of them stay in the system and alleviate the following process. This is the reason why Q_i depends on the air content, while Q_b does not.

However, this discrepancy is only a part of the common problem. Above, we discussed the cavitation number from its origin: this quantity must reflect some critical state of the system, in dependence on external conditions. Here we meet another side of the problem: how unified are the parameters of the critical state?

Indeed, cavitation is a partly irreversible process (see the previous section), and the degree of irreversibility strongly depends on the conditions at which cavitation starts. If the abrupt change in pressure is caused by the variation of external conditions, then many details will be important to the process (time of the process, first of all), and not only the magnitude values.

---

[5] See Section 2.3 below.

In commonly, it looks like we are dealing with a system, the properties of which may strongly depend on poorly controlled parameters, such as the initial composition of a fluid, fluctuations in the incoming flow, etc. For certain conditions, for the same systems (especially—closed systems, like a pump's tract), it is possible to obtain the critical parameter more or less appropriately. But in open systems, in undefined conditions—like for breaking bottles, for example, when the strike is difficult to reproduce—any attempts to determine a single critical parameter go rough.

Thus, generally, cavitation number can be used with caution. The result obtained for the given system can be poorly translated to a close relative of this system. The estimation $Q_{cr} \sim 1$ works well indeed for most problems, and, sometimes, it is difficult to obtain the clarification for it.

2.2 Cavitation on wings

2.2.1 How wings work

The main principle of a hydrofoil (underwater wing) is the same as of an airfoil (air wing). A wing is a body of a special shape that allows it to interact with an incident flow in such a manner that the lifting force arises. Due to the structure of the flow past such an object, the pressure above the wing is lower, and below the wing it is higher than the 'normal' pressure in an undisturbed medium. Thus, moving forward, the wing also tries to lift up. Note, of course, that in a common case the 'lifting force' can be directed somewhat arbitrarily, not necessarily opposite to the acceleration of gravity: for instance, for a helicopter tail rotor, or for an underwater device which moves in its special direction.

Usually, of course, the body (that is, the wing) moves in a (comparatively) immobile medium, but the whole picture can be represented in that manner as the flow moves at velocity $\vec{u}_\infty$ onto a stationary wing (see Fig. 2.2.1), like during aerodynamic tests.

As it shown in Fig. 2.2.1, the drag force is always defined along the direction of the incident flow (axis x is chosen along the velocity $\vec{u}_\infty$). Correspondingly, the lifting force is defined orthogonally, along the axis y.

These forces can be expressed with the value $\rho u_\infty^2/2$. Adopting some value of the effective area of the wing S, one may write

$$F_L = C_L S \frac{\rho u_\infty^2}{2};$$

(2.2.1)

$$F_D = C_D S \frac{\rho u_\infty^2}{2}.$$

(2.2.2)

Quantities C_L and C_D are termed the lift and the drag coefficient respectively. In some special cases, these coefficients can be found analytically (see Chapter 4), but usually this operation is impossible. In a common case, these coefficients depend on the shape of a wing and on the Reynolds number,[6] their determination, actually, is the main problem of the airfoil construction.

The method of creating the lifting force described above can be used in any medium—both in gases and in liquids. There are no additional questions for a wing in air: a flow creates the lifting force for aircrafts, and does not provoke any special effects for them, except that usually we want a higher C_L at a lower C_D. A wing in a liquid is a different case. As we mentioned above, the whole principle of wing functioning implies a low-pressure zone above the wing. We already know what may 'low-pressure zone' mean in a liquid: it may cause cavitation.

Thus, the work of a wing in water may be accompanied by cavitation. Before we discuss how to describe it and how to avoid it, we discuss the common principles of cavitation of such type.

---

[6] The Reynolds number is the dimensionless velocity: $Re = ud/v$, where u is velocity, d is the spatial scale, v is the kinematic coefficient of viscosity.

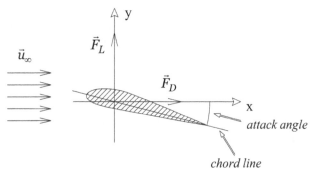

Fig. 2.2.1. The lifting force F_L and the drag force F_D of a wing.

2.2.2 Qualitative analysis of cavitation on a wing

In this section, we consider cavitation in the flow past a body—cavitation that is caused by pressure redistribution around an object moving in a liquid. From the notions made above, it is clear that the 'dangerous zone' for a wing—the region where the risk of cavitation is high—is located at the upper side of a hydrofoil (see our remark concerning the 'upper side' above); cavitation may arise in this low-pressure zone.

The usual form of such type of cavitation is bubble cavitation: bubbles appear from small nuclei in the region where pressure is above the saturation value (see Chapter 1). Possibly, these bubbles are not as small as they are shown in Fig. 2.2.2—that depends on certain conditions. It should be noted that, strictly, the configuration of the low-pressure zone in a cavitating flow differs from the one in a single-phase flow, because cavitation leads to redistribution of pressure; in addition, for a moving object the zone of cavitating bubbles may be wider than the region of low pressure—cavitating bubbles may penetrate high-pressure regions, where they collapse.

Another cavitation pattern arises when some bubbles merge into a larger cavern. This cavern may be attached to the foil (Fig. 2.2.3), or not. The last case is shown in Fig. 2.2.4, when the giant cavern is formed, covering all the foil; it is called supercavitation or developed cavitation.

Here we return to the question that was discussed in Section 1.3—about stability of the cavitating cavern. Let us consider developed cavitation, and try to describe it with stationary equations. Thus, we are dealing with a giant stationary cavern formed in vicinity of a solid surface; fluid flows around this cavern, as it is shown in Fig. 2.2.4.

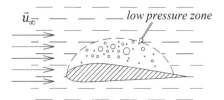

Fig. 2.2.2. Bubble cavitation in vicinity of a hydrofoil.

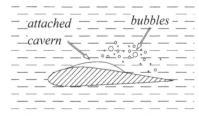

Fig. 2.2.3. Some cavitation bubbles are merged, but the cavern is attached to the foil.

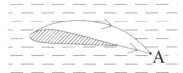

Fig. 2.2.4. Supercavitation: All the bubbles are merged into a giant bubble in the low-pressure zone.

For the streamline along the cavern, on that boundary, there must be

$$p + \frac{\rho u^2}{2} = const. \tag{2.2.3}$$

Inside the cavern, the pressure–pressure of gas—is constant, that is, the fluid at the interface is under constant pressure. From the equation (2.2.3), it follows that if $p = const$, then $u = const$ also. However, at point A (at the rear stagnation point) velocity must be zero.

This contradiction is the Brillouin paradox. Strictly, this means that the flow considered above is impossible, and we have to consider the full, time-dependent problem. It is exactly the matter that was discussed in Section 1.3: stability of a cavitating cavern is a very special case. On the other hand, physicists do not surrender so easily, and some special schemes were proposed to get out of the difficulty.

We return to this matter in Chapter 4, where hydrodynamics of a cavitating flow will be discussed. Hence, we will follow this way gradually: in Chapter 1, we discussed the common principles, here we see some certain results which, however, must be maintained by some correlations.

2.2.3 Physical description of cavitation on a wing

Let us consider a flow past a body. Denote the pressure at infinity as p_∞ and velocity of the incoming flow as u_∞. We are interested in the emergence of cavitation at some point where pressure is p and velocity is u. To describe the beginning of cavitation, we may use the cavitation number introduced in the previous section

$$Q = \frac{p_\infty - p_s}{\rho u_\infty^2 / 2}. \tag{2.2.4}$$

On the other hand, for the considered flow, the Bernoulli equation (2.2.3) can be written as

$$p_\infty + \frac{\rho u_\infty^2}{2} = p + \frac{\rho u^2}{2}. \tag{2.2.5}$$

From this equation, we may withdraw the expression for the so-called pressure coefficient:

$$C_p = \frac{p - p_\infty}{\rho u_\infty^2 / 2} = 1 - \left(\frac{u}{u_\infty}\right)^2, \tag{2.2.6}$$

which represents the dimensionless pressure at the given point.[7] To understand this parameter for our purposes, we may note that the lower the pressure at the given point, the lower the pressure coefficient (or the higher the negative value of pressure coefficient). Thus, the most dangerous point, with respect to cavitation occurrence, corresponds to the minimum pressure p_{min} on the foil, and the corresponding minimum pressure coefficient there is

---

[7] In some references, this value is introduced with the opposite sign, i.e., as $C_p = \frac{p_\infty - p}{\rho u_\infty^2 / 2}$. This is more convenient in view of the sign of the results: one may note that quantity $-C_p$ is actually represented on all the figures.

$$C_p^{min} = \frac{p_{min} - p_\infty}{\rho u^2 / 2}. \tag{2.2.7}$$

Cavitation starts on the given hydrofoil if the minimum pressure on it is lower than the saturation pressure: $p_{min} \leq p_s$, that is, comparing (2.2.7) with (2.2.4), if

$$-C_p^{min} \geq Q. \tag{2.2.8}$$

If cavitation takes place, then the negative pressure coefficient at least is equal to the cavitation number, or higher.

Thus, we have the clear condition for the occurrence of cavitation. For the given parameters of the incoming flow p_∞ and u_∞, and for the given saturation pressure $p_s(T)$, one may find the cavitation number Q. Then, obtaining from the calculations the pressure distribution past a foil, we get C_p at any point. Thereby, we may check whether cavitation can be observed at any point or not by using the inequality (2.2.8).

In this manner, we see the way to find out the possibility of cavitation: one may operate with the pressure coefficient, analyzing its value and comparing it with cavitation number. Indeed, the pressure coefficient plays a special role in such problems. In its turn, as we see from (2.2.6), C_p is determined by the velocity at the given point. Thereby, to calculate C_p, one has to determine the velocity field past a foil. Then, one may calculate the pressure coefficient with (2.2.6), or use a simplified approach which can be useful for analytical operations. For instance, in the case when a hydrofoil creates a small disturbance in the velocity field, so that

$$u_x = u_\infty + u_x', \ u_y = u_y', \ |u_x'|, |u_y'| \ll u_\infty \tag{2.2.9}$$

(axes correspond to Fig. 2.2.1), one may note that the pressure coefficient can be calculated in a linearized form as

$$C_p = -\frac{2u_x'}{u_\infty} - \left(\frac{u_x'}{u_\infty}\right)^2 - \left(\frac{u_y'}{u_\infty}\right)^2 \approx -\frac{2u_x'(x)}{u_\infty}. \tag{2.2.10}$$

Here we emphasize that the pressure coefficient depends on the coordinate x on the chord line. The last two terms in the first equation are negligible since they are of an order of $\sim z^2$ in comparison with the first term $\sim z$ for $z \ll 1$; the last condition follows from (2.2.9).

In a common case, the pressure coefficient can be calculated numerically or can be analyzed only qualitatively. We choose the second way, since it explains all the features of the problem without excessive expressions.

First, let us consider the distribution of the pressure coefficient along the wing. For the flow pattern at zero angle of attack, we see that the coordinate $x = 0$ coincides with the stagnation point, where $u = 0$. Correspondingly, here $C_p = 1$ (see (2.2.6)). For a larger x, velocity increases, but at a different rate for the upper side of the foil and for the lower one: at the upper side, the rate of increase is higher, as well as the absolute values of velocities are higher too. The corresponding dependence of $C_p(x)$ for the dimensionless x (ratio to the chord length) is shown in Fig. 2.2.5. For non-zero attack angles, the stagnation point does not correspond to zero on the chordwise, and the upper curves will be displaced higher, in general, i.e., for normal foils, so to say (see again Fig. 2.2.5).

Then, we may also construct another dependence: C_p from the attack angle α for different coordinates x; the result is presented in Fig. 2.2.6.

From the consideration presented above, we understand that it is sufficient if cavitation would begin at any single point, that is enough to create the danger zone in vicinity of the wing. Thus, this is enough for us, if on the line at a constant attack angle we may find a single value C_p^{min} corresponding to the condition (2.2.8)—this fact would mean the beginning of cavitation at the given attack angle.

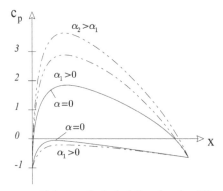

Fig. 2.2.5. Distribution of the pressure coefficient on the hydrofoil surface for different angles of attack α. Upper curves correspond to the upper surface of a wing; lower curves—to the lower one.

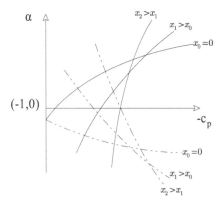

Fig. 2.2.6. Pressure coefficient as the function of the attack angle for different coordinates on a wing. Solid curves—upper surface, dashed curves—lower surface.

Thus, for the goal to establish the beginning of cavitation, we can only be interested in a single value of the pressure coefficient, of the minimum value $C_p^{\min}(\alpha)$, which is reached at some point on the foil. Actually, as we see, this value corresponds to the most right-hand point on the set of curves in Fig. 2.2.6, corresponding to the set of all the values of x from 0 to 1. From this point of view, the dependence $C_p^{\min}(\alpha)$ is an envelope of the family of curves plotted in Fig. 2.2.6 for all $x \in [0,1]$. Note, however, that usually this envelope can be constructed without very high values of x—this fact means that usually cavitation starts far from the trailing edge.

The dependence $C_p^{\min}(\alpha)$ plotted on the graph is called the cavitation bucket diagram, or simply the pressure envelope, see Fig. 2.2.7.

Let us assume that the cavitation number for the given flow, that is, for the given p_∞, u_∞ and $p_s(T)$, is $Q = 2$. Then, we look at the cavitation bucket diagram for our foil presented in Fig. 2.2.7. In this diagram, we see that the values satisfying the condition (2.2.8) correspond to attack angles $\alpha > 3°$ or $\alpha < -2.5°$. Thus, the ranges of attack angles $\alpha > 3°$ and $\alpha < -2.5°$ correspond to cavitation modes, while the range $-2.5° < \alpha < 3°$ corresponds to the flow free from cavitation.

Here we consider some common examples of hydrofoil. In (Eppler 1990), one may find some cavitation diagrams specific for real hydrofoils. The common features of such diagrams are the same: as a rule, the risk of cavitation increases with the increase of the attack angle. For some values of the cavitation number ($Q < Q_{cr}$ in Fig. 2.2.7), there are no safe attack angles: any flow pattern corresponds to a cavitation mode.

As we see from Fig. 2.2.6 and 2.2.7, the pressure coefficient $C_p^{\min}$ changes very rapidly at small values of x, i.e., at the leading edge. This case corresponds to the occurrence of cavitation at high values of attack angles, and also at high values of the cavitation number. On the contrary, at small

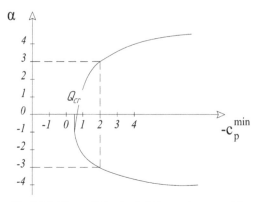

Fig. 2.2.7. The cavitation bucket diagram (an example).

values of Q cavitation begins far away from the leading edge, and in this range C_p^{min} is a weak function on α.

After cavitation begins, the pressure distribution along the wing tends to be flattened out: the dependencies of $C_p(x)$ become smoother than for a single-phase flow. The physics of a cavitating flow is very difficult, and we will not discuss it there. On the other hand, for practical purposes, the main problem is rather to avoid cavitation than to describe the process of the foil's destruction properly.

From this consideration, the common way to avoid cavitation on hydrofoils becomes clear. Of course, the main purpose of a wing is to create the lifting force. This statement is correct for underwater wings too. But, since the work of such devices is restricted by the cavitation process, we have to find a compromise between the lifting force and the durability of use. In other words, if we sacrifice the lifting force by changing the hydrofoil geometry, so that the pressure at the upper side will be decreased much less, then we can stop being afraid of cavitation in that region. The reverse side of this medal is the loss in lifting force because of that very insufficient pressure decrease.

Another way to get away from cavitation is to follow cavitation diagrams. They show us, first of all, the attack angles which are appropriate for a non-cavitating flow. The second choice, actually, is also hidden on cavitation diagrams: reducing u_∞ (which is, actually, the speed of the moving object), we get higher cavitation number which shifts the dangerous attack angles. On the other hand, here we come close to the general principle: no flow, no cavitation, but this is not a suitable thing too.

2.2.4 Cavitation on rotating wheels

Despite some differences in geometry, cavitation on a body during rotational motion, not at the translational one, is based on the same physical principles. Again, starting from the Bernoulli equation, we finish at the decreased pressure at the region of the enhanced velocity.

This enhanced velocity can originate in different places of such machines. For instance, on a rotating wheel, the maximum velocity (ωR) is reached at the periphery of the rotating disk. For this reason, a rotating spiral drill produces cavitating bubbles on its edges.

Usually, such rotational devices are designed for the same properties as wings—to create additional pressure for some purposes: propeller screws are required to move a ship, impeller blades pump fluids, and so on. In these cases, the increased velocity is observed in the vicinity of special aerodynamic (to be exact, hydrodynamic) elements, such as a turbine blade, which is, actually, a hydrofoil in its origin.

Therefore, all the principles discussed above are suitable for rotating devices. Details about these machines can be found in (Carlton 2012, Tupper 2013); we do not discuss such special matters here.

However, one apparatus stands apart. A centrifugal pump is a machine which creates pressure by rotating an impeller, and cavitation in it can be considered in the same manner as for a common hydrofoil. However, another point is of interest: to work, this pump must draw a fluid in, producing a low-pressure zone inside it. Thus, the pump walks together with a cavitation danger, and consideration of pumps—despite the ordinary cavitation mechanism inside them—demands special attention because of the unusual source of cavitation at their entrance.

2.3 Cavitation in pumps

2.3.1 Centrifugal pumps

To consider cavitation in pumps, we will cover a certain type of them—a centrifugal pump. This does not restrict in any way the common principles of the problem.

A centrifugal pump is shown in Fig. 2.3.1.

The main part of a centrifugal pump is an impeller, the rotation of which creates low pressure at the input (upstream) flange and high pressure at the output (downstream) flange. Thus, the pump engulfs the fluid from the suction line and forces it to the pressure line.

Drawing a fluid into a pump requires low pressure inside it. If the pressure is too low, cavitation occurs at the impeller.

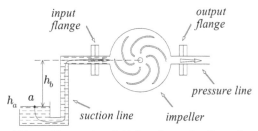

Fig. 2.3.1. A centrifugal pump. This machine receives fluid from the suction line and pushes it to the pressure line. An impeller creates lower pressure at the input flange and high pressure at the outer one with its blades. A reservoir with a fluid may be placed both below (on this figure) and above.

2.3.2 Pressure head at a suction line

Let us determine the pressure at the input flange. For certainty (again), we will assume that the pump vessel is located below the pump (relative to the axis of the impeller). This is not obligatory: in some cases, the reservoir can be located above the pump.

Using the Bernoulli equation, we may read for the line $a - b$, that is, from the surface of the pump vessel to the pump entrance:

$$p_a + \rho g h_a + \frac{\rho u_a^2}{2} = p_b + \rho g h_b + \frac{\rho u_b^2}{2}. \tag{2.3.1}$$

Choosing the reference line at the surface of the reservoir, we have $h_a = 0$. In this approach, the height h_b coincides with the middle of the input pipe; the difference between the height of the axis of the impeller and h_b can be taken into account with the adopted margin.

The pressure p_a, in a general case, is not only an atmospheric pressure: some extra pressure can be created in the reservoir. Velocity u_b cannot be equal to zero in principle: the condition $u_b = 0$ would mean zero mass discharge at the entrance to the pump.

As for the velocity u_a, this is a more complicated matter. This quantity may be put to zero, but may not, depending on the certain system of the suction line. The net mass discharge must be equal both at section a (with the area of cross-section S_a) and b (of cross-section S_b); for an incompressible fluid, it gives

$$u_a S_a = u_b S_b. \tag{2.3.2}$$

Thus, the first reason to put $u_a \approx 0$ is the condition $S_a >> S_b$. Another opportunity is the case when the level of the fluid in the reservoir is maintained artificially in some way. Anyway, below we set $u_a = 0$; in some special cases, one may return to this point and obtain a refined result.

In all the conditions, we have the relation for the pressure at the upstream flange:

$$p_b = p_a - \rho g h_b - \frac{\rho u_b^2}{2}. \tag{2.3.3}$$

We may remind that h_b would be negative, if the pump vessel is located above the impeller.

However, the pressure p_b is not the minimal pressure in the pump. To engulf the fluid inside the pump, it creates additional depressurization inside it due to the rotation of the impeller, see Fig. 2.3.2.

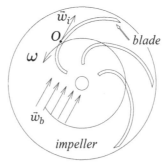

Fig. 2.3.2. The flow past an impeller's blade. Due to rotation, the blade is flown over by the fluid coming to the pump (into the input flange); the corresponding relative velocity is $\vec{w}_b$. At some point O at the upper side of the blade, the pressure is minimum, and the corresponding relative velocity there is $\vec{w}_i$. If the pressure at point O is equal or below the saturation one, cavitation begins. In its common characteristics, this is a particular case of cavitation on a hydrofoil.

Thus, if the fluid has the initial pressure p_b and relative velocity w_b, then in vicinity of the impeller's blade, the relative velocity of which is w_i, the minimum pressure obtains another Bernoulli equation:

$$p_b + \frac{\rho w_b^2}{2} = p_{\min} + \frac{\rho w_i^2}{2}. \tag{2.3.4}$$

Consequently, the pressure at the input flange can be represented with the minimum pressure as

$$p_b = p_{\min} + \frac{\rho w_b^2}{2}\left(\frac{w_i^2}{w_b^2} - 1\right). \tag{2.3.5}$$

The goal of our consideration is to establish the conditions at which the minimal pressure in the pump will exceed the saturation pressure, i.e., the limiting case for this minimal pressure is $p_{\min} = p_s$. With this information, we may take a closer look at the term in the brackets in (2.3.5). From the same equation (2.3.4), we see that this term can be expressed as

$$\frac{w_i^2}{w_b^2} - 1 = \frac{p_b - p_{\min}}{\rho w_b^2 / 2} = \frac{p_b - p_s}{\rho w_b^2 / 2} = C, \tag{2.3.6}$$

that the cavitation number for a wing (in the given case—for the impellers blade). Here we obtain the expression for the cavitation number—exactly in the form discussed above (in Section 2.2, expression 2.2.4 is the same).

Finally, we have the theoretical limiting condition for the absence of cavitation or, at another angle, the condition of the inception of beginning of cavitation:

$$p_s + C \frac{\rho w_b^2}{2} = p_a - \rho g h_b - \frac{\rho u_b^2}{2}. \tag{2.3.7}$$

From (2.3.7), we may find the restriction for height h_b—the maximum value at which cavitation in the pump is absent:

$$h_b = \frac{p_b - p_s}{\rho g} - \frac{u_b^2}{2g} - C \frac{w_b^2}{2g}. \tag{2.3.8}$$

With (2.3.8), we obtained the answer for height, and, structurally, this answer looks like $h = p/\rho g$ (see more details concerning the last two terms below). In hydraulics, the comparable value is the ratio of pressure to the product of density by acceleration due to gravity.

Let us take a look at the equation (2.3.8). The last two terms, containing velocities in the pump, indeed, depend on the operating mode of the pump. Combined, these terms can be defined as the pressure head required for the given mode, that is, the relation (2.3.8) can be rewritten as

$$h_b = \frac{p_b - p_s}{\rho g} - h_p, \tag{2.3.9}$$

where, obviously, $h_p = \frac{u_b^2}{2g} + C \frac{w_b^2}{2g}$.

Let us establish the expression for h_p. The pressure head required by the pump must depend on the rotation frequency ω [s^{-1}], on the volume discharge Q [m^3/s], and on the gravitational acceleration g, [m/s^2]. It is possible to combine these variables into a multiplex with dimensions in meters, corresponding to the head:

$$h_p \sim \omega^a Q^b g^c. \tag{2.3.10}$$

However, one may see that this combination cannot give a certain answer yet, since we have two equations (obtained from the conditions for the power of seconds and meters in the final answer for the pressure head) for three variables a, b, c:

$$[\text{for meters}] \; 3b + c = 1, \tag{2.3.11}$$

$$[\text{for seconds}] \; -a - b - 2c = 0. \tag{2.3.12}$$

Thereby, from (2.3.11) and (2.3.12), we may get many combinations for parameters a, b, c. We need some additional hint. Such a hint can be obtained with different approaches: we may consider the physical process of rotation, find connections between Q and ω, or try to establish the connection between the head and the rotation frequency, etc. But the simplest way is to notice that, due to the nature of the pressure head, the acceleration g and head h_p can be figured in an expression only in combination $g h_p$. Actually, this may be seen directly from the definition of h_p: this equation is the relation of the squares of velocities (which depend only on the frequency of rotation and the volume discharge) to the gravity acceleration. Thus, we have the third additional condition:

$$[\text{from the hint}] \; c = -1. \tag{2.3.13}$$

With (2.3.13), the rest of equations (2.3.11) and (2.3.12) gives

$$a = 4/3, \; b = 2/3. \tag{2.3.14}$$

Thus, we have for the pressure head

$$h_p = N^{-4/3} \frac{\omega^{4/3} Q^{2/3}}{g},$$ (2.3.15)

where the dimensionless parameter N, which must appear obligatory, is the suction specific speed:

$$N = \frac{\omega \sqrt{Q}}{\left(gh_p\right)^{3/4}}.$$ (2.3.16)

Suction specific speed is the parameter of the given pump, which depends on its type;[8] this is the known parameter for the pump you are dealing with.

Thus, now we can calculate the required pressure head for the pump; then we may return to the determination of the maximum height h_b, which can be used without cavitation inside the pump on its blades.

Following this way, first, we may notice that in practice, in real life, pressure losses exist. The pressure at the input flange will be smaller than the value predicted by the Bernoulli equation because of the friction processes; thus, the equivalent quantity h_l must be included in the equation (2.3.9), which now has the form

$$h_b = \frac{p_a - p_s}{\rho g} - h_p - h_l.$$ (2.3.17)

The new term h_l must be determined for the given suction system: this is the parameter of that contour, not of the pump itself.

Then, we may note that, indeed, the condition (2.1.17) means not the absence of cavitation, but rather its beginning. Thus, especially if we are currently considering technological appliances, some margin is needed. This margin can be of two kinds: as an additional absolute value (usually from 0.5 m to 1 m for water), or as a relative one: it can be from 3% to even 25% margin that must prevent cavitation for some unexpected cases (and, of course, for some errors in calculations, especially of the pressure loss).

The first such unexpected case is temperature. Saturation pressure strongly depends on temperature; this is not a big problem for room temperatures, since at these parameters $p_s \ll p_a$ anyway; but this matter may be important for high temperatures. For instance, at temperatures $\sim 80°C$, the saturation pressure of water is ~ 0.5 atm. This fact restricts the pressure head significantly.

2.3.3 Net positive suction head

In this section, we briefly translate the 'physical' results obtained in the previous section into the 'engineering' language.

The head, as it was mentioned above, is the pressure head—the corresponding pressure divided by product ρg. Thus, the net positive suction head (NPSH) corresponds to the pressure in the suction line. It may be of two kinds:

✓ NPSH-R—net positive suction head required;
✓ NPSH-A—net positive suction head available.

---

[8]   Sometimes this parameter is also referred to as the cavitation specific speed; specific speed itself is the quantity calculated with (2.3.16) but with the head created by the pump H instead of the required head h_p. Note that the suction specific speed of the pump can be calculated with various dimensions of parameters in (2.3.16); it is also often that the acceleration g figured out in this expression is represented as its numerical value (taken in the corresponding power).

NPSH-R is the parameter of the pump; in our previous consideration, it corresponds to the quantity h_p—this is the head that must be provided to the pump from the suction side:

$$\text{NPSH-R} = h_p. \tag{2.3.18}$$

In its turn, to output this head, the pump must get the corresponding head from the suction side.

NPSH-A is the parameter of the whole suction system. If NPSH-R is the head that must be supplied, NPSH-A is the head which can be supplied from the given suction line. In our previous consideration, the suction head available is

$$\text{NPSH-A} = \frac{p_a - p_s}{\rho g} - h_b - h_l. \tag{2.3.19}$$

Again, we notice that h_b may be of the opposite sign.

As it follows from the previous consideration, there must be

$$\text{NPSH-A} > \text{NPSH-R} \tag{2.3.20}$$

by some value, otherwise cavitation may surprise us.

2.3.4 Pump cavitation

Finally, we may bring together all our knowledge about cavitation in a pump. In this section, we are not interested in the useful work of the pump, in the head created by it, etc. We only consider the processes at the suction side of the pump.

To output the given pressure and the given discharge, the pump needs a certain value of pressure in the suction line—at the input flange. In engineering terminology, this is the net positive suction head—the pressure at the suction side that must support the high pressure which this pump creates at the pressure line. The given value of NPSH-R will be set when we turn on the centrifugal pump, and the impeller starts to rotate at the required velocity (at the angular velocity that has been set).

This NPSH-R of the pump must be maintained by the external suction system; the corresponding parameters are called the net positive suction head available—NPSH-A. The fluid enters the pump from some outer reservoir which can be placed at some height relative to the pump (to the axis of its impeller): either higher than the pump or somewhat lower than it. The last configuration may be determined by the requirement of the whole pumping system's operation: it is often necessary to deflate a fluid from somewhere. Thus, the pump vessel (that reservoir) is located somewhere below, and the pump is mounted where it can be possible; the example is the pumping of water from a cellar after a minor flood.

The minimum pressure that can be reached for the given NPSH-R is

$$p_{min} = p_a - \rho g h_b - \rho g \cdot \text{NPSH-R} - \rho g h_l. \tag{2.3.21}$$

This is the expression (2.3.17) rewritten with all the consequent nuances (with the suction head required and pressure losses); the negative sign at h_b corresponds to the location of the pump reservoir below the pump's impeller. If this pressure falls below the saturation pressure p_s, cavitation begins inside the pump.

A centrifugal pump contains an impeller (precisely, this the main part of the pump overall); the blades of the impeller create low-pressure zones and, if the pressure there drops below p_s, these zones generate cavitation, i.e., huge bubbles that will destroy the pump gradually.

We may also note that cavitation may appear inside a pump of any construction, not only in a centrifugal pump. In simple words, to engulf a fluid from the outer source, pump must create a lower pressure inside it. If this pressure is too low, below p_s, then cavitation begins in this pump.

2.3.5 Thoma's cavitation factor

Above, we gave a representation of cavitation in pumps, but did not introduce any cavitation number. Actually, as we see, this is not obligatory, but, for the sake of order, we must mention that for such systems such parameter can also be used. The corresponding form of the cavitation number is called the Thoma cavitation factor; it is introduced in a form similar to (2.1.7), through the pressure difference:

$$Q_{TH} = \frac{p_{in} - p_s}{p_{in} - p_{out}}. \tag{2.3.22}$$

For the given case, the pressure difference in the denominator is proportional to the total pressure head of the pump H, while the numerator has a familiar form: it is the difference between the pressure at the pump's entrance $p_a - \rho g h_b$ and the saturation pressure p_s. Thus, using the representation with pressure heads, we have:

$$Q_{TH} = \frac{\Delta h}{H}, \tag{2.3.23}$$

where $\Delta h = \dfrac{p_a - p_s}{\rho g} - h_b$.

2.4 Cavitation effects

2.4.1 Erosion vs. evaporation

Despite the fact that, generally, the term 'effect' means both positive and negative consequences, the cavitation effects can be considered only in the second—harmful—meaning. As we discussed in Section 1.5, all the positive effects of cavitation are indeed its destructive effects which are properly applied.

In the previous chapter, we have already mentioned that the main effect of cavitation on a surface is erosion. Erosion is the mass carryover under an external mechanical influence. We all see erosion in natural conditions; probably, the most famous example is soil erosion. In nature, the main source of erosion is wind, which may carry sand particles, or water, which washes shores. All these sources of erosion are more sparing that the cavitation erosion, which can damage a metal surface noticeably in a couple days of work.

During the erosion process, the mass is carried away in the form of macroscopic particles, not separate atoms. Nevertheless, the mechanics of erosion can be explained in the same way as the carryover by atoms during the evaporation process.

Evaporation process differs from boiling: this process takes place in any external conditions. If the temperature in non-zero (i.e., always, as it is promised by the third law of thermodynamics), then atoms[9] on the surface are distributed on their kinetic energies (as promised by Maxwell), so some atoms have higher kinetic energy than others. If this kinetic energy exceeds the potential energy that holds this atom on the surface, then this atom escapes its site.

There may be a second type of evaporation—evaporation induced by an external flow; for instance, by the condensation flux. This process looks like erosion, at some view angle. Indeed, some external energy must be applied to separate a piece of a surface, regardless of what it is: a single atom or a chunk of a solid surface. This circumstance is clear, of course; the main distinction between erosion and induced evaporation is less obvious: self-healing.

During evaporation, a single atom accepts the energy from an external source instantly. Then it has very little time to be detached (if its new energy is sufficient), otherwise this additional energy will dissipate in the medium, that is, it will be transferred to other particles. This looks like

---

[9]   We refer everywhere to atoms, but there may be molecules too; this is not crucial.

self-healing of the surface at atomic spatial scales: if the external disturbance did not cause evaporation immediately, then this disturbance did not contribute to evaporation.

On the contrary, large-scale impacts leave traces forever, as usual. If a single strike of a macroscopic particle (or of a droplet, or of a collapsing bubble) did not produce the destruction, never mind. Other hits will finish the process, so sooner or later the surface will be damaged. From some point of view, large-scale disturbances are poisonous hits: they inflict small damages, which accumulate with time and produce the departure of mass. When the next piece of surface detaches, this is the result of not only the final strike, but also the result of all the previous hits that loosen the surface.

Thus, the rate of erosion must vary with time, at least at the initial stage (while there is something to break down). Consecutive impacts produce more and more weakening of the surface, and later the second phase begins: intense erosion at increasing rate. At this second stage, erosion has the avalanche character, covering all new areas.

Then, when the surface is eroded enough, and many caverns are formed on it, the character of the process changes. Large caverns are some kinds of concentrators of cavitation, because those caverns usually contain additional gas. Thus, these caverns tend to grow. On the other hand, cavitation is not caused by the gaseous phase as it is, but by the collapsing gaseous phase. Because that additional gaseous phase prevents intense bubble collapse, the rate of cavitation erosion decreases. Moreover, a highly damaged surface changes the type of flow in its vicinity, producing more vortexes there; this factor affects the destruction caused by cavitation and increases it, generally, twofold. At small spatial scales, we should expect the decrease in the rate of cavitation erosion: such irregular small-scale flow 'washes' the bubbles formed close to the surface away from it. On the other hand, such huge vortex structures help to form large-scale cavitation vortexes, the effect of which may be macroscopic and strong. However, the last type of influence is not erosion at all. Summing up, in other words, we may conclude that after everything that could be easily destroyed has been destroyed, the rate of erosion decreases.

Generally, the curve of cavitation erosion rate can be represented as in Fig. 2.4.1. We should perceive it rather like a qualitative representation, not the attempt at reconstruction of a certain common dependence. The main reason to deny such an attempt is a suspicion that the general dependence of the erosion rate on time does not exist. In (Preece and Macmillan 1977), one may find the discussion concerning various experimentally determined—we should emphasize this point—dependencies of the erosion rate by different authors. As always, every experimental curve can be explained well; the difference between diverse experimental data can be explained much worse. Surely, the reasoning given above must be considered only as a generalized consideration.

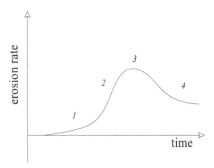

Fig. 2.4.1. General dependence of the rate of cavitation erosion on time. Section *1* corresponds to the absence of erosion (impacts do not cause the mass carryover); then erosion appears, increases (*2*), reaches its maximum (*3*) and fades (*4*).

2.4.2 Schemes of cavitation erosion treatments

Since erosion is the most vivid manifestation of cavitation, this process is investigated separately. The main purpose is to explore the influence of cavitation on different materials in the same conditions.

In this way, we may find which materials are more resistant to cavitation and, taking their structure into account, try to understand the methods to minimize the consequences of cavitation.

There may be many such schemes; we briefly explain only three of them.

The first type is presented in Fig. 2.4.2. This type corresponds to the setup for investigation of acoustic cavitation (see Chapter 6); all the photos of cavitation erosion in this book are obtained on this setup. In such experiments, it is convenient to explore the cavitation damage in 'pure conditions': all the parameters are well-controlled and well-reproducible. The ultrasonic waveguide hits a liquid with great energy; these impacts produce a lot of small cavitation bubbles the collapse of which damages the surface of the waveguide. In these experiments, we were interested in the special mechanisms of the cavitation effect (except for ones listed below in Section 2.4.3). The results can be found in Chapter 5.

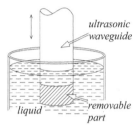

Fig. 2.4.2. Erosion test for acoustic cavitation. The removable part is fixed on the ultrasonic waveguide that vibrates in the liquid at an ultrasonic frequency.

The second type of cavitation tests can be implemented for pipe conditions. As we see in Fig. 2.4.3, these are the experiments which actually reproduce a cavitating flow in a pipe. A special removable section is mounted at the cavitation region; its examination allows one to define such an important parameter as erosion rate, among others.

Another type of experimental setup corresponds to erosion in pumps, cavitation mills, etc. The main element of this scheme is the rotating disk which produces cavitation on its heterogeneities (for an example of a cavitation mill, see Fig. 1.5.2 above). Thus, for the erosion test this disk is equipped with special expandable elements. Again, these elements can be made of various materials with surfaces processed with different methods.

The results of erosion experiments show that erosion is weaker for soft materials (Pearsall 1972) (for lead, for example). Thus, one may assume that the mechanisms of mass departure are connected with the fragile character of destruction. However, the best result was obtained for stainless steel.

The absolute values of the erosion rate differ for different conditions. Possibly, for some estimations, we may point the value of ~ 1–10 micrometers per minute (Preece and Macmillan 1977), that is, every minute several micrometers of the surface vanish. Of course, in other conditions the erosion rate may differ significantly from this range.

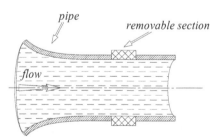

Fig. 2.4.3. Erosion test for cavitation in a pipe.

2.4.3 Mechanisms of impact

Erosion is caused by a mechanical impact on a surface. From the general point view, the source of this impact is, obviously, cavitation. However, the certain nature of these impacts is not clear.

At first, one may suppose that it is connected with the influence of pressure that arises around the collapsing bubble (Bai and Bai 2019). The physical details of this process are considered in Chapter 7.

However, this mechanism is not the only one. Today, the main process which causes the destruction of materials during cavitation is supposed to be the hits from microjets formed in collapsing bubbles (Fig. 2.4.4). Such a strange shape of bubbles collapsing near a solid wall can be easily explained (see Chapter 4), and exactly microjets are mainly supposed to be responsible for the surface destruction.

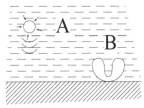

Fig. 2.4.4. Two ways for a bubble to destroy a surface. A—the distant way: an oscillating bubble produces a pressure wave that acts on a surface. B—themelee way: collapsing near the surface, a bubble generates a microjet which impacts a solid surface.

2.4.4 Increase in hydrodynamic resistance

Cavitation erosion is the most dangerous consequence of cavitation, but it should not be forgotten that this is not the only effect which cavitation has. Besides the mass carryover, cavitation may also influence the flow structure, which leads to not destructive, but undesired effects.

Partially, this matter was discussed in Section 2.1. The local hydraulic resistance increases in every case of abrupt deviation of the channel configuration and, for some of such deviations, cavitation plays a role in the friction enhancement.

However, those were standard cases, so to say (like a flow in an orifice). In special cases, when cavitation arises in pipes, the hydrodynamical resistance increases in another manner: because of appearance of a two-phase flow. Simply, another sort of medium is now in the pipe, and that fact affects the pressure loss. Below, we consider a flow in a pipe and calculate the additional friction that arises after cavitation begins.

Let us assume that a two-phase flow can be described as a homogeneous system but with effective properties. In such a model, we consider a uniform 'quasi-fluid' with uniform properties; this proposition means that bubbles move exactly with the same velocity as the liquid phase, so no additional terms in friction appear due to relative motion of phases.

The void fraction—the ratio of volume of the gaseous phase V_g to the total volume of the gas–liquid mixture $V_m = V_l + V_g$—can be defined as[10]

$$\varphi = \frac{V_g}{V_l + V_g}.$$
(2.4.1)

Thereby, the fraction of the liquid phase can be expressed as

$$\frac{V_l}{V_l + V_g} = 1 - \varphi.$$
(2.4.2)

---

[10] Below, we use expressions for the differential pressure where, strictly, another quantity must be used: $\beta = V_g/V_m$—the ratio of volume discharges. In the absence of a phase slip, as we adopted, $\varphi = \beta$.

The density of the gas–liquid mixture ρ_m is the ratio of the total mass to the total volume:

$$\rho_m = \frac{\overbrace{\rho_l V_l}^{m_l} + \overbrace{\rho_g V_g}^{m_g}}{V_m} = \rho_l(1-\varphi) + \rho_g\varphi = \rho_l - \varphi\Delta\rho, \tag{2.4.3}$$

where

$$\Delta\rho = \rho_l - \rho_g. \tag{2.4.4}$$

Then, the differential pressure can be expressed for two cases—for the flow of a single-phase fluid with density ρ_l and for the flow of a mixture with density ρ_m:

$$\Delta p_l = \xi_l \frac{\rho_l u_l^2}{2}, \tag{2.4.5}$$

$$\Delta p_m = \xi_m \frac{\rho_m u_m^2}{2}. \tag{2.4.6}$$

The comparison between (2.4.5) and (2.4.6) must be provided for the same discharge

$$\rho_l u_l = \rho_m u_m; \tag{2.4.7}$$

that is, we have a transition from the single-phase flow to the cavitating flow in the same pipe.

The factors ξ_l and ξ_m define the friction of the fluid at the pipe's wall. If we assume that for both cases—for the initial single-phase flow and for the cavitating flow—the liquid phase is in contact with the wall, then we obtain that

$$\xi_l = \xi_m. \tag{2.4.8}$$

Thus, we have for the ratio of (2.4.6) and (2.4.5), with (2.4.7) and (2.4.8)

$$\frac{\Delta p_m}{\Delta p_l} = \frac{1}{1 - \varphi\Delta\rho/\rho_l}. \tag{2.4.9}$$

Since $\varphi\Delta\rho > 0$, then $\Delta p_m/\Delta p_l > 1$—the differential pressure increases in the cavitating flow. In other words, cavitation itself causes additional friction in a flow, even without any additional effects.

2.4.5 Abnormal modes of operation

This matter is related to the previous subject—enhancement in hydraulic resistance. However, the scale and effect of this phenomenon are essentially higher: if small bubbles may increase friction in a channel, as we see above, then a huge bubble, the size of which is comparable to the inner diameter of the volume distorts the entire flow pattern.

Among other dangerous consequences of cavitation, this effect must not be forgotten. Cavitation produces a new phase—sometimes of a small scale, but sometimes much larger, as we see on examples of supercavitation on wings. Simultaneously, due to the appearance of these new objects in a flow, the pressure distribution in the flow may be changed significantly, and this fact may dramatically influence the very flow. This effect can be especially strong in closed volumes—in pumps, hydro turbines, etc.—where the path of fluid particles is long in comparison with the one in a pipe. When a huge cavitation cavern appears in such a closed space, the damage of the whole setup is possible.

This mode, sometimes directly connected with the so-called cavitation vortex, can be considered only in a full, non-stationary hydrodynamical statement. On a qualitive level, we may assume that this giant cavern will pulsate due to many reasons: both because of interfacial process and because of the non-stationary pressure redistribution in the volume. Moreover, the rotation of the wheel (an impeller or some other aggregate in this volume) will amplify the frequency corresponding to its rotation. This process may be stopped by the resonant damage of the construction.

Here, we may also mention another, more humble effect. When collapsing, cavitation bubbles produce pressure waves and, in a common case, the amplitude of these pulsations may be high (see Chapter 7). This is a non-stationary effect that cannot be detected by numerical simulations of stationary fluid dynamics equations, but it affects the flow pattern too.

Conclusion

Cavitation causes great problems for technological devices. For this reason, many engineers assume that technical application is the main part of the whole 'cavitation science': the main purposes of investigation of cavitation are to predict its occurrence and, if it is impossible to prevent it, to minimize the consequences. In this section, we explained the basics of the application of cavitation in engineering.

Simplifying the reasons, we may consider the origin of cavitation as follows. Due to the Bernoulli equation, there must be

$$p + \frac{\rho u^2}{2} = const$$

along the streamline. Thus, the higher the velocity, the lower the pressure at that point. If the pressure drops below the saturation pressure p_s, then thermodynamic conditions correspond to the instability of the liquid phase, and cavitation begins. Travelling into regions of higher pressure, these bubbles collapse. For many, these processes are the essence of cavitation.

We should also point that, actually, any decrease of pressure leads to some growth of air bubbles located in the liquid, as well as the subsequent pressure increase causes the collapse of those bubbles. Due to the similarity in common features (increase in the gaseous phase), this process can be considered as some sort of cavitation too, but the 'true cavitation', when pressure drops below p_s, is a much more intense process. The growth of air bubbles is too restricted to compete with vapor bubbles appearing at pressures below p_s.

The main parameter which helps us to describe the cavitation inception is the cavitation number

$$Q = \frac{p - p_s}{\rho u^2 / 2}.$$

When this parameter is about unity, cavitation begins. Unfortunately, this quantity cannot be considered as the single and unique measure for cavitation: in different conditions, cavitation starts at different values of Q. But, at least, this parameter helps to roughly estimate the dangerous parameters for the given flow.

A more serious analysis of cavitation inception on wings can be conducted with cavitation diagrams. These diagrams, created for certain types of hydrofoils, predict cavitation inception in the dependence of the attack angle and, consequently, the parameters of the flow at which cavitation begins at any angle of attack.

Cavitation in pumps is the direct menace, since the whole principle of operation of this device implies the suction of fluid into it which, in its turn, is achieved by creating low-pressure inside the pump. This low-pressure zone is a necessary region in a pump (otherwise it cannot receive a fluid at all), but the question is—how low is this pressure? If the pressure inside a pump drops below the saturation value, then cavitation inside it produces its usual set of consequences. We considered, for certainty, a centrifugal pump: in this apparatus, cavitation indeed is a kind of cavitation on hydrofoils: the main processes take place on the impeller blades, and they are similar to cavitation on wings.

However, for pumps the common situation is not too bad, since there exist developed methods to calculate all the factors that influence the beginning of cavitation.

The main consequence of cavitation for appliances is erosion of their elements, but cavitation effects as a whole—just like the reasonings that lead to erosion—are very versatile. Cavitation influences the entire structure of a flow, and, sometimes, may destroy the technological device completely.

References

Bai, Y. and Q. Bai. 2019. *Subsea Engineering Handbook*. Gulf Professional Publishing.

Bergant, A. and A. R. Simpson. 1999. Pipeline column separation flow regimes. *Journal of Hydraulic Engineering*, 125: 835–48.

Brennen, C. E. 1994. *Hydrodynamics of Pumps*. Oxford University Press.

Carlton, J. S. 2012. *Marine Propellers and Propulsion*. Butterworth-Heinemann.

Cherkassky, V. M. 1985. *Pumps, Fans, Compressors*. Mir publisher: Moscow.

Eppler, R. 1990. *Airfoil Design and Data*. Verlag: Springer Berlin Heidelberg. https://doi.org/10.1007/978-3-662-02646-5_6.

Ferrari, A. 2017. Fluid dynamics of acoustic and hydrodynamic cavitation in hydraulic power systems. *Proceedings of the Royal Society A: Mathematical, Physical and Engineering Sciences*, 473: 20160345. https://doi.org/10.1098/rspa.2016.0345.

Idelchik, I. E. 1986. *Handbook of hydraulic resistance*. Hemisphere Publishing Corp.: Washington.

Martynov, S. B., Mason, D. J. and M. R. Heikal. 2006. Numerical simulation of cavitation flows based on their hydrodynamic similarity. *International Journal of Engine Research*, 7: 283–96. https://doi.org/10.1243/14680874JER04105.

Nurick, W. H. 1976. Orifice Cavitation and its effect on spray mixing. *Journal of Fluids Engineering*, 98: 681–7. https://doi.org/10.1115/1.3448452.

Payri, R., C. Guardiola, F. J. Salvador and J. Gimeno. 2004. Critical cavitaion number determination in diesel injection nozzles. *Experimental Techniques*, 28: 49–52. https://doi.org/10.1111/j.1747-1567.2004.tb00164.x.

Payri, R., F. J. Salvador, J. Gimeno and O. Venegas. 2013. Study of cavitation phenomenon using different fuels in a transparent nozzle by hydraulic characterization and visualization. *Experimental Thermal and Fluid Science*, 44: 235–44.

Pearsall, I. S. 1972. *Cavitation*. Mills and Boon Limited: London.

Pease, D. C. and L. R. Blinks. 1947. Cavitation from solid surfaces in the absence of gas nuclei. *The Journal of Physical Chemistry*, 51: 556–567.

Preece, C. M. and N. H. Macmillan. 1977. Erosion. *Annual Review of Materials Science*, 7: 95–121. https://doi.org/10.1146/annurev.ms.07.080177.000523.

Sebestyen, Gy., A. Szabo, A. Verba and M. Rizk. 1984. The characteristic of cavitating flow and the results of noize investigation. *Periodica Polytechnica Mechanical Engineering*, 29: 29–49.

Tupper, E. C. 2013. *Introduction to Naval Architecture*. Butterworth-Heinemann.

Washio, S. 2014. *Singular properties of flow separation as a cause of cavitation inception*. *In*: Recent Developments in Cavitation Mechanisms. Woodhead Publishing. https://doi.org/10.1533/9781782421764.159.

CHAPTER 3

Pressure
Positive and Negative

3.1 What is pressure?

3.1.1 Pressure at first glance

During cavitation, the growth of bubbles is caused by an abrupt depression in a fluid. When the outer pressure drops, the inner pressure in a bubble pushes the liquid up, if the pressure there is lower than inside a cavern. Usually, it is assumed that the pressure inside a bubble corresponds to the saturation pressure at the given temperature; this a reasonable approach given the lack of a better hypothesis.

Thus, the liquid pressure at cavitation is low. It is interesting, nonetheless, how small the pressure in the liquid can be. Can the pressure be zero? Or can it even be negative?

These questions are very interesting from many points of view. To answer them, we must start from the basic notions.

Everybody knows the simplest definition of pressure from school. As we know from mechanics, pressure p is a force acting on a unit of an area:

$$p = \frac{force}{surface\ area}.$$

(3.1.1)

For example, in this manner we are able to calculate the pressure of a human standing on a scale.

Initially, to be exact, historically, thermodynamics used the same understanding of pressure. Pressure acting on a surface does a work on a body, causing the normal displacement of the surface, which leads to deviation in volume dV, that is, the corresponding work is pdV. This term appears in the first law of thermodynamics, which can be transformed into the second law[1] with representation for heat $\delta Q = TdS$; formulated with the free Helmholtz energy F, this equation can be written as

$$dF = -SdT - pdV.$$

(3.1.2)

Thereby, one may say that

$$p = -\left(\frac{\partial F}{\partial V}\right)_T,$$

(3.1.3)

that is, pressure is a partial derivative from the Helmholtz energy at constant temperature. Moreover, the relation (3.1.3) can be considered as a definition of pressure. Represented with specific values $f = F/m$ and $v = V/m = \rho^{-1}$, we have

$$p = -\left(\frac{\partial f}{\partial v}\right)_T = \rho^2\left(\frac{\partial f}{\partial \rho}\right)_T$$

(3.1.4)

---

[1] The second law of the thermodynamics can be formulated in many ways. From some point of view, if entropy appears in the heat balance equation, this equation now represents the second law of thermodynamics.

If we find, as it's done in statistical physics, the free energy as a function of density and temperature, then equation (3.1.4) gives us the equation of state—the correlation between pressure, density and temperature in the form of

$$p = p(\rho, T). \tag{3.1.5}$$

One may notice that we have already gone quite far away from the consideration of some 'forces' and 'surfaces': the last relation does not contain such elements. This whole consideration is so clear that it may seem boring. But now hydrodynamics shakes us up.

3.1.2 The stress tensor

Hydrodynamics of viscous fluids uses such a physical quantity as the momentum flux density tensor $\prod_{ij}$ (Landau and Lifshitz 1987); with this quantity, the Navier–Stokes equation can be written as

$$\frac{\partial \rho u_i}{\partial t} = -\frac{\partial \Pi_{ik}}{\partial x_k}. \tag{3.1.6}$$

Here

$$\Pi_{ik} = \rho u_i u_k - \sigma_{ik}, \tag{3.1.7}$$

where the stress tensor consists of two parts—of the pressure one and of the viscous term

$$\sigma_{ik} = -p\delta_{ik} + \tau_{ik}; \tag{3.1.8}$$

here δ_{ik} is the Kronecker symbol.

It should be emphasized that the normal component of the viscous stress tensor $\tau_{ii} \neq 0$. Therefore, we see that the normal component of stress is not pressure, as one may expect from (3.1.1); it is $p - \tau_{ii}$. So, it is time to return to the question: what is pressure?

Thermodynamics, that had been applied in the previous section, describes only equilibrium systems, reversible processes.[2] Thus, the equation (3.1.2) and everything that follows from it, up to the equation of state (3.1.5), concerns only equilibrium processes. In other words, with the correlation (3.1.5), we may calculate only the equilibrium pressure, neglecting all dissipation processes that take place in the medium.

On the other hand, the part τ_{ii} corresponds exactly to viscous, dissipative forces. These forces act on a surface area in the normal direction too; so, the direct definition of pressure includes these ones. However, usually under pressure we mean a function that obeys the thermodynamic equation of state, and if so, then this 'thermodynamic pressure' is not exactly pressure.

This is a typical situation for thermodynamics. When you translate a problem into thermodynamics language, you lose all the in equilibrium details of it. The problem that you formulate in thermodynamic terms is 'passed through the equilibrium filter'; this fact should not ever be ignored.

3.1.3 Saturation pressure

As we see above, thermodynamic pressure is the equilibrium pressure itself. However, there also exists a special equilibrium-equilibrium pressure—the saturation pressure.

The condition of thermodynamic equilibrium of the two phases in a single-component system is (see Appendix A)

$$\varphi_1(T, p) = \varphi_2(T, p). \tag{3.1.9}$$

---

[2] Exactly, there exists a direction of thermodynamics called 'irreversible thermodynamics', which was very popular in the mid-XX century and causes mixed feelings now. In this sense, classical thermodynamics is limited to a couple of wordings about inequalities for irreversible processes (such as the Clausius inequality).

Here pressure and temperature are equal in both phases; the case of non-equal pressures is considered in the next section.

In each phase, pressure, temperature and density are connected by the equation of state (3.1.5). Generally, to define the pressure in a phase (liquid or vapor), one must determine two parameters—density and temperature. But for two phases which are in equilibrium, we have the additional connection between temperature and pressure (3.1.9). Thus, the saturation pressure p_s is an unambiguous function of T, and at saturation—for the phase equilibrium—only temperature defines the pressure in the system. Note that this sentence is correct for any phase transition: for solid–gas, solid–liquid and liquid–gas.

In the particular case, when we have a bubble in a liquid at a given temperature, we may expect the pressure inside the bubble to correspond to the saturation pressure (with important correction on the surface tension, see the next section). Thereby, for cavitation the pressure in a liquid must drop lower than p_s—in this case, bubbles would grow.

The saturation pressure for water is shown in the Fig. 3.1.1. As we see, for low temperatures, the saturation pressure is quite low; at first glance, it is even difficult to imagine how the pressure in the liquid can drop by one or even on two orders of magnitude (in comparison with the normal atmospheric pressure) to provoke cavitation.

But cavitation exists, as we know, since our pressure can be far from our imagination—see below.

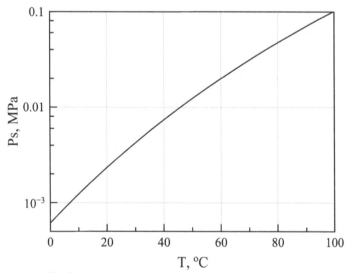

Fig. 3.1.1. Saturation pressure for water; data from the NIST.

3.1.4 Pressure at microlevel

Now we can turn our attention to the molecular level of the problem. On this level, pressure is the influence of impacting particles: atoms or molecules of the media strike a surface (real or imaginary one) and give it their impulse. The velocity of a molecule consists of two components: the average velocity of the flux $\bar{u}$ and the chaotic velocity component v'. The last one is connected with temperature: its mean value is $\overline{u'^2} \sim T$ (for instance, in 3D case $\overline{u'^2} = 3kT/m$, where k is the Boltzmann constant, m is the mass of the molecule).

Then, using the velocity distribution function $D(u')$, we may calculate the pressure as the integral

$$p = 2\int_0^\infty mu'^2 D(u')\,du'. \tag{3.1.10}$$

For the equilibrium distribution function, (3.1.10) gives the relation that contains temperature T; thus, we see that $p \sim \overline{u'^2}$, i.e., the chaotic motion of molecules provides the pressure indeed.

If the velocity distribution function is unknown, then one may obtain the corresponding function from numerical simulations, considering the motion of atoms interacting with each other (the Lennard–Jones potential is the most popular correlation). In other words, one may obtain the results using pure microlevel mechanics.

There also exists another way to get the correlation for pressure in a mechanical system, where each particle has coordinates $\vec{r}$. In its complete form, the virial theorem states that

$$2\overline{K} - \underbrace{\overline{\sum_i \vec{r_i} \frac{\partial U}{\partial \vec{r_i}}}}_{\tilde{U}} - 3pV = 0, \tag{3.1.11}$$

where the upper line denotes the mean value on time, U is the potential energy of interaction of particles (atoms or molecules) and K is the kinetic energy of the system:

$$K = \sum_i \frac{m_i u_i^2}{2}. \tag{3.1.12}$$

In accordance with (3.1.11), one may find the pressure in a mechanical system. Here, pressure is a uniform force per unit of area that acts on the surface bounding the volume V. The key word in this term is 'uniform'; all parts of the surface bounding the volume must be stricken by molecules with equal probability. Generally, this means that the number of particles in a system must be large, otherwise one may experience interesting problems, when the pressure would depend on the selected volume or even on its form. Of course, in a real system this condition is fulfilled; these notes concern only numerical simulations.

Thus, mechanics also lets us determine the pressure in a system and, besides, in various ways.

3.1.5 Negative pressure as it is

In sum, we have seen many definitions of pressure prior to this page, but the equation (3.1.11) is the most suitable to explain the nature of negative pressure in media. From (3.1.11), we have for pressure

$$p = \frac{2\overline{K}}{3V} - \frac{\overline{U}}{3V}. \tag{3.1.13}$$

First, we should point that the first term corresponds to a perfect gas and provides the Clapeyron equation. Indeed, in this case $\overline{U} = 0$ (one of possible definitions of a perfect gas but not the best),[3] the kinetic energy of molecules $\overline{K}$ $-NkT$, and we have from (3.1.13), using the concentration function $n = N/V$, the Clapeyron relation

$$p = nkT. \tag{3.1.14}$$

It is clear that the pressure of a perfect gas can only be positive. To obtain the negative value, the second term in the right-hand must be (1) non-zero, (2) to be exact, with the positive sign, (3) larger than the perfect-gas pressure $2\overline{K}/3V$. That is, the interaction of molecules must be significant for the medium to stretch so much that the pressure inside it becomes sub-zero. In other words, negative pressure can be reached only in dense media—in solids and liquids.

---

[3]   The definition of perfect gas as a substance with no interaction between particles inside it requires some theoretical model, which is difficult to analyze in its fundamentals and is inconvenient in practical use. The definition of perfect gas as a medium obeying the Clapeyron equation—despite, possibly, looking strange when you read it for the first time—is easy to verify and provides no additional questions.

Now we come to the problem that can be formulated in two different ways:

- ✓ above, we stated that if the pressure inside a liquid becomes sufficiently low (but positive), cavitation begins: caverns start to grow everywhere in the liquid, and so on; thus, how can we reach not a very low pressure, but the negative pressure at all?
- ✓ how stable is the state with negative pressure?

To answer these questions (this question, to be exact), we have to consider some thermodynamic aspects of phase transitions.

3.1.6 Metastable states

States of a substance are not divided into stable and unstable ones. Between these two extremes, there also exists an intermediate one—the state that is unstable not in relation to any small disturbance (which corresponds to the unstable state), but in relation to disturbance of a finite amplitude. Such—a metastable—system can withstand small disturbances, but cannot resist the strong ones. Metastable states are limited in their parameters: from the common condition of thermodynamic stability, a state can be stable only if

$$\left(\frac{\partial p}{\partial V}\right)_T < 0. \tag{3.1.15}$$

Thus, when the derivative (3.1.15) becomes zero, such a state is absolutely unstable. The states that satisfy this condition correspond to spinodal curve on the phase diagram.

Let us take a look at the phase diagram of a one-component substance on pressure–temperature coordinates; see Fig. 3.1.2. Here curves AQ corresponds to sublimation (solid–gas phase transition), QB corresponds to melting (solid–liquid phase transition), QC corresponds to boiling (liquid–gas phase transition). Q is the triple point—the only point where all three phases can coexist; C is the critical point of this substance.

Along each curve, the slope is defined by the Clapeyron–Clausius relation:

$$\frac{dp}{dT} = \frac{L}{T(v_n - v_o)} , \tag{3.1.16}$$

where L is the enthalpy (latent heat) of the phase transition; $L > 0$ when the old phase (more ordered) with specific volume v_o transits to a new phase (less ordered) with specific volume v_n at temperature T.

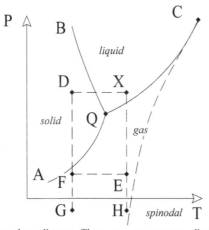

Fig. 3.1.2. The pressure–temperature phase diagram. There are many ways to displace from the point X; some ends correspond to the solid state, some ones to the gas.

As we see, the slope of curves is determined by the difference in specific volumes. For phase transitions, liquid–gas and solid–gas $v_n > v_o$ (the specific volume of the gas is larger than the specific volume of the condensed medium); melting is an ambiguous case. For most substances, the density of the solid state is higher than the density of the liquid, that is, $v_n > v_o$, and the slope of the curve QB is positive too. However, for some substances, water for example, the density of the solid is less than the density of the liquid (ice floats on water, as we know), so the curve QB has a negative slope, as it shown in Fig. 3.1.2.

It is useful to compare the slope of curves AQ, QB and QC for water in vicinity of the triple point, where $T_Q = 273$ K. For water, the sublimation enthalpy $L_{AQ} = 2834$ kJ/kg, the melting enthalpy $L_{QB} = 333$ kJ/kg, the vaporization enthalpy $L_{QC} = 2501$ kJ/kg. The specific volumes: of solid (ice) $v_s = 0.00109$ kg/m³, of liquid $v_l = 0.00100$ kg/m³, of gas $v_g = 206$ kg/m³; all data taken from the NIST. Therefore, we have

$$\left.\frac{dp}{dT}\right|_{AQ} = 50\,\text{Pa/K}, \left.\frac{dp}{dT}\right|_{QB} = -1.4\cdot10^7\,\text{Pa/K}, \left.\frac{dp}{dT}\right|_{QC} = 44\,\text{Pa/K}. \tag{3.1.17}$$

Thus, in simple words, the melting branch QB is almost vertical in comparison to the sublimation curve and the vaporization one. Note that usually such a diagram is presented only by its upper part; for our consideration, the lower zone—of negative pressure—has a crucial value. The region in vicinity of zero temperature is shown schematically.

Let us imagine that initially we have a liquid with parameters corresponding to the point X. What would happen if we (a) decrease temperature at constant pressure, i.e., go to the point D, or (b) drop the pressure at constant temperature, i.e., go to the point E?

The point D corresponds, normally, to the solid state. Thus, normally, one should expect that after cooling we necessarily obtain the solid state. However, this is not exactly the case. It is insufficient to organize the external parameters corresponding to a new phase; to switch to a new state, the old phase must also receive some initial disturbance—get the starting seed of the new phase. If this seed is absent, then the old phase would exist for a long time (for example, water may exist at temperatures below 0°C), how long—it depends on supercooling; actually, with time, sooner or later, spontaneous fluctuations would do their job, transiting the old phase to a new one. Finally, one may say that the liquid is supercooled at the point D.

Now consider the point E; note that for such experiments (see below), we actually may have to deal with the isothermal pressure drop—not with the abrupt adiabatic depression,[4] as it takes place during cavitation. Parameters at the point E correspond to the gas phase obligatory, since the slope of the sublimation curve AQ is positive. By analogy, the liquid tends to boil at the point E, but it has to have some initial nucleus inside it—if so, the liquid turns to gas, but if it will wait for a seed, it would remain in the condensed phase. Thus, the liquid at the point E is superheated: yes, we can overheat a liquid at constant temperature—it is a wonder of thermodynamics, but not the last one.

Then, let us take a look at the point F, which corresponds to cooling and depression of the liquid with initial parameters at the point X. Considering the weak slope of the curve AQ, we may conclude that the point F corresponds rather to a gas than to a solid, if only the pressure in F is lower than $p_Q = 611$ Pa at the triple point (besides, the opposite case is not impossible, of course; it can be realized for strong cooling and weak depression). Thus, the point F is corresponding to a superheated condensed medium: looking at the phase diagram, one may assume that this is an overheated solid, but considering the origin of the point F—we reach this point starting from X—one must say that the point F corresponds to a superheated liquid, despite the fact that the temperature in F is lower than at the initial point X. Sic.

As at the point E, the medium at the point F tends to become a gas. The absence of disturbances—the absence of seeds of a vapor phase—is the only reason why the medium is still condensed.

---

[4] See Chapter 1; sometimes, these conditions mean the same.

Then, if depression is strong, we get into the region of negative pressure (point *H*—depression, and point *G*—cooling and depression). From the common thermodynamic sense, the magnitude of disturbance that is required to transform into a new phase depends on the distance from the saturation line; keep also in mind that any disturbance at the spinodal curve leads to destabilization of the system.

Anyway, both points *G* and *H* are metastable—this is the answer to the second question given at the end of the previous section. A liquid with parameters at point *G* (or *H*) can only exist at special, very delicate conditions. By the way, the type of a phase that would appear if disturbances are strong enough is also a question. The distinction, for example, between points *E* and *H* is that the point *E* corresponds to the stable gas phase, while point *H*, at its negative pressure, corresponds to nothing. The final—stable—state, which appears after some perturbations, may be both a gas (on the upper side in Fig. 3.1.2, at $p > 0$) and a condensed phase (liquid or a solid)—which depends on the process that will establish the 'normal', positive pressure in the system. The most possible way: pressure increases, a cavity arises, so the two-phase system appears—that state corresponds to the point on the saturation curve.

To the first question—why cavitation cannot prevent the further decrease in pressure when it reaches the value of the saturation pressure, i.e., on the curve *QC*—the answer is obvious. Above, we described cavitation as a process of cavity growth: the pressure inside a cavity corresponds to *QC*, the outer pressure wanes, so the pressure inside a cavity becomes higher than in the surrounding liquid, and the cavity increases in size. Now, there is no cavity. To reach negative pressure, the liquid must be very clear from any additional objects.

3.1.7 How to reach negative pressure

Reviews of old methods can be found in (Imre2007, Hayward 1971). Actually, such experiments are known from the mid of the XVII century: Huygens was the first to investigate the possibility of the liquid stretch; later F.M. Donny reproduced these experiments and obtained some quantitative estimations of the negative pressure he produced.

The scheme of the experiment by Donny is presented in Fig. 3.1.3.

Usually, the liquid in both section of a U-shape tube has the same level. At least, it is correct for an open tube, when the outer pressure is equal in both parts. On the other hand, if the external pressure in section *B* is higher than in the section *A* by the value

$$\Delta p = p_B - p_A, \tag{3.1.18}$$

then this pressure difference can support the column of the height Δh of liquid with density ρ:

$$\Delta p = \rho g \Delta h. \tag{3.1.19}$$

The pressure difference (3.1.18) can be reached in two ways. In the normal situation, when $p_B > p_A > 0$, additional pressure is caused by a pump. However, in Donny's experiments, the section *B* was vacuumed, so $p_B = 0$. Thus, the liquid in the section *A* is at negative pressure: $p_A < 0$.

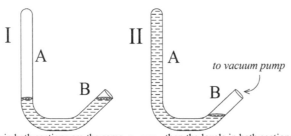

Fig. 3.1.3. (I) If pressures in both sections are the same, $p_A = p_B$, then the levels in both sections are the same. (II) Donny's experiment to reach negative pressure in a U-shaped pipe. The level of the liquid in section A is higher because the pressure in section B is higher, including the case $p_B = 0$ and $p_A < 0$.

The experiment was organized as follows. Initially, when the section *A* was placed in a horizontal position, this part of the tube was completely filled with liquid—without any free volume there. Then the tube was turned vertically; in this position, the liquid column is balanced by the atmospheric pressure. Then we start to pump out the gas from section *B*; finally, we obtain the case depicted in the Fig. 3.1.3. The liquid at negative pressure is supported by vacuum.

In such experiments, Donny reached negative pressure of about –0.1 atm. Later, Osbourne Reynolds repeated these experiments with mercury; he achieved pressures of up to –3 atm.

Another method, which allows to reach higher values of negative pressure (by modulus), was used by M. Bertelot, see Fig. 3.1.4.

Take a capillary, fill it with liquid, and then solder its upper end. Inevitably, an air bubble would remain in the upper part of the capillary, but it must be as small as possible. Then the soldered capillary is heated; during this process, the air bubble is absorbed by the expanding liquid. After that, the liquid that fills the capillary completely, begins to cool down. In this process, there is no gaseous phase in the capillary (until, of course, the negative pressure does not become too high), the liquid adheres to the walls and stretches. The negative pressure that was achieved by this method was around –30 atm.

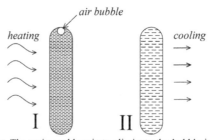

Fig. 3.1.4. The Bertelot experiment. The main problem is to eliminate the bubble in the soldered capillary; then, with a cooling process, it is possible to reach significant negative pressure in the capillary.

With time, the Bertelot scheme became classic: at that time, the 'world record' for negative pressure in water (–148 atm) was achieved with a variation of this method (Henderson and Speedy 1980, Speedy 1982).

Besides the static methods, there are also dynamic ones. Some of them are described in (Hayward 1971); their principle is based on the short-moment influence on a liquid. It may be the tearing of bars 'glued' together with a thin film of water from each other, or rotating rollers in fluid, or even an explosion of nitroglycerine. In (Ando et al. 2012), the greatest value of tensile strength was obtained by a laser-induced shock wave: –600 atm was achieved; in short, a shock wave in a vessel consists of stages of compression and tension, and the last one can be accompanied by cavitation. In details, shock waves are discussed in chapter 5.

The simplest and the most productive way to investigate the dynamic effects in a liquid is the ultrasound impact. We will discuss this method in Chapter 6, devoted to acoustic cavitation.

Today, as in many other fields of physics, the trend was displaced from experiments to numerical investigations, including, especially, the molecular dynamics simulation; a good illustration is the collection (Imre 2002), where many problems are considered with the numerical simulation approach. We return to numerical simulations in Section 3.3.3.

Meanwhile, negative pressure not only can be discovered in numerical simulations or examined in complicated experiments, it can be observed in nature: the mechanism of water transport in xylems is possible because the pressure inside it is negative (Steudle 1995). This may look strange, but this is the only reason that can explain the how a liquid (tree sap) travels dozens of meters from the roots to the treetops. The mechanism of creation of negative pressure in trees includes specific properties of live cells (as far as we can judge, dead materials cannot provide such an effect), evaporation from the leaf which creates the initial pressure gradient that leads to tension in the xylem system, and—possibly, surprisingly for physicists—a negligible role of the capillary forces: the surface tension cannot lift a liquid to such height (up to 100 meters) in such capillaries (of diameter ~10 μm).

Well, biologists examine what they want, while physicists do what they can. We cannot explain the water transport in living mater, but at least we may consider capillary forces closely.

3.2 The Laplace condition

3.2.1 Surface tension

We can define the concept of surface tension in different ways (as always, or, at least, in general case).

The first way begins in thermodynamics. According to the general thermodynamic approach, a system may do different kinds of work, each of them corresponds to the change of a certain thermodynamic parameter, i.e., the correlation for this kind of work is

$$\delta A = X dx, \tag{3.2.1}$$

see Appendix A. The simplest kind of work corresponds to the expansion: $\delta A^{vol} = p dV$; this work is conditioned by the variation in volume of the body. However, this is not the only case: except the change in volume, the body may change its surface area. Generally, this change demands some sort of work which would correspond to the change in the surface area Σ:[5]

$$\delta A^{surf} = -\sigma d\Sigma, \tag{3.2.2}$$

Here σ is the surface tension itself. We choose the negative sign in (3.2.2) because we expect this work to be done against the system: one may recall that (3.2.1) represents the work of a system, not the work performed on a system; thus, we expect that surface tension tends to shrink the body, not to stretch it out; otherwise, bodies would behave differently. Thereby, in this interpretation, surface tension is the proportionality coefficient for the work done to change the surface area.

From this point of view, we may write the first law of thermodynamics for a body that can change both its volume and its shape (i.e., surface area) in the form

$$\delta Q = dU + p dV - \sigma d\Sigma. \tag{3.2.3}$$

The second way also follows from thermodynamics and draws from the works by Gibbs. How many phases may we point out in a 'water + vapor' system? By definition, a thermodynamic phase is the system with homogeneous properties;[6] it is obvious that liquid and vapor are separate phases. It is not so obvious that when we see a combination of a liquid and its vapor, we actually see three thermodynamic phases instead of two: one additional is the surface layer—the boundary layer that separates the condensed phase and the gaseous one, see Fig. 3.2.1. In this layer, which has a width of around several intermolecular distances in the liquid, the properties of the medium differ both from the (bulk of) liquid and from the gas. For liquid, this region is too lax and irregular; for gas, it is too dense. Thus, one may conclude that this is a special thermodynamic phase; usually, it is called the 'Gibbs phase'. Since the width of this phase is very small, we may neglect its volume completely: this is a 2D object in our 3D word.

Thus, we have a phase of zero volume and, consequently, zero mass, etc. If so, what are the parameters to describe the state of this phase? For this surface phase, as for any other one, we may introduce the free energy (the Helmholtz energy):

$$F_\Sigma = \sigma\Sigma. \tag{3.2.4}$$

---

[5]   We have to use this strange symbol for the surface area, because, as usual, in vicinity of thermodynamics, all the 'normal' letters are occupied: S for entropy, F for free energy.

[6]   Not to confuse a phase and an aggregate state: the second one is a particular case of the first one.

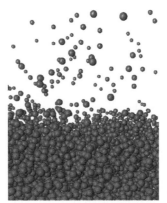

Fig. 3.2.1. The surface of a liquid—the boundary between liquid and vapor; this is a screenshot from a molecular dynamics simulation. This boundary differs both from liquid and vapor—this a special phase.

So, we see that surface tension is the free energy of a surface (which has the meaning of a special phase) per the unit of surface area, which is another meaning of σ. Note that, of course, we obtain nothing principally new here: $dF_\Sigma = -dA^{surf}$.

We should remember this representation of surface tension—by considering the new specific phase—in view of the future consideration of sonoluminescence (Chapter 9), where we will discuss probable explanations of this specific light emission.

Close to this explanation of surface tension, we may provide an additional theory. In the bulk of a condensed medium, the potential energy of a single atom does not depend (in average) on a certain place of this atom in the medium. Another case in vicinity of the surface: here, an atom has less neighbors; thus, its potential energy is lower. In other words, near the surface the potential energy (in thermodynamics terminology—the inner energy) is lower; thus, this discrepancy in energy—the energy defect—related to the surface area is the surface tension.

The third way starts from mechanics and, partially, from observations. It is an experimental fact that we can fill a glass with water to a level that is slightly higher than the height of this glass (see Fig. 3.2.2). This might be possible because of the surface tension forces—the forces that act along the surface (see Fig. 3.2.3). The surface tension itself is the force dF that arises at elementary width of the surface dl per that width:

$$\sigma = \frac{dF_l}{dl}. \tag{3.2.5}$$

All the definitions of surface tension mean the same and, as we see below, give the same equation for the balance of forces at the surface.

Fig. 3.2.2. A glass of water held by surface tension.

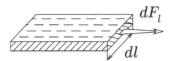

Fig. 3.2.3. Surface tension from the mechanical point of view.

3.2.2 The Laplace equation in hydrodynamics

Let us consider a phase equilibrium in a two-phase system (see Fig. 3.2.4) placed in a thermostat at constant pressure.

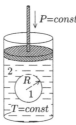

Fig. 3.2.4. A two-phase system at conditions p = *const*, T = *const*. It may be a drop of liquid in a vapor, or a bubble in a liquid—it does not matter.

This equilibrium in such a system corresponds to the minimal Gibbs energy (see Appendix A)

$$\Phi = F + pV = \underbrace{m_1 f_1 + m_2 f_2 + \sigma \Sigma}_{F} + p\underbrace{\left(m_1 v_1 + m_2 v_2\right)}_{V}, \tag{3.2.6}$$

that is, there must be

$$\delta\Phi = 0. \tag{3.2.7}$$

To calculate the variation of Φ, one must determine the independent argument of this function. Generally, this is an arbitrary set; we choose specific volumes of both phases v_1 and v_2, and also the mass of the 'spherical phase' m_1. Consequently, we have instead of (3.2.7)

$$\left(\frac{\partial\Phi}{\partial v_1}\right)_{v_2,m_1} \delta v_1 + \left(\frac{\partial\Phi}{\partial v_2}\right)_{v_1,m_1} \delta v_2 + \left(\frac{\partial\Phi}{\partial m_1}\right)_{v_1,v_2} \delta m_1 = 0, \tag{3.2.8}$$

where lower indexes denote constants for these derivatives; we omit temperature among these indexes because it is common for all the derivatives. Since the variations of these arguments are arbitrary, one may conclude that individual derivatives in (3.2.8) must be equal to zero:

$$\left(\frac{\partial\Phi}{\partial v_1}\right)_{v_2,m_1} = 0, \left(\frac{\partial\Phi}{\partial v_2}\right)_{v_1,m_1} = 0, \left(\frac{\partial\Phi}{\partial m_1}\right)_{v_1,v_2} = 0. \tag{3.2.9}$$

Therefore, for derivatives in (3.2.8) from (3.2.9) we have

$$\left(\frac{\partial\Phi}{\partial v_1}\right)_{v_2,m_1} = m_1 \frac{\partial f_1}{\partial v_1} + m_1 p + \sigma\left(\frac{\partial\Sigma}{\partial v_1}\right)_{m_1} = 0; \tag{3.2.10}$$

$$\left(\frac{\partial\Phi}{\partial v_2}\right)_{v_1,m_1} = m_2 \frac{\partial f_2}{\partial v_2} + m_2 p = 0; \tag{3.2.11}$$

$$\left(\frac{\partial\Phi}{\partial m_1}\right)_{v_1,v_2} = f_1 + p v_1 - f_2 - p v_2 + \sigma\left(\frac{\partial\Sigma}{\partial m_1}\right)_{v_1} = 0. \tag{3.2.12}$$

The derivative from the Helmholtz energy on the volume is pressure with reverse signs (keep in mind that temperature is assumed to be constant too), i.e.,

$$\frac{\partial f_1}{\partial v_1} = -p_1, \frac{\partial f_2}{\partial v_2} = -p_2. \tag{3.2.13}$$

Derivatives from the surface area Σ are more complicated. For the spherical shape of the phase number 1, we have for the surface area Σ and for the volume V:

$$\Sigma = 4\pi R^2, \quad V = v_1 m_1 = \frac{4}{3}\pi R^3. \tag{3.2.14}$$

Thus, we see that $\Sigma(R)$ while $R(v_1 m_1)$. Using these correlations, we get

$$\left(\frac{\partial \Sigma}{\partial v_1}\right)_{m_1} = \frac{d\Sigma}{dR}\left(\frac{\partial R}{\partial v_1}\right)_{m_1} = 8\pi R\frac{m_1}{4\pi R^2} = \frac{2m_1}{R}, \tag{3.2.15}$$

$$\left(\frac{\partial \Sigma}{\partial m_1}\right)_{v_1} = \frac{d\Sigma}{dR}\left(\frac{\partial R}{\partial m_1}\right)_{v_1} = 8\pi R\frac{v_1}{4\pi R^2} = \frac{2v_1}{R}. \tag{3.2.16}$$

Thereby, from (3.2.10–12) we obtain the equations for pressure in each phase. In phase 2, from (3.2.11):

$$p_2 = p, \tag{3.2.17}$$

This equation is obvious: of course, the pressure in the 2nd phase must be equal to the external pressure, since this phase 'supports' the piston. Such equations are a verifying test rather than a scientific result.

For the 1st phase we get the correlation, which can be written with respect to (3.2.10) as

$$p_1 = p_2 + \frac{2\sigma}{R}, \tag{3.2.18}$$

This is the Laplace equation. This correlation shows that the pressure in the 'convex' phase is higher than the outer pressure; note again that this relation is correct both for a drop in a vapor and for a bubble in a liquid.

To finish this consideration, we obtain the third equation from (3.2.12); it is

$$\underbrace{f_1 + \left(p + \frac{2\sigma}{R}\right)v_1}_{\varphi_1(p_1,T)} = \underbrace{f_2 + pv_2}_{\varphi_2(p_2,T)}. \tag{3.2.19}$$

This is an interesting relation from the thermodynamic point of view. It implies the standard condition $\varphi_1 = \varphi_2$ for the phase equilibrium of two media, but takes unequal pressures in the phases into account.

Let us return to the Laplace equation. In accordance with it, the pressure inside a bubble of radius R is higher than the pressure of the surrounding liquid by $2\sigma/R$. For water, $\sigma \approx 0.07$ N/m, that is, in a bubble of $R = 1$ mm the pressure jump is around 140 pascals. This is, surely, not an impressive value, considering the fact that the common pressure level in our atmosphere is $\sim 10^5$ Pa. To 'feel' the pressure difference caused by surface tension, much smaller bubbles are needed, of radius lesser by two or three orders of magnitude.

The thermodynamic derivation of the Laplace equation holds all the disadvantages that thermodynamics brings in itself. The obtained relation concerns only an equilibrium case, when the phase 1 has equilibrium (spherical) form. For common case, we need a relation that can be applied for surfaces of any conceivable form.

3.2.3 The Laplace equation in mechanics

Here we obtain a more common form of the Laplace correlation: the surface may not be spherical, but a condition of a kind of (3.2.18) must exist anyway.

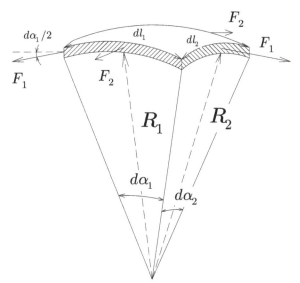

Fig. 3.2.5. The calculation of the surface tension forces acting on an elementary surface area.

Let us consider an elementary surface area between two phases (Fig. 3.2.5). Using the 'mechanical' definition of surface tension from the equation (3.2.5), we calculate the force that acts on that elementary area.

As we see, we have four forces: two equal forces at any direction which we index by numbers 1 and 2; these forces are

$$F_1 = \sigma dl_2, \; F_2 = \sigma dl_1. \tag{3.2.20}$$

The corresponding projections of these forces on the vertical axis are determined by the sinuses of angles $d\alpha_1$ and $d\alpha_2$:

$$F_1^n = F_1 \sin\frac{d\alpha_1}{2} \approx F_1 \frac{d\alpha_1}{2}, \; F_2^n = F_2 \sin\frac{d\alpha_2}{2} \approx F_2 \frac{d\alpha_2}{2}, \tag{3.2.21}$$

since the angles are small.

Thereby, the total force that acts on the elementary area in vertical direction is the doubled forces (3.2.21) (doubled because, as we see in Fig. 3.2.5, we have one force on two opposite sides), which give with (3.2.20)

$$F^n = 2F_1^n + 2F_2^n = \sigma dl_2 d\alpha_1 + \sigma dl_1 d\alpha_2. \tag{3.2.22}$$

Then, to calculate the pressure that acts on this surface, we must divide the normal force (3.2.22) by the area of that elementary surface

$$\Sigma = dl_1 dl_2. \tag{3.2.23}$$

As we see from Fig. 3.2.5, the lengths are connected with angles as

$$dl_1 = R_1 d\alpha_1, \; dl_2 = R_2 d\alpha_2. \tag{3.2.24}$$

Thus, we obtain the pressure difference at the surface with respect to surface tension

$$\Delta p = \frac{F^n}{\Sigma} = \sigma\left(\frac{1}{R_1} + \frac{1}{R_2}\right). \tag{3.2.25}$$

So, we get the Laplace equation again, this time for the common shape of the surface. The pressure difference Δp acts in the direction of the convex phase; thus, to compensate it the pressure in this phase (under the surface in Fig. 3.2.5) must be higher than the one in the concave phase (above the surface in Fig. 3.2.5).

Radiuses R_1 and R_2 are the main curvature radiuses of the surface at the given point. For a sphere, $R_1 = R_2 = R$, and we get the Laplace equation in the form of (3.2.18). For cylindrical geometry, one radius is $R_1 = R$ (the cylinder radius) while the other one is $R_2 = \infty$, so we have the equation in the form of $\Delta p = \sigma/R$.

Note that the reverse radiuses are called curvatures: $k_{1,2} = R_{1,2}^{-1}$; the mean curvature of the surface at the given point is $k = k_1 + k_2$. Thus (3.2.25), can be written in a very short and nice form:

$$\Delta p = \sigma k. \tag{3.2.26}$$

In common case, when the surface at the given cross-section can be represented as the equation $z = z(x)$, the curvature at this cross-section (i.e., for this curve) can be represented as

$$k(x) = \frac{z''}{\left(1 + z'^2\right)^{3/2}}, \tag{3.2.27}$$

where dashes mean the derivative on x.

It should be understood that, in our sense, the curvature (so, the radius too) must be positive; thus, a modulus or something else is required in (3.2.27). For an almost flat surface, when the curvature is small, one may neglect the derivative in the denominator in (3.2.27), so we get a curvature that coincides with the second derivative.

Consequently, for the almost flat surface that is defined by the equation $z = z(x, y)$, one has a corresponding second derivative for every term of the curvature, and in the sum

$$k = k_1 + k_2 = \frac{\partial^2 z}{\partial x^2} + \frac{\partial^2 z}{\partial y^2}. \tag{3.2.28}$$

Again, one should be cautious with the sign in this equation.

3.2.4 Some notes on surface tension

The first question that sometimes arises: does surface tension exist for a flat surface? The answer is positive: surface tension exists in this case (see Fig. 3.2.3 and equation 3.2.5); however, it doesn't create a pressure difference at both sides of the surface—for a flat surface $k = 0$, so, $\Delta p = 0$.

Then, surface tension reasonably depends on temperature; for water, $\sigma = 0.07288$ N/m at 20°C, and $\sigma = 0.06948$ N/m at 40°C. This character of dependence can be expected: we feel that cold water is 'more tense' than the hot one.

The temperature dependence of surface tension presents us with an interesting problem from the thermodynamics point of view. We all know, but sometimes do not remember, that the function δQ in equation (3.2.3) can be represented in a form satisfying the second law of thermodynamics

$$\delta Q = TdS \tag{3.2.29}$$

if and only if the Frobenius condition is satisfied: the differential form

$$\delta Z = X_1 dx_1 + X_2 dx_2 + X_3 dx_3 \tag{3.2.30}$$

has an integrating multiplier if

$$X_1 \left(\frac{\partial X_3}{\partial x_2} - \frac{\partial X_2}{\partial x_3} \right) + X_2 \left(\frac{\partial X_1}{\partial x_3} - \frac{\partial X_3}{\partial x_1} \right) + X_3 \left(\frac{\partial X_2}{\partial x_1} - \frac{\partial X_1}{\partial x_2} \right) = 0. \tag{3.2.31}$$

In other words, for a system described by (3.2.3), entropy exists if

$$p \left(\frac{\partial \sigma}{\partial U} \right)_{V,\Sigma} = \left(\frac{\partial \sigma}{\partial V} \right)_{U,\Sigma} + \underbrace{\left(\frac{\partial p}{\partial \Sigma} \right)_{U,V}}_{0} + \sigma \left(\frac{\partial p}{\partial U} \right)_{V,\Sigma}. \tag{3.2.32}$$

We put the second term in the right side of (3.2.32) as zero, because we cannot imagine any physical reasons for dependence $p(\Sigma)$. We may also assume that the first term in the right-hand side of (3.2.32) is equal to zero (for that we have some reasons, but no strict proof, because σ depends on the curvature, see below; for a spherical bubble, this derivative vanishes, but in general case this is a questionable item). Anyway, the other two terms—the derivatives on the inner energy—leave us with a problem, especially in view of evident dependences $U(T)$ and $\sigma(T)$. Since not all the derivatives in (3.2.32) vanish, the equation like $\sigma/p = f(U)$ follows from this relation, the correctness of which should be examined. Indeed, the existence of entropy for a system with surface tension should be explored carefully.

Also, surface tension depends on the size; to be exact, it depends on the curvature of the surface. If the curvature is small (i.e., the surface is almost flat), one may represent surface tension as the series

$$\sigma = \sigma\big|_{k=0} + \frac{d\sigma}{dk}\bigg|_{k=0} \cdot k. \tag{3.2.33}$$

Denoting $\sigma\big|_{k=0} = \sigma_0$—surface tension of a flat surface and representing $\dfrac{d\sigma}{dk}\bigg|_{k=0} = -\sigma_0\delta$, where the Tolman length δ is supposed to be a constant, we get the scale dependence of surface tension

$$\sigma = \sigma_0(1 - \delta k). \tag{3.2.34}$$

A more common form of (3.2.34) called the Gibbs–Tolman relation (Tolman 1949) usually represents the radius of a drop (or a bubble): $k = 2/r$. The spatial scale $\delta \sim 1$ nm; this fact should be taken into account when we want to consider more complicated forms of (3.2.34). Indeed, as we see, the spatial scales on which surface tension depends on the size of a system is about δ, but the interatom distance in a condensed medium is of the same order. All the approaches considered here are based on the continuous medium representation (i.e., when the elementary volume contains many atoms); thus, the case $r \sim \delta$ is questionable, while $r \ll \delta$ is almost impossible, so the relation (3.2.34) seems to be a good equation to stop at. Note also that the Tolman length is not a constant: it depends on temperature (Rekhviashvili 2011).

One may think that we show prejudice against surface tension, remarking only on its problems. To compensate, we may note that surface tension is one of the few physical quantities that can be measured using only with a cam in the simplest experiment, and the accuracy of this experiment is around several percent—not bad for a thermophysics practice.

Surface tension holds a liquid in a capillary; it also holds a droplet that hangs on the capillary (see Fig. 3.2.6) if only the droplet is not too heavy. A very large (and, therefore, a very massive) droplet would detach from the capillary; thus, gradually increasing the droplet mass (pushing it away from the syringe's needle), we may fix the limit droplet weight that can be balanced by the capillary forces.

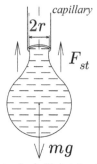

Fig. 3.2.6. A droplet hangs on a capillary. Surface tension holds it, balancing out the gravity forces, until the droplet becomes too massive.

To be exact, when a droplet hangs on a capillary (see Fig. 3.2.6), it is held by the surface tension forces, which can be defined in accordance with (3.2.5)

$$F_{st} = 2\pi r\sigma, \tag{3.2.35}$$

where r is the radius of the 'neck'—the narrowest part of the liquid column that holds the droplet. In the simplest way, this can be assumed as the diameter of the capillary. This force balances out the gravity forces that act on the droplet:

$$F_g = mg, \tag{3.2.36}$$

as long as it is possible: when a droplet becomes too heavy, it tears away from the liquid inside the capillary. Thereby, we may determine the surface tension in the experiment as

$$\sigma = \frac{mg}{2\pi r}. \tag{3.2.37}$$

The droplet mass can be determined either by measuring its weight (preferably) or, if you only have a cam as a measuring instrument, by measuring the size of the falling droplet (and with known density). In Fig. 3.2.7, one may see the process of the droplet detachment.

The last matter that must be mentioned here: what is the capillary constant? In various problems, we consider gravity forces ρg and capillary forces $2\sigma/r$. The spatial scale at which these two forces are comparable is called the capillary radius

$$a = \sqrt{\frac{2\sigma}{\rho g}}. \tag{3.2.38}$$

For water at room temperatures, $a \approx 3.5$ mm.

Fig. 3.2.7. The detachment of the water droplet. From this series of photos, knowing the scale, i.e., the capillary radius, one may define the surface tension of water. By the way, here the inner capillary radius is 1 millimeter, so we have $\sigma = 0.07$ N/m; this value is close to the reference data.

3.3 Rupture of liquid

3.3.1 Hard rupture

The first thing that comes to mind about the term 'rupture of liquid' is that a liquid stretches so much that, when the applied stress is sufficiently strong, it breaks in two parts, like a dissected solid.

This phenomenon can be imagined, as any other, but actually it leads to unbelievable values of physical parameters. Particles in liquids are attracted to each other by the interatom forces; the most frequently used expression is the Lennard–Jones potential which describes the potential energy $\varphi(r)$ of the interaction of two atoms

$$\varphi(r) = 4\varepsilon\left[\left(\frac{\sigma}{r}\right)^{12} - \left(\frac{\sigma}{r}\right)^{6}\right], \tag{3.3.1}$$

where r is the distance between two atoms, σ is the spatial scale of the interaction (at $r = \sigma$ the potential energy $\varphi = 0$) and ε has a sense of a typical energy of interaction (the minimal value of

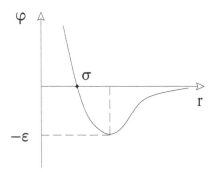

Fig. 3.3.1. The Lennard–Jones potential.

$\varphi = -\varepsilon$, which also corresponds to the equilibrium point since here $d\varphi/dr = 0$ is reached at the distance $r_0 = \sigma 2^{1/6}$); see also Fig. 3.3.1.

Derived for the interaction of simple atoms like ones of rare gases, the potential (3.3.1) is used wherever it is possible and not, because it is so simple and clear: initially, the second term was obtained for the attraction of two dipoles at a large distance (conceivably, each atom turns its neighbor into a dipole, so two atoms couple as two dipoles), and the first one represents, a strong, a very strong repulsive force at small distances—there is no any analytical foundation for this term, seriously. All can we say is that we used the expression (3.3.1) ourselves to describe the interaction of Cu and Ar atoms.

For many non-metallic elements, especially for atoms of rare gases, for which the expression (3.3.1) was constructed, the parameter σ is around several angstroms, and $\varepsilon \sim 0.01 - 0.1$ eV.[7] Thus, the force that acts between two atoms is around $f \sim \varepsilon/\sigma$, and, if the interparticle distance in the medium is $\sim \sigma$ too (in liquid it is slightly more, of course, but we realize the estimations on an order of magnitude). Thus, the limiting stress—the force per the unit of area—is a quantity of an order $\tau \sim \varepsilon/\sigma^3$, that is, to show the parameters we mean in calculations

$$\tau \sim \frac{10^{-21} \; J}{\left(3 \cdot 10^{-10}\right)^3 \; m^3} \sim 10^7 \, \text{Pa}. \tag{3.3.2}$$

That is, we see that the theoretical values of the limiting stress are around hundreds of atmospheres; this is rather the lower limit value of τ, since we choose the minimal estimation of ε; for $\varepsilon \sim 0.1$ eV, we would get the value greater by an order than (3.3.2). For water, because, as we have already pointed out, this potential is used for any substance, we have $\varepsilon \sim 0.03$ eV, $\sigma \sim 2.7$ Å (Lin 2004) and $\tau \sim 2.4 \cdot 10^8$ Pa. As we saw in Section 3.1, the value calculated in (3.3.2) correlates with the 'record' experiment results obtained in the most delicate experiments.

By the way, we can make a special note on the method to obtain such macro-parameters from the micro-ones. For instance, according to such an approach, the surface tension can be estimated as ε/σ^2. Let's make 'more precise estimations', so to say: for argon $\varepsilon = 0.011$ eV, $\sigma = 3.3$ Å, that is, the surface tension can be estimated as 0.016 N/m; the experimental value is ~ 0.01 N/m at 90–100 K. Not the worst consistency if we take into account the minimal labor intensity of the work done. Such methods work because usually the coefficients that arise in such formulae are about unity; on the other hand, this is not a common law and the case of a multiplier $(4\,\pi)^3$ is not excluded, of course, as Einstein wittily noticed almost a century ago, while he proclaimed this approach to general physical problems.

Thus, we see that, indeed, to rupture a liquid, theoretically, great forces must be applied. However, we know that usually the forces that cause the rupture are much smaller, at least by an

---

[7]   1 eV (electron-volt) = $1.6 \cdot 10^{-19}$ J.

order, or even two or three: such results can be easily obtained in not-so-delicate experiments. Thus, one may conclude that usually some factors play a role to deviate a practical case from the ideal one.

3.3.2 Soft rupture: The role of nucleation

Let us take a closer look at our understanding of the liquid rupture. At first glance, we may consider the rupture of a liquid as the rupture of a solid: indeed, there are no nuances pointed out in Section 3.1, which differentiates the problem statement specifically for a liquid, not for a solid. From this approach, there is no difference between tearing a liquid or a solid piece—to break up some material, one needs to apply some force that exceeds the strength of that material, only the value of that strength varies for a liquid and a solid.

Meanwhile, we may reasonably assume that the rupture of a liquid differs from the rupture of solids not only because a liquid has different qualitative parameters (like limiting tensile strength or something); it is almost evident that due to a much higher mobility of atoms in a liquid, some special processes are possible in liquids. Experiments confirm this statement: during melting, the strength of liquid tin is lower than the strength of the solid one by an order of magnitude (Kanel 2015).

Let us avoid that difficulty and formulate the physical picture of the rupture of a liquid that will be close to experiments.

First, we should note that the huge negative pressure we discussed above—of around hundreds or even thousands of atmospheres—almost cannot be reached in practice. At normal conditions, it is impossible to stretch a liquid at all—bubbles occur in a liquid not even at negative, but at low positive pressures. In accurate experiments, when bubbles are removed from the liquid, it is possible to reach negative pressures of approximately 1–10 atm. In very accurate experiments, when all more or less valuable bubbles are removed from the liquid, the negative pressure of ~ 100 atm can be achieved.

However, the following question is important: even if bubbles were removed at the initial stage of the experiment (like in the Berthelot scheme, see Section 3.1), why cannot they appear again, during the main stage of the experiment, in the process of stretching the liquid?
To answer this question, we must consider the types of bubble nucleation in a liquid.

Generally, a bubble may appear inside the bulk of a liquid (homogeneous nucleation) or at the boundary of a liquid (heterogeneous nucleation); in the last case, it can be walls of the vessel or fine impurities in the liquid itself. Of course, for the thorough experiments discussed here impurities can be ignored; the nucleation on a smooth wall can also be negligible—it depends, finally, on the properties of the surface surrounding the liquid, and can be achieved very well.

However, homogeneous nucleation cannot be as small as we want (to be exact, the heterogeneous one too: all that we can reach is that the nucleation at the wall will be of the same probability as that in the liquid). The nature of nucleation in a liquid has probabilistic nature, so it cannot be avoided completely.

Atoms (or molecules—it does not matter here) move inside the liquid, so in some regions, locally, at very short spatial scales, the liquid becomes denser, and in some regions, thinner than average. Of course, we are interested in the latter objects—the low-density regions in the liquid. These regions represent, actually, a gap in the liquid, and the formation of them limits the tensile strength of the liquid: under stretching, a liquid does not break like a rubber band at limiting stress; when an external force exceeds the limiting strength of the liquid, bubbles arise inside it. Because the strength limit is connected with nucleation, it is sometimes called the 'cavitation limit'.

Let us consider the formation of a bubble in a thermostat (see Fig. 3.3.2). Assuming that not only temperature, but also the inner energy in the system remains constant when a small bubble arises here, one may write directly

$$T \Delta S = -L, \tag{3.3.3}$$

that is, the deviation of entropy is connected to the work that was spent on the formation of that nucleus.

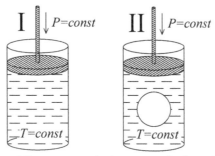

Fig. 3.3.2. Bubble formation in a vessel. Spontaneous nucleation accompanies the work spent on forming the bubble and the corresponding change in entropy.

The minimal work to form a nucleus of radius R consists of two parts (see Section 3.2): volume and surface must be formed (Skripov 1974). If the pressure inside a nucleus (in a gas phase) is p_1 and in the liquid is p_2, then the corresponding work that is required to form the volume is

$$L_V = \frac{4\pi R^3}{3}(p_2 - p_1).$$
(3.3.4)

The work that must be spent to form the spherical surface[8]

$$L_S = 4\pi R^2 \sigma_s.$$
(3.3.5)

In Chapter 7, you will see a more common and stricter approach to nucleation, where bubble dynamics is considered.

Then, we may get the total work from (3.3.4) and (3.3.5) considering the Laplace condition at the bubble surface

$$p_1 = p_2 + \frac{2\sigma_s}{R},$$
(3.3.6)

so, we have the expression for the work

$$L = L_V + L_S = \frac{4\pi R^2 \sigma_s}{3} = \frac{16\pi\sigma_s^3}{3(p_1 - p_2)^2}.$$
(3.3.7)

The last equation is used more widely, because it does not contain the radius of the nucleus.
Then, we may consider the Boltzmann interpretation of entropy:[9]

$$S = k \ln W.$$
(3.3.8)

Here W is the thermodynamic probability—the number of ways in which the given state can be described. W is not a 'regular' probability ω which expresses the relative number of 'successful cases' and is represented by a fraction (while W is a large integer number). However, we understand that these probabilities must be connected, so $\omega \sim W$.

If the thermodynamic probability at the initial state, without a nucleus, with entropy S_0 is W_0, and at the final state with entropy $S = S_0 + \Delta S$ is W, then we see that

$$\Delta S = k \ln \frac{W}{W_0}.$$
(3.3.9)

---

[8] Here we use notation σ_s for surface tension to distinguish from the parameter of the Lennard–Jones model, which was used in this section too.

[9] By the way, the first one who applied this equation in practical sense was Max Planck, and the Boltzmann constant k was defined for the first time exactly by Planck in his work devoted to black body radiation.

Thus, the probability of the state where the nucleus was formed can be written with relation (3.3.3) as

$$w = A \exp\left(\frac{\Delta S}{k}\right) = A \exp\left(-\frac{L}{kT}\right), \tag{3.3.10}$$

where L is determined by (3.3.7), A is the normalizing constant.

Thereby, the probability of appearance of a nucleus of radius R depends on temperature: the higher the temperature, the higher the corresponding probability. This is not a surprising result, of course, and it can be expected from various considerations: that is why the experiments concerning rupture of liquid contain a cooling phase.

Nevertheless, we now stand on the thermodynamics way. Let us see what we can obtain directly from the thermodynamic formalism.

3.3.3 Thermodynamics of rupture

Many theoretical approaches can be based on thermodynamics. Thermodynamics is a common science; thus, it may provide the explanation for many phenomena; however, probably, this explanation may have certain special characters: for someone, it may seem too general and too formalistic. Anyway, this is a good starting point for almost every problem.

Let us take a look at the negative pressure and at the rupture of a liquid abstractly. From a common sense, the given medium obeys some equation of state.[10] Then, we may introduce the van der Waals equation as the expression that 'corrects' the Clapeyron equation for (a) the interaction of molecules, (b) the restriction in volume since molecules occupy some volume themselves:

$$\left(p + \frac{a}{v^2}\right)(v - b) = RT, \tag{3.3.11}$$

$R = 8.314$ J/(mol·K) is the universal gas constant, v is a specific volume (per mole), and a and b are some parameters. Note that this equation can also be written with $R = R/\mu$—the individual gas constant—ratio of the universal gas constant to the molar mass of the substance, with redefining the parameters a, b, and v, which now become the volume perunit of mass.

Parameters a and b can be obtained in various ways (or even can be determined for separate ranges of parameters, but this is an extreme case and an almost forbidden trick); the good method is to connect them with pressure at the critical point p_c and temperature at the critical point T_c

$$a = \frac{27}{64} \frac{R^2 T_c^2}{p_c}, \; b = \frac{RT_c}{8 p_c}. \tag{3.3.12}$$

By the initial presumption, the equation of state (3.3.11) describes both the liquid phase and the gaseous one. Like the Lennard–Jones potential, the van der Waals equation is used widely for qualitative analysis, because it is simple and clear; unfortunately, it does not provide suitable results in calculations: qualitatively, the predictions of the van der Waals equation leave much to be desired.

Nevertheless, one may try to use this equation to analyze negative pressure (Temperley 1947). Isotherms for the van der Waals equation are presented in Fig. 3.3.3. Note again (see Section 3.1) that the stability zone is restricted by the condition $(\partial p/\partial v)T < 0$, i.e., isotherms cannot include 'rising sites'. That is, the stability region is restricted by the condition

$$\frac{2a}{v^3} - \frac{RT}{(v-b)^2} \le 0; \tag{3.3.13}$$

---

[10] Note that the derivation of the equation of state is not the subject of thermodynamics, this is the outsource matter for this science. The equation of state can be obtained with statistical physics, or with physical kinetics, or from the experimental data, etc. A special method for obtaining such an equation is to write some common expression for pressure as a function of density and temperature, and then to find the approximation constants for it.

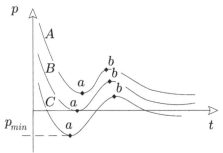

Fig. 3.3.3. Isotherms for the van der Waals equation: the left part corresponds to a liquid; the right part, to a vapor. Points *a* and *b* on each isotherm restrict the instability regions; the set of such points compose a spinodal. Isotherm *A* is an isotherm of a common form; isotherm *B* touches the abscissa axis and corresponds to temperature T_m (see in text); isotherm *C* corresponds to temperature when negative pressure can be observed in the liquid phase. The minimal pressure that can be achieved in liquid phase corresponds to the point *a* on the given isotherm.

from the equality in (3.3.13), we may determine two volumes that correspond to two extremums on a subcritical isotherm (i.e., the isotherm that corresponds to temperature $T < T_{cr}$).

Now, let us notice that, theoretically, we may find the temperature that corresponds to the extremum at $p = 0$ (see Fig. 3.3.3); the abscissa of this point corresponds to

$$v = 2b, \tag{3.3.14}$$

while the corresponding temperature, calculated from (3.3.11) at $p = 0$ and (3.3.14), is

$$T_m = \frac{a}{4bR}. \tag{3.3.15}$$

Thereby, temperatures that may correspond to negative pressure must be $T < T_m$. For instance, for water

$$a = 0.5537 \text{ (N·m}^4\text{)/mol}^2, \ b = 30.5 \text{ cm}^3\text{/mol, (Lide 1998)} \tag{3.3.16}$$

so, we have the theoretical limit for the existence of negative pressures in water:

$$T_m = 546 \text{ K.} \tag{3.3.17}$$

Then, continuing the thermodynamic approach, we may determine the limiting value of negative pressure at the given temperature. For temperatures $T < T_m$, one may note that the minimum value of pressure corresponds to the left extremum of the corresponding isotherm (see Fig. 3.3.4) at $v = v_l$ determined by the equation (3.3.13):

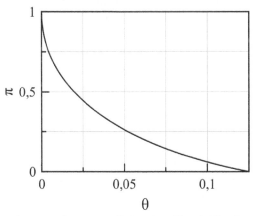

Fig. 3.3.4. The module of the maximum negative pressure that can be achieved at the given temperature: the van der Waals approach. Units are dimensionless, see in text.

$$p = \frac{a(v_l - 2b)}{v_l^3}. \tag{3.3.18}$$

Of course, since for $T < T_m$ the corresponding volume $v_l < 2b$, the pressure calculated with (3.3.18) is sub-zero. On the other hand, from the meaning of the parameter b, $v_l \geq b$, so the theoretical limit for negative pressure is

$$p_{min} = -\frac{a}{b^2}, \tag{3.3.19}$$

which gives for water at (3.3.16) the value

$$p_{min} \approx -6 \cdot 10^8 \text{ Pa}. \tag{3.3.20}$$

Note that the value p_{min} is too theoretical, so to say, since it corresponds to zero temperature, as it follows from (3.3.11). To understand more realistic values, we may consider the volume $v_l = 1.4b$ that corresponds to temperature 255 K, and get for minimum pressure $-1.3 \cdot 10^8$ Pa—the value of the same order, anyway. Again, as we see, these values correlate with the simplest approach presented in the Section 3.3.1: we continue to advertise that primal approach.

The results obtained above can be generalized. Using dimensionless variables

$$\theta = \frac{T}{T_0} = \frac{TRb}{2a}, \pi = -\frac{p}{p_0} = -\frac{pb^2}{a}, v = \frac{v}{b}, \tag{3.3.21}$$

i.e., π is the module of the maximum negative pressure that can be achieved at temperature θ, equations (3.3.13) and (3.3.18) can be represented as

$$\theta = \frac{(v-1)^2}{v^3}, \pi = \frac{2-v}{v^3}. \tag{3.3.22}$$

From some point of view, the dimensionless volume parametrizes the pressure–temperature function. The dependence $\pi(\theta)$ is shown in Fig. 3.3.4.

Since the van der Waals equation suits poorly for practical calculations, one should try to use other forms of the equation of state. That is, using the same methodology, we may try to investigate the sub-zero parts of isotherms, identifying the extremum in that region with the limiting value of negative pressure.

For example, in (Malyshev2015) the Redlich–Kwong equation was used:

$$p = \frac{RT}{v-b} - \frac{a}{\sqrt{T}v(v+b)}; \tag{3.3.23}$$

it is easy to notice that the Redlich–Kwong equation corrects the Van der Waals equation like, in its turn, the latter corrects the Clapeyron equation. Again, in (3.3.20) v is the volume per single mole, and parameters a and b can be connected to the critical pressure p_c and the critical temperature I_c:

$$a = \frac{0.4275 R^2 T_c^{2.5}}{p_c}, b = \frac{0.08664 RT_c}{p_c}. \tag{3.3.24}$$

Again, one can determine the volume corresponding to the extremum $(\partial p / \partial v)_T = 0$, which gives

$$\frac{a(b+2v)(v-b)^2}{(v+b)^2} = RT^{3/2}, \tag{3.3.25}$$

and then find pressure from (3.3.23); we cannot obtain a short equation like (3.3.18), so we do not write the long one; it does not make much sense to rewrite huge similar constructions of various kinds. In (Malyshev 2015), with Redlich–Kwong equation the value -370 atm was obtained for argon.

The main problem with empirical equations of state is that they can only be derived for certain conditions. The van der Waals equation has no fundamental character, and the constants chosen to fit the critical point cannot provide a good enough agreement with the experimental data even at the 'normal domain', that is, for non-extremal values of temperature and at positive pressure. For instance, the van der Waals equation gives the compressibility factor at the critical point

$$z_c = \frac{p_c v_c}{RT_c} = \frac{3}{8} = 0.375 \tag{3.3.26}$$

for any substance, while indeed this value differs for various substances and for most of them belongs to the range $0.2 - 0.3$ (closer to 0.2).

Thus, it is difficult to expect that such an approximate equation may give us the exact values of negative pressure. The calculations of the kind presented above may rather provide the estimations of p_{min}, but these estimations can be obtained even with a much simpler implementation—only through the parameters of interatom interaction—which give, as we saw above, quite reasonable values.

Consequently, in order to obtain the precise value of the extremal pressure that may exist in a liquid, one should continue to consider the matter at the atomic level.

3.3.4 Numerical experiment: Theory

Today, the physical science is divided in three groups:

- ✓ experiment,
- ✓ theory,
- ✓ numerical simulations.

Actually, if we do not take ancient Greeks and medieval philosophers into account, these branches appeared in that sequence: experiments demanded explanation and generalization, so theoretical physics appeared (approximately, in the late XIX century); numerical simulations were born a century later from two parents: the impossibility to obtain analytical solutions for the mathematical equations of physics and the growing power of computing systems.

The relationships between these three families are interesting. The common view on physics, which we heard repeatedly, especially from our fellow theorists, was that this is an experimental science. The experiment prevails, while the theory is some non-obligatory side of physics. Another opinion dominates, which may look strange, in engineering. 'The golden rule: an experiment means nothing until it is confirmed by calculations'. Actually, this situation is a more serious variation of the known joke by Einstein: 'A theory is something nobody believes, except the person who made it; an experiment is something everybody believes, except the person who made it'. When we face an experimental result, there are four ways before us:

- ✓ consider this result as a particular case of a more common class of phenomena, and the existing theory explains this result well; example: bubble oscillations;
- ✓ try to consider this result as a particular case of a more common class of phenomena, but the existing theory cannot explain this result well; example: rupture of a liquid;
- ✓ leave it on its own—this experimental fact replenishes the collection of other experimental facts; this is suitable when this result has intrinsic value, especially of practical character, while the theory is evidently too complicated and, as far as we can judge, may provide only qualitative explanation in the near future, not the qualitative one; example: cavitation erosion;
- ✓ to ask ourselves: do we know physicsalittle?; example: sonoluminescence.

Experimenters, especially today, love the results of the third kind; numerical simulators prefer the second one; theorists admire the first one and particle physics; and all like, at special conditions, the fourth variant. Anyway, all three groups of scientists can happily coexist in harmony. For example, this book is written by the representatives of all three specialties: one of us is an experimenter, another one—a numerical simulator, and the third one is a theorist.

In short, to generalize, we have to theorize, but the tools available to theorists are restricted by two important factors: (a) the complexity of the mathematical problem from one side, and (b) the wish of non-theorists to get the 'exact' results, from another side. For a complicated problem, a theorist may give an approximate expression, but in particular cases more is needed. Thus, the hour of numerical simulations is coming.

Sometimes, numerical simulations are also called the numerical experiment: if we are absolutely assured in the correctness of the given physical equation, then we assume that this equation and the natural object described by this equation are identical. This point of view is questionable for computational fluid dynamics, for example, considering all the problems with hydrodynamics equations. However, the method of molecular dynamics (MMD) may claim the title of 'numerical experiment' more reasonably.

MMD considers the dynamics of a mechanical system: particles interact with each other by some potential (for example (3.3.1)), so we have a system of equations of a kind of

$$m_i \frac{d\dot{x}_i}{dt} = -\sum_j \frac{\partial \varphi_{ij}\left(r_{ij}\right)}{\partial \vec{r}_{ij}}, \qquad 3.3.27)$$

where r_{ij} is the distance between particles i and j. Of course, external forces can also be added to (3.3.27). By formulating the equation (3.3.27) for each of N particles, we may explore the dynamics of this ensemble, obtaining macroscopic characteristics from the microscopic ones. For instance, temperature may be calculated with the mean kinetic energy of the particles:

$$\frac{3}{2} kT = \frac{1}{N} \sum_{i=1}^{N} \frac{mu_i^2}{2}. \qquad (3.3.28)$$

Indeed, here only the chaotic component of velocity $\vec{u}_i' = \vec{u}_i - \dfrac{1}{N} \sum_{i=1}^{N} \vec{u}_i$ must be used, but, practically, since the average velocity is much smaller, velocity u_i is used in (3.3.4) directly: it does not look good, but does not give a noticeable error.

Usually, the number of particles $N \sim 10^3 - 10^7$ can be considered: that depends on the computing power. This is a small amount of particles anyway, if we take into account that any cubic centimeter of a perfect gas at 10^5 Pa and 300 K contains $2.4 \cdot 10^{19}$ particles. Practically, to obtain the information about a homogeneous system, the number of particles $\sim 10^4$ is sufficient. On the other hand, sometimes a complicated system must be explored: a cavity in a liquid is the example. If so, the dynamics of many more particles must be calculated.

Considering the fact that, indeed, all that we see around us consists of interacting atoms, one may assume that MMD provides the complete approach to a real problem. Many particles must be considered, but, at least theoretically, MMD gives the complete description of a physical problem.

Unfortunately, the situation is not as shiny. We do not know the exact expression for the potential of the interaction between two atoms, at least for one simple reason: generally, a third atom distorts the pair-potential function. Of course, we may repeat that some parts of potentials are only a model: they are not obtained using the strict theoretical approaches.

Thus, as for the exact quantitative results, MMD may not provide the detailed information. Especially, this matter concerns the simulation of molecular systems (i.e., systems which consist of molecules, not atoms). Anyway, MMD may give interesting quantitative results, the nature of which is undisputed. Of course, MMD should not be over criticized: in comparison with the empirical equations of state, the results of numerical simulations are quite good.

3.3.5 Numerical simulation results

The history of the MMD simulations of a bubble in a liquid is not very long, due to, among others, very prosaic reasons. One may model an arbitrarily small drop: it may even consist of 100 atoms, for example, and the researcher would call it a 'drop' anyway. However, to model a hole in the condensed phase (note: a hole which contains some atoms inside it in the gaseous phase), we have to consider many more particles in this simulation. For a proper consideration, if the radius of a bubble is 10 nm, then the layer of the surrounding liquid must be of the same order, so the considered volume is $(4R)^3$, i.e., more than 10^4 nm^3. A single particle occupies a volume roughly equal to σ^3, where $\sigma \sim 0.3$ nm is the parameter of the Lennard–Jones potential. Thus, to simulate such an object, we should consider $\sim 10^6$ atoms. To solve this problem, very powerful computers are required. Of course, a bubble may be of a smaller size, but not significantly. Anyway, we cannot do with a thousand or two of particles.

The first work where the problem of formation of a bubble at negative pressure was considered is (Kinjio1998). Almost 11000 particles were considered—a good result for the XX century computers. Later, the number of particles increased significantly: in (Bazhirov 2008) and (Baidakov 2014), the number of particles reached half a million.

Generally, the results of a molecular simulation confirm the experimental results (or, its results are confirmed by experiments). For instance, let us take a look at the results of (Malyshev 2015), where the rupture of a pure liquid was investigated with MMD

The time dependence (smoothed, of course, since simulations give the experiment-like non-monotonous graphs) of pressure is shown in Fig. 3.3.5: when a liquid is being stretched, the pressure inside it initially decreases, reaches the minimal value that can be interpreted as the rapture strength, and then suddenly increases.

In (Malyshev 2016), the influence of gas inclusions on the rupture strength was considered. The liquid argon + neon mixture was considered. In accordance with the previous consideration, a bubble appears in a stretched liquid; note that the bubble is neither 'argon' nor 'neon': it contains atoms of both components.

The minimal pressure depends on the concentration of neon atoms:

$$C_{Ne} = \frac{N_{Ne}}{N_{Ar} + N_{Ne}}. \tag{3.3.29}$$

In (Malyshev 2016), it was obtained that the dependence of the rupture strength $|p_{min}|$ has linear dependence on C_{Ne}; approximately, it can be expressed by the relation

$$|p_{min}| = 367 - 785 \cdot C_{Ne} \text{ [atm]}, \tag{3.3.30}$$

that is, in absence of neon the minimal negative pressure is –367 atm, while for $C_{Ne} = 0.2$ this value decreases down to –210 atm.

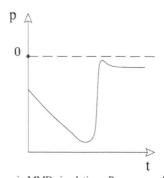

Fig. 3.3.5. The time dependence of pressure in MMD simulations. Pressure reaches the minimal value, and then increases abruptly.

The dependence (3.3.22) can be explained by the properties of neon in comparison with argon: for *Ne*, the parameter ε in the Lennard–Jones potential is more than three times lower than for *Ar*. In other words, atoms of neon interact more weakly than atoms of argon while, as we see, the strength of the liquid depends on ε. Thus, when the concentration of such weakly interacting atoms becomes significant (of an order of a percent or more), they influence the total properties of the medium. We should emphasize the absence of the 'subtle' effects of impurities: we see that the influence manifests clearly when the number of 'impurity' atoms becomes comparable with the number of 'basic' particles.

Thus, the rupture of a liquid indeed accompanies the formation of a new phase in it. Numerical MMD simulations can clarify the picture of that process, but, on the other hand, we should not expect too much from it. To date, the available spatial and time scales for MMD are too short to make definite conclusions for macroscopic scales; for instance, the results for the rate of formation of nuclei are far from the classical macroscopic theory: the difference is around several orders of magnitude (Baidakov2014). For such spatial scales, many macroscopic parameters almost lose physical sense, surface tension being an example. Even the term 'bubble surface' needs a special definition, when you look at the heap of moving atoms.

Undoubtedly, molecular dynamics simulation has a great future, but not for a while.

Conclusion

This chapter concerns the pressure in a liquid and its connection to the pressure in the gas phase—the Laplace equation.

First, the pressure in a liquid can be defined in various ways, and the 'normal force per unit of area' is not the best possible formulation, since pressure, as we usually understand it, is also a thermodynamic function. Thus, 'pressure' is an equilibrium parameter from this point of view, and some controversy may be found in hydrodynamics, because it also operates the non-equilibrium (dissipative, viscous) forces.

Pressure can be negative, which corresponds to the tension of a medium. Negative pressure cannot exist if particles of a medium do not interact with each other, so there is no point in expecting negative pressure in a perfect gas. However, in liquids, the negative pressure can be obtained: like solids, liquids can be stretched; moreover, liquids also can be ruptured, but this process in liquids differs significantly from the rupture of solids.

In liquids, a new—gaseous—phase tends to be formed at low pressures. The process of nucleation, by the way, prevents the observation of a stretched liquid at domestic conditions (or in poorly-organized experiments): cavitation, that is the formation and development of the gaseous phase in a cold liquid, stops the further extension. Under a huge negative stress, a liquid tears not like a solid, breaking in two parts, but stops the stretching by producing gaseous bubbles inside it. Thus, cavitation plays a role of a 'fuse' here: cavitation begins, and the stretching process ends.

The force that prevents the formation of gas in a liquid, that is, tries to prevent cavitation, is surface tension. To make a bubble in a liquid, one must spend some work to form its surface, and this work is not negligible. The energy barrier of surface tension stops cavitation for the time being, but not in perpetuity: at certain stretch, sooner or later, the liquid will be broken. To understand this process in detail, numerical simulation is a suitable tool: today, we are able to model a physical system by modeling the motion of particles that make up this system. In common, such numerical simulations give reasonable qualitative illustrations and some quantitative estimations; however, the great difference in spatial and temporal scales takes its toll, and some macroscopic parameters cannot be determined directly from numerical simulations.

On the other hand, the surface tension σ has an intrinsic matter. It leads to the connection between pressures in a gas and in a liquid—the Laplace equation. In accordance with it, the pressure in the convex phase is greater by value $k\sigma$, where k is the curvature of the surface; for a sphere of radius *r*, for example, the curvature is $k = 2/r$. This equation will be used in every following chapter of this book.

References

Ando, K., A. Q. Liu and C.-D. Ohl. 2012. Homogeneous nucleation in water in microfluidic channels. *Physical Review Letters*, 109: 044501-1–5. https://doi.org/10.1103/PhysRevLett.109.044501.

Azouzi, M. E. M., C. Ramboz, J.-F. Lenain and F. Caupin. 2013. A coherent picture of water at extreme negative pressure. *Nature Physics*, 9: 38–41. https://doi.org/10.1038/nphys2475.

Babb, S. E. 1963. Parameters in the simon equation relating pressure and melting temperature. *Reviews of Modern Physics*, 35: 400–13. https://doi.org/10.1103/RevModPhys.35.400.

Baidakov, V. G. 2016. Spontaneous cavitation in a Lennard-Jones liquid: Molecular dynamics simulation and the van der Waals-Cahn-Hilliard gradient theory. *The Journal of Chemical Physics*, 144: 074502-1–11. https://doi.org/10.1063/1.4941689.

Baidakov, V. G. and K. S. Bobrov. 2014. Spontaneous cavitation in a Lennard-Jones liquid at negative pressures. *The Journal of Chemical Physics*, 140: 184506-1–11. https://doi.org/10.1063/1.4874644.

Bazhirov, T. T., G. E. Norman and V. V. Stegailov. 2008. Cavitation in liquid metals under negative pressures. Molecular dynamics modeling and simulation. *Journal of Physics: Condensed Matter*, 20: 114113-1–11. https://doi.org/10.1088/0953-8984/20/11/114113.

Bertrand, C. E. and M. A. Anisimov. 2011. Peculiar thermodynamics of the second critical point in supercooled water. *The Journal of Physical Chemistry B*, 115: 14099–111. https://doi.org/10.1021/jp204011z.

Caupin, F., A. Arvengas, K. Davitt, M. E. M. Azouzi, K. I. Shmulovich, C. Ramboz, D. A. Sessoms and A. D. Stroock. 2012. Exploring water and other liquids at negative pressure. *Journal of Physics: Condensed Matter*, 24: 284110-1–7. https://doi.org/10.1088/0953-8984/24/28/284110.

Chitnelawong, P., Sciortino, F. and P. H. Poole. 2019. The stability-limit conjecture revisited. *The Journal of Chemical Physics*, 150: 234502-1–10. https://doi.org/10.1063/1.5100129.

Davitt, K., A. Arvengas and F. Caupin. 2010. Water at the cavitation limit: Density of the metastable liquid and size of the critical bubble. *Europhysics Letters*, 90: 16002-1–6. https://doi.org/10.1209/0295-5075/90/16002.

Green, J. L., D. J. Durben, G. H. Wolf and C. A. Angell. 1990. Water and solutions at negative pressure: raman spectroscopic study to –80 Megapascals. *Science*, 249: 649–52. https://doi.org/10.1126/science.249.4969.649.

Harvey, A. H. 2019. *Properties of Ice and Supercooled Water*. CRC Press.

Hayward, A. T. J. 1971. Negative pressure in liquids: can it be harnessed serve man? *American Scientist*, 59: 434–43.

Henderson, S. J. and R. J. Speedy. 1980. A Berthelot-Bourdon tube method for studying water under tension. *Journal of Physics E: Scientific Instruments*, 13: 778–82. https://doi.org/10.1088/0022-3735/13/7/019.

Imre, A., K. Martinás and L. P. N. Rebelo. 1998. Thermodynamics of negative pressure in liquids. *Journal of Non-Equilibrium Thermodynamics*, 23: 351–75. https://doi.org/10.1515/jnet.1998.23.4.351.

Imre, A. R., H. J. Maris and P. R. Williams. 2002. *Liquids Under Negative Pressure*. Springer. https://doi.org/10.1007/978-94-010-0498-5.

Imre, A. R. 2007. *How to Generate and Measure Negative Pressure in Liquids?* In: Rzoska S. J. and V. A. Mazur. (eds.). Soft Matter under Exogenic Impacts. NATO Science Series II: Mathematics, Physics and Chemistry, vol. 242. Springer. https://doi.org/10.1007/978-1-4020-5872-1_24.

Kanel G. I., A. S. Savinykh, G. V. Garkushin and S. V. Razorenov. 2015. Dynamic strength of tin and lead melts. *Journal of Experimental and Theoretical Physics Letters*, 102: 548–551. https://doi.org/10.1134/S0021364015200059.

Kashchiev, D. 2020. Nucleation work, surface tension, and Gibbs–Tolman length for nucleus of any size. *The Journal of Chemical Physics*, 153: 124509-1–11. https://doi.org/10.1063/5.0021337.

Kinjo, T. and M. Matsumoto. 1998. Cavitation processes and negative pressure. *Fluid Phase Equilibria*, 144: 343–350.

Kuksin, A. Yu., G. E. Norman, V. V. Pisarev, V. V. Stegailov and A. V. Yanilkin. 2010. A kinetic model of fracture of simple liquids. *High Temperature*, 48: 511–7. https://doi.org/10.1134/S0018151X10040085.

Landau, L. D. and E. M. Lifshitz. 1987. *Fluid Mechanics*. Oxford: Pergamon Press.

Lide, D. R. 1998. *Handbook of Chemistry and Physics*. CRC Press.

Lin, D. T. W. and C.-K. Chen. 2004. A molecular dynamics simulation of TIP4P and Lennard-Jones water in nanochannel. *ActaMechanica*, 173: 181–94. https://doi.org/10.1007/s00707-004-0134-x.

Mahmud, K. A., F. Hasan, M. I. Khan and A. Adnan. On the molecular level cavitation in soft gelatin Hydrogel. *Scientific Reports*, 10: 9635-1–13. https://doi.org/10.1038/s41598-020-66591-9.

Malyshev, V. L., D. F. Marin, E. F. Moiseeva, N. A. Gumerov and I. Sh. Akhatov. 2015. Study of the tensile strength of a liquid by molecular dynamics methods. *High Temperature*, 53: 406–412. https://doi.org/10.1134/S0018151X15020145.

Malyshev, V. L., D. F. Mar'in, E. F. Moiseeva and N. A. Gumerov. 2016. Influence of gas on the rupture strength of liquid: Simulation by the molecular dynamics methods. *High Temperature*, 54: 640–4. https://doi.org/10.1134/S0018151X16030123.

Netz, P. A. and F. W. Starr. Static and dynamic properties of stretched water. *The Journal of Chemical Physics*, 115: 344–8. https://doi.org/10.1063/1.1376424.

Rehner, P., Aasen, A. and Ø. Wilhelmsen. 2019. Tolman lengths and rigidity constants from free-energy functionals—General expressions and comparison of theories. *The Journal of Chemical Physics*, 151: 244710-1–13. https://doi.org/10.1063/1.5135288.

Rekhviashvili, S. S. and E. V. Kishtikova. 2011. On the size dependence of the surface tension. *Technical Physics*, 56: 143–6. https://doi.org/10.1134/S106378421101021X.

Roedder, E. 1967. Metastable superheated ice in liquid-water inclusions under high negative pressure. *Science*, 155: 1413–7. https://doi.org/10.1126/science.155.3768.1413.

Scripov, V. P. 1981. Low-temperature asymptotic-behavior and thermodynamic similarity of melting lines. *High Temperature*, 19: 66–72.

Sekine, M., K. Yasuoka, T. Kinjo and M. Matsumoto. 2008. Liquid–vapor nucleation simulation of Lennard-Jones fluid by molecular dynamics method. *Fluid Dynamics Research*, 40: 597–605. https://doi.org/10.1016/j.fluiddyn.2007.12.012.

Simon, F. and G. Glatzel. 1929. Bemerkungen zur Schmelzdruckkurve. *Zeitschrift für anorganische Chemie*, 178: 309–16. (in German) https://doi.org/10.1002/zaac.19291780123.

Skripov, V. P. 1974. *Metastable Liquids*. Wiley: New York.

Speedy, R. J. 1982. Stability-limit conjecture. An interpretation of the properties of water. *The Journal of Physical Chemistry*, 86: 982–91. https://doi.org/10.1021/j100395a030.

Steudle, E. 1995. Trees under tension. *Nature*, 378: 663–664. https://doi.org/10.1038/378663a0.

Temperley, H. N. V. 1947. The behavior of water under hydrostatic tension: III. *Proceedings of the Physical Society*, 59: 199–208. https://doi.org/10.1088/0959-5309/59/2/304.

Tolman, R. C. 1949. The effect of droplet size on surface tension. *The Journal of Chemical Physics*, 17: 333–7. https://doi.org/10.1063/1.1747247.

Tsuda, S., S. Takagi and Y. Matsumoto. 2008. A study on the growth of cavitation bubble nuclei using large-scale molecular dynamics simulations. *Fluid Dynamics Research*, 40: 606–15. http://dx.doi.org/10.1016/j.fluiddyn.2008.02.002.

Viet M. H., P. Derreumaux and P. H. Nguyen'. 2016. Nonequilibrium all-atom molecular dynamics simulation of the bubble cavitation and application to dissociate amyloid fibrils. *The Journal of Chemical Physics*, 145: 174113-1–7. https://dx.doi.org/10.1063/1.4966263.

Zheng, Q., D. J. Durben, G. H. Wolf and C. A. Angell. 1991. Liquids at large negative pressures: water at the homogeneous nucleation limit. *Science*, 254: 829–32. https://doi.org/10.1126/science.254.5033.829.

CHAPTER 4

Hydrodynamics of Cavitation

4.1 The hydrodynamic description of irrotational flow

4.1.1 Potentials

Hydrodynamics, from the mathematical point of view, is a very complicated science. Hydrodynamic formulation represents a system of non-linear partial differential equations. Their integration is a very difficult problem.

However, some methods may facilitate the integration of hydrodynamic equations for special but practically important cases.

At first, let us take a closer look at the continuous equation for an incompressible fluid:

$$div\vec{u} = 0. \tag{4.1.1}$$

This condition means that the velocity vector $\vec{u}$ is the curl of some vector potential $\vec{\Psi}$:

$$\vec{u} = rot\,\vec{\Psi}. \tag{4.1.2}$$

Indeed, $div(rot\,\vec{\Psi}) \equiv 0$ for any vector $\vec{\Psi}$. The construction (4.1.2) may even be useful for a 3D flow, but it is especially helpful for the flat flow. In this case, the vector $\vec{u}$ has only two projections u_x and u_y, while the vector $\vec{\Psi}$ has only one component $\Psi_z \equiv \psi(x, y)$, called in this case a stream function. The components of the velocity vector are connected with ψ via (4.1.2) as

$$u_x = \frac{\partial \psi}{\partial y},\, u_y = -\frac{\partial \psi}{\partial x}. \tag{4.1.3}$$

Lines $\psi = const$ correspond to stream lines; the velocity vector is tangent to these curves. It is clear from representation

$$d\psi = \frac{\partial \psi}{\partial x}dx + \frac{\partial \psi}{\partial y}dy = -u_y dx + u_x dy, \tag{4.1.4}$$

which gives

$$\frac{dx}{u_x} = \frac{dy}{u_y}. \tag{4.1.5}$$

Another 'useful tool' is the scalar potential. If we know that the flow is irrotational, that is, vorticity

$$rot\,\vec{u} = 0, \tag{4.1.6}$$

thus, remembering that $rot(grad\varphi) \equiv 0$, we may deduce from (4.1.6)

$$\vec{u} = grad\varphi, \tag{4.1.7}$$

which means for the flat flow that

$$u_x = \frac{\partial \varphi}{\partial x}, \, u_y = \frac{\partial \varphi}{\partial y}. \qquad (4.1.8)$$

For incompressible liquid from (4.1.1), we have the Laplace equation for φ

$$\Delta \varphi = 0. \qquad (4.1.9)$$

The boundary conditions for velocity of a viscous fluid are:

(a)   impermeability, which means zero normal projection of velocity v_τ;

(b)   sticking, which means zero tangential components of velocity of the fluid relative to a solid.

   The last condition, by the way, means that irrotational flow may only exist in a viscous fluid in a domain with immobile boundaries, otherwise the integral of vorticity must be

$$\int\limits_{(F)} rot \vec{u} d\vec{F} = \oint\limits_{(L)} \vec{u} d\vec{l} \neq 0, \qquad (4.1.10)$$

where L is the contour restricted F, and equation (4.1.4) does not take place. If we consider a perfect, i.e., non-viscous, fluid we have no condition (b). Below, through the entirety of this chapter, we will only consider perfect liquids.

   Solving the hydrodynamic problem in variables φ, ψ can be much more convenient than the solution directly for function u, which will be demonstrated in following sections. The correlation (4.1.9) represents a simple linear differential equation (and, by the way, a well-known one) in contrast to the non-linear Navier–Stokes equation.

From (4.1.3) and (4.1.8) we see that

$$\frac{\partial \varphi}{\partial x} = \frac{\partial \psi}{\partial y}, \, \frac{\partial \varphi}{\partial y} = -\frac{\partial \psi}{\partial x}. \qquad (4.1.11)$$

   Taking a derivative from the fist equation (4.1.11) on y, taking a derivative from the second one on x, and subtracting one result from another, we see that the stream function also satisfies the Laplace equation (4.1.9). Thus, φ and ψ are 'symmetric' functions. Note that this is true only for flat geometry; for cylindrical geometry, the relation is more complicated.

   Equations (4.1.11) represent the Cauchy–Riemann conditions, which means that for the complex potential

$$\omega = \varphi + i\psi \qquad (4.1.12)$$

of the complex argument

$$z = x + iy. \qquad (4.1.13)$$

the following correlation is true

$$\frac{d\omega}{dz} = \frac{\partial \omega}{\partial x} = \frac{\partial \omega}{\partial (iy)} \qquad (4.1.14)$$

This equation follows from (4.1.11) directly.

Following this way, one may also introduce a complex velocity[1]

---

[1]   There are different descriptions for the complex velocity in literature. In a straightforward approach, complex velocity can be defined as $v = u_x + iu_y$; in this case, we see that the derivative $d\omega/dz$ does not give the velocity v, but its conjugate function v^*. Another approach defines velocity in the same line, but with another designations: the conjugate velocity is denoted as v, while the complex velocity function is $v = u_x + iu_y$. We follow the simplest way, as it seems to us: complex velocity is the derivation of a complex potential on a complex coordinate.

$$v \equiv Ae^{-i\theta} = \frac{d\omega}{dz} = \frac{\partial\varphi}{\partial x} + i\frac{\partial\psi}{\partial x} = u_x - iu_y, \qquad (4.1.15)$$

where absolute value $A = \sqrt{u_x^2 + u_y^2}$ coincides with the modulus of the velocity vector, and θ determines the angle relative to axis x.

4.1.2 Notes about complex numbers

Representation with complex numbers may possibly look peculiar. Here we briefly recall the main properties of such objects. Those who are familiar with this elementary material may omit this section.

A complex number is indeed a 'flat vector'. One may represent a complex number in different forms

$$z = x + iy = A(\cos\theta + i\sin\theta) = Ae^{i\theta}. \qquad (4.1.16)$$

The first special object in (4.1.16) is i—an imaginary unit. Everywhere we will only use representation with $i^2 = -1$ (that is, we only have to deal with elliptic complex numbers—special comment for advanced readers). Part $x = \mathrm{Re}\, z$ of the complex number is called a 'real part', part $y = \mathrm{Im}\, z$ is called an 'imaginary part'. The complex number modulus

$$A = \sqrt{x^2 + y^2} \qquad (4.1.17)$$

arises from the definition

$$A^2 = |z|^2 = zz^* = \underbrace{(x + iy)}_{z}\underbrace{(x - iy)}_{z^*}, \qquad (4.1.18)$$

where z^* is the conjugate number for z.
The complex number argument θ provides

$$\mathrm{tg}\theta = \frac{y}{x}. \qquad (4.1.19)$$

Figure 4.1.1 illustrates the complex number issues. This is, actually, an adequate representation of the vector with components x, y.

The representation on a complex plane allows us to consider both components of coordinates and velocity simultaneously. As we will see in following sections, this,very abstract, method simplifies the solution of many problems.

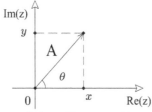

Fig. 4.1.1. Complex number on a plane is an equivalent of the vector with components x and y; also shown are its modulus A and argument θ.

4.1.3 The hodograph of velocity

Complex velocity can be represented on a complex plane as a corresponding vector (originated from the origin, so to say); a hodograph is the line formed by the ends of this vector taken at different moments of time or at different spatial points which correspond to various values θ.

For some problems, a hodograph of velocity can be easily constructed (see Section 4.3) by virtue of elementary considerations based on the geometry and boundary conditions; this approach may significantly simplify a solution.

4.1.4 Electro-hydrodynamic analogy

From the point of view of physicists, mathematics is a language of science. Mathematicians, by the way, are not content with such a modest role, but this story is about something else. Then, problems of physics translated into mathematical language may look similar and, if so, one can apply to these, different in their physical nature, problems the same methods of analysis.

Probably, the best illustration is the partial differential equation

$$\frac{\partial Y}{\partial t} = a\Delta Y, \tag{4.1.20}$$

which is named by all mathematicians and almost all theoretical physicists as the 'diffusion equation' despite the fact that this equation for $Y = T$ (temperature) describes heat conduction; for this case, a is the heat diffusivity. In physical language, it is quite difficult to describe such a parameter as temperature as a diffusing quantity—by parity of reasoning with the stochastic motion of particle which leads, finally, to the real diffusion equation. However, since the heat conduction equation and the diffusion equation look similar, one may apply all the methods developed for the diffusion process to the heat conduction problem: for example, techniques like mathematics of random walks, Brownian motion, etc. Questions like 'At this description, for the heat conduction we deal with a Brownian motion, then, with a Brownian motion of what?' do not have any sense. If we were able to construct an analogy, we may apply it formally, if this application is productive.

After this deviation, let us return to hydrodynamics. The Laplace equation (4.1.9) for the hydrodynamic potential φ is identical to the equation for the electrostatic potential. Thus, we have a formal analogy between electrostatics and hydrodynamics (by the approximation used in this chapter), on a level of mathematical description, but this is quite enough.

Let us develop this analogy (Lamb 1993). The electric field may be caused by point charges or by dipoles (the same charge but with opposite signs pushed apart on distance l_d). For example, the electric field of the point charge is

$$\varphi(r) = \frac{A}{r}, \tag{4.1.21}$$

where a includes a charge as some dimension coefficients (e.g., $(4\pi\varepsilon_0)^{-1}$); the electric field of the point dipole, i.e., the dipole with $l_d \to 0$, is

$$\varphi(r) = \frac{B\cos\theta}{r^2}, \tag{4.1.22}$$

where θ is the angle between the dipole axis and the vector $\vec{r}$ directed from the dipole at the given point, B contains dipole moment and dimension constants.

In accordance with the Laplace equation, if we consider it as the equation for electric potential, the electric field is created only at the boundary—there is no free charge density that can appear in the right-hand side of (4.1.9), transforming the Laplace equation to the Poisson equation. Thus, the electrostatic problem can be reduced to finding the electric potentials from sources like (4.1.21) or (4.1.22) at the boundary of the medium.

The same goes for a hydrodynamical problem: as we see from our analogy, the hydrodynamic potential in a fluid can be calculated if we can define potentials from the given distribution of sources; however, the sources of another form are more useful in this case. But its 'physical nature' does not matter: we can solve problems with such terms, this is the main thing.

However, potentials like (4.1.21) and (4.1.22) are not very convenient in hydrodynamics. Other potentials, i.e., other solutions of the Laplace equation, are much more practical.

The first important solution of the Laplace equation suitable for a hydrodynamical problem is very simple:

$$\varphi(x, y) = u_{x0}x + u_{y0}y, \tag{4.1.23}$$

which corresponds to the flow at constant velocity with components (u_{x0}, u_{y0}). Such flow can also be determined by the stream function, which can be obtained with (4.1.9)

$$\psi(x, y) = -u_{y0}x + u_{x0}y. \tag{4.1.24}$$

Corresponding complex potential (4.1.10) is

$$\omega(z) = u_{x0}x + u_{y0}y - iu_{y0}x + iu_{x0}y = vz. \tag{4.1.25}$$

Then, the Laplace equation has a solution in the form

$$\varphi(r) = \frac{C}{2\pi}\ln r, \tag{4.1.26}$$

where $r = \sqrt{(x-x_0)^2 + (y-y_0)^2}$ is measured from the point (x_0, y_0), with corresponding stream function and complex potential:

$$\psi(\theta) = \frac{C\theta}{2\pi}, \tag{4.1.27}$$

$$\omega(z) = \frac{C}{2\pi}\ln z, \tag{4.1.28}$$

where θ is a polar angle on the plane (x, y) (counted from axis x). The potential (4.1.26) corresponds to the point source in (x_0, y_0) with radial component of velocity

$$u_r = \frac{C}{2\pi r}, \tag{4.1.29}$$

i.e., streamlines are directed from the point (x_0, y_0) outward (or inward for negative C, i.e., for a sink), see Fig. 4.1.2.

On the other hand, if there is a point source, then we may compose a system 'two point sources of different signs at distance $2h$', which is sometimes called a 'dipole' in hydrodynamics literature. This dipole should not be confused with an electrostatic dipole (4.1.22)—those are different constructions; electrostatic dipole consists of two sources that create the potential (4.1.21), while the hydrodynamic dipole consists of two sources of the form (4.1.26).

Thereby, we consider our hydro-dipole (4.1.3), which creates at point O a potential from its two point sources: from the positive end (with $+C$ in (4.1.26)) and from the negative one (with $-C$ in (4.1.26)):

$$\varphi(r) = \frac{C}{2\pi}\ln r^+ - \frac{C}{2\pi}\ln r^-. \tag{4.1.30}$$

Approximately, for $h \ll r$ (see Fig. 4.1.3), we get

$$r^+ \approx r\left(1 + \frac{hx}{r^2}\right), \; r^- \approx r\left(1 - \frac{hx}{r^2}\right) \tag{4.1.31}$$

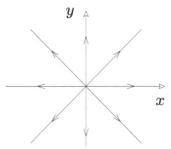

Fig. 4.1.2. The streamlines for the point source (4.1.26).

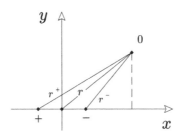

Fig. 4.1.3. Calculating the potential of a dipole.

(these equations can be obtained with the cosine theorem and an expansion on a small parameter h/r), so we get for (4.1.30)

$$\varphi = \frac{\overbrace{2Ch}^{D}x}{2\pi r^2} = \frac{Dx}{2\pi r^2}. \tag{4.1.32}$$

Here, of course, parameter D can be called the dipole moment. The corresponding stream function and the complex potential are

$$\psi = -\frac{Dy}{2\pi r^2}, \tag{4.1.33}$$

$$\omega = \frac{D}{2\pi z}. \tag{4.1.34}$$

At last (for now), the Laplace equation gives us a function

$$\varphi(\theta) = \frac{E\theta}{2\pi} = \frac{E}{2\pi}\arcsin\frac{y - y_0}{r}. \tag{4.1.35}$$

which corresponds to the stream function and the complex potential

$$\psi(r) = -\frac{E}{2\pi}\ln r, \tag{4.1.36}$$

$$\omega(z) = \frac{E}{2\pi i}\ln z. \tag{4.1.37}$$

The potential (4.1.35) corresponds to the vortex: streamlines are circular (see Fig. 4.1.4), and the velocity has only θ-projection

$$u_\theta = \frac{E}{2\pi r}. \tag{4.1.38}$$

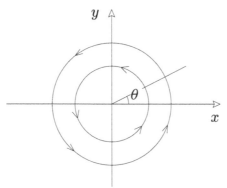

Fig. 4.1.4. The streamlines for the source (4.1.35).

In (4.1.35), parameter E is the circulation along the streamline of radius r:

$$\oint_{(l)} u_\theta dl = \int_0^{2\pi} \frac{E}{2\pi r} r d\beta = E \equiv \Gamma, \tag{4.1.39}$$

using the standard definition of circulation.

Note that the potential (4.1.35) is a non-periodic function: when angle increases on θ, φ increases on the value of circulation Γ.

One may note that the potential φ and the stream function ψ are 'crossed' for the two last cases. Generally, this is a common case for a pair $\varphi - \psi$: as we know, each of these pairs represents a solution of the Laplace equation (again, for a flat problem).

4.1.5 The dynamics of the cavitation cavern: What we have to deal with

Unfortunately, the description of the cavitating flow is more complicated even in comparison with traditional 'single-fluid' problems. For the continuous fluid, we have a single type of boundary conditions—on solids restricting the flow; sometimes the conditions on a free surface can also be defined, for instance for open water, but the position of this surface is determined initially. Another matter is cavitation: here we not only have a free surface, but a free surface whose position and shape must also be determined within the solution.

Actually, hydrodynamics of cavitation concerns two groups of problems:

I. The dynamics of a single cavitation bubble. In this chapter, we consider the common hydrodynamic description of this problem (right in the next section), while the dynamics of the spherical oscillating cavitation bubble will be considered in Chapter 7 specially devoted to this problem.

II. The physical description of the cavitating flow: evolution of the developed (or super-) cavitation, when all bubbles are merged into a giant cavity. This happens behind the bodies moving in the liquid: a typical example is hydrofoils, which will be considered in Sections 4.3 and 4.4.

In addition, in the final section of this chapter, we consider such special questions as waves and instabilities on the liquid–gas interface.

For all aforementioned problems, the fluid can usually be considered as

✓ non-viscous

✓ incompressible

✓ the flow can be considered as irrotational.

Because of the last two circumstances, one may use the convenient description with potentials φ, ψ and ω, as it was stated above.

Due to the first assumption, we only have the condition of zero normal velocity on solid: there is no sticking, and the fluid may flow freely along the solid surface. Thus, the only condition for the solid boundary is

$$u_n = \frac{\partial \varphi}{\partial n} = 0. \tag{4.1.40}$$

Then, for the potential flow we have the Lagrange–Cauchy integral that reads

$$\frac{\partial \varphi}{\partial t} + \frac{u^2}{2} + \frac{p}{\rho} = f(t), \tag{4.1.41}$$

where $f(t)$ is an arbitrary function. Equation (4.1.41) is the generalization of the Bernoulli integral and is sometimes called the same; see also Appendix B: indeed, for the stationary flow $\partial \varphi / \partial t = 0$ and $f = const$. Despite the external similarity, there is one significant distinction between the Bernoulli equation derived for the potential flow and the common Bernoulli integral: to obtain the Bernoulli correlation, we must somehow eliminate the term $u \times rotu$. In the common case, it is possible along the streamline, since it is perpendicular to u and gives no projection on the streamline. On the contrary, for the potential flow this term is identical to zero everywhere. Thus, the Bernoulli correlation for the common case is fulfilled only along the streamline (i.e., with different constants in the right-hand side of it), while a similar relation for the potential flow is common for all the fluid mass.

Thus, for the stationary case we have equation

$$\frac{u^2}{2} + \frac{p}{\rho} = const \tag{4.1.42}$$

which leads to an interesting thing concerning the boundary conditions on the cavern. If we neglect the surface tension forces, which is possible for a cavitation cavern which is giant in comparison with the capillary scales, then the pressure along the cavern is constant: $p = const$. Also, for the stationary cavern—the cavern of the stationary form, whose boundaries are immobile—we have the condition (4.1.42). In sum, from all these statements, we derive that the tangent projection of velocity is constant along the cavity, i.e.,

$$u_\tau = \frac{\partial \varphi}{\partial \tau} = const \tag{4.1.43}$$

Below we consider the first important case of the hydrodynamics description of the cavitation cavern.

4.1.6 Example: The collapse of a cavity near a solid wall

We begin with the phenomenon that was referred to in the first chapter: the collapse of a bubble near a solid surface.

The collapse of a spherical cavity is a classic problem for cavitation. Many particular cases of it will be considered in Chapter 7, where the bubble will be assumed as spherical. The simplicity of a spherical bubble lets us consider many effects like the influence of surface tension, alternating external pressure, dependence of the internal pressure on temperature, etc.

In this section, we omit all remaining complications, but take into account another matter—non-sphericity of a bubble. At first, this problem was considered by Plesset and Chapman (Plesset and Chapman 1970, 1971), and their results found confirmation in experiments.

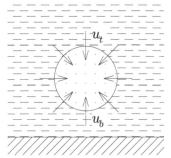

Fig. 4.1.5. The collapse of a bubble near a solid wall: the principal scheme of asymmetry. The collapsing velocity at the top of the bubble u_t can be much greater than the velocity u_b at the bottom, i.e., close to the wall; directly on the wall, the normal component of velocity is equal to zero.

Let us consider the collapse of the initial spherical cavity in the liquid in vicinity of the solid wall (see Fig. 4.1.5). Because of the collapse of the bubble, the surrounding fluid must flow toward the cavity, that is, the collapse must be accompanied by the flow of fluid.

Further, we assume that the fluid is incompressible. Thus, when the cavity disposes far from the wall, these boundary restrictions do not take effect on the process, and the continuity equation (4.1.1) can be written for fluid at the condition of spherical symmetry:

$$\frac{1}{r^2}\frac{\partial}{\partial r}\left(r^2\frac{\partial\varphi}{\partial r}\right)=0, \tag{4.1.44}$$

and $\varphi(r)\sim r^{-1}$, etc.

However, a solid wall near the cavity dramatically changes the scene. Directly on the wall, the fluid flow toward a bubble is distorted; to be exact, the flow is absent directly on the wall—normal projection of velocity $u_n=0$. Note that here, like everywhere in this section, we will consider non-viscous fluid; thus, tangent components of velocity may exist.

It is quite understandable on qualitative level, how bubble dynamics will change in such circumstances. When the cavity walls move (e.g., collapse) at velocity u, the fluid around it must also flow at velocity u. This is possible for the upper side of the cavity, but the bottom side of the bubble is close to the wall, where $u_n=0$, and u cannot significantly exceed u_n in this area. Thus, under the pressure gradient, the upper side of the bubble will collapse much faster than the underside: the top of the cavity will collapse at velocity determined by the difference of external and internal pressures, while the bottom of the cavity would be almost immobile. In other words, if a bubble lays on a solid surface, its bottom would not collapse at all: the fluid flow cannot support this process at this point.

Below, we consider the fundamental principles that allow us to analyze this problem theoretically.

The problem must be defined with the same velocity potential φ, but we cannot further consider the spherical-symmetric case. Now, the problem becomes 2D at least (we will not over complicate the problem here), so the cylindrical coordinates are much more convenient. For coordinates (r, z), the Laplace equation can be written in the form

$$\frac{\partial^2\varphi}{\partial z^2}+\frac{1}{r}\frac{\partial}{\partial r}\left(r\frac{\partial\varphi}{\partial r}\right)=0. \tag{4.1.45}$$

Also, we have the Lagrange–Cauchy integral (4.1.41). Far from the cavity, one may assume the absence of a flow, that is, at infinity $\partial\varphi/\partial t=0$, $u=0$ and we see that (4.1.41) becomes

$$\frac{\partial\varphi}{\partial t}+\frac{u^2}{2}+\frac{p}{\rho}=\frac{p_\infty}{\rho}. \tag{4.1.46}$$

On the free surface of the fluid, at the cavity wall, the pressure in liquid is equal to the pressure of gas in cavity p_g, if we neglect the surface tension, that is, the boundary condition for φ is

$$\frac{\partial \varphi}{\partial t}\bigg|_{r_s, z_s} = \overbrace{\frac{p_\infty - p_g}{\rho}}^{\Delta p} - \frac{u^2}{2}, \tag{4.1.47}$$

where $u^2 = u_r^2 + u_z^2 = \left(\dfrac{\partial \varphi}{\partial r}\right)^2 + \left(\dfrac{\partial \varphi}{\partial z}\right)^2$.

The key hardship: for this problem, as well as for other problems too, we have two difficulties for boundary corresponding to a cavity (a) its shape is unknown and must be determined with other parameters during the solution (b) the boundary moves. Thus, the solution of the problem must include the procedure of determination of the interface, which was done in (Plesset and Chapman 1971) in the following manner. The position r_s, z_s of the interface can be calculated with components u_r, u_z and differential equations

$$\dot{r}_s = u_r\big|_{r_s, z_s}, \quad \dot{z}_s = u_z\big|_{r_s, z_s}. \tag{4.1.48}$$

The total derivative of velocity potential φ at the free surface is

$$\frac{d\varphi}{dt}\bigg|_{r_s, z_s} = \frac{\partial \varphi}{\partial t} + u_r \frac{\partial \varphi}{\partial r} + u_z \frac{\partial \varphi}{\partial z} = \frac{\Delta p}{\rho} + \frac{u^2}{2}, \tag{4.1.49}$$

which was obtained with (4.1.8).

The problem with such a statement was solved numerically in (Plesset and Chapman 1971) and (Voinov and Voinov 1976), where the results correlated with experimental data from Chapter 1.

In (Plesset 1971), the equation (4.1.45) was solved numerically at each time step. With obtained φ, the velocities u_r and u_z were calculated, so that the displacement of the boundary is calculated with (4.1.48) as $r_s \Delta t$, $z_s \Delta t$; the new value of φ—the boundary conditions for the next time step—were obtained for the new boundary position with $\Delta \varphi$ from (4.1.49) written through finite differences, i.e., with replacement $d\varphi/dt \to \Delta\varphi/\Delta t$.

Plesset and Chapman (Plesset and Chapman 1971) observed the formation of a jet in the collapsing bubble directed to the wall; see Fig. 4.1.6 where the process is shown schematically. The jet formed at the upper side of the bubble accelerated up to significant speed: the magnitude of the jet speed was ~ 100 m/s.

In (Voinov and Voinov 1976), authors denied the spherical shape and considered the bubble as ellipsoid; they found out that even a small deviation from sphericity causes great increase in the jet speed: if a bubble would flatten at the direction to the solid surface by 10%, the formed jet would be thinner by five times and two times faster.

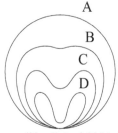

Fig. 4.1.6. The shape of a collapsing bubble near a solid wall. A—initial (spherical shape); B—collapsed, slightly deformed bubble; C—jet is formed; D—jet strikes the solid surface.

Thus, the jet formed in the bubble remembers the condition of its formation; its parameters cannot be calculated precisely by analytical correlations. But, at least, we may consider the process qualitatively.

At the first stage, before the jet forms (shape B on Fig. 4.1.6; this is actually the jet-ancestor—this bubble deformation produces the jet), its speed can be estimated with the Bernoulli integral approximately as $u_0 \sim \sqrt{2\Delta p / \rho} \sim 14\,\text{m/s}$ (for $\Delta p = 10^5$ Pa and $\rho = 10^3$ kg/m³); note that this estimation correlates with numerical results from Plesset and Chapman. The area of cross-section of this jet S_0 is bigger by an order of magnitude than the area of the full upper part of the bubble; we put $S_0 \sim R_{bubble}^2$, because we find it difficult to separate exactly that 'upper part'; by the way, for such calculations, generally speaking, there is no need to hold '2' in the correlation for velocity: we did it only to show the origin of that correlation.

Then, when the jet spreads to the surface, from the condition of the constant discharge $uS = const$, we may estimate the speed at the moment of the impact on the solid surface (shape D on Fig. 4.1.6): $u \sim u_0 S_0/S$. For such a fat jet that is pictured on Fig 4.1.6, the jet diameter is less than the initial radius by about three times, i.e., the speed increases up to ~ 100 m/s.

Note that the real form of the jet depends on many factors, including instabilities at the liquid–gas interface.

4.2 The problem of a flow past a body

4.2.1 Physical formulations

In the following sections, we consider a developed cavitation, when all bubbles are merged into the macroscopic cavern past a moving body (see Fig. 4.2.1). This is a special case of the two-phase flow with separated phases.

To solve the problem, we must set it correctly. The job of a theoretical physicist is similar to the work of a sculptor: they both take a material—mathematical equations or marble—and cut off all unnecessary parts. A hydrodynamic description may be of any level of complexity, but we want to have such a description that can give us some certain results. So, simplification is the key point for a successful solution.

Foremost, we will consider a flat two-dimensional case, mainly of the flat geometry. This is enough to obtain results of sufficient quality in order to compare them with the experimental data.

Then, we will consider a stationary case, which is a more questionable choice. This assumption works well for a single-phase flow, but it is a bit desperate for a two-phase one. The cavern shown on Fig. 4.2.1 is non-stationary indeed; its rear side cannot be depicted in static without special tricks set out in Section 4.3.5, where different schemes of the cavern's 'closure' are discussed.

Nevertheless, we represent this phenomenon as the stationary irrotational flow. Hence, the problem turns into the physics described in Section 4.1: a flat flow over a complicated boundary consisting of a solid contour and a free surface.

The boundary conditions abide the statement. Considering a perfect fluid, we get the condition of impermeability on the solid (line *ab* on Fig. 4.2.1). For the stationary cavern at constant pressure, we have the condition of the constant tangent velocity along the free surface.

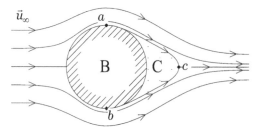

Fig. 4.2.1. The cavitating flow past a body. Fluid flows around the complicated region consisting of the part *ab* on the solid body *B* and of the part *bca* on the cavern *C*.

Note another useful feature that we will use below. Because of its nature, the stream function is constant along the streamline. At the same time, the surface of the cavern corresponds to the streamline (corresponding to lines *ac* and *bc*). Thus, if one somehow knows the stream function as a function of coordinates ψ (x, y), then from the equation $ψ(x, y) = const$ one may determine the shape of the cavern.

Theréby, we have the proper physical statement of the problem. However, this is not enough, since the mathematical implementation may be different too.

4.2.2 Mathematical apparatus

The problem of the cavitating flow is a generalization of the classical problem of the flow past a body, the basic results of which were obtained by such classics as Helmholtz, Kirchhoff, Joukowski, etc. There may be different methods to describe it, and several methods will be used below.

The common issue of these methods is the use of hydrodynamic potentials, of course, but details differ. Some results can be obtained directly with φ and ψ, while for others the complex potential ω is needed.

The suitable way is to represent the potential field with the sum of potentials corresponding to elementary 'sources' of potentials; for instance, to represent the external flow the linear potential (4.1.23) is very convenient. Then we can add other potentials using the property of superposition of hydrodynamic potentials, the sources of which are placed on the boundaries. Directly on the boundaries potentials have singularities, but this is not a significant inconvenience, since we are interested only in the velocity filed, i.e., the potentials, in the fluid.

The main method of solution for such problems is to use the complex potential ω and the complex velocity v, that is, to use complex plane $z(x, y)$. For most cases, one may map this plane to another plane $t(ξ, η)$, where boundary has a convenient form: for the example considered in the next section, the arbitrary contour transforms into a circle. In plane t, the solution can be obvious; for example, one may simply compose it from the potentials corresponding to the flow around a cylinder: the circle on 2D corresponds to the cylinder the axis of which is directed toward the picture.

Then, when we find the solution in the complex plane $t(ξ, η)$, we have to translate it to the initial plane v; this part of the problem is the most difficult.

4.2.3 The simplest consideration: A non-cavitating flow

Since in the following sections we consider problems concerning cavitating flows, it is useful to remember the single-phase case of the flow past a wing. Besides other reasons, we explain some techniques concerning the complex potentials, complex velocities, maps, etc. On the other side, this is a classical solution, and you are probably familiar with it.

The problem of the flow around a body reduces to the map that transforms the physical space $z(x, y)$ to the space $t(ξ, η)$ (see Fig. 4.2.2); the airfoil corresponds to the dashed area on them. The main problem is to define the corresponding map which converts points from the points on plane t away from the circle to the points on plane z for the outer (non-dashed) area. The corresponding map is

$$z = f(t) \tag{4.2.1}$$

For this transform, there exist two additional conditions:

1) $ξ \rightarrow \infty$ corresponds to $z \rightarrow \infty$,

2) the direction of vector $\vec{u}_\infty$, determined by the attack angle α, is the same on both planes.

Denote the complex potential on the plane z as $ω(z)$, corresponding complex potential on the t-plane is $ω(t)$; quantities on the corresponding plane will be denoted by the appropriate indexes.

At first, we consider t-plane and, respectively, the flow past a cylinder. The potential of the corresponding flow at zero circulation can be obtained as the sum of potentials of the dipole and of the external flow:

$$\omega(t) = u_\infty t + \frac{u_\infty R^2}{t}. \tag{4.2.2}$$

Corresponding streamlines are shown on Fig. 4.2.3. If we are interested in it, the complex velocity in this flow is

$$\upsilon = \frac{d\omega}{dt} = u_\infty \left(1 - \frac{R^2}{t^2}\right). \tag{4.2.3}$$

Now, if we want to 'turn on' non-zero circulation on the contour, then we may simply add the corresponding potential (4.1.35) to (4.2.3) and get

$$\omega(t) = u_\infty \left(t + \frac{R^2}{t}\right) + \frac{\Gamma}{2\pi i} \ln t. \tag{4.2.4}$$

The stagnation points are the points of zero velocity on the contour. Finding them for potential (4.2.4) is a trivial exercise: for $t = Re^{i\theta}$, we obtain that the distribution of the velocity on the cylinder is

$$|\upsilon| = \left|2u_\infty i e^{-i\theta} \sin\theta\right| = 2u_\infty |\sin\theta|, \tag{4.2.5}$$

that is, points A and B on the contour, corresponding to $\theta = \pi$ and $\theta = 0$, respectively, are the stagnation points. For the common potential, the answer depends on the value of circulation Γ. We have

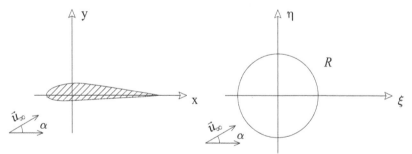

Fig. 4.2.2. The map from the coordinates (x,y) to (ξ,η). Initial flow past a body of an arbitrary shape can be represented as the flow past a cylinder; we only need to find a corresponding map that transforms one set of coordinates into the other.

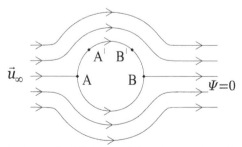

Fig. 4.2.3. The flow past a cylinder for case $\Gamma = 0$. Points A and B are the stagnation points (points of zero velocity). Points A' and B' correspond to stagnation points for $\Gamma \neq 0$.

$$v = \frac{d\omega}{dt} = u_\infty \left(1 - \frac{R^2}{t^2}\right) + \frac{\Gamma}{2\pi i t}, \tag{4.2.6}$$

and from the condition $v = 0$, we obtain the quadratic equation (that is obvious) and its solution

$$t = \frac{i\Gamma}{4\pi u_\infty} \pm \sqrt{R^2 - \frac{\Gamma^2}{16\pi^2 u_\infty^2}}. \tag{4.2.7}$$

The simplest case corresponds to the positive root expression in (4.2.7): for this case, two values of t have the same imaginary part (i.e., the same positive coordinate η) and two real parts— coordinates ξ—differing by signs. This case corresponds to points A' and B', placed on the cylindric contour (see Fig. 4.2.3).

In all these expressions, the quantity u_∞ is the velocity of the incoming flow that is directed along the axis x. If the incoming flow has angle α with axis x (the attack angle), then the complex potential must be represented with a complex velocity

$$\omega = v_\infty t + \frac{v_\infty^* R^2}{t} + \frac{\Gamma}{2\pi i} \ln t \tag{4.2.8}$$

Now we return to our initial problem—the flow past a body with an arbitrary contour. Above, we obtained the solution on t-plane; because we cannot yet make any certain inference about the correspondence of velocities and circulations on z-plane and t-plane, we rewrite the complex potential on t-plane with the following designations:

$$\omega(t) = v_{t\infty} t + \frac{v_{t\infty}^* R^2}{t} + \frac{\Gamma_t}{2\pi i} \ln t. \tag{4.2.9}$$

Complex velocities in both planes v_z, v_t are connected as

$$v_t = \frac{d\omega}{dt} = \frac{d\omega}{dz}\frac{dz}{dt} = v_z f', \tag{4.2.10}$$

where $f' = df/dt \equiv dz/dt$ is the derivative of the map (4.2.1). Particularly, for the velocity at infinity, we have the relation

$$v_{t\infty} = v_{z\infty} \underbrace{\frac{df}{dt}\Big|_{t=\infty}}_{k_\infty} = k_\infty v_{z\infty}. \tag{4.2.11}$$

For the circulation, we have with (4.2.3)

$$\Gamma_t = \mathrm{Re} \oint v_t \, dt = \mathrm{Re} \oint v_z f' \, dt = \mathrm{Re} \oint v_z \, dz = \Gamma_z, \tag{4.2.12}$$

that is, the circulations remain the same after map $z \to t$. Generally, we may get a solution for any Γ; however, the Joukowski–Chaplygin postulate states that the velocity on the trailing edge must be finite (exactly, it is equal to zero, that is, there must exist a stagnation point on the trailing edge). This definition determines the certain value of circulation.

Thus, the expression for the complex potential on z-plane has a similar form:

$$\omega(z(t)) = k_\infty \left(v_{t\infty} t + \frac{v_{t\infty}^* R^2}{t} \right) + \frac{\Gamma_z}{2\pi i} \ln t. \tag{4.2.13}$$

The general form of the map $t \rightarrow z$ with function f can be expressed as

$$f(t) = k_{\infty} t + \sum_{n=0}^{\infty} \frac{k_n}{t^n} \tag{4.2.14}$$

with coefficients

$$k_n = \frac{1}{2\pi i} \oint_{C_t} f(t) t^{n-1} dt. \tag{4.2.15}$$

The first and the simplest transformation function $f(t)$—the Joukowski–Chaplygin transform—corresponds to coefficients

$$k_1 = c^2/2, \; k_{\infty} = 1/2; \; k_n = 0, \text{ for any } n \neq 1, \infty \tag{4.2.16}$$

and gives

$$z = \frac{1}{2}\left(t + \frac{c^2}{t}\right), \text{ or } \frac{z-c}{z+c} = \left(\frac{t-c}{t+c}\right)^2. \tag{4.2.17}$$

This map corresponds to the representation of the circle of radius c on t-plane as the line segment on z-plane. Indeed, for $t = ce^{i\theta}$ we have

$$z = \frac{c}{2}\left(e^{i\theta} + e^{-i\theta}\right) = c\cos\theta. \tag{4.2.18}$$

In other words, when we follow the circle on t-plane, looking over all angles θ from 0 to 2π, we get values z from c to... c, including $-c$ for $\theta = 2\pi$.

Thus, we come close to the flow past a plate. However, we are interested in the cavitating flow, not in a simple single-phase flow.

Thus, we are finishing the explanation of the complex-map-skills, and getting back to the main problems of this chapter and of this book.

4.3 The cavitating flow around a body

4.3.1 Initial assumptions

The first assumption involves the incompressible fluid approach. Actually, this is a common suggestion that is used almost everywhere in this book; so we will not discuss it in this section.

Below, we will consider the so-called weightless fluid. Possibly, this is not a very suitable term; exactly, this means that in the Navier-Stokes equation, which can be written for these purposes in a simple form,

$$\frac{\partial \vec{u}}{\partial t} + \vec{u}\nabla\vec{u} = -\frac{1}{\rho}\nabla p + \nu\Delta\vec{u} + \vec{g} \tag{4.3.1}$$

one may neglect the last term on the right-hand side. This term can be assumed negligible if it is much lower than the second term on the left-hand side, which, estimated by an order of magnitude, is $\sim U^2/d$, where U is the typical velocity and d is the spatial scale. Thus, we need to fulfill the condition for the Froude number Fr

$$Fr = \frac{U^2}{gd} \gg 1. \tag{4.3.2}$$

Then, we will neglect, as above, the viscosity term – the second term on the right side of (4.3.1), and also consider the stationary case. Despite the fact that the Navier–Stokes equation is now reduced to the simplest version of the Euler equation (and gives, by the way, the Bernoulli integral), we will not consider exactly that: as in previous sections, the solution will be presented with hydrodynamic potentials.

The final assumption: we will consider an irrotational flow, i.e., we can apply all the implementation provided in Section 4.1.

4.3.2 Boundary conditions

We will consider a flat flow, i.e., 2D-geometry in Cartesian coordinates.

In comparison to the consideration from Section 4.2, we have to deal with a more complicated case (see Fig. 4.3.1). For the cavitating flow around a body, we have two types of boundary conditions: liquid–solid and liquid–vapor.

As for the liquid–solid boundary, we will use here, of course, the same condition as above: normal component of velocity must be zero, that is,

$$\frac{\partial \varphi}{\partial n}\bigg|_{l-s} = 0. \tag{4.3.3}$$

The liquid–vapor interface demands another condition along with itself. A common assumption for a gas cavern is the constant, homogeneous pressure inside it: indeed, one may expect no gas flow around the cavern; therefore, $\nabla p = 0$. If so, from the Bernoulli integral

$$\frac{\rho u^2}{2} + p = const, \tag{4.3.4}$$

we see that if $p = const$ along the streamline corresponding to the boundary of the cavern, then $u^2 = const$; for the absence of the normal component of velocity, i.e., for a stationary cavern, this means that

$$\frac{\partial \varphi}{\partial \tau}\bigg|_{l-g} = const. \tag{4.3.5}$$

In other words, the fluid flows along the cavern at a constant velocity.

If we only have these two types of boundary conditions, the problem is simplified in terms of the complex velocity.

For the free surface (liquid–gas interface), the constancy of u^2 means that the modulus V of the complex velocity

$$v = u_x - iu_y = Ve^{-i\theta} \tag{4.3.6}$$

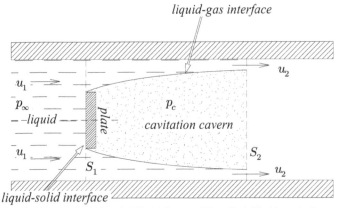

Fig. 4.3.1. The cavitating flow. There are two types of boundary conditions here. Simple reasonings provide estimations of the correlation between the cavitation number and geometry.

remains constant. That is, the hodograph of the velocity corresponding to the streamline along the free surface of the fluid represents the arc of a circle at constant radius.

For a plane, i.e., for boundaries in the form of straight lines, the condition of impermeability $u_n = 0$ means that $u_x/u_y = const$: the velocity has a constant direction, that is, $\theta = const$ in (4.3.6). In other words, the hodograph represents a radial line in this case.

Summing up, we have two types of lines on the t-plane that correspond to interfacial boundaries for the flow past a plate:

(1) radial lines for solid boundaries;

(2) arcs for free surfaces.

4.3.3 The flow past a plate

Let us consider a flat flow past a plane barrier. For this case, as we have seen above, we know a lot about this kind of flow in advance.

At first, if the flow is enclosed in a pipe (see Fig. 4.3.1), one may provide simple estimations concerning the properties of the flow. Let us denote the velocity at the plate as u_1, the area of cross-section here is S_1; the corresponding parameters past a plate in the cavitation area are u_2 and S_2. Due to incompressibility, the flux must be the same in both cross-sections:

$$u_1 S_1 = u_2 S_2. \tag{4.3.7}$$

To use this relation, i.e., to replace $\int u dS \rightarrow uS$, the flow must be uniform. This is obvious for the initial flow at velocity u_1, where any velocity profile can be set (note that we consider a perfect fluid which does not stick to the solid, and, therefore, velocity can be u_1 right on the wall). For the stream behind a barrier, it is not as obvious, and we consider a well-established velocity profile, where u_2 is uniform too.

If the pressure in the liquid ahead of the plate is p_∞, and pressure in the cavern is p_c, then, from (4.3.4) we have

$$\frac{\rho u_1^2}{2} + p_\infty = \frac{\rho u_2^2}{2} + p_c. \tag{4.3.8}$$

If the pressure in the cavern is lower than at the entry, $p_c < p_\infty$, then $u_2 > u_1$ (according to (4.3.8)) and $S_2 < S_1$ (according to (4.3.7)). Thus, the cavitation cavern has a form shown on Fig. 4.3.1, and we have from equations (4.3.7)–(4.3.8) for the cavitation number

$$Q = \frac{p_\infty - p_c}{\rho u_1^2 / 2} = \left(\frac{u_2}{u_1}\right)^2 - 1 = \left(\frac{S_1}{S_2}\right)^2 - 1. \tag{4.3.9}$$

Then we consider the case when the flow runs onto a plate at an angle α (see Fig. 4.3.2) in an infinite medium. Choose the axis x along the plate and axis y directed to the plate. Let the width of the plate be L.

The incoming flow splits at the stagnation point O into two streams. The x coordinate of this point is evident for $\alpha = \pi/2$: in this case, point O lies in the middle of the plate; in other cases, the position of O must be determined additionally.

Below, we will use the complex potential ω and the complex velocity v. As it was discussed above, the velocity on the free surface is constant; thus, we denote the modulus of v on the free streamline as V_F. At the edge points on the plate, consequently, $V_A = V_B = V_F$.

As it was described above, for this case the hodograph contains simple parts corresponding to interfaces. Thus, we will solve the following—the inverse problem: knowing the velocity, we try

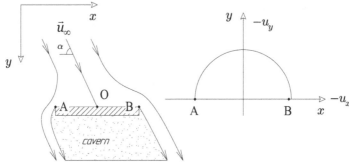

Fig. 4.3.2. The flow around a plate. The plate is placed along the axis x, the attack angle is α. Point O is the stagnation point: here $u_x = u_y = 0$. The hodograph is also shown; the horizontal line AB corresponds to the plate.

to find the coordinates in accordance with (4.1.1) using the auxiliary function T (here we follow (Birkhoff and Zarantonello 1957)):

$$z = \int \frac{d\omega}{\upsilon} = \int \frac{f(T)dT}{\upsilon}, \tag{4.3.10}$$

where T has a common form

$$T = \frac{a\left(\tilde{\upsilon}^{2n}+1\right)+2b\tilde{\upsilon}^n}{c\left(\tilde{\upsilon}^{2n}+1\right)+2d\tilde{\upsilon}^n}, \tag{4.3.11}$$

with real coefficients a, b, c, d, and $ad < bd$; here we use $\tilde{\upsilon} = \upsilon/V_F$, which is a normalized complex velocity; with such definition, $\tilde{\upsilon}_A = -1$, $\tilde{\upsilon}_B = 1$.

It can be proven (Birkhoff and Zarantonello 1957) that the common form of the function $f(T)$ for such type of flow is

$$f(T) = C\frac{\prod\limits_i (T - A_i)}{\prod\limits_k (T - B_k)}, \tag{4.3.12}$$

where A_i and B_k are real.

It is required that at the stagnation point O, where the flow splits on the solid surface in two different flows, and where the velocity $\upsilon = 0$ (see Fig. 4.3.2), our auxiliary function $T = 0$. Thus, in (4.3.11) $a = 0$ and $bc > 0$, and (4.3.11) can be represented for $n = 1$ as

$$T = \frac{2h\tilde{\upsilon}}{\tilde{\upsilon}^2 - 2D\tilde{\upsilon}+1}, \tag{4.3.13}$$

where parameters are connected as

$$h = b/c, \ D = -2 \ d/c. \tag{4.3.14}$$

For simplicity, the function (4.3.12) can be written as a linear function on T: $f(T) = T$, so

$$\omega = \frac{T^2}{2} = \frac{2h^2\tilde{\upsilon}^2}{\left(\tilde{\upsilon}^2 - 2D\tilde{\upsilon}+1\right)^2}, \tag{4.3.15}$$

and (4.3.10) now can be taken apart

$$z = \frac{\omega}{V_F \tilde{v}} + \int \frac{\omega d\tilde{v}}{V_F \tilde{v}^2} = \frac{\omega}{V_F \tilde{v}} + \int \frac{2h^2 d\tilde{v}}{V_F \left(\tilde{v}^2 - 2D\tilde{v} + 1\right)^2}. \tag{4.3.16}$$

At infinity $v = e^{-i\alpha}$ (see Fig. 4.3.2), there must be $\omega = \infty$, and the denominator in the integral (4.3.15) must be equal to zero:

$$e^{-i\alpha} \left(e^{-i\alpha} - 2D + e^{i\alpha}\right) = 2e^{-i\alpha} \left(\cos\alpha - D\right) = 0, \tag{4.3.17}$$

thus, $D = \cos\alpha$.

Integrating (4.3.16) on the complex plane, we get

$$z = \frac{2h^2 \tilde{v}}{V_F \left(\tilde{v}^2 - 2\tilde{v}\cos\alpha + 1\right)^2} + \frac{h^2}{V_F \sin^2\alpha} \left(\frac{\tilde{v} - \cos\alpha}{\left(\tilde{v}^2 - 2\tilde{v}\cos\alpha + 1\right)} - \frac{i}{2\sin\alpha} \ln\frac{\tilde{v} - e^{i\alpha}}{\tilde{v} - e^{-i\alpha}}\right). \tag{4.3.18}$$

On a plate, the complex velocity has only a real component u_x; thus, here we may consider only the real form of (4.3.16). Introducing a function

$$\chi = \tilde{v} - \cos\alpha, \tag{4.3.19}$$

so that $\tilde{v}^2 - 2\tilde{v}\cos\alpha + 1 = \chi^2 + \sin^2\alpha$, integration of (4.3.16) there (on a plate) gives the coordinate along the plate

$$z_p(\tilde{v}) = \frac{2h^2 \tilde{v}}{V_F \left(\chi^2 + \sin^2\alpha\right)^2} + \frac{h^2}{V_F \sin^2\alpha} \left(\frac{\chi}{\chi^2 + \sin^2\alpha} + \frac{1}{\sin\alpha} \arctan\frac{\chi}{\sin\alpha}\right). \tag{4.3.20}$$

Relation (4.3.20) may also contain a constant that defines the exact position of some point of the plate; we omit this constant.

Then, with (4.3.20) one may find the coordinates of all interesting points on a plate. The left edge—point A:

$$z_A = z_p(-1) = -\frac{h^2}{2V_F \left(1 + \cos\alpha\right)^2} - \frac{h^2}{V_F \sin^2\alpha} \left(\frac{1}{2} + \frac{\pi - \alpha}{2\sin\alpha}\right). \tag{4.3.21}$$

The stagnation point O:

$$z_O = z_p(0) = -\frac{h^2}{V_F \sin^2\alpha} \left(\cos\alpha + \frac{\pi/2 - \alpha}{\sin\alpha}\right). \tag{4.3.22}$$

The right edge—point B:

$$z_B = z_p(1) = \frac{h^2}{2V_F \left(1 - \cos\alpha\right)^2} + \frac{h^2}{V_F \sin^2\alpha} \left(\frac{1}{2} + \frac{\alpha}{2\sin\alpha}\right). \tag{4.3.23}$$

To whom who follows, we may note that

$$\arctan\left(\frac{1 + \cos\alpha}{\sin\alpha}\right) = \frac{\pi - \alpha}{2}, \arctan\left(\frac{1 - \cos\alpha}{\sin\alpha}\right) = \frac{\alpha}{2}. \tag{4.3.24}$$

Now we can connect the only unknown parameter h in our equations with the width of the plate L:

$$L = z_B - z_A = \frac{h^2}{2V_F} \underbrace{\left(\frac{4}{\sin^4 \alpha} + \frac{\pi}{\sin^3 \alpha} \right)}_{s(\alpha)}, \tag{4.3.25}$$

Thus, constant h, which we heroically dragged through the entire derivation, is

$$h = \sqrt{\frac{2LV_F}{s(\alpha)}}. \tag{4.3.26}$$

Generally, there are two ways in theoretical physics. The first one, which was the mainstream throughout the XX century, prescribes to write all equations in dimensionless form where correlations do not contain any dimension parameters. However, such relations are very difficult for practical analysis and for calculations; actually, they need a decoding procedure to get anything useful from them. Thus, we prefer the (not very popular) second way: to write full equations; if someone needs to generalize them, it can be easily done *a posteriori*.

Then, one may find the coordinate of the stagnation point relative to the left edge of the plate. Thus, we get

$$\Delta z_O = z_O - z_A = \frac{h^2}{2V_F} \underbrace{\left(\frac{1}{(1+\cos\alpha)^2} + \frac{1-2\cos\alpha}{\sin^2\alpha} + \frac{\alpha}{\sin^3\alpha} \right)}_{p(\alpha)}. \tag{4.3.27}$$

or, as a relative value with (4.3.25)

$$\frac{\Delta z_O}{L} = \frac{p(\alpha)}{s(\alpha)}. \tag{4.3.28}$$

The position of the stagnation point is shown on Fig. 4.3.3.

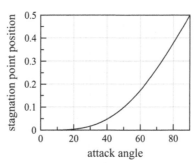

Fig. 4.3.3. The position of the stagnation point (relative to the width of the plate) in dependence of the attack angle α. Position is measured from the left edge of the plate (see Fig. 4.3.2).

To calculate the total force acting on the plate from the flow, we have to take the integral

$$F = \int_{(L)} \left(p_f - p_c \right) dz. \tag{4.3.29}$$

Here F is a force per unit of length in the direction perpendicular to Fig. 4.3.2; p_f is the pressure from the fluid (acting from the upper side on Fig. 4.3.2), p_c is the pressure in the cavitation cavern (acting from the bottom side on Fig. 4.3.2).

In accordance with the Bernoulli integral,

$$p_f - p_c = \frac{\rho V_F^2}{2} - \frac{\rho |v|^2}{2}, \tag{4.3.30}$$

where v is the velocity of the flow on the plate. Since the complex velocity has only a real component on the plate (corresponding to the x-value), we may replace $|v|^2 = v^2$. With (4.3.30), we have for (4.3.29)

$$F = \frac{\rho V_F^2}{2} \int_{z_A}^{z_B} \left(1 - \tilde{v}^2\right) dz. \tag{4.3.31}$$

Then, we can rewrite the differential in (4.3.31) as $dz = \dfrac{\partial z}{\partial \omega} d\omega$ and rewrite (4.3.31) in the same manner as (4.3.16):

$$F = \frac{\rho V_F}{2} \int_{(L)} \left(\tilde{v}^{-1} - \tilde{v}\right) d\omega = \frac{\rho V_F}{2} \left[\omega\left(\tilde{v}^{-1} - \tilde{v}\right)\Big|_{(L)} + \int_{(L)} \omega\left(1 + \tilde{v}^{-2}\right) d\tilde{v} \right]. \tag{4.3.32}$$

Integrals over the plate surface in variable $\tilde{v}$ imply integration from $\tilde{v} = -1$ to $\tilde{v} = 1$. So, the first integral in (4.3.32)

$$\omega\left(\tilde{v}^{-1} - \tilde{v}\right)\Big|_{-1}^{1} = 0, \tag{4.3.33}$$

since the bracket gives zero on both ends.

The second integral is more complicated. With ω from (4.3.15), the integral

$$\int_{-1}^{1} \frac{\tilde{v}^2 \left(1 + \tilde{v}^{-2}\right) d\tilde{v}}{\left(\tilde{v}^2 - 2\tilde{v}\cos\alpha + 1\right)^2} \tag{4.3.34}$$

(we take the constant $2h^2$ off from the integral) splits in two:

$$\int_{-1}^{1} \frac{\tilde{v}^2 \, d\tilde{v}}{\left(\tilde{v}^2 - 2\tilde{v}\cos\alpha + 1\right)^2} = \frac{\left(2\cos^2\alpha - 1\right)\tilde{v} - \cos\alpha}{2\sin^2\alpha\left(\tilde{v}^2 - 2\tilde{v}\cos\alpha + 1\right)}\Bigg|_{-1}^{1} + \frac{1}{2\sin^3\alpha}\arctan\frac{\tilde{v} - \cos\alpha}{\sin\alpha}\Bigg|_{-1}^{1} =$$

$$= -\frac{1}{2\sin^2\alpha} + \frac{\pi}{4\sin^3\alpha}; \tag{4.3.35}$$

$$\int_{-1}^{1}\frac{d\tilde{\upsilon}}{\left(\tilde{\upsilon}^2-2\tilde{\upsilon}\cos\alpha+1\right)^2}=\frac{\tilde{\upsilon}-\cos\alpha}{2\sin^2\alpha\left(\tilde{\upsilon}^2-2\tilde{\upsilon}\cos\alpha+1\right)}\bigg|_{-1}^{1}+\frac{1}{2\sin^3\alpha}\arctan\frac{\tilde{\upsilon}-\cos\alpha}{\sin\alpha}\bigg|_{-1}^{1}=$$

$$=\frac{1}{2\sin^2\alpha}+\frac{\pi}{4\sin^3\alpha}. \tag{4.3.36}$$

Finally, with the help of (4.3.24), we get for the whole integral (4.3.32)

$$F=\frac{\pi\rho V_F h^2}{2\sin^3\alpha}=\frac{\pi\rho V_F^2 L}{s(\alpha)\sin^3\alpha}. \tag{4.3.37}$$

Thus, dividing by the width of the plate L, one may obtain the expression for the mean pressure on a plate (it is obvious and does not need to be written; keep in mind that F is defined per unit of length). Note that for a streamline on a free surface, due to the Bernoulli integral

$$\underbrace{p_\infty+\frac{\rho u_\infty^2}{2}}_{\text{at infinity}}=\underbrace{p_c+\frac{\rho V_F^2}{2}}_{\text{at interface}}, \tag{4.3.38}$$

consequently, we know that

$$V_F^2=u_\infty^2\left(1+\underbrace{\frac{p_\infty-p_c}{\rho u_\infty^2/2}}_{q}\right)=u_\infty^2\left(1+q\right). \tag{4.3.39}$$

Thus, the average force acting on the plate per unit of area is

$$P=\frac{\pi\rho u_\infty^2\left(1+q\right)}{s(\alpha)\sin^3\alpha}. \tag{4.3.40}$$

Function P determines the modulus of force that acts on the plate in normal direction. Calculating its projections on the direction of the initial flow and on the perpendicular to it, one may find the drag coefficient C_D and the lift coefficient C_L:

$$C_D=\frac{P\sin\alpha}{\rho u_\infty^2/2}=\frac{2\pi\left(1+q\right)}{s(\alpha)\sin^2\alpha},C_L=\frac{P\cos\alpha}{\rho u_\infty^2/2}=\frac{2\pi\left(1+q\right)\cos\alpha}{s(\alpha)\sin^3\alpha}. \tag{4.3.41}$$

Functions $C_D(\alpha)$ and $C_L(\alpha)$ are shown in Figs. 4.3.4 and 4.3.5 for $q=0$. The drag coefficient has a maximum at $\alpha=90°$, the lift coefficient reaches its maximum value at the attack angle $\alpha\approx38°$.

The problem solved in this section is sometimes called 'the Kirchhoff problem'. Indeed, the basic principles of the solution were given by Kirchhoff about one and a half century ago. It is very impressive that one can obtain the full solution of such a complicated problem; moreover, calculations with the relations obtained in this section are in good agreement with the experimental results, see (Birkhoff and Zarantonello 1970).

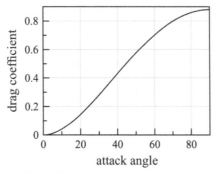

Fig. 4.3.4. The drag coefficient C_D as the function of the attack angle α. Here we put $q = 0$.

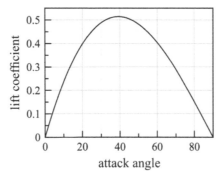

Fig. 4.3.5. The lift coefficient C_L as the function of the attack angle α. Here we put $q = 0$.

4.3.4 The model forms of the cavitation cavern

Previously,we considered a quite optimistic form of the cavitation cavern (see Fig. 4.3.2): we were not interested in the form of the cavitation cavern—we assumed it was infinite. Strictly, this solution suits only the case where $q = 0$.

However, it should be understood that the real form of the cavern is qualitatively different. The flow past a plate must be closed beyond the cavern; this is evident from our intuition, but our intuition does not predict whether this closure is caused by the gravity of forces or not.

In any way, if we want to consider a more realistic flow pattern, we have to change the boundary conditions—the geometry of the considered problem.

Perhaps, the first scheme that comes to mind is to make the flow pattern more natural as shown in Fig. 4.3.6. Here we see the closed cavern, as it seems to be at first glance.

In Fig. 4.3.6, we see that the flow behind a cavern must converge to the point A—the second stagnation point in addition to the one on the solid surface—point O. The total velocity at point A must be zero, same as at point O. However, point A lies on a free surface where, as we discussed above, the velocity must be constant. Thus, we come to a contradiction: the stationary pattern shown in Fig. 4.3.6 does not exist.

Traditionally, physicists do not like non-stationary problems, so in order to find solutions for this crux, we will stand by the static character of the flow and try to invent other images of the flow.

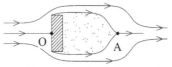

Fig. 4.3.6. An impossible cavitation scheme with two stagnation points: O—the real stagnation point at the solid surface, C—the non-existent stagnation point on the free surface.

The first correction is the Efros–Gilberg–Rock (EGR) (Ivanov1980) scheme shown in Fig. 4.3.7. The principal element of this pattern is the inverse jet at the end of the cavern, where the fluid flows back to the plate.

Fig. 4.3.7. The Efros–Gilbarg–Rock pattern of the cavitating flow.

In this theoretical scheme, the ends of the inverse streamlines continue beyond the plate on another sheet of the Riemann surface. The second stagnation point is located in the fluid. Such re-entrant jets are observed in experiments as well as second stagnation point; however, these objects are non-stationary. There is no need to explain that the disappearance of the inverse flow at the plate—we mean disappearance in our normal, physical space—is a fantastic construction. Despite such a special mathematical trick, the EGR-model provides reasonable results for the forces acting on the plate.

A close relative to such schemes is description of the cavity closure with the help of exotic mathematical apparatus given in (Gurevich 1965). It can be shown that the closure condition

$$\oint_c dz = 0 \qquad (4.3.42)$$

can be satisfied if the complex potential at the closure point (where it can be set $\omega = 0$) corresponds to the singularity of the complex velocity as

$$\ln \upsilon \sim -\frac{1}{\sqrt{\omega}}. \qquad (4.3.43)$$

Of course, such singularities exist only in theoretical models, not in nature.

Another modification of the cavitation pattern is presented in Fig. 4.3.8. It is relative to the EGR-pattern on another basis.

Fig. 4.3.8. The Kuznetsov pattern of the cavitating flow.

Here the inverse flow is closed on two plates, where the velocity of the flow monotonically fades from V_F to zero. The main advantage of this scheme is that we have no non-physical sink of the re-entrant flow. From the other side, not every engineer or physicist is able to realize this scheme in his mind because it is only a mathematical construction. Note that the Kuznetsov pattern does not give good results for the forces (i.e., pressures) for an arbitrary attack angle.

The Joukowski pattern (Fig. 4.3.9), from some point of view, is a reversed Kuznetsov scheme.

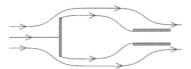

Fig. 4.3.9. The Joukowski pattern: No inverse flow.

The flow pattern on Fig. 4.3.9 does not contain re-entrant jets in any form. It may seem that, considering the fact that such jets really exist, such a scheme of cavitation is inappropriate. However, it is not that bad at all, since this pattern is clear, simple and allows one to determine the shape of the cavern (by variation of the plates' position).

The generalized Ryabushinski pattern is shown in Fig. 4.3.10; here we use the terminology from (Ivanov 1980) where any type of cavitation scheme closed on the solid finite body was called with this term.

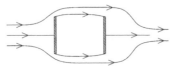

Fig. 4.3.10. The generalized Ryabushinski scheme. The closed surface has finite size.

The scheme in Fig. 4.3.10 looks clearer than other diagrams that contain infinite plates, crossed plates, singularities, etc. However, it should not be forgotten that the second body in this scheme is also fictional. Particularly, to calculate the drag force one needs to 'omit' it and only integrate the pressure on the real obstacle.

A more detailed analysis of various forms of cavitation patterns can be found in (Birkhoff and Zarantonello 1970, Gurevich 1979, Ivanov 1980).

4.3.5 The integral methods

The potentials introduced in Section (4.1.4) can be applied for hydrodynamic problems. In the bulk of a liquid, where the flow can be considered irrotational, the fluid can be described with the potential listed above. The sources of these potentials must be defined on the boundary contours: on solid surfaces or on free surfaces. Moreover, because the Laplace equation is linear, one may compose different potentials for a given problem.

For the cavitating flow, it is convenient to use the source of the form (4.1.35) for the complicated contour formed by the solid surface and the cavitation cavern: for the last kind of potential, it is natural to satisfy the condition (4.1.39) for non-zero circulation on it. On this boundary contour, the elementary sources $d\Gamma = \omega(s)ds$ are placed continually, where s is a coordinate along the contour, $\omega(s)$ is the density of the sources along that complicated contour 'profile + cavern'; each such elementary source creates a potential of a form (4.1.24) $\sim \theta/2\pi$, i.e., produces a part of elementary potential at the point (x, y)

$$d\varphi = \frac{\theta(l, x, y)\, d\Gamma}{2\pi},$$ (4.3.44)

θ is counted out from the given point on the contour. If the given point on the contour is determined by coordinates (x_0, y_0), then the correlation (4.3.44) describes the potential at the distance $r = \sqrt{(x - x_0)^2 + (y - y_0)^2}$ from this point. Correspondingly, the potential created by the whole contour is the sum of all the elementary sources, i.e., the integral of (4.3.44):

$$\varphi(x, y) = \int \frac{w(s)\theta(s, x, y)\, ds}{2\pi}.$$ (4.3.45)

Note that the intensity density $\omega(s) = d\Gamma/ds$ also has a meaning of the local tangent velocity u_s along the contour. Despite the fact that potential (4.1.35) has a singularity at the very point of the location of the source, this singularity can be eliminated with various methods, and we obtain that $\oint u_s ds = \oint w(s)ds = \Gamma.$

To describe the full flow, we must add the potential of external flow of a form (4.1.23)—we can do so due to the superposition of the potential. Directing axis x along the incoming flow, we get for the total potential:

$$\varphi(x, y) = \frac{1}{2\pi} \int w(s) \theta(s, x, y) \, ds + u_\infty x, \qquad (4.3.46)$$

To satisfy the boundary condition $u_n = 0$ on the contour, we can take a derivative of (4.3.46); the second term gives an obvious function, while the first term produces the tangent velocity $u_c \sim r(l,x,y)^{-1}$ that is directed perpendicularly to the line connecting two points on the contour: the given point and the 'source' point (see Fig. 4.3.11). Finally, for the normal projection of velocity

$$u_n = -\frac{1}{2\pi} \int w(s) \frac{\sin\gamma}{r} \, ds + u_\infty \sin\chi = 0. \qquad (4.3.47)$$

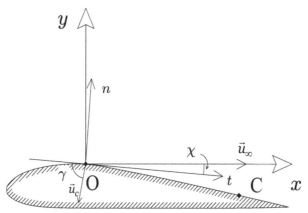

Fig. 4.3.11. Velocities at point O on hydrofoil. Axis n is normal to the profile at point O, t is the tangential axis. In accordance with (4.3.47), here we have two components: from the first 'vortex' term of (4.3.47) and $\vec{u}_\infty$ from the second one. Let us consider only a single part of the 'vortex' term created only by the potential located at point C; the corresponding part of velocity is denoted as $\vec{u}_c$, note that $\vec{u}_c \perp (OC)$; $|OC| = r$. χ is the angle between axes x and t, γ is the angle between $\vec{u}_c$ and $-t$.

Note that the sign of $\sin\gamma$ varies: if point C lies left to point O, the velocity $\vec{u}_C$ is directed up. Using (4.3.54), we may, generally, calculate the velocity on a contour $\omega(s) = u_s$ for a given geometry of a contour, i.e., for the given function $\gamma(s)$, which can be calculated for the function $y(s)$ determining the hydrofoil. However, for the cavitating flow the problem is harder, because the contour also includes the free surface which is unknown *a priori* and must be determined itself.

Then, introducing the contour with the distributed circulation on it (see Section 4.1), we may write the equation for the stream function of a cavitating flow which consists of parts corresponding to the incoming flow (4.1.34) and to the contour 'solid + cavern':

$$\psi(x, y) = \frac{1}{2\pi} \int w(s) \ln r(s, x, y) \, ds + u_\infty y. \qquad (4.3.48)$$

Since the stream function remains constant along the streamline, we may arbitrarily set this constant to zero along the one corresponding to the contour, obtaining equation for points on contour (x_c, y_c)

$$\frac{1}{2\pi} \int w(s) \ln r(s, x_c, y_c) \, ds + u_\infty y_c = 0. \qquad (4.3.49)$$

4.3.6 The partial cavitation

Generally, there may be two types of cavitation on the wing, shown in Fig. 4.3.12. The first type—further we will call it as 'full cavitation'—is similar to the cases discussed above: the cavitation cavern is closed in the fluid beyond the obstacle. The second one, the 'partial cavitation', corresponds to a case when the cavern finishes at the wing itself.

Fig. 4.3.12. Two types of cavitating flow. A: full cavitation, when the cavern is closed behind the wing; note that the form of the cavern near point *b* is nominal—see above about the different schemes of closure. B: partial cavitation, when the cavern begins and finishes at points *a* and *b* located on the wing.

Let us consider a cavitating flow in the vicinity of a separation point *a*, where the cavitation cavern begins or, in other words, where the flow detaches from a wing. In a particular case, when the wing is not smooth so that the tangent to its profile contains a breakpoint, it is supposed that such a breakpoint is the separation point (Ivanov 1980). Otherwise, for a smooth hydrofoil, the situation is more interesting, because we do not see any 'evident' candidates for such type of a point.

The condition for the separation point is named after Brillouin–Villat (Birkhoff and Zarantanello 1957). The initial thesis states that the pressure must reach its minimum at the cavitation cavern beyond the wing: indeed, otherwise cavitation would take place somewhere else. If so, the free streamlines must convex toward a cavity, following the negative pressure gradient. On the other hand, a free streamline cannot penetrate the wing, i.e., the curvature of the hydrofoil cannot be exceeded by the curvature of the cavern (that is, by the curvature of the free streamline). In the simplest formulation, the Brillouin–Villat principle states that the curvature of the cavern at the separation point *a* is equal to the curvature of the wing.

Point *b* is more interesting. Here we see the breakdown of the principle stated above: for the pattern pictured in Fig. 4.3.12B, the free streamline enters the wing at a finite angle. Moreover, for this case, point *b* is a stagnation point; for this reason, here $v_b = 0$. On the contrary, at a free surface the total velocity must be constant everywhere; denote this constant value as V_F in accordance with previous sections. Thus, the velocity has a break at point *b*: from V_F at the streamline to zero at the wing.

Of course, this nodus follows from the presumption about the stationarity of the process. In practice, such problems are solved by comparing the results obtained with a given theoretical method to the experimental data (Ivanov 1980).

Note that the non-stationarity can be taken into account in different ways. Of course, it is possible to use the Lagrange–Cauchy equation instead of the Bernoulli integral, and the additional term $\partial\varphi/\partial t$ provides all the non-stationary features; this is a very difficult way because the form of the cavitation cavern is a self-consistent function. However, it is possible to use another way. We may consider the set of the quasi-stationary cases which differ (slightly) by a different form of the cavitation cavern in its rear part. Then, we average the results obtained in that quasi-stationary states, i.e., get, for example, the equation (4.3.51) for function $\bar{\psi}(x, y)$, which contains the mean value $\overline{\omega(s) \ln r}$ under the integral. The last term, of course, does not equal to the production $\overline{\omega(s)}$ and $\overline{\ln r}$, where each factor corresponds to stationary functions: the distribution of the elementary circulation on the contour and, actually, the form of the contour itself (the relative position *r* depends on the cavern profile). Thus, actually, the equation (4.3.59), which determines function $\omega(s)$ for stationary case (in our new terminology), would contain an additional term that describes the correlation between ω and $\ln r$.

The simplest way to solve the problem for a partial cavitating flow is to use the fact that the cavern may be sufficiently 'thin' to use the linearization of a problem. For the geometry presented in Fig. 4.3.12B, all sizes along the axis y are much smaller than along the axis x. This means that along the cavern

$$\left| y' \right| = \left| dy_c / dx_c \right| = \varepsilon \ll 1 \tag{4.3.50}$$

and that the x-projection of the velocity on the free streamline is almost V_F (to be exact, V_F is the constant modulus of velocity, which is directed along the cavern); to be exact, the horizontal projection of velocity is $V_F + \tilde{u}_x$, the vertical projection is $\tilde{u}_y$, and $\left| \tilde{u}_x \right|$, $\left| \tilde{u}_y \right| \ll V_F$. Let us find the form of such (i.e., almost flat) cavern that satisfies the condition $\omega(s) = V_F = const$.
We will use (4.3.50), where $r^2 \approx (x_c - s)^2$, and obtain

$$y_c = -\frac{V_F}{2\pi u_\infty} \int_a^b \ln \left| s - x_c \right| ds. \tag{4.3.51}$$

Even in the form of an integral, one may see that the condition (4.3.51) cannot be satisfied at the edges, i.e., at boundaries where $x \sim a$ or $x \sim b$. However, in the middle of that range the answer may be reasonable. We get:

$$y_c = \frac{V_F}{2\pi u_\infty} \left[b - a - \ln \left(b - x_c \right)^{b-x_c} \left(x_c - a \right)^{x_c - a} \right]. \tag{4.3.52}$$

This relation, describing the shape of the cavitation cavern, is correct in the middle of the range $[a, b]$ in the form of (4.3.50): otherwise, the linearization cannot be applied.

4.3.7 The modern trends

In this entire section, we restricted ourselves with only qualitative and analytical approach of the developed (or, as it is also called today, the 'super') cavitation. This approach arose at the dawn of hydrodynamics and reached its peak in the middle of the XX century (see the works of Wu (Wu 1962), Tulin(Tulin 1964), Terentiev (Terentiev 1977), Ivanov (Ivanov 1980), etc.). That was the era when theoretical physics walked next to mathematics not only in the field of astrophysics, but everywhere. Many results were obtained not by the physicists but also by mathematicians.

We paid a significant attention to this approach because, as we think, this allows one to understand every step of the solution. In contrast to the black-box numerical simulations, analytical (or semi-analytical) methods offer clear receipts: we know what the result is made of. We may discuss and change some assumptions, replace one element of the scheme with another one, etc. That was an open game, but today, this game is over.

Now computers can reach incredible power. Note that for semi-analytical solutions (like solving the integral equations presented above), computers played some role too, but now their computational speed allows to solve the Navier-Stokes equation immediately.[2] The progress in computer hardware was one of two tremendous factors that turned theoretical investigations into pure numerical simulations based directly on the Navier-Stokes equation, not on some consequences of it (like integral correlations for the velocity potential). The second factor that promotes the change in trend is the extreme complexity and abstraction of analytical representation. Enter potential→enter a complex potential→proceed to other complex variables→find the map: this is a long and shady way. It is not surprising that once the computers allowed to simulate the problem directly, it began to be solved directly right away.

---

[2]   One would say 'directly', and this term can be applied even in its strict form too: modern computers allow 'direct numerical simulations', i.e., calculation of the turbulent flow without special models of it.

Among other advantages, numerical simulations allow to solve complicated problems: studying a turbulence (Park and Rhee 2012) for a flow around hydrofoil (Chen et al. 2016), considering 3D cavitation (Roohi et al. 2016), including investigation of the cavitating flow past a body of submarine shape (Shang 2013), or designing the control system for limiting cavitation for hydrofoils (Capurso et al. 2018), etc.

On the other hand, the last problem can also be considered with the analytical apparatus. Thus, in the next chapter, we are going to return to our preferable complicated-abstract approach.

4.4 Cavitation diagrams: Theory of reconstruction

4.4.1 General principles

In Chapter 2, we discussed the cavitation diagrams. Here, in this section, we return to this problem from a 'more scientific' point of view; we mean, we will consider this problem armed with theoretical physics apparatus, which implies a lot of mathematics. Like other sections in this chapter, this one is quite difficult.

Let us consider the profile of a solid body surrounded by the incoming flow at the attack angle α. As it was discussed in Chapter 2, the cavitation diagram shows the minimal pressure coefficient $F(\alpha)$ (defined with a reversed sign) in the dependence of the attack angle α:

$$F(\alpha) = \frac{p_\infty - p_{\min}(\alpha)}{\rho u_\infty^2 / 2};$$

(4.4.1)

where p_∞ is the pressure at infinity in the incoming flux, u_∞ is the velocity of the free stream flow, ρ is density, and $p_{\min}$ is the minimal pressure on the surface.

From the standard theory, cavitation occurs if the pressure at the given point is lower than the critical value, which is usually taken as the saturation pressure p_s, that is, the cavitation number

$$Q = \frac{p_\infty - p_s}{\rho u_\infty^2 / 2}$$

(4.4.2)

must be exceeded.

Formulating in a slightly different manner, one may conclude that the absence of cavitation corresponds to in equation

$$F(\alpha) < Q.$$

(4.4.3)

Note that the design of airfoils for a common purpose usually includes the design of the pressure diagrams, i.e., many diagrams have already been designed for airfoils—for wings of airplanes, for example. But, from the methodological point of view, there is (almost) no difference between the hydrodynamic description of a plane in the air and a submarine in the water (the only difference may lie in compressibility of the media). Thus, some pressure diagrams can be borrowed from a related area—aerodynamics, i.e., diagrams for airfoils may be comparable for hydrofoils. For example, the book by (Eppler 1990) is very useful, despite the fact that the authors doubted in this, since 'there exist computer programs which allow the design of airfoils which are very good, if not optimally adapted to the requirements resulting from certain special applications' (written in 1990). Of course, again from this book, we may state that (1) modern computer codes allow one to simulate many hydrodynamic problems on a completely different level than in the late 20th century, (2) old-fashioned methods are indispensable for understanding the qualitative picture of phenomena. Thus, a book like (Eppler 1990) and methods of simplifying calculations have a great value even today. Taking into account the Bernoulli integral for an incompressible fluid

$$p + \frac{\rho u^2}{2} = p_\infty + \frac{\rho u_\infty^2}{2},$$

(4.4.4)

we reformulate (4.4.3) for the function

$$f(\alpha) = \frac{u}{u_\infty} = \sqrt{1 + F(\alpha)},\qquad(4.4.5)$$

finding $f(\alpha)$ so that

$$f(\alpha) > W,\qquad(4.4.6)$$

where $W = \sqrt{1 + \dfrac{p_\infty - p_s}{\rho u_\infty^2 / 2}}$ is the critical value corresponding to the beginning of cavitation.

For instance, all the results in (Maklakov 1997) were derived for function $f(\alpha)$; below, we follow this way.

4.4.2 Hydrofoil design

In this section, we discuss the following problem: we want to find a hydrofoil which would satisfy the given velocity distribution. This is the so-called inverse problem of hydrodynamics: usually, we have a direct one—to calculate the velocity field in the vicinity of the object.

Here, as everywhere in this chapter, we consider analytical representations only. Today, the solution of this problem can be obtained with numerical simulations, but the 'old-school technique' deserves some attention.

Let us consider a flow of a perfect incompressible fluid around a body (see Fig. 4.4.1). We are interested in the pressure dependence on x—the dimensionless coordinate measured along the chord of that hydrofoil, or, using representation (4.4.5), the dependence of the relative velocity $u/u_\infty = \tilde{u}(x)$. The value $x = 0$ corresponds to the stagnation point, value $x = 1$ corresponds to the trailing edge. The attack angle $\alpha = 0$ corresponds to the zero 'lift force'; quotes remind that we are obtaining results for a ship, not for an airplane—vector g (acceleration of gravity) can be directed perpendicular to hydrofoil.

We will use the same method as everywhere in this chapter. Let us consider a conformal mapping of the complex field $z(x, y) \leftarrow t(\xi, \eta)$ (Lighthill 1945). The value $z = 1$ (i.e., $x = 1$, $y = 0$) corresponds to $t = 1$ (i.e., $\xi = 1$, $\eta = 0$). The given point on a circle is defined by the parameter θ (see Fig. 4.4.1), i.e., $t = e^{i\theta}$.

Corresponding map

$$z = z(t),\qquad(4.4.7)$$

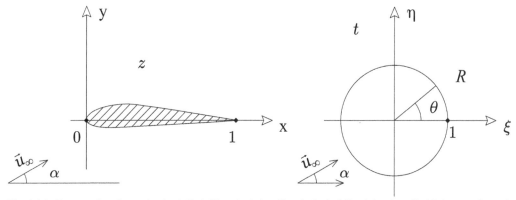

Fig. 4.4.1. Two complex planes $z(x, y)$ and $t(\xi, \eta)$. Plane (x, y) describes the hydrofoil as it is; plane (ξ, η) is its map. Our point is to find the map $z(t)$ based on the desired conditions for a certain (or any) attack angle α.

which establishes the correlation between two complex variables, also obeys the rule

$$\left|\frac{dz}{dt}\right| \to 1 \text{ as } t \to \infty, \tag{4.4.8}$$

or, in simpler words, $z \to \infty$ as $t \to \infty$.

The dimensionless complex potential ω for a flow past a hydrofoil, which satisfies the Joukowsky condition

$$\frac{d\omega}{dt}\bigg|_{t=1} = 0, \tag{4.4.9}$$

is

$$\omega = \omega_0 \left(te^{-i\alpha} + \frac{e^{i\alpha}}{t} + 2i\sin\alpha \ln t \right). \tag{4.4.10}$$

Here α is the attack angle, ω_0 is a constant.

Thus, for the complex velocity

$$\upsilon = \frac{d\omega}{dz} = \frac{1}{z'(t)}\frac{d\omega}{dt}, \tag{4.4.11}$$

where $z'(t)$ is the derivative of (4.4.7), we get, remembering that $2\sin\alpha = e^{i\alpha} - e^{-i\alpha}$:

$$\upsilon = \frac{\omega_0 (t-1)}{z'}\left(\frac{e^{-i\alpha}}{t} + \frac{e^{i\alpha}}{t^2} \right). \tag{4.4.12}$$

The velocity vector is normalized with the parameter ω_0 to provide unity at infinity.

The function $z(t)$ can be represented as $z(t) = e^{-i\alpha} z_0(t)$, where the part $z_0(t)$ corresponds to the map for $\alpha = 0$, i.e., corresponds to zero lift force. Therefore, since $t = e^{i\theta}$, the equation (4.4.12) can be represented as

$$\upsilon = \frac{4\omega_0 i e^{-i(\theta+\alpha)}}{z_0'\left(e^{i\theta}\right)}\sin\left(\frac{\theta}{2}\right)\cos\left(\frac{\theta}{2}-\alpha\right), \tag{4.4.13}$$

and the modulus

$$|\upsilon|^2 = \underbrace{\frac{16\omega_0^2}{|z_0'|^2}\sin^2\left(\frac{\theta}{2}\right)\cos^2\left(\frac{\theta}{2}-\alpha\right)}_{s^2(\theta)}. \tag{4.4.14}$$

Thus, as we see, the absolute value of velocity depends on the attack angle only as a cosine. This means that two values $|\upsilon(\theta, \alpha)| = V(\theta, \alpha)$ for different attack angles α_1 and α_2 are connected with each other as

$$\frac{V(\theta,\alpha_1)}{|\cos(\theta/2-\alpha_1)|} = \frac{V(\theta,\alpha_2)}{|\cos(\theta/2-\alpha_2)|}. \tag{4.4.15}$$

As it follows from (4.4.15), if we know a function $V(\theta, \alpha)$ for one value of the attack angle, then we may calculate it for another.

From (4.4.14), one also may obtain the differential equation for $V(\theta, \alpha)$ (Maklakov 1997). Differentiating on α, we have

$$\frac{\partial V(\theta, \alpha)}{\partial \alpha} = V(\theta, \alpha) \tan\left(\frac{\theta}{2} - \alpha\right). \tag{4.4.16}$$

Equations (4.4.15) and (4.4.16) are related—they establish the correlation between velocity and the attack angle.

The hydrofoil and the velocity function are directly connected through the function $s(\theta)$:

$$s(\theta) = \frac{4\omega_0 \left|\sin(\theta/2)\right|}{\left|z_0'(\theta)\right|}, \tag{4.4.17}$$

that is, the inverse problem—to find a map $z_0(e^{i\theta})$ for the given $V(\theta, \alpha)$, i.e., for the given distribution of velocity on coordinate θ at a certain attack angle—can be easily solved. With the known $V(\theta, \alpha)$ we have

$$s(\theta) = \frac{V(\theta, \alpha)}{\left|\cos(\theta/2 - \alpha)\right|}, \tag{4.4.18}$$

and, consequently, we know the real part

$$\varsigma(\theta) = \mathrm{Re}\big(\zeta(t)\big) = \ln\left|z_0'(t)\right| = \ln s(\theta) - \ln\big(4\omega_0 \sin(\theta/2)\big) \tag{4.4.19}$$

of the complex function $\zeta(t) = \ln z_0'(t)$ of the complex argument $t = e^{i\theta}$. Reconstructing the function $\zeta(t)$ with the Schwarz integral

$$\zeta(t) = \frac{1}{2\pi} \int_0^{2\pi} \varsigma(\theta) \frac{e^{i\theta} + t}{e^{i\theta} - t} d\theta + iC, \tag{4.4.20}$$

(where C is a real constant), we can finally get $z_0(t)$ with

$$z_0(t) = \int_1^t e^{\zeta(\upsilon)} d\upsilon. \tag{4.4.21}$$

Note that the function $V(\theta, \alpha)$ must satisfy a special condition: to the boundary condition

$$\int_{-\pi}^{\pi} \ln V(\theta, \alpha) d\theta = 0 \tag{4.4.22}$$

and to the closure condition

$$\int_{-\pi}^{\pi} e^{i\theta} \ln V(\theta, \alpha) d\theta - 2\pi i e^{i\alpha} \sin \alpha = 0. \tag{4.4.23}$$

Thus, the given problem is solved. For the given attack angle α, we want to obtain the function $V(\theta, \alpha)$ which satisfies the required conditions; this is provided by the function (4.4.21) with all intermediates.

However, if we want to consider cavitation diagrams exactly (or some analogs, see below), we need additional calculations. Suppose that we choose a function $V(\theta, \alpha)$, which provides the absence of cavitation at the very given attack angle α and, finally, find the hydrofoil $z_0(t)$, i.e., the function that describes the shape of the profile. But, at least, we have to be sure that cavitation will be absent at any other attack angle.

We must develop the mathematical apparatus further.

4.4.3 The cavitation functions

As we stated above, we are interested in functions $F(\alpha)$ or $f(\alpha)$ (in short, one may call them cavitation functions), which correspond to the extremal value of pressure at the given attack angle. It must be the minimum value of pressure, defined by the function $F(\alpha)$, or the maximum velocity, determined by the function $f(\alpha)$. Indeed, both cavitation functions provide the required result, but it is more convenient to deal with function $f(\alpha)$.

The connection between $f(\alpha)$ and $V(\theta, \alpha)$ is almost obvious: $f(\alpha)$ corresponds to the maximum value of $V(\theta, \alpha)$ that is reached at a certain value $\theta_m(\alpha)$, i.e., for any given attack angle α, there exists a coordinate θ_m where the function $V(\theta, \alpha)$ has a maximum value denoted as $f(\alpha)$. In mathematical language, function $f(\alpha)$ is an envelope for function $V(\theta, \alpha)$.

Let us take a closer look at correlation (4.4.15). Choosing $\theta = \theta_m(\alpha)$ there, we may rewrite it by replacing $\alpha_1 \rightarrow \beta$, $\alpha_2 \rightarrow \alpha$

$$\frac{V\left(\theta_m, \beta\right)}{\left|\cos\left(\theta_m / 2 - \beta\right)\right|} = \frac{V\left(\theta_m, \alpha\right)}{\left|\cos\left(\theta_m / 2 - \alpha\right)\right|}. \tag{4.4.24}$$

But, as we stated above, $V(\theta_m, \alpha) = f(\alpha)$; therefore, we get

$$V\left(\theta_m, \beta\right) = f\left(\alpha\right)\left|\frac{\cos\left(\theta_m / 2 - \beta\right)}{\cos\left(\theta_m / 2 - \alpha\right)}\right|. \tag{4.4.25}$$

Function (4.4.25), still determined at the single coordinate θ_m, corresponds to the maximum value at the given α. However, we may define (4.4.25) for any θ by redefining α in this relation: now $\alpha_m(\theta)$ corresponds to θ, where function $V(\theta, \alpha)$ reaches its maximum value $V(\theta, \alpha_m(\theta)) = f(\alpha_m(\theta))$. If so, we have

$$V\left(\theta, \beta\right) = f\left(\alpha_m\right)\left|\frac{\cos\left(\theta / 2 - \beta\right)}{\cos\left(\theta / 2 - \alpha_m\right)}\right|. \tag{4.4.26}$$

By analogy, the differential representation (4.4.16) can be rewritten with $f(\alpha)$ and $f'(\alpha) = \partial V(\theta_m, \alpha) / \partial \alpha$:

$$\frac{f'\left(\alpha\right)}{f\left(\alpha\right)} = \tan\left(\frac{\theta_m}{2} - \alpha\right), \tag{4.4.27}$$

and from (4.4.27) we get the connection between $\theta_m(\alpha)$ and α

$$\theta_m\left(\alpha\right) = 2\left(\alpha + \arctan\frac{f'\left(\alpha\right)}{f\left(\alpha\right)}\right), \tag{4.4.28}$$

Also, (4.4.27) has formal solutions:

$$f\left(\alpha\right) = f\left(\alpha_0\right)\exp\left[\int_{\alpha_0}^{\alpha} \tan\left(\frac{\theta_m\left(\alpha\right)}{2} - \alpha\right) d\alpha\right]. \tag{4.4.29}$$

With (4.4.5), the expressions obtained in this section establish correlations directly for cavitation functions. For instance, with these equations one may determine a hydrofoil for a desired function $f(\alpha)$.

On the other side, we must mention some problems with such solution approach (Maklakov 1997): to apply this scheme, the function (4.4.28) must give in result (a) continuous and (b) unambiguous set of coordinates θ_m. In simple words, it is possible that for a certain function $f(\alpha)$ at some α, the expression (4.6.28) gives more than one result for θ_m while, possibly, some values of θ_m do not correspond to any α. In the last case, $\theta_m(\alpha)$ will be a piecewise function and solution will be impossible.

The answer to these doubts lies in the area of pure mathematics; for details, see (Maklakov 1997), where conditions for the existence of a solution are given.

4.5 Waves and instabilities at a liquid–gas interface

4.5.1 Stationarity

The study of instabilities plays a special role in physics. Traditionally, physicists prefer to investigate stationary constructions even if they understand that the real phenomenon has clear non-stationary properties. An excellent example is given in Section 4.3: the closure of a cavitation cavern is a principally non-stationary process, but the desire to use a simplified mathematical apparatus (exactly—the apparatus that may provide some clear, understandable results) leads to static patterns, which must be sometimes supported by very unusual theoretical models: see various schemes for the cavern closure.

The same scenes take place in many areas of physics, except for, probably, classical mechanics, but the finding of motion integrals plays a big role there too. Everywhere we try to develop a static theory: obtain stationary solutions of differential equations or even construct special approximate models (for instance, the theory of the big sister of cavitation—boiling—consists of many very rough models; what else, if the subject is so tough?) to get a qualitative and, if lucky, quantitative picture of the phenomenon.

Except for some special (really static) cases, there may be two variants of a developed static theory: it can be absolutely inappropriate or it can have a restricted area of application.

An inappropriate static theory develops when volatility becomes an inherent property of the system. The clearer examples may be found in economics or population dynamics theories: in both cases, changeability is the key feature of the investigated object. It is impossible to explain the population sizes of wolves and hares in the forest as well as the share prices on the market with a static theory. There cannot be a correct static theory for a clock; to be exact, the static theory applied for the working clock gives two correct values per day.

Of course, for the old, well-known static physical models we have another way out. These theories are adequate at special ranges of (a) time scale, (b) ruling parameters. Under the 'time scale', we not only mean a time scale of the considered physical process, but also a time scale of our conceptions. With time, after development, a static theory may get out of order and become non-static. The best instance is the Universe theory: in accordance with absolutely solid scientific views (different at various times), our Universe has been (1) static, (2) expending with constant speed, (3) possibly, collapsing with time, (4) possibly, oscillating, (5) currently expanding with acceleration (but maybe not, in accordance with modern doubts concerning interpretations of measurements). Note that all these conceptions altered during the course of a single century, and we suppose that no one knows which properties of the Universe will be discovered in future.

Returning to the usual interpretation of the term 'time scale', we can consider a lot of examples. We may even imagine a static theory for weather, but in most of the world this theory would only be valid for a single season. We may develop a static theory for ocean currents, but, as we know, these currents will change or disappear with time. Not to stray too far from the topic: one may invent a static theory for the circulation of water flowing out from a bathtub, but this theory will start working after some time from the beginning (when that famous vortex[3] would form) and until the destruction of that vortex at the very end.

The restriction from the system parameters is much more complicated. This subject deserves a special topic.

4.5.2 Stability

As we tried to explain in the previous section, stationarity exists only in our scientific imagination. The static pattern arises from a non-static mathematical equation: everything is changing, and variables that denote physical quantities change too.

For instance, we have to deal with the simplest dynamic system for a single quantity x:

$$\frac{dx}{dt} = \mu x - x^3. \tag{4.5.1}$$

As we tend to act everywhere, we (initially) are not interested in the dynamics of that system, we want to know the 'final answer'—the stationary solution. If so, we see that the stationary state of that system, i.e., when $dx/dt = 0$, appears for x_0, for which

$$\mu x_0 - x_0^3 = 0. \tag{4.5.2}$$

From this equation, we obtain three roots for x_0:

$$x_0 = 0 \text{ or } x_0 = \pm\sqrt{\mu}. \tag{4.5.3}$$

Someone might not be ready for so many roots. Does this fact mean that all three solutions are equally feasible? In reality, no.

Let us try to find a stability of the point, i.e., how deviation from the stationary points (4.5.3) behaves. Consider variable $x(t) = x_0 + u(t)$, where u depending on time is a small deviation from the point x_0. If $|u(t)|$ increases with time, then any, arbitrarily as small as we wish, deviation would discard x from x_0 forever. The nature of deviation can be anything: for calculations, it can be rounding errors; in real life, if such a system describes a certain quantity (for instance, the population size with hard intraspecific competition), there may exist some random variations of that parameter that lie beyond the main model (for instance, occasional excessive fertility).

Thus, we have to return to the non-stationary equation, at least to its simplified form. Inserting $x = x_0 + u$ in (4.5.3), we have

$$\frac{du}{dt} = \mu x_0 + \mu u - \left(x_0 + u\right)^3. \tag{4.5.4}$$

On the presumption that u is sufficiently small, we may expand the cubic term in (4.5.4) and restrict this expression only to linear terms:

$$\left(x_0 + u\right)^3 \approx x_0^3 + 3x_0^2 u, \tag{4.5.5}$$

and with (4.5.2) we get

$$\frac{du}{dt} = \underbrace{\left(\mu - 3x_0^2\right)}_{a} u \tag{4.5.6}$$

---

[3] For educational purposes only and because myths die hard. The vortex formed in a bathtub (or in other objects: we remember how Bart Simpson found out that he was in Australia) has nothing to do with the Coriolis forces. These forces are very weak and very 'slow'; they show an effect only for extended periods of time. The direction of the vortex formed in a bathtub depends on initial conditions: initial motion of fluid in the bath itself, etc.; it depends even on the movement when the plug is pulled out.

The solution of (4.5.6) is the exponent function $u = u_0 \exp(at)$, and we realize, according to the previous consideration, that u increases with time if $a > 0$, u decreases if $a < 0$

Thus, for the root $x_0 = 0$, we have $a = \mu$, and this point is stable for $\mu < 0$ and unstable for $\mu > 0$.

For the roots $x = \pm \sqrt{\mu}$, we obtain $a = -2\mu$, and these roots are stable for $\mu > 0$; note that for $\mu < 0$, these roots are not unstable – they do not exist.

Finally, we see that for different values of the ruling parameter μ, there are different solutions for (4.5.1) realization; knowing this we understand the existing, i.e., stable, values. Unstable values x_0 cannot be observed: both in physical nature and in mathematical calculations. Practically, we will observe either $x_0 = 0$ for $\mu < 0$ or $x_0 = \pm \sqrt{\mu}$ for $\mu > 0$ (in this case, one of the two points is realized depending on initial conditions).

For system (4.5.1), we see that the parameter μ rules the dynamics of the system: the stable point depends on μ, which can be seen in Fig. 4.5.1, where stable points are shown. On this diagram, the point $\mu = 0$ is special: at this point, the type of stable point changes—before $\mu = 0$ there is a single stable point $x_0 = 0$, after that point there are two stable points $x_0 = \pm \sqrt{\mu}$. Such a change—the change in the type of stable solutions—is called a bifurcation, so μ is also called a bifurcation parameter, and point $\mu = 0$ is a bifurcation point.

We understand two important things from this canonic example (that was the so-called pitchfork bifurcation):

✓ a stationary solution does not necessarily mean a stable solution; from the practical point of view, one may say that unstable solutions do not exist, implying that such solutions cannot be observed in real cases;

✓ the stability of a stationary solution depends on the parameters of the system: at some values of parameters, there may be one type of stable solution, at other values—another one; here we must note that the theory is mainly developed for a single ruling (bifurcation) parameter.

Now, it is time to widen our conceptions. Except for stationary points, there may also exist other types of system dynamics at infinite time (corresponding objects in the phase space—a space, constructed with all the variables of the dynamic system) called attractors: it can be a periodic motion (called cycle with a single frequency and torus with two or more frequencies) or an aperiodic one—a very special type of attractor corresponding to a chaotic system. In simple words, a chaotic system represents a system with a property of strong dependence on initial condition.

Let us consider a dynamic system at two close initial conditions: $x(0)$ and $x(0) + \delta x$. There may be three variants for the trajectories corresponding to these initial conditions:

✓ trajectories remain at the same distance as at the origin;

✓ trajectories merge with time;

✓ trajectories scatter with time.

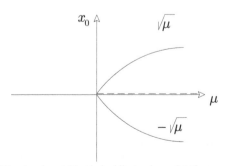

Fig. 4.5.1. Stable (solid lines) and unstable (dashed line) values of stationary points of a system (4.5.1).

The last case corresponds exactly to a chaotic system, and the scattering of initially close trajectories depicts a strong dependence on initial conditions. Indeed, this is a sad story for any mathematical description of such a system. Due to many reasons, there would always be many disturbances (see above), and each of these disturbances lead the system away along a special path (in comparison with the absence of any disturbance or with another disturbance). In a poetic representation, that would mean that a butterfly may cause a hurricane on another continent by the flap of its wings:[4] weather, i.e., an open hydrodynamic system on a large scale, is an example of a chaotic system. Seriously, such theoretical descriptions are assumed to be incorrect (the so-called Hadamard's lemma) until the scientists discover that many systems obey the chaotic rules, and, therefore, we would have to adapt to live with them together (possibly, until another theoretical apparatus, for absolutely other quantities,would be discovered).

In general, stability is an exhaustible theme. All the more surprising that any explanation of it is not complete without an illustration of a ball on a relief (Fig. 4.5.2).

As we clearly realize, the ball is placed at a stable position for the case shown in Fig. 4.5.2A: after displacement, it will return to the initial position (at the bottom of the hollow), and stop there eventually because of friction, by the way. The case depicted in Fig. 4.5.2B corresponds to an unstable position: after displacement, we lose this ball.

It is not clear how to interpret what the patterns shown in Fig. 4.5.2 have in common with a real situation. What is a ball, what is a hollow or a hill, and what is gravity (which indeed rules this whole situation), in translation into a certain physical problem?

In a physical system, the equilibrium (i.e., the stationary point) appears when different forces acting on this system come to a balance. As we know, equilibrium can be stable or unstable: that becomes evident when, due to accidental reasons, one of the forces would increase or decrease its value. What can we expect from other forces? If other forces change their intensity so that they compensate the new value of that initial 'rebel' force, the system would return to the initial equilibrium state—that is an example of a stable equilibrium. If the variation in the single force cannot be balanced out with other forces, the system would leave this unstable equilibrium state.

A common principle that establishes the behavior of a system in a stable equilibrium state is the Le Chatelier—Brown principle. It is a non-trivial task to formulate this principle in such a manner that would provide us with something new in addition to the definition of the state of stable equilibrium (which is a state to which the system returns after perturbations). Possibly, the most successful formulation is: 'For a system in the state of stable equilibrium, an external influence causes the forces in that system to aspire to return this system to the state of equilibrium'. At least, this wording states that some special forces must appear in the system after the external influence.

A   B

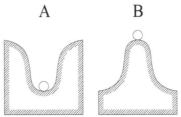

Fig. 4.5.2. Illustration of a stable state (A) and of an unstable one (B).

---

[4]   In reality no, of course. Not a single hurricane has ever been caused by the wings of a butterfly. First of all, all possible states of a system must belong to the same aperiodic attractor. Then, for such type of attractor, every disturbance leads to a new turn in the fate of the system, and every consequence of every turn still belongs to the same attractor. In short, nothing new can exist in a chaotic system: all that has to happen in this system, will happen.

Definitely not every system must be stable and, therefore, the Le Chatelier principle cannot be applied to every system. However, take into account the fact that unstable states do not appear as a stationary (or equilibrium) state: the biggest thing to expect from such a state is that it can be an intermediate state in some dynamical process. Thus, if we obscure the stationary state, we can be sure that this state is stable for some external disturbances at least;[5] therefore, this is a system to which the Le Chatelier principle can be applied.

4.5.3 Turbulence

The word 'hydrodynamics' is present in the title of this chapter, the word 'instability' is present in the title of this section. Together, these two words combine into the term 'turbulence'. We cannot avoid this matter even though our goal in this section is located in a slightly different field.

Hydrodynamic instabilities are undoubtedly the most famous instabilities in physics, and turbulence is one of the most complicated phenomena, despite its domestic nature. Everyone saw the water flow, and any person in the street would likely say that he does not understand how to describe this phenomenon in a scientific language. Alas, a scientist has to echo this statement, except for, probably, our colleagues engaged in numerical simulations and several mathematicians. Turbulence—a disordered fluid flow, where every liquid particle moves on a complicated trajectory (we consciously avoid the term 'chaoticity', see below)—is very hard to describe in mathematical language, and, therefore, on the level that allows quantitative prediction of physical quantities. Turbulence occurs in pipes when the Reynolds number

$$\mathrm{Re} = \frac{ud}{v} \qquad (4.5.7)$$

(u is the pipe diameter, v is the kinematic viscosity, $u = 4G/\rho\pi d^2$ is the velocity of the fluid at the discharge G) exceeds the critical number $\mathrm{Re}_{cr} \sim 2300$; the certain value is not unknown but does not exist: the transition depends on many additional factors. For instance, for a water tap of diameter $d = 2$ cm, with dynamic viscosity of cold water $\mu = 10^{-3}$ Pa·s, we obtain the value of the critical discharge $G = 0.036$ kg/s. With such a discharge, a glass of 200 ml volume will be filled in 5.5 seconds: if the glass is being filled faster than in 5.5 s, the flow in the water tap is turbulent. Turbulence is complicated in two aspects:

✓ we have no clear representation of what is turbulence;

✓ we have no theory that predicts the transition from laminar flow to a turbulent one.

These matters are connected, but not as closely as one may assume.

As for the representation of turbulence, today we may observe a very interesting picture. In every area of science and of engineering connected to this scientific area, different specialists use different methods and deal with different representations of this phenomena, etc. This is quite common: for instance, the gas discharge concerns a wide range of specialists from electrical engineers to pure plasma physicists. But the corresponding range of specialists in turbulence is much wider: from technologists to mathematicians. Initially, the laminar–turbulent transition was formulated through the jump of hydraulic resistance, but with time the definition (the exact one) of that process shifted to formulations based on various conceptions for the nature of turbulence. For instance, in hydrodynamics of the middle of the XX century, the formulations that used words like 'casual' and 'chaotic' as synonyms were very popular. Then—we omit the complete history—turbulence began to be interpreted as a chaotic motion in a strict (not verbal, not domestic) sense of it, i.e., as a system which has a strong dependence on initial conditions and a system which can be described with the strange (aperiodic) attractor. However, this 'chaotic' interpretation did not penetrate from

---

[5]   System can be stable for an infinitesimal disturbance but be unstable for perturbations with a finite magnitude. In almost all areas of physics, such states are called 'metastable'.

the community of mathematicians and pure[6] physicists to the more 'applied' specialists. Note, by the way, that there is not enough experimental data to make a solid conclusion on this matter to date. Now we have, among others, two large categories of specialists in turbulence: one of them is absolutely sure that turbulence is chaos and uses these terms synonymously, while the other category does not even understand what the term 'chaos' exactly means and, if they find out the sense of that term, how it concerns turbulence at all. In some areas, it happens that specialists use slightly different language, but it is rare when the language is different this much, up to the opposite sense. Indeed, the discussion about turbulence with different groups of scientists is very exciting.

Therefore, if one admits that turbulence is an example of chaotic motion, then we have to construct a theory of transition to it from laminar flow. The mathematical mechanism developed by Ruelle, Takens and Newhouse explains the fundamental principles of such a transition, but it cannot be directly applied to hydrodynamic equations, i.e., there are no theories that may derive from the Navier–Stokes equations the critical Reynolds number which corresponds to the transfer to turbulence after several Hopf bifurcations (the Hopf bifurcation is another type of bifurcation when the number of independent frequencies for the attractor increases by one: after the Hopf bifurcation, from the stable point we get the stable cycle, from the cycle—torus, etc.).

Fortunately, our direct goals do not involve the problem of turbulence; further, we restrict our consideration only to some examples of hydrodynamic instability for particular cases.

4.5.4 Long gravitational waves in a shallow fluid

Let us begin with the type of waves on the surface of fluid, when the wavelength λ is much larger than the depth of the liquid h.

At first, we derive the continuity equation in a novel manner. Consider the fluid in a channel (see Fig. 4.5.3); the width of channel is b, the depth of liquid is h, so the cross-section of the liquid layer is $S = bh$.

The continuity equation reflects the fact that the fluid mass cannot disappear or appear from nowhere. The 'common' continuous equation (Landau and Lifshitz 1987) establishes the correlation for the density and velocity of the fluid based on this circumstance. For an incompressible fluid (where $\rho = const$) in a channel of variable geometry, we may obtain a relation for that geometry.

In the given cross-section from x to $x + dx$, the incoming mass flux is $\rho u S$, where u is the single (x-) component of velocity. Since the outcoming flux is $\rho u S + \partial(\rho u S)$, the mass change in this cross-section during time interval ∂t is $\partial m = - \rho \partial(Su) \, \partial t$; if $\partial(Su) > 0$, then fluid leaves the volume and

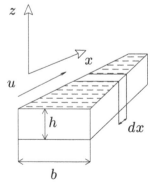

Fig. 4.5.3. The channel with variable width.

$\partial m < 0$. On the other hand, this means that the mass here becomes $\partial m = \rho \partial V = \rho \partial S \partial x$. Equating these relations for ∂m, we get the continuity equation

$$\frac{\partial S}{\partial t} + \frac{\partial (Su)}{\partial x} = 0. \tag{4.5.8}$$

From some points of view, this equation represents the conservation law for the cross-section area of the flow—sic. This equation can be applied for many problems, but further we apply it for the channel of a constant width $b = const$, so the only reason for the variation of the cross-section area is the depth of liquid h. In its turn, variation in the fluid level is caused by the wave on the surface: $h = h_0 + \xi(t, x)$, where h_0 is the depth in the absence of the wave, $\xi(t, x)$ is the variation of fluid level caused by the wave. Finally, we have

$$\frac{\partial \xi}{\partial t} + h_0 \frac{\partial u}{\partial x} + u \frac{\partial \xi}{\partial x} = 0. \tag{4.5.9}$$

Since we consider long waves, the last term in (4.5.9) can be neglected: the term $\partial \xi / \partial x \sim a/\lambda$ where $a << h_0$ is the wave amplitude, and the third term in the right-hand side of (4.5.9) is much less than the second one.

Then, all that is left from the Navier–Stokes equation for x-projection of velocity is

$$\frac{\partial u}{\partial t} = -\frac{1}{\rho} \frac{\partial p}{\partial x}, \tag{4.5.10}$$

while for z-projection, we have the Pascal equation:

$$-\frac{1}{\rho} \frac{\partial p}{\partial z} - g = 0, \, p = p_0 + \rho g (\xi - z), \tag{4.5.11}$$

where p_0 corresponds to pressure at the surface, i.e., at $z = \xi$.
Thus, the pressure gradient along axis x is

$$\frac{\partial p}{\partial x} = \rho g \frac{\partial \xi}{\partial x}, \tag{4.5.12}$$

and (4.5.10) turns into

$$\frac{\partial u}{\partial t} = -g \frac{\partial \xi}{\partial x}. \tag{4.5.13}$$

Taking the derivative on time from (4.5.9), and using (4.5.13), we obtain the wave equation

$$\frac{\partial^2 \xi}{\partial t^2} - gh \frac{\partial^2 \xi}{\partial x^2} = 0, \tag{4.5.14}$$

from where we see that the wave speed is

$$c = \sqrt{gh_0}. \tag{4.5.15}$$

This result means that the wave speed depends on the average fluid level in the channel.

However, some questions to this solution remain. First, it is not clear where exactly the assumption about shallow fluid has been used. Second, we may note that our approach looks somewhat strange, indeed. In a wave moving along the axis x, the potential φ must behave accordingly, i.e., φ must be a function on x. This means that, in accordance with the equation $\Delta \varphi = 0$, this potential must also depend on z (except for the case when function $\varphi(x)$ is linear, but our case differs: for the wave, there must be the function $\varphi \sim \cos x$), but we did not consider this dependence.

Actually, these two assumptions are connected to each other. We are not interested in the dependence on axis z simply because the fluid is shallow: the flow is too thin to take into consideration any function $\varphi(z)$.

Further in this section, we obtain more common solutions and make sure that the correlations are proper too.

4.5.5 Gravitational-capillary waves

Previously, we considered long waves with a very small curvature. For such deformations of the surface, capillary forces are negligible: the magnitude of these forces is proportional to the curvature $K = 1/R_1 + 1/R_2$, where R_1 and R_2 are the main radiuses of curvature at the point.

Now we discuss a more common problem. Further, we will consider a fluid of an infinite depth, i.e., the case $h_0 \rightarrow \infty$; thereby, it is logical to translate the origin closer to the liquid surface: now $z = 0$ corresponds to the undisturbed surface; still $z = \xi(x,t)$ determines the interface, and now the bottom is placed at $z = -\infty$.

For our problem, we have another condition on the surface for pressure: in the previous section, we assumed that at the interface the pressure in the liquid is equal to the pressure in the gas: $p = p_0$. Now we have to take into account the Laplace pressure jump at the surface, which leads to a moderate curvature:

$$p = p_0 - \sigma\left(\frac{\partial^2 \xi}{\partial x^2} + \frac{\partial^2 \xi}{\partial y^2}\right) \tag{4.5.16}$$

We still have to deal with a unidirectional flow (along the axis x), that is, we will omit the second term in the brackets in (4.5.16).

Then, we will consider an irrotational flow in a perfect fluid, that is, we may use the potential φ for the velocity $u = \partial\varphi/\partial x$ along the channel. Then, the Lagrange–Cauchy integral along any streamline is

$$\frac{\partial \varphi}{\partial t} + \frac{p}{\rho} + gz = const = \frac{p_0}{\rho}, \tag{4.5.17}$$

that is, on the free surface, with (4.5.16) we have

$$\rho\frac{\partial \varphi}{\partial t} + \rho g\xi - \sigma\frac{\partial^2 \xi}{\partial x^2} = 0. \tag{4.5.18}$$

Also, at the surface we may put for the vertical component of velocity

$$u_z = \frac{\partial \varphi}{\partial z} = \frac{\partial \xi}{\partial t}, \tag{4.5.19}$$

and, taking the derivative on time from (4.5.18) with respect to (4.5.19), we get

$$\rho\frac{\partial^2 \varphi}{\partial t^2} + \rho g\frac{\partial \varphi}{\partial z} - \sigma\frac{\partial^3 \varphi}{\partial z \partial x^2} = 0. \tag{4.5.20}$$

Then, the potential φ in a fluid must obey the Laplace equation

$$\Delta\varphi = 0. \tag{4.5.21}$$

with boundary condition (4.5.20).

The solution for equation (4.5.21) must fade at $z \to -\infty$; consequently, we choose for it the function in form $\varphi \sim e^{kx} \cos(kx + C)$. Keeping in a mind that at the surface we must have some wave-type correlation, we put $C = -\omega t$, so that

$$\varphi = Ae^{kz} \cos(kx - \omega t).$$ (4.5.22)

Thus, we have from (4.5.20) with function (4.5.22) an additional restriction for parameters k and ω: they cannot be set arbitrarily, they obey the correlation

$$\omega^2 = gk + \frac{\sigma}{\rho} k^3.$$ (4.5.23)

A correlation of such type is called the dispersion equation. Relation (4.5.23) gives correlations in two limited cases: for the gravitational wave ($\sigma = 0$)

$$\omega^2 = gk,$$ (4.5.24)

for capillary waves ($g = 0$)

$$\omega^2 = \frac{\sigma}{\rho} k^3.$$ (4.5.25)

Let us estimate the boundary between these two types of waves. As we see, to consider both terms (not to neglect one of them) there must be

$$g \sim \frac{4\pi^2 \sigma}{\rho \lambda^2};$$ (4.5.26)

here we use the wavelength $\lambda = 2\pi/k$ instead of the wave number to get clearer results. From (4.5.26) for water where $\rho \sim 10^3$ kg/m^3, $\sigma \sim 10^{-1}$ N/m, and we obtain the wavelength $\lambda \sim 2\pi\sqrt{\sigma/\rho g} \sim 2$ cm. Shorter waves can be considered as capillary waves (ripples), and longer waves are gravitational; the dependence on λ^2 helps to divide the range in two areas and make the 'transition' part of the wavelength range narrower.

The wave speed can be obtained with a common relation (4.5.23):

$$c = \frac{\partial \omega}{\partial k} = \frac{1}{2\omega}\left(g + \frac{3\sigma k^2}{\rho}\right).$$ (4.5.27)

It is clear that (4.2.27) does not turn into (4.5.15), of course, even for pure gravitational waves; for $\sigma \to 0$ we have

$$c = \sqrt{g/4k},$$ (4.5.28)

i.e., the speed depends on the wavelength $\lambda = 2\pi/k$ but not on the depth of the liquid which was set to infinity for this problem.

The common solution for gravitational waves in the liquid of arbitrary depth h can be obtained in the same manner. We still have the Laplace equation (4.5.21), but now we must use its common solution instead of (4.5.22)

$$\varphi = \left(Ae^{kz} + Be^{-kz}\right)\cos(kx - \omega t).$$ (4.5.29)

For this case, we have an additional boundary condition: at the bottom, the normal velocity must be zero, that is

$$\left.\frac{\partial \varphi}{\partial z}\right|_{z=-h} = 0,$$ (4.5.30)

that is, in (4.5.28)

$$B = Ae^{-2kh},$$ (4.5.31)

$$\varphi = \underbrace{2Ae^{-kh}}_{C}\cosh\left[k\left(z+h\right)\right]\cos\left(kx-\omega t\right).$$ (4.5.32)

where $\cosh x = (e^x + e^{-x})/2$ is a hyperbolic cosine.
Then, from the boundary condition (4.5.20) at $z = 0$ with $\sigma = 0$, we obtain

$$\omega^2 = kg\tanh \underbrace{kh}_{s},$$ (4.5.33)

where hyperbolic tangent $\tanh x = \sinh x/\cosh x$.
The wave speed (4.5.27) now turns into

$$c = \frac{\partial \omega}{\partial k} = \frac{\sqrt{g}}{2\sqrt{k\tanh s}}\left(\tanh s + \frac{s}{\cosh^2 s}\right).$$ (4.5.34)

Thus, for a long wave in a fluid of finite depth $s = kh \ll 1$, we have $\tanh s \approx 2s$, and we get $c = \sqrt{gh}$, i.e., correlation (4.5.15). For a finite wavelength and infinite depth $s = kh \gg 1$, the second term from the brackets in (4.5.33) vanishes, and we get $c = \sqrt{g/2k}$ which coincides with the relation (4.5.28).

4.5.6 Solitons

Solitons—solitary waves—are special objects in mathematical physics. First of all, we must specify that here we discuss only the solitons for waves on the surface of a fluid, while solitary waves, generally, can appear in almost any physical problem: optics, plasma physics, etc.

The surface waves described above were harmonic functions: the raise of the fluid in one area was compensated by the drop in other neighboring area. Solitons have special forms. In their most radical appearance, it may be a single liquid hill racing on the water surface—the so-called killer-wave. Put yourself in the place of a sailor: you rest after a hard day's work under the sun, you look at the sea and, suddenly, you see a single crest of a wave that rushes to your ship. What do you think at this moment? Of course, not every solitary wave is a killer wave (rather the opposite: usually, solitons are peaceful).

The science of solitons has partly difficult fate. Not all scientists believed in solitons in past, and, interestingly, they still do not believe in some special kinds of them today. Some colleagues doubt that killer-waves in the open ocean exist, because the formation condition of them in such a medium remains unclear despite many evidences and numerical simulations (Dyachenko and Zakharov 2008). Fortunately, the hardest times are behind and, in general, the concept of solitons is undisputed now.

Solitons were discovered by John Scott Russell in 1834 by direct observation (Newell 1985). He saw from the shore how horses pull up a barge on a river (the Union channel) and, suddenly, the water mass gathered at the ship' bow separated and moved forward ahead of the ship. Impressed, Russell devoted his life to investigation of this phenomenon. Initially, his ideas were criticized (see some citations in (Rayleigh 1876)), but finally the scientific community admitted Russell's correctness.

Theoretically, solitons were described by Boussinesq (1871), Rayleigh (1876) and, in the most complete form, by Korteweg and de Vries (1895).

A solution in the form of a solitary wave can be obtained, approximately, in the same way as for other problems considered in this section. However, for this soliton we must use non-linear representation.

Following Rayleigh, whose work was devoted to justification of Russell's empirical investigations, we should consider the full Bernoulli integral along the streamline corresponding to the liquid surface, which for the stationary case is

$$\frac{\rho u_x^2}{2} + \frac{\rho u_z^2}{2} + \rho g z + p = const = P, \tag{4.5.35}$$

where directions of axes correspond to Fig. 4.5.3 and velocities can be calculated with the stream function ψ:

$$u_x = \frac{\partial \psi}{\partial z}, u_z = -\frac{\partial \psi}{\partial x}. \tag{4.5.36}$$

Because of (4.5.36), in (4.5.35) one may replace $u_z^2 = u_x^2 z'$, where $z' = dz/dx$ is the derivative of the dependence $z(x)$ determining the shape of the streamline along the surface, i.e., the soliton shape. The solution for the Laplace equation for ψ can be represented as expansion

$$\psi(x, z) = \sum_{n=0}^{\infty} \frac{(-1)^n z^{2n+1} f^{(2n)}(x)}{(2n+1)!} = zf - \frac{z^3}{6} f'' + ..., \tag{4.5.37}$$

where $f^{(2n)}$ is the derivative of order $2n$, so that

$$u_x = \frac{\partial \psi}{\partial z} = \sum_{n=0}^{\infty} \frac{(-1)^n z^{2n} f^{(2n)}}{(2n)!} = f - \frac{z^2 f''}{2} + ... \tag{4.5.38}$$

$$u_z = -\frac{\partial \psi}{\partial x} = -\sum_{n=0}^{\infty} \frac{(-1)^n z^{2n+1} f^{(2n+1)}}{(2n+1)!} = -zf' + \frac{z^3 f'''}{6} + ... \tag{4.5.39}$$

On the surface, i.e., on the free streamline, there must be

$$\psi = const = \Psi. \tag{4.5.40}$$

Thus, we see the common solution path for the soliton shape. From (4.5.35) we have

$$z'^2 = \frac{2(P - p - \rho g z)}{\rho u_x^2} - 1, \tag{4.5.41}$$

where u_x is determined with (4.5.38) at condition (4.3.40). For instance, if we suppose that the function $f(x)$ is slow—we may neglect all derivatives of f in (4.5.37) and (4.5.38)—then we have from (4.5.37) and (4.5.40) $f = \Psi/z$, from (4.5.38) $u_x = \Psi/z$, and the soliton shape is defined by the integral

$$x = \pm \int \frac{dz}{\sqrt{\dfrac{2(P - p - \rho g z)z^2}{\rho \Psi^2} - 1}}. \tag{4.5.42}$$

Of course, this is not the best expression for the soliton shape that can be obtained with this approach. However, it is a good illustration of what can be done based on the simplest approach that we use in this section.

Today, the science about solitary waves is (a) developed very well, (b) leaves the acute phase, which seems inevitable for every scientific area that becomes popular in a wide scientific community (as it happened with the catastrophe theory, chaos, fractals, etc.), when corresponding objects have been found everywhere—where they are and where, in fact, they are not. If you are interested in that theme, we recommend to begin from the monograph by Alan C. Newell (Newell 1985).

4.5.7 The Helmholtz–Kelvin instability

Now we leave the waves behind and return to instabilities which are even more impressive than solitary waves. We begin with the Helmholtz–Kelvin instability that occurs when one fluid glides over the surface of another one. Below, we will consider only the interaction between a gas and a liquid, despite the fact that this theory can be applied to any fluids with different densities.

Actually, everyone saw this instability, if not in life, then at least in pictures where beautiful spiral waves on the surface of water were portrayed; our imitation can be seen in Fig. 4.5.4: surely, it falls short of Hokusai prints.

Analyzing the phenomenon of instability, firstly, it is necessary to distinguish the 'waves' caused by instabilities from the regular waves on the surface: the last ones are ordinary periodic functions (if they are not solitons, of course), while instability leads to irregular complicated forms at the interface.

The Helmholtz–Kelvin instability takes place at the boundary of two different fluids with a tangent break of velocity at their interface. As we know, perfect (non-viscous) fluids are able to move along a boundary, including a boundary in the form of adjacent liquid, at the velocity different from the velocity of that boundary; such velocity difference is called a tangent break. Thus, if (a) a fluid can flow over another fluid at its own velocity; (b) the fluid motion causes the pressure variation that affects the fluid flow, then one may expect that this system might have a positive feedback, which means, in physical language, exactly an instability.

The mathematical apparatus to explore instabilities is a developed scheme that can be used for any problem—this instrument has been polished through decades. The scheme looks as follows (some designations are symbolic).

✓ Find the solution S of the system of mathematical equations that describes the problem. It may be a stationary solution, a cycle, a wave, etc.

✓ Apply a noise N to this solution as a periodic function with small amplitude, i.e., introduce a function of a form $Ae^{i(kx-\omega t)}$ (or something like that). Here k is the wave number of the disturbance, ω is the cyclic frequency of the disturbance.

Fig. 4.5.4. A typical pattern of a developed Helmholtz–Kelvin instability.

✓ Then we insert the function $S + N$ into our system. Using the fact that (initially, by assumption) $N << S$, we may linearize the system—obtain a uniform system of linear algebraic equations for magnitudes of the noise A.

✓ The system of linear equations to determine the noise magnitudes A contains k (appeared from spatial derivatives) and ω (appeared from time derivatives).

✓ The uniform system of linear algebraic equations has a solution for A if and only if its determinant D is equal to zero.

✓ From condition $D = 0$, we obtain a correlation that connects ω and k—the dispersion equation. Thus, we have the function $\omega(k)$, probably, in implicit form, but it doesn't matter.

✓ From the correlation $\omega(k)$, we may see whether ω has an imaginary part with a positive sign or not. If yes, i.e., such part exists and is equal to iF, then we see that the noise has a multiplier $\sim Ae^{Ft}$. Thus, the initial disturbance increases with time and our system is unstable under conditions when $F > 0$; usually that depends on the wavenumber k.

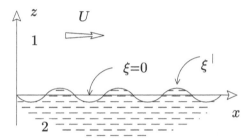

Fig. 4.5.5. The problem statement for studying the Helmholtz–Kelvin instability. Initially, the upper phase of liquid 1 flows on the surface of liquid 2 which is immobile; the surface is flat. Then we have a disturbance at the interface, which is accompanied by deviations in velocity and in pressure.

Now let us apply that implementation to the Helmholtz–Kelvin instability.

We consider two liquids (see Fig. 4.5.5); the system of mathematical equations for each liquid consists of the Euler equation, which will be linearized further

$$\frac{\partial \vec{u}}{\partial t} + \vec{u}\nabla\vec{u} = -\frac{\nabla p}{\rho} \tag{4.5.43}$$

and of the Poisson equation for pressure, which follows from the linearized Euler equation for incompressible fluid (divergence for it gives zero on the left-hand side):

$$\Delta p = 0. \tag{4.5.44}$$

In addition, we have to take into account that on the surface, the normal component of velocity of the fluid is connected to the displacement of the interface:

$$u_z = \frac{d\xi}{dt} = \frac{\partial \xi}{\partial t} + \vec{u}\nabla\xi. \tag{4.5.45}$$

This system of equations accommodates for the simplest stationary solution: in both liquids, pressures are the same and constant

$$p_1 = p_2 = const, \tag{4.5.46}$$

and only x-projection of velocity in the first liquid is non-zero:

$$u_{x1} = U, u_{x2} = u_{z1} = u_{z2} = 0. \tag{4.5.47}$$

Then we add disturbances to all these quantities in the form of $e^{i(kx-\omega t)}$; the corresponding quantities are signed with tildes. Note that it is sufficient to consider only z-projections of velocity, so

$$\tilde{u}_{z1,2} = \tilde{u}_{01,2}e^{i(kx-\omega t)}, \tilde{p}_{1,2} = \tilde{P}_{1,2}(z)e^{i(kx-\omega t)}, \tilde{\xi} = \tilde{\xi}_0 e^{i(kx-\omega t)}, \tag{4.5.48}$$

(we omit index z at amplitudes $\tilde{u}$).

Since the pressure satisfies the Laplace equation, we find that

$$\tilde{p}_1 = \tilde{p}_0 e^{-kz}e^{i(kx-\omega t)}, \quad \tilde{p}_2 = \tilde{p}_0 e^{kz}e^{i(kx-\omega t)}, \tag{4.5.49}$$

since pressure must vanish at infinity. Note that we set equal pressures in both liquids at $z = 0$: it is possible because we do not take into account both gravity and capillary forces.

Thus, the Navier–Stokes equation takes the form

$$\frac{\partial \tilde{u}_{z1}}{\partial t} + U\frac{\partial \tilde{u}_{z1}}{\partial x} = -\frac{1}{\rho_1}\frac{\partial \tilde{p}_1}{\partial z}, \quad \frac{\partial \tilde{u}_{z2}}{\partial t} = -\frac{1}{\rho_2}\frac{\partial \tilde{p}_2}{\partial z}, \tag{4.5.50}$$

where we take into account different densities of liquids.

By inserting representations (4.5.48) and (4.5.49) here, from (4.5.50) we obtain shortened exponents everywhere

$$\tilde{u}_{01}i\rho_1(Uk-\omega) - \tilde{p}_0 k = 0, \tag{4.5.51}$$

$$-\tilde{u}_{02}i\rho_2\omega + k\tilde{p}_0 = 0 \tag{4.5.52}$$

Then, from the boundary conditions in both liquids

$$\tilde{u}_{z1} = \frac{\partial \tilde{\xi}}{\partial t} + U\frac{\partial \tilde{\xi}}{\partial x}, \tilde{u}_{z1} = \frac{\partial \tilde{\xi}}{\partial t}. \tag{4.5.53}$$

we obtain two correlations

$$\tilde{u}_{01} = \tilde{\xi}_0 i(Uk-\omega), \tilde{u}_{02} = -\tilde{\xi}_0 i\omega, \tag{4.5.54}$$

that give us the third equation

$$\tilde{u}_{01}\omega + \tilde{u}_{02}(Uk-\omega) = 0. \tag{4.5.55}$$

Equations (4.5.51), (4.5.52) and (4.5.55) form a system of uniform linear algebraic equations to find $\tilde{u}_{01}$, $\tilde{u}_{02}$ and $\tilde{p}_0$. But, as we know, non-trivial solutions of this system exist only in case of zero determinant:

$$\begin{vmatrix} i\rho_1(Uk-\omega) & 0 & -k \\ 0 & -i\rho_2\omega & k \\ \omega & (Uk-\omega) & 0 \end{vmatrix} = 0. \tag{4.5.56}$$

From (4.5.56), we obtain the equation

$$\rho_1(Uk-\omega)^2 + \rho_2\omega^2 = 0. \tag{4.5.57}$$

from which we get the dispersion relation

$$\omega = Uk \frac{\rho_1 \pm i\sqrt{\rho_1\rho_2}}{\rho_1 + \rho_2}. \tag{4.5.58}$$

Thus, the problem is solved. All that remains is to interpret the results.

First, we see that at any k there exists a positive imaginary part of ω. This means that such a flow, shown in Fig. 4.5.5, is always unstable. The disturbances at the interface (4.5.58) increase, so do the additional components of velocity: components that are normal to the surface appear.

It is wrong to think that the flow will hold the pattern described by relations (4.5.48), i.e., without the tangent projection of velocity. Of course, instability will cause the flow with a full set of velocity projections; that is clear even from the continuous equation (see above).

It is important to understand the reason behind the Helmholtz–Kelvin instability. Firstly, the difference in densities does not play a role: even for $\rho_1 = \rho_2$, (4.5.58) gives the same result. On the other hand, in a homogeneous fluid it is more difficult to create a tangent break that initiates all the processes.

If $U = 0$, (4.5.57) gives nothing, but it is interesting whether we must hold the convective term $\vec{u}\,\nabla\,\vec{u}$ in the Navier–Stokes equation, or we may neglect it there and consider U only at the boundary condition (4.5.45)? Technically, it would mean consideration with the irrotational approach (see Section 4.1). The answer is negative: if we omit the convective term in equation (4.5.43), i.e., put $U = 0$ in the first element of the matrix, we would finally obtain only the real part of ω, which corresponds to regular oscillations of the surface with initial amplitudes. If the role of the non-linear term in the Navier-Stokes would be negligible, such instability could not exist.

Thus, the Helmholtz–Kelvin instability principally has a non-linear nature. The qualitative description of this process in (Batchelor 1970) and (Drazin 1970) is based on the vortex structures arising during this instability; we used a more common and formal explanation.

Also, note that gravity forces do not play any role in this instability: we did not consider gravity at all; the 'upper liquid' corresponded only to a picture—we may swap liquids with the same result.

Intuitively, we expect that viscosity would change the corollary. Viscosity damps the processes, so one may think that instability would be quenched by it. However, it is not so: viscosity changes the beginning of the Helmholtz–Kelvin instability, but does not consume the phenomenon itself.

4.5.8 Common case. The Rayleigh–Taylor instability

The instability considered in the previous section does not consider gravity and capillary forces. Now we take into account both of these elements.

Still, we have the Navier–Stokes equations as a main undisturbed flow as

$$\frac{\partial \vec{u}}{\partial t} + \vec{u}\nabla\vec{u} = -\frac{\nabla p}{\rho} + \vec{g}, \tag{4.5.59}$$

which gives for undisturbed flow and flat surface:

$$u_{x1} = U_1,\ u_{x2} = U_2,\ u_{z1} = 0,\ u_{z2} = 0, \tag{4.5.60}$$

$$p_1 = P_1 - \rho_1 gz,\ p_2 = P_2 - \rho_2 gz,\ \xi = 0, \tag{4.5.61}$$

where $P_1 = P_2$, because for the flat surface at $\xi = z = 0$, the pressures must be equal.

In (4.5.60) and (4.5.61), we take into account different horizontal velocities in both liquids: due to gravity forces, we may expect to obtain asymmetry in our problem relative to swapping liquids. Then we turn on disturbances of a kind $\sim Ae^{i(kx-\omega t)}$, i.e., place into (4.5.59) the sum of the stationary values (4.5.60) and (4.5.61), and those disturbed values are marked by tildes. Thus, the corresponding equation for z-projections of velocity in two liquids becomes:

$$\frac{\partial \tilde{u}_{z1}}{\partial t} + U_1 \frac{\partial \tilde{u}_{z1}}{\partial x} = -\frac{1}{\rho_1} \frac{\partial \tilde{p}_1}{\partial z}, \quad \frac{\partial \tilde{u}_{z2}}{\partial t} + U_2 \frac{\partial \tilde{u}_{z2}}{\partial x} = -\frac{1}{\rho_2} \frac{\partial \tilde{p}_2}{\partial z}. \tag{4.5.62}$$

and, since the pressures still satisfy the Laplace equation, so that

$$\tilde{p}_1 = \tilde{p}_{01} e^{-kz} e^{i(kx - \omega t)}, \quad \tilde{p}_2 = \tilde{p}_{02} e^{kz} e^{i(kx - \omega t)} \tag{4.5.63}$$

we get from (4.5.62), (4.5.63) the analogues of (4.5.51):

$$\tilde{u}_{01} i \rho_1 (U_1 k - \omega) - \tilde{p}_{01} k = 0, \tag{4.5.64}$$

$$\tilde{u}_{02} i \rho_2 (U_2 k - \omega) + \tilde{p}_{02} k = 0, \tag{4.5.65}$$

where $\tilde{u}_0$, $\tilde{p}_0$ are the amplitudes, like in the previous section; note that now $\tilde{p}_{01} \neq \tilde{p}_{02}$ due to the Laplace pressure jump on the surface.
On the surface, there must be conditions (4.5.44):

$$\tilde{u}_{z1} = \frac{\partial \tilde{\xi}}{\partial t} + U_1 \frac{\partial \xi}{\partial x}, \quad \tilde{u}_{z2} = \frac{\partial \tilde{\xi}}{\partial t} + U_2 \frac{\partial \xi}{\partial x} \tag{4.5.66}$$

which give

$$\tilde{u}_{01} = \tilde{\xi}_0 i (U_1 k - \omega), \quad \tilde{u}_{02} = \tilde{\xi}_0 i (U_2 k - \omega). \tag{4.5.67}$$

The second condition at the interface now is much more complicated than a simple equality of pressures. At the surface, the entire pressure must have a jump due to capillary forces:

$$P_1 + \tilde{p}_1 - \rho_1 g \tilde{\xi} = P_2 + \tilde{p}_2 - \rho_2 g \tilde{\xi} + \sigma \frac{\partial^2 \tilde{\xi}}{\partial x^2}. \tag{4.5.68}$$

Here P_1 and P_2 are cut, and we have for amplitudes

$$\tilde{p}_{01} - \rho_1 g \tilde{\xi}_0 = \tilde{p}_{02} - \rho_2 g \tilde{\xi}_0 - \sigma k^2 \tilde{\xi}_0. \tag{4.5.69}$$

Now we have five equations (4.5.64), (4.5.65), (4.5.67) and (4.5.69) for five variables $\tilde{u}_{01}$, $\tilde{u}_{02}$, $\tilde{p}_{01}$, $\tilde{p}_{02}$, $\tilde{\xi}_0$. As before, we may find the determinant 5x5 of that system, but, for the sake of brevity and variety, we choose another way: we express all other quantities with $\tilde{\xi}_0$.
We have

$$\tilde{p}_{01} = \frac{i \rho_1 (U_1 k - \omega)}{k} \tilde{u}_{01} = \frac{i \rho_1 (U_1 k - \omega)}{k} \underbrace{i \tilde{\xi}_0 (U_1 k - \omega)}_{\tilde{u}_{01}} = -\frac{\rho_1 (U_1 k - \omega)^2}{k} \tilde{\xi}_0, \tag{4.5.70}$$

$$\tilde{p}_{02} = -\frac{i \rho_2 (U_2 k - \omega)}{k} \tilde{u}_{02} = -\frac{i \rho_2 (U_2 k - \omega)}{k} \underbrace{i \tilde{\xi}_0 (U_2 k - \omega)}_{\tilde{u}_{02}} = \frac{\rho_2 (U_2 k - \omega)^2}{k} \tilde{\xi}_0. \tag{4.5.71}$$

Then we replace $\tilde{p}_{01}$ and $\tilde{p}_{02}$ in (4.5.69) with (4.5.70) and (4.5.71) and divide the obtained equation by $\tilde{\xi}_0$:

$$\rho_1 (U_1 k - \omega)^2 + \rho_2 (U_2 k - \omega)^2 = k \left[(\rho_2 - \rho_1) g + \sigma k^2 \right]. \tag{4.5.72}$$

Thus, the cyclic frequency is

$$\omega_{1,2} = \frac{k(U_1\rho_1 + U_2\rho_2) \pm \sqrt{k(\rho_1+\rho_2)\left[(\rho_2-\rho_1)g + \sigma k^2\right] - k^2\rho_1\rho_2(U_1-U_2)^2}}{\rho_1+\rho_2}. \tag{4.5.73}$$

We are interested in the under-root expression: if the value of function

$$f = k(\rho_1+\rho_2)\left[(\rho_2-\rho_1)g + \sigma k^2\right] - k^2\rho_1\rho_2(U_1-U_2)^2 \tag{4.5.74}$$

is negative, then, in accordance with (4.5.73), ω would have a positive imaginary part. If the first term in (4.5.74) is equal to zero, any non-zero velocity in any liquid (note that we finally obtain symmetry concerning the velocity swap; case $U_1 = U_2$ gives no tangent break at all) provides instability.

Then, if $\rho_2 > \rho_1$, that means that the lighter liquid is above; we see that the tangent break must have a finite value to cause an instability:

$$(U_1-U_2)^2 > U_{cr}^2 = \frac{(\rho_1+\rho_2)\left[(\rho_2-\rho_1)g + \sigma k^2\right]}{k\rho_1\rho_2}. \tag{4.5.75}$$

The dependence of U_{cr} on $\lambda = 2\pi/k$ for water ($\rho_2 \approx 10^3$ kg/m^3, $\sigma \approx 10^{-1}$ N/m) in air ($\rho_1 \approx 1$ kg/m^3) is presented in Fig. 4.5.6. The minimal value of U_{cr}

$$U_{cr}^{min} = \sqrt{2\frac{\rho_1+\rho_2}{\rho_1\rho_2}\sqrt{\sigma(\rho_2-\rho_1)g}} \approx 8\,\text{m/s}. \tag{4.5.76}$$

corresponds to the wavelength

$$\lambda = 2\pi\sqrt{\frac{\sigma}{(\rho_2-\rho_1)g}} \approx 2\,\text{cm}; \tag{4.5.77}$$

we already obtained this value when we analyzed the critical value needed to separate gravitational and capillary waves. Partially, this result explains why we do not always see the beautiful waves from Fig. 4.5.4 (or from Japanese prints): a quite decent wind is needed to provide such waves.

Hence, gravitational and capillary forces partially dampen the instability caused by tangent breaks, but not only these ones. It is clear that such a '$\sigma - g$ combination' tends to damp any oscillations of the surface. Moreover, continuing to generalize, the quantity g may not be the gravity itself, the acceleration might be caused by external reasons—by external motion, the only proviso:

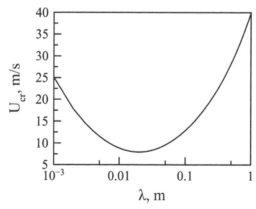

Fig. 4.5.6. Critical velocity to induce the Helmholtz–Kelvin instability of the corresponding wavelength in water: normal stratification—water under air

itself, the acceleration might be caused by external reasons—by external motion, the only proviso: this acceleration must be directed from the gas to the liquid. Domestic example: you fill a glass up to the brim, and, when you carry it, water creates a bulge on top of the glass. To avoid splashing, you, instinctively, raise the glass up swiftly; by this intuitive motion, you produce the very additional acceleration directed at the liquid phase from the gaseous one, and, therefore, suppress the instability exactly according to the rigorous scientific theory. Such cases demonstrate that physical intuition is intrinsic to all the beings; however, this is not the most impressive example. For instance, elephants are waving ears: their ears, among other functions, are heat exchangers, and elephants know that the heat transfer coefficient is higher for a forced convection than for a free one. Elephants are clever creatures.

Above, we considered the 'normal case'—when a lighter liquid is placed onto a heavier one. However, the opposite variant is not impossible; exactly, for the cavitation problem—when the gaseous phase is located inside the liquid one—this case is quite real.

If gas and liquid are swapped, so that a heavy liquid is placed over the light gas, in (4.5.74) the difference $(\rho_2 - \rho_1) < 0$, and gravity aims to develop the instability, not prevent it. First, we consider the case without any tangent velocities, i.e., for $U_1 = U_2 = 0$. This pure case is called the Rayleigh–Taylor instability.

Such type of instability depends, as we see from function (4.5.74), on the sign of the term

$$(\rho_2 - \rho_1)g + \sigma k^2; \qquad (4.5.78)$$

if it is negative, then instability takes place. In simple words, gravity forces aspire to break the interface, to push it through, while capillary forces try to prevent it. Withal, the magnitude of capillary forces depends on the spatial scale: the greater the scale, the lesser the capillary force is. Thus, long-wave disturbances that correspond to a small value of k may develop: for such waves, gravity defeats capillarity. The exact value of the critical wavelength is defined, again, by the correlation (4.5.77), but with the negative sign in the density difference, i.e., one must replace in (4.5.77) $(\rho_2 - \rho_1) \rightarrow (\rho_1 - \rho_2)$ with the same estimation value.

Again, here g is not exactly gravity but any acceleration directed from the liquid mass to the gas. The Rayleigh–Taylor instability may be observed in domestic conditions through an experiment similar to the one that was described above. Fill the glass, not up to the brim this time, and then abruptly drop it down. The waves that you will see correspond to the instability caused by such great scientists as Rayleigh and Taylor.

Finally, we consider the case when the heavier liquid is above and tangent velocity exists.

For this problem, we have the same answer—the correlation (4.5.75), we only have to swap densities in it. The result is shown in Fig. 4.5.7.

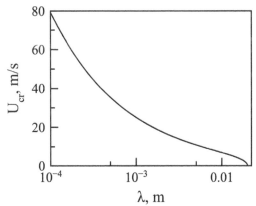

Fig. 4.5.7. Critical velocity to induce the Helmholtz–Kelvin instability of the corresponding wavelength in water: reverse stratification—water above air; wavelength of 2 cm corresponds to the pure Rayleigh–Taylor instability at zero velocity.

Conclusion

Hydrodynamics is a difficult science in itself. Alas, for two-phase flows, in comparison with a single-phase one, the problem doubles, and not because we have to describe twice the number of phases. Indeed, usually the flow in a gas is not taken into account, and only hydrodynamics of a liquid must be considered. The difficulty lies in another matter: generally, the liquid–gas boundary is unknown.

A common approach to obtaining results for the irrotational flow of an incompressible fluid is to use hydrodynamic potentials: the scalar potential φ and the stream function ψ. The first one always obeys the Laplace equation

$$\Delta\varphi = 0,$$

which, contrary to the Navier–Stokes equation, is linear. Moreover, to get a solution for a particular problem, one may combine 'standard' potentials caused by the sources placed at the boundaries of the flow (on solids or on the free surface of the fluid).

Many problems can be considered in language of hydrodynamic potentials. However, the cavitating flow past a body is a much more complicated problem than the classical case described in the aerodynamics works by Joukowski, Chaplygin, etc. Using the same methods as for the stationary single-phase flow, we meet significant difficulties in describing the cavitation cavern which is really non-static.

On the other hand, some problems for the cavitating flow can be solved 'up to numbers'; the classic (to be exact, slightly modified) solution for the cavitating flow past a plate is an excellent example of how analytical solutions can be advanced, if they can overwhelm the mathematical difficulties.

The same methods that allow to describe the flow around the body may be helpful for the construction of cavitation diagrams. Initially designed for airfoils, this methodology allows to reconstruct profiles which provide the given cavitation number.

A special part of hydrodynamics is the representation of waves and instabilities. The interface, so to say, is the subject of both these processes. There may be several types of 'regular' waves—gravitational, capillary, or their mix—which can be described within the same method of hydrodynamic potentials, and the special type of it—the solitary wave (soliton). Of course, soliton can also be treated within the same approach.

The Helmholtz–Kelvin instability occurs when one fluid moves over another at finite velocity. This is a common case for a two-phase flow, and the simplest consideration predicts instability at any value of the relative velocity, which would actually mean that the surface is always unstable. However, if we take into account other elements of the problem, at least gravity and surface tension, then we would find out that these additional factors partially damp the instability.

To be exact, at the beginning of this section, we considered a part of the common thread throughout the entire book—the collapse of a bubble. For very simple boundary conditions at the bubble–liquid interface, we may consider a very important particular part of the common problem— the collapse near a solid wall. In this geometry, the fluid flow around a bubble—which must be accompanied by the shrinking of this bubble, since the fluid must replace the compressible gaseous phase—is asymmetric. In vicinity of a solid wall, the velocity of a fluid is around zero (the velocity is equal to zero directly on the wall), that is, the outer side (relative to the wall) of the bubble collapses much faster. Moreover, such collapse is accompanied by the formation of a jet that moves from the outer side of the bubble to the wall and hits it at velocity $\sim k\sqrt{\Delta p/\rho}$, where Δp is the difference between the pressure in liquid and inside the bubble, $k \sim 10$ or even higher. Like other problems considered in this chapter, this special kind of collapse can also be considered with the same theoretical approach.

References

Ahn, B.-K., T.-K. Lee, H.-T. Kim and C.-S. Lee. 2012. Experimental investigation of super cavitating flows. *International Journal of Naval Architecture and Ocean Engineering*, 4: 123–31. https://doi.org/10.3744/JNAOE.2012.4.2.123.

Amromin, A. L. and A. N. Ivanov. 1982. Determination of the position of separation points of a cavitation cavity from a body with account of fluid viscosity and capillarity. *Doklady Akademii Nauk SSSR*, 262: 823–6 (in Russian).

Batchelor, G. K. 1970. *An introduction to fluid dynamics*. Cambridge Univ. Press.

Birkhoff, G. and E. H. Zarantonello. 1957. *Jets, Qakes, and Cavities*. New York: Academic Press.

Braun, M. J. and W. M. Hannon. 2016. Cavitation formation and modelling for fluid film bearings: A review. *Proceedings of the Institution of Mechanical Engineers, Part J: Journal of Engineering Tribology*, 224: 839–63. https://doi.org/10.1243/13506501JET772.

Butuzov, A. A. 1967. Artificial cavitation flow behind a slender wedge on the lower surface of a horizontal wall. *Fluid Dynamics*, 2: 56–58. https://doi.org/10.1007/BF01015141.

Capurso, T., G. Menchise, G. Carama, S. M. Camporeale, B. Fortunato and M. Torresi. 2018. Investigation of a passive control system for limiting cavitation inside turbomachinery under different operating conditions inside turbomachinery under different operating conditions. *Energy Procedia*, 148: 416–23. https://doi.org/10.1016/j.egypro.2018.08.103.

Chen, Y., X. Chen, Z. Gong, J. Li and C. Lu. 2016. Numerical investigation on the dynamic behavior of sheet/cloud cavitation regimes around hydrofoil. *Applied Mathematical Modelling*, 40: 5835–57. http://dx.doi.org/10.1016/j.apm.2016.01.031.

Cumberbatch, E. and T. Y. Wu. 1961. Cavity flow past a slender pointed hydrofoil. *Journal of Fluid Mechanics*, 11: 187–208. https://doi.org/10.1017/S0022112061000469.

Drazin, P. G. 1970. Kelvin–Helmholtz instability of finite amplitude. *Journal of Fluid Mechanics*, 42: 321–35. https://doi.org/10.1017/S0022112070001295.

Dyachenkoa, A. I. and V. E. Zakharov. 2008. On the formation of freak waves on the surface of deep water. *Journal of Experimental and Theoretical Physics Letters*, 88: 356–9. https://doi.org/10.1134/S0021364008170049.

Eppler, R. and D. M. Somers. 1980. *Computer Program for the Design and Analysis of Low-Speed Airfoils*. Hampton: NASA Langley Research Center.

Eppler, R. 1990. *Airfoil Design and Data*. Verlag: Springer Berlin Heidelberg. https://doi.org/10.1007/978-3-662-02646-5.

Gurevich, M. I. 1965. *Theory of Jets in Ideal Fluids*. Academic Press.

Guzevsky, L. G. 1975. Calculation of Axisymmetric Flows with Free Surfaces. *Doklady Akademii Nauk SSSR*, 225: 269–71 (in Russian).

Guzevsky, L. G. 2006. Method of boundary integral equations of solving plane and axisymmetric Ryabushinskii's problems. *Vychislitelnye Tekhnologii*, 11: 68–81 (in Russian).

Hoerner, S. F., W. H. Michel, L. W. Ward and T. M. Buermann. 1954. *Hydrofoil Handbook. Volume I, Design of Hydrofoil Craft*. New York: Bath. Iron Works Corp. by Gibbs and Cox, inc.

Ivanov, A. N. 1980. *Hydrodynamics of Developed Cavitating Flows*. L: Sudostroenie.

Jiang, Y., Bai, T. and Y. Gao. 2017. Formation and steady flow characteristics of ventilated supercavity with gas jet cavitator. *Ocean Engineering*, 142: 87–93. http://dx.doi.org/10.1016/j.oceaneng.2017.06.054.

Katz, J. 1984. Cavitation phenomena within regions of flow separation. *Journal of Fluid Mechanics*, 140: 397–436. http://dx.doi.org/10.1017/S0022112084000665.

Kim, D.-H., Park, W.-G. and Ch.-M. Jung. 2013. Numerical simulation of cavitating flow past axisymmetric body. *International Journal of Naval Architecture and Ocean Engineering*, 4: 256–66. http://dx.doi.org/10.32478/IJNAOE-2013-0094.

Kirchhoff, G. 1869. Zur Theorie freier Flüssigkeitsstrahlen. *Journal für Mathematik*, 70: 289–298. https://doi.org/10.1515/crll.1869.70.289.

Korteweg, D. J. and G. de Vries. 1895. XLI. On the change of form of long waves advancing in a rectangular canal, and on a new type of long stationary waves. *Philosophical Magazine Series 5*, 39: 422–43. https://doi.org/10.1080/14786449508620739.

Lamb, H. 1993. *Hydrodynamics*. Cambridge University Press.

Landau, L. D. and E. M. Lifshitz. 1987. *Fluid Mechanics*. Oxford: Pergamon Press.

Lavrentev, M. A. and B. V. Shabat. 1973. *Hydrodynamics problems and their mathematical models*. Moscow: Nauka (in Russian).

Lighthill, M. J. 1945. *A New Method of Two-dimensional Aerodynamic Design*. Reports and Memoranda No. 2112. London: His Majesty's Stationery Office.

Lin, C. C. 1955. *The theory of hydrodynamic stability*. Cambridge: University Press.

Maklakov, D. V. 1997. *Nonlinear problems of hydrodynamics of potential flows in uncertain boundaries*. Yanus-K (in Russian).

McNown, J. S. and C.-S. Yih. 1953. *Free-Streamline Analyses of Transition Flow and Jet Deflection*. Iowa: University Iowa City.

Newell, A. C. 1985. *Solitons in Mathematics and Physics*. SIAM.

Park, S. and Rhee H. 2012. Computational analysis of turbulent super-cavitating flow around a two-dimensional wedge-shaped cavitator geometry. *Computer & Fluids*, 70: 73–85. http://dx.doi.org/10.1016/j.compfluid.2012.09.012.

Plesset, M. S. and R. B. Chapman. 1970. *Collapse of an initially spherical vapor cavity in the neighborhood of a solid boundary*. Report No. 85–49. California: California Institute of Technology.

Plesset, M. S. and R. B. Chapman. 1971. Collapse of an initially spherical vapour cavity in the neighbourhood of a solid boundary. *Journal of Fluid Mechanics*, 47: 283–290. https://doi.org/10.1017/S0022112071001058.

Pykhteev, G. N. 1960. Cavitation flow of an ideal incompressible fluid in a slot. *Journal of Applied Mathematics and Mechanics*, 24: 213–20. https://doi.org/10.1016/0021-8928(60)90156-8.

Rayligh, M. A. 1876. XXXII. On waves. *Philosophical Magazine and Journal of Science*, 1: 257–79.

Roohi, E., Pendar, M.-R. and A. Rahim. 2016. Simulation of three-dimensional cavitation behind a disk using various turbulence and mass transfer models. *Applied Mathematical Modelling*, 40: 542–64. http://dx.doi.org/10.1016/j.apm.2015.06.002.

Ruelle, D. and F. Takens. 1971. On the nature of turbulence. *Communications in Mathematical Physics*, 20: 167–192.

Shang, Z. 2013. Numerical investigations of supercavitation around blunt bodies of submarine shape. *Applied Mathematical Modelling*, 37: 8836–45. http://dx.doi.org/10.1016/j.apm.2013.04.009.

Terentiev, A. G. 1977. Inclined entry of a thin body into incompressible liquid. *Izvestia AN SSSR MZhG*, 5: 16–24 (In Russian).

Tulin, M. P. 1964. Supercavitating flows. Small perturbation theory. *Journal of Ship Research* 8: 16–37. https://doi.org/10.5957/jsr.1964.8.1.16.

Voinov, O. V. and V. V. Voinov. 1975. Numerical method of calculating nonstationary motions of an ideal incompressible fluid with free surfaces. *Doklady Akademii Nauk SSSR*, 221: 559–62 (in Russian).

Voinov, O. V. and V. V. Voinov. 1976. Scheme of collapse of a cavitation bubble near a wall and formation of a cumulative jet. *Doklady Akademii Nauk SSSR*, 227: 63–6 (in Russian).

Wu, T. Y.-T. 1962. A wake model for free-streamline flow theory. Part 1. *Journal of Fluid Mechanics*, 13: 161–81. https://doi.org/10.1017/S0022112062000609.

Wu, T. Y.-T. 1964. A wake model for free-streamline flow theory. Part 2. *Journal of Fluid Mechanics*, 18: 65–93. https://doi.org/10.1017/S0022112064000052.

Wu, T. Y.-T. 1972. Cavity and Wake Flows. *Annual Review of Fluid Mechanics*, 4: 243–84. https://doi.org/10.1146/annurev.fl.04.010172.001331.

CHAPTER 5

Hydraulic Shocks

5.1 What is a hydraulic shock?

5.1.1 Common physical nature of hydraulic shock

Contrary to some other physical terms,[1] the term 'hydraulic shock' means exactly what the combination of these two words means: an impact caused by the liquid mass. Slightly translating this word combination into physical language, one may expand this to 'the change in pressure caused by the change in velocity of a fluid'.

Then, we may continue with this expanded definition. The first thing that comes to mind when we analyze this phrase is that the force caused by the impact is directed along velocity. For instance, a water mass impinges on a solid surface; during this process, velocity (to be exact, momentum $\vec{p}$) diminishes so the corresponding difference is the force as it is due to the Newton law:

$$\vec{F} = \frac{d\vec{p}}{dt}.\tag{5.1.1}$$

This, in fact, is a hydraulic shock, of course. This type of hydraulic shock causes destruction of ships, coastal structures, etc. However, this is not the only manifestation of hydraulic shock.

The thing is that the pressure in a liquid acts in all directions. Let us consider a flow moving in a pipe at the direction $\vec{x}$ (along the axis of the pipe, surely), and along this axis the flow velocity is undergoing some change. Due to this change, the pressure changes too. But, since the pressure acts in all directions, the forces caused by this change in velocity would act not only along the axis $\vec{x}$: the walls of the pipe would also feel that change in velocity. Actually, the abrupt manipulation with the liquid discharge in the pipe may cause deformation of the pipe walls; don't try it at home with the plumbing—despite the small chances of success, results can be astonishing, especially if your plumbing is quite old.

Thus, one can determine two phenomena that are called 'hydraulic shock':

1. The phenomenon of a sharp change in pressure in a pipe.
2. The impact of a liquid mass impinging on a solid surface.

The physical nature of these phenomena is similar, but the appearance of them is quite different. Further, we discuss both these matters in details.

5.1.2 Hydraulic shock in pipes

For better understanding, we begin with a stationary description. In general, we discussed this question in the first chapter; here we mention the basic principles.

----

[1] A good example of a confusing term is the 'black body'. The optical properties of the body are not connected to its color as directly as one may think: snow emits light as a black body, because it does it in the infrared range.

In a pipe of variable diameter, pressure is lower at the narrower cross-section. Indeed, pressure is connected to velocity through the Bernoulli correlation, in the simplest case

$$p + \frac{\rho u^2}{2} = const.$$ (5.1.2)

Velocity is connected to the discharge G that is a constant for any cross-section of pipe S

$$G = \rho u S = const.$$ (5.1.3)

The discharge is constant along the pipe: equal liquid mass must flow through any cross-section of the pipe per second.

Consequently, in a stationary case at condition (5.1.2), pressure depends on the cross-section as

$$p + \frac{G^2}{2\rho S^2} = const.$$ (5.1.4)

Thus, if $S\downarrow$ or $G\uparrow$, then $p\downarrow$ and vice versa; in previous chapters, we discussed how this phenomenon leads to cavitation. For the static case, we have an ordinary situation that cannot surprise: with equations (5.1.3) and (5.1.4), one may describe the flow in a channel of variable cross-section. However, the problem can be much more interesting if the change in S is abrupt, that is, pressure is abrupt too. A non-static case demands a completely different level of consideration.

Below, we consider the description of the flow in a pipe caused by an abrupt pressure variation. Despite the fact that this method has a wider area of application, we will consider a certain particular problem.

If we have a finite Δp at the given point, i.e. on the length $\Delta x \rightarrow 0$, this point would be the source of instant disturbance. This additional component of the flow may be described with the velocity potential φ, because pressure does not produce vorticity. Consequently, we have for the longitudinal component of velocity

$$u = \frac{\partial \varphi}{\partial x}.$$ (5.1.5)

Disturbances occur at the sharp heterogeneity of the pipe and lead to deviations in pressure and, consequently, the corresponding deviations in density, which are small in comparison to the normal density of the liquid, i.e., $\Delta \rho / \rho \ll 1$. These perturbations propagate along the pipe as waves of the velocity potential, i.e., in such a wave, the velocity potential is the function of a form

$$\varphi = \varphi (x \pm at),$$ (5.1.6)

where a is the speed of the wave propagation; the sign of the second term will be chosen below based on physical reasoning. In the next section, we derive the function strictly: it follows from the wave equation, which can be derived for φ of the irrotational flow at small deviations of density; here we want to concentrate on the common physical meaning.

Thus, velocity (5.1.5) is the same function on coordinates and time as the potential (5.1.6), and from the Navier-Stokes equation, neglecting the non-linear term $\vec{u} \nabla \vec{u}$,

$$\frac{\partial u(x+at)}{\partial t} = -\frac{1}{\rho}\frac{\partial p}{\partial x},$$ (5.1.7)

we may also conclude that $p(x + at)$. In other words, the speed a is the propagation speed of pressure in the pipe. Moreover, one can find the solution for this equation using the self-similar variable[2] $\xi = x + at$, with what we have

---

$$\frac{\partial u}{\partial t} = \frac{du}{d\xi}\frac{\partial \xi}{\partial t} = a\frac{du}{d\xi}, \quad \frac{\partial p}{\partial x} = \frac{dp}{d\xi}\frac{\partial \xi}{\partial x} = \frac{dp}{d\xi}. \tag{5.1.8}$$

Now we see the reason to choose the sign '+' in the following statement: only in this case, we get the acceleration of the fluid at a positive pressure gradient.

Now equation (5.1.7) transforms into

$$a\frac{du}{d\xi} = \frac{1}{\rho}\frac{dp}{d\xi}, \tag{5.1.9}$$

and, after integration, we have the relation

$$\Delta p = -\rho a \Delta u. \tag{5.1.10}$$

Possibly, it should be explained what do the differences in (5.1.10) exactly mean. Here $\Delta p = p(\xi_1) - p(\xi_2)$, for example, can be defined at two different self-similar coordinates, which means, particularly, the difference at the same coordinate but at two different instants, or the difference at two points at the same moment of time.

Thus, according to (5.1.10), the change in pressure is connected to the change in velocity through the density of the fluid ρ and the speed of propagation of disturbance a: if $\Delta u = u_2 - u_1 < 0$, i.e., if the fluid flow suddenly drags, then pressure increases: $\Delta p = p_2 - p_1 > 0$.

This picture can be easily imagined. The fluid flows in a pipe, then someone pulls down the gate valve harshly. If one does it sufficiently fast, then a hydraulic shock can be expected: pressure will increase abruptly as it follows from equation (5.1.10).

On the other hand, another question is important: what do 'harshly' or 'abruptly' exactly mean? At what rate the valve must be closed for these terms to be correct? If we close the valve smoothly, we should expect a qualitatively different picture: no shocks, and the flow will be stopped without any dramatic events.

We will consider these questions in the next section; now we must discuss some special matters, concerning the speed of shock propagation a.

At first glance, speed a is the speed of sound in the liquid

$$c_S = \sqrt{(\partial p / \partial \rho)_s}. \tag{5.1.11}$$

However, it is not so. It should be taken into account that the sound, we mean the small disturbances of the medium, propagates not only in the liquid, but also in the walls of the pipe. The problem with finding the parameter a is much more complicated. Below, we will follow the original work of Joukowsky with some simplifications in the explanation.

Let us consider a fluid flow in a pipe (see Fig. 5.1.1) of radius r and the area of cross-section $S = \pi r^2$. Initially, the fluid flows at velocity u_0, then one suddenly lowers a gate somewhere downstream, so velocity drops to value $u < u_0$. This perturbation propagates upstream at velocity a; we consider two cross-sections at distance $\Delta l = a\Delta t$, where Δt is the time of hydraulic shock: at this time, velocity diminishes on $u_0 - u$, and pressure increases by the corresponding value in accordance with (5.1.10).

At first, we may note that the fluid would accumulate in the volume $S\Delta l$, because the fluid flows in it at a higher velocity than it flows out. Exactly, the additional volume is

$$\Delta V_1 = (u_0 - u)S\Delta t. \tag{5.1.12}$$

This additional volume must be compensated somehow; one can imagine two ways for it. The first and the simplest way is that the walls of the pipe widen: the radius increases on Δr, and the cross-section area increases on ΔS, so the corresponding increase in volume is

$$\Delta V_2 = \Delta S \Delta l. \tag{5.1.13}$$

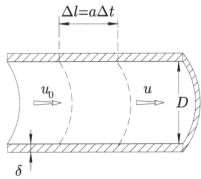

Fig. 5.1.1. Hydraulic shock. Velocity of the fluid at the end of the region of length Δl is less than at the entry. Fluid tends to accumulate in this region, shrinking and widening the walls of the pipe. The velocity of shock propagation a differs from the speed of sound in the fluid.

The second way to compensate the increased volume is the increase in density of the fluid: we may stuff the fluid tighter in the pipe. So, this second way is based on the compressibility of the liquid. If the mass m is located in the volume $V = S\Delta l$, then after the increase of density by value $\Delta\rho$, the volume of the liquid will be decreased by volume

$$\Delta V_3 = \frac{m}{\rho} - \frac{m}{\rho + \Delta\rho} \approx \frac{m}{\rho^2}\Delta\rho = \frac{S\Delta l \Delta\rho}{\rho}. \tag{5.1.14}$$

Thereby, we have the condition

$$\Delta V_1 = \Delta V_2 + \Delta V_3, \tag{5.1.15}$$

which can be rewritten taking into account that $a = \Delta l/\Delta t$

$$u_0 - u = a\frac{\Delta S}{S} + a\frac{\Delta\rho}{\rho}. \tag{5.1.16}$$

Let us analyze the first term on the right-hand side of (5.1.16). It can be expressed with radius

$$\frac{\Delta S}{S} = \frac{2\pi r \Delta r}{\pi r^2} = \frac{2\Delta r}{r}. \tag{5.1.17}$$

On the other hand, the relative extension is related to the stress σ in the pipe wall, which, in its turn, is connected to the additional pressure Δp:

$$\frac{\Delta r}{r} = \frac{\sigma}{E} = \frac{1}{E}\underbrace{\frac{\Delta p D}{2\delta}}_{\sigma}; \tag{5.1.18}$$

the corresponding relation $\sigma(\Delta p)$ is suitable for thin-walled pipes, in which the wall width δ is much lower than the pipe diameter $D = 2r$; E is the Young's modulus.

The second term on the right-hand side (5.1.16) can be expressed with the corresponding change in pressure Δp in the liquid. Because the process can be considered as adiabatic, we have

$$\Delta\rho = \left(\frac{\partial\rho}{\partial p}\right)_s \Delta p = \frac{\Delta p}{c_s^2}. \tag{5.1.19}$$

Finally, we may combine all the correlations into one: using (5.1.10) for $(u_0 - u) = -\Delta u$ on the left-hand side of (5.1.16), (5.1.18) and (5.1.19), we have

$$\frac{\Delta p}{\rho a} = a\frac{\Delta p D}{E\delta} + a\frac{\Delta p}{\rho c_s^2}. \tag{5.1.20}$$

Finally, by canceling Δp, we get the expression for the speed a:

$$a = \frac{c_s}{\sqrt{1 + \dfrac{\rho D c_s^2}{E\delta}}}. \tag{5.1.21}$$

Thus, the speed of the hydraulic shock propagation is equal to the acoustic velocity c_s only if $E \to \infty$, i.e., the pipe absolutely cannot deform.

It is interesting that sometimes an equation of a form (5.1.21) can be written differently. The derivative

$$K^{-1} = \beta = \frac{1}{\rho}\left(\frac{\partial \rho}{\partial p}\right)_s = \frac{1}{\rho c_s^2}. \tag{5.1.22}$$

is the adiabatic compression ratio, while K is the compression modulus that can be defined not only for a solid, but for a liquid too.[3] If so, then (5.1.21) can be represented by only the parameters of elasticity of fluid and solid wall:

$$a = \frac{\sqrt{K/\rho}}{\sqrt{1 + \dfrac{KD}{E\delta}}}. \tag{5.1.23}$$

In this variant, the equation has a more 'symmetric' form; however, we prefer (5.1.21) anyway, at least because the acoustic velocity can be found easier in reference data than the elastic modulus for fluids.

The fact that the speed of sound can be represented in two different ways (through the equation of state or with the elastic modulus) demands more detailed consideration; it will be considered in Section 5.2, where we clarify what sound is.

5.1.3 Water hammer from the technical point of view

This loud term—war hammer—implies exactly the same phenomenon as hydraulic shock; it is often used in technical literature and in popular science.

Let us consider a large reservoir and a pipeline diverted from it (Fig. 5.1.2). The reservoir creates constant pressure at the end A of the pipeline; it is so big that any phenomenon in the pipeline cannot disturb it. The pipeline is equipped with a valve at the end B. Below we consider the processes that occur when we suddenly shut the valve and finally, as we promised above, we translate the term 'suddenly' into technical language.

After the valve has been shut, the following processes takes place in the pipeline.

1. The valve is shut. At the end B of the pipeline the fluid stops, pressure increases by Δp in accordance with (5.1.10). Hydraulic shock propagates at velocity a from point B to point A: this is the wave of increased pressure. From point B to the point of the current position of the hydraulic shock front, the fluid is immobile.

2. The hydraulic shock wave reaches point A. Now we have two states of liquid: the compressed liquid at high pressure in the pipeline (right-hand relative to point A) and the liquid at initial parameters within the reservoir (left-hand to point A). At this moment, the fluid is immobile throughout the whole pipeline. The time to reach this state is

$$\tau = L/a, \tag{5.1.24}$$

---

[3] For a common case, the equation (5.1.22) is simplified: for solids, the acoustic velocity is different for different directions; see Section 5.2 and relations (5.2.36–39) there. However, for liquids this correlation is quite suitable.

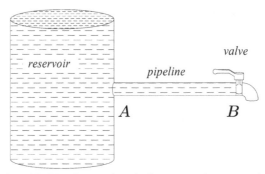

Fig. 5.1.2. Water hammer in a pipeline connected to a reservoir.

where *L* is the length of the pipeline—the distance between points A and B.

3. It is clear that the state described in item 2 is unstable. The compressed fluid will flow from the pipeline back to the reservoir. During this process, pressure in the pipeline decreases. This process—the leaking of fluid accompanied with the decreasing of pressure—can be represented and described as a wave of quenching: this wave moves from point A to point B at velocity *a*. Up to time 2τ, we have the phase when the fluid flows away from the pipeline to the reservoir—from point B toward point A; fluid retreats from the valve.

4. If the fluid detaches the valve (at point B) completely or not, anyway, it creates a low-pressure zone in vicinity of the valve. Thus, we have an area of decompression in vicinity of the valve; this area produces a decompression wave that propagates from point B to point A again at velocity *a*. This process ends at time 3τ.

5. At the moment 3τ, the reservoir contacts with the immobile fluid at point A, and the pressure in the reservoir is higher than the pressure in the pipeline. Therefore, the fluid begins to flow from the reservoir to the pipeline, accompanying the wave that restores the initial pressure (like in the reservoir) from point A to point B.

6. Thus, at time 4τ the fluid flowing through the pipeline meets the closed valve and produces a new hydraulic shock: fluid stops, pressure rises, etc. At this moment, we return to the item 1.

The cycle 1–6 can be repeated many times, until the dissipation processes stops it. It behaves like a string: we compress it, and after the release this string does not return to the initial state and freezes in it, but moves further to the tension state, then shrinks again, and so on and so on. Only friction can stop this process.

Stages 1–3, when the pressure in vicinity of the valve is higher than in the reservoir by value Δ*p*, are sometimes called the 'shock phase'. As we understand, this phase lasts for 2τ = 2*L/a*. Usually, this time scale is used to predict the valve closing time: to produce a hydraulic shock, one must close it faster than 2τ. In practice, however, the multiplier is not 2 but smaller, usually it is chosen as ~ 1.5. Note that the whole description presented here is somewhat idealized, so in certain experiments some discrepancy with the theory can be easily observed (Kodura 2016, Muhammad 2019, Lema 2012). The finite shutting time, for example, affects all the characteristics of the water hammer process, not only the magnitude values of pressure.

Let us estimate a water hammer in a pipe of length *L* = 2 m and of diameter *D* = 2 cm with walls of width δ = 1 mm. For steel, $E \sim 2 \cdot 10^{11}$ Pa; acoustic velocity in water c_s = 1500 m/s, density ρ = 10^3 kg/m³. For such conditions, $\dfrac{\rho D c_s}{E \delta} \sim 0.2$, so $a \sim 1.4 \cdot 10^3$ m/s. The time to close a valve to cause the shock is ~3 ms—this time is too short to reach it practically. On the other hand, if someone can accomplish this, then for the velocity of water 1 m/s one can achieve a water hammer at Δ*p* ~ 14 atm. This is a decent value, especially if we take into account that the velocity produced is quite frugal.

Water hammer may damage the pipeline, but more often it contents itself with loud acoustic vibrations. Of course, as it follows from the consideration, it is mostly dangerous for long pipes, where the value of time τ is practically achievable.

Note that water hammer can be positive or negative, named not after the consequences, which are always negative, but for the reason that causes it. A water hammer that is caused by shutting a valve is called a positive hydraulic shock. This shock was considered above in details. Another way to cause the cavitation is to open the valve abruptly; this is a negative hydraulic shock. It can be explained by analogy.

5.1.4 Cavitation during water hammer

As it was considered above, a hydraulic shock consists of two phases (which, in fact, alternate): the stage when pressure in the pipe is high and the stage when this pressure is low—even lower than the outer pressure.

The second case may provoke cavitation, if the pressure is sufficiently low. In (Kalkwijk and Kranenburg 1971), one may find the old theoretical description of this problem; in (Geng et al. 2017, Jansson 2017), the modern results of numerical simulations of this phenomenon are presented.

Generally, cavitation induced by water hammer does not significantly differ from cavitation induced by any other reason, except for the spatial scale: generally, the whole pipe length can be covered by cavitation.

5.1.5 Hydraulic shock from a jet

The hydraulic impact from a jet is a direct and, in a manner of speaking, a natural demonstration of the term 'water hammer'. Here our main point is to find the pressure of a jet on an obstacle.

We will consider the scheme presented in Fig. 5.1.3. A cylindrical column of water falls on a solid surface. Generally, due to the impact, both bodies—liquid and solid—would deform. Below, we neglect the deformation of the solid surface, since (i) to take it into account we need more additional information about it, (ii) as we saw above, such approach gives only a correction to the main term corresponding only to the compression of the liquid.

The proper correlation can be obtained based on the conservation laws. The liquid column falls on the surface, stops completely and compresses; as we agreed above, the surface remains undeformed.

Thus, all the kinetic energy of the liquid column was spent on work to compress it. From this equality, we will find the pressure that has been reached in the liquid.

If the volume of the column is V, then for the liquid of density ρ moving at velocity u we have its kinetic energy as

$$E_{kin} = \frac{\rho V u^2}{2}.$$

(5.1.25)

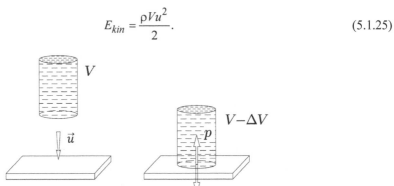

Fig. 5.1.3. A column of liquid falls on a solid surface. As a result of the impact, its volume decreases by $\Delta V \ll V$, and the excess pressure appears.

The work done to compress this liquid column is

$$A = -\int p\,dV \tag{5.1.26}$$

This work is positive, since V decreases.

Thereby, to calculate the work we need to find the pressure function for this process. Assuming that the impact is abrupt (which is quite reasonable: this is not a gentle landing of the column on the surface, this is a shock), we will consider this process adiabatic. Thus, in (5.1.26)

$$dV = \left(\frac{\partial V}{\partial p}\right)_s dp = \underbrace{\frac{1}{V}\left(\frac{\partial V}{\partial p}\right)_s}_{-\beta_s} V\,dp, \tag{5.1.27}$$

where β_s is the adiabatic compressibility.

With (5.1.27), we have for work, assuming $p = 0$ before the fall:

$$A = \int_0^p p\beta_s V\,dp. \tag{5.1.28}$$

Here we may consider $\beta_s = const$ and $V = const$ (actually, the main reason to use such parameters is to assume them as constants) since the compression of the liquid is small anyway. Therefore, the work is

$$A = \frac{\beta_s V p^2}{2}. \tag{5.1.29}$$

We have an alternative representation for compression ratio, taking into account the definition (5.1.11) for the speed of sound:

$$\beta_s = \frac{1}{\rho}\left(\frac{\partial \rho}{\partial p}\right)_s = \frac{1}{\rho c_s^2}. \tag{5.1.30}$$

Finally, from the equality

$$E_{kin} = A \tag{5.1.31}$$

with (5.1.25), (5.1.29) and (5.1.32) we have for pressure

$$p = \rho c_s u. \tag{5.1.32}$$

Some closing notes:

The relation (5.1.32) is obtained for a liquid that is being compressed: thus, despite the operation with integral (5.1.28), despite neglecting the value of this compression, the compression as a physical process is crucial to understand the impact of liquid.

Then, we neglect the deformation of the solid surface. From a common sense, it is clear that if we take into account the solid too, then not all kinetic energy would be spent on the compression of the liquid, so pressure would be lower than according to (5.1.32).

In this chapter, we examine how hydraulic shock is connected to cavitation. Above, we saw that a hydraulic shock in a pipe may cause cavitation, but there may also exist an opposite case where cavitation causes a hydraulic shock, however, on smaller spatial scales. Indeed, as we will see below, the main danger of cavitation is hydraulic shocks caused by microjets, considered in the previous chapter and to be considered further in the following sections.

5.2 Shock waves in a liquid

5.2.1 Shock wave: Definition

Let us consider a motion of a piston in a tube filled with some medium,a gas, which is a common substance to be examined in such problems, or a liquid. As long as the motion of the piston is slow, we may expect that this motion of the piston would cause a uniform motion of that medium along the tube, neglecting friction. The simplest example is an outflow from a medical syringe, which can be easily observed in domestic conditions. Yes, this is a more interesting object than it seems at first glance. For instance, if you are a champion athlete, can you push the piston hard enough to get a different flow mode of the outflow?

To understand the physical formulation of this thought (or, perhaps, real) experiment, we have to flesh out the meaning of the term 'slow motion' used above. Slow relative to what?

Small perturbations propagate in the medium at the speed of sound. Exactly, sound is a small perturbation of the medium in its nature. The speed of sound has a sense of velocity of small adiabatic (isentropic) disturbances:

$$c_s = \sqrt{\left(\frac{\partial p}{\partial \rho}\right)_s},$$ (5.2.1)

which means that one takes the derivative of pressure on density at constant entropy, which is equivalent for the reversible adiabatic process. For a perfect gas, the equation of state is

$$p = \rho \tilde{R} T,$$ (5.2.2)

where $\tilde{R} = R/\mu$, $R = 8.314$ J/mol·K is the universal gas constant, μ is molar mass, we get

$$c_s = \sqrt{\gamma \tilde{R} T},$$ (5.2.3)

where $\gamma = C_p/C_V$ is the adiabatic exponent: the ratio of isobaric and isochoric heat capacities (see all the details in the next section).

Now let us take a look at the relation (5.2.3) from another angle. The gas constant R is related to the Boltzmann constant k and the Avogadro number N_A as $R = kN_A$, that is, because $\mu = m_1 N_A$ (where m_1 is the mass of a single particle), we have

$$c_s = \sqrt{\gamma \frac{kT}{m_1}}.$$ (5.2.4)

As we see, the speed of sound is roughly equal to the mean velocity of a particle in its thermal motion: $c_s \sim v_T = \sqrt{8kT/\pi m}$. The time scale of establishing of a new state in the medium is roughly equal to the ratio of the mean free path of the molecule to the velocity: l_{mfp}/v_T; in other words, external influence cannot spread, on molecular level, at any speed; the speed of propagation of this influence is limited by the velocity of molecules in a medium, and this velocity is roughly equal to the speed of sound.

Thus, when you push the piston of a syringe, this push travels in the medium at the speed of sound, and the medium transfers this influence freely: it contracts slightly on macroscopic scale, which leads to a negligible variation in temperature.

Now let us push the piston at the speed that exceeds the speed of sound in this medium.

The piston pushes a single molecule of the medium. This averaged molecule—a molecule that has the average velocity—cannot get to its neighbor in time to transmit its energy, it cannot escape from the piston yet. Thus, the energy of that molecule that is still being pushed by the piston will increase: not to imagine that the piston directly drags a molecule, but this process looks similar.[4] To be exact, the piston pushes a large number of molecules in front of it, compressing and heating this ensemble. Thus, the new state in the small volume of the medium appears in front of the piston: denser and hotter. The formation of this new state will continue until the molecules of this new state will be able to leave the volume in vicinity of the piston.

Leaving that tiny volume near the piston, 'hot' molecules will heat the next small area of the medium. Note that these 'escaping' molecules have higher velocities than molecules in the non-disturbed medium, so the next small part of the medium will be significantly changed before it transmits this influence further.

In simple words, the initial medium is completely surprised when the flow of fast molecules from the perturbed part smashes into it, and this initial medium does not have enough time to transfer this perturbation further to another non-disturbed area before it gains the parameters that correspond to the disturbed part.

Finally, we may conclude that there must be two regions in front of the piston moving at velocity higher than the speed of sound in the initial medium: the first region is fully perturbed and propagates toward the undisturbed medium at velocity w. We may get some very rough estimations for the velocities in such a tube: denoting the regions by numbers as it is shown on Fig. 5.2.1, we may conclude that $w \sim c_{s2} > u_2 > c_{s1}$. To obtain more accurate results, we have to consider the complete hydrodynamics problem.

In accordance with previous reasoning, we may consider two areas, denoted by 1 (non-disturbed area) and 2 (the fully-perturbed area). The boundary between them is the shock wave: on this interface, parameters change abruptly.

Then, we will consider a flow in a coordinate system where the shock wave is stationary, with corresponding velocities denoted by tildes. In this system, the medium 1 moves toward the shock front at velocity $u_1 = w$, and the medium 2 runs away from it at velocity $\tilde{u}_2 = w - u_2$, where u_2 is the velocity of the piston and, correspondingly, of the disturbed medium on the laboratory coordinate system.

At the shock wave, all the quantities—mass flux, momentum flux, energy flux—must be conserved. This means that

$$\rho_1 \tilde{u}_1 = \rho_2 \tilde{u}_2, \tag{5.2.5}$$

$$p_1 + \rho_1 \tilde{u}_1^2 = p_2 + \rho_2 \tilde{u}_2^2, \tag{5.2.6}$$

$$h_1 + \frac{\tilde{u}_1^2}{2} = h_2 + \frac{\tilde{u}_2^2}{2}. \tag{5.2.7}$$

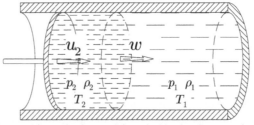

Fig. 5.2.1. The shock wave. The area 1 is steady, the area 2 is pushed by the piston at velocity u_2. The shock wave moves from left to right at velocity w. Below, we move on to the coordinate system where the shock wave is steady.

---

4   In such considerations, it is useful to remember that the piston, in its turn, also consists of molecules. Actually, we deal with the interaction 'molecules (of piston)—molecules (of gas)', not with the interaction 'continuous wall—molecules of gas'.

The last relation follows from the assumption of adiabaticity of the process (see Appendix A). Note that adiabaticity does not necessarily mean isentropic conditions: if the process is irreversible, then entropy increases; we make a certain conclusion in Section (5.2.4).

To obtain the final relations from equations (5.2.5–7), we must know some additional information about the medium—the equation of state. In addition, we have to discuss some special matters concerning, at first glance, well-known objects.

5.2.2 Two views on the adiabatic exponent

Almost everyone knows that the Poisson adiabat

$$pV^\gamma = const, \tag{5.2.8}$$

(where $\gamma = C_p/C_V$) is suitable only for a perfect gas, i.e., for gas with equation of state corresponding to (5.2.1). Then, we can show you a trick below. Follow the hands.

The first law of thermodynamics states, written with internal energy U:

$$\delta Q = dU + pdV. \tag{5.2.9}$$

The same written with enthalpy H:

$$\delta Q = dH - Vdp. \tag{5.2.10}$$

Then, for adiabatic process $\delta Q = 0$, and we obtain, dividing one equation by the other:

$$-\frac{V}{p}\left(\frac{\partial p}{\partial V}\right)_S = \underbrace{\left(\frac{\partial H}{\partial U}\right)_S}_{\kappa}. \tag{5.2.11}$$

The symbol 'S' in the derivative means that entropy remains constant during the process—the reversible adiabatic process. Then, from the obtained equation

$$\frac{dp}{p} + \kappa\frac{dV}{V} = 0 \tag{5.2.12}$$

we get the condition for adiabatic process

$$pV^\kappa = const. \tag{5.2.13}$$

Note that we did not use any assumptions for the substance, e.g., a perfect gas. Was this a mistake?

Above, following from (5.2.12) to (5.2.13), we suppose that $\kappa = const$. For a common case, it can be wrong; to ascertain it, we have to define a clearer sense of this parameter.
From (5.2.9) with evident

$$dU = \underbrace{\left(\frac{\partial U}{\partial T}\right)_V}_{C_V} dT + \left(\frac{\partial U}{\partial V}\right)_T dV \tag{5.2.14}$$

and not very evident

$$C_p = \left(\frac{\delta Q}{dT}\right)_p = C_V + \left[\left(\frac{\partial U}{\partial V}\right)_T + p\right]\left(\frac{\partial V}{\partial T}\right)_p, \tag{5.2.15}$$

which follows from (5.2.9) and (5.2.14) and gives

$$\left(\frac{\partial U}{\partial V}\right)_T + p = \left(C_p - C_V\right)\left(\frac{\partial T}{\partial V}\right)_p, \tag{5.2.16}$$

we may rewrite the first law of thermodynamics in the form

$$\delta Q = C_V\, dT + \left(C_p - C_V\right)\left(\frac{\partial T}{\partial V}\right)_p dV. \tag{5.2.17}$$

This equation suits for every substance: we did not make any assumptions about it. Here we may replace

$$dT = \left(\frac{\partial T}{\partial p}\right)_V dp + \left(\frac{\partial T}{\partial V}\right)_p dV. \tag{5.2.18}$$

and get for the adiabatic process

$$C_V\left(\frac{\partial T}{\partial p}\right)_V dp + C_p\left(\frac{\partial T}{\partial V}\right)_p dV = 0, \tag{5.2.19}$$

$$\left(\frac{\partial p}{\partial V}\right)_S = -\underbrace{\frac{C_p}{C_V}}_{\gamma}\left(\frac{\partial p}{\partial V}\right)_T. \tag{5.2.20}$$

Again, this correlation suits any substance, not only a perfect gas. Thus, we see that adiabatic exponent in its usual definition—parameter γ—defines the distinction between the adiabatic derivation and the isothermal one. To obtain the Poisson equation (5.2.9) from (5.2.20), we must have

$$\left(\frac{\partial p}{\partial V}\right)_T = -\frac{p}{V}, \tag{5.2.21}$$

and this equation is correct only for a perfect gas. In a common case, we have (5.2.11) with

$$\kappa = -\frac{C_p}{C_V}\frac{V}{p}\left(\frac{\partial p}{\partial V}\right)_T = -\gamma\frac{V}{p}\left(\frac{\partial p}{\partial V}\right)_T, \tag{5.2.22}$$

which follows with the help of (5.2.20). Thus, in the relation (5.2.13) we have to consider not the adiabatic exponent as it is, but the averaged value of (5.2.22).

Let us compare the adiabatic exponents γ and κ for water. At pressure 10^5 Pa and temperature 300 K, the parameter $\gamma \approx 1.01$, while $\kappa \approx 1.7 \cdot 10^4$; such huge difference arises due to high value of derivative $(\partial p/\partial V)_T$ which is $\sim 10^{12}$ Pa·kg/m³.

One may think that (5.2.13) with (5.2.22) gives us some usual recipes to use the Poisson-like equation for an arbitrary substance, not only for perfect gases. However, from the practical point of view, this approach has narrow area of use. For dense gases and, *a fortiori*, for liquids, the parameter κ may vary by an order of magnitude or even more in temperature ranges of about $\sim 10^2$ K. If so, any averaging does not make much sense, and the relation (5.2.13) would give a huge error from the real dependence $p(V)$ in adiabatic process.

Nevertheless, this approach to the Poisson equation with mean value of κ instead of the adiabatic exponent γ may be useful for some intermediate cases, when the gas is 'semi-perfect' so to say.

5.2.3 Two views on the speed of sound

One may think that these two views were presented above—from thermodynamics and from molecular physics. But we are going to explore another matter.

The wave process implies that some quantity A propagates in accordance with the law

$$\frac{1}{c^2}\frac{\partial^2 A}{\partial t^2} - \frac{\partial^2 A}{\partial x^2} = 0. \tag{5.2.23}$$

Actually, here may even be the 'plus' sign, it does not matter.

So, let us realize what quantity obeys (5.2.23) during the sound propagation. The sound, it is easy to check, arises when some source generates a local medium compression—the local increase in density $\rho' \ll \rho_0$ and pressure $p' \ll p_0$; index '0' corresponds to the parameters of a non-disturbed medium. We produce such a compression when we speak, for instance. Then this local compression spreads in the medium, and we call this process the 'sound propagation'.

We will consider the problem as an irrotational flow of liquid (or gas), with hydrodynamic potential φ (see the previous chapter). Thus, the velocity can be represented as

$$u = \frac{\partial \varphi}{\partial x}. \tag{5.2.24}$$

We will consider a one-directional flow. In the continuity equation

$$\frac{\partial \rho}{\partial t} + \frac{\partial (\rho u)}{\partial x} = 0 \tag{5.2.25}$$

we must use density $\rho = \rho_0 + \rho'$, and, since the deviation ρ' is small while the main pressure p_0 is constant, we get

$$\frac{\partial \rho'}{\partial t} + \rho_0 \frac{\partial u}{\partial x} = 0, \tag{5.2.26}$$

that is, with (5.2.24)

$$\frac{\partial \rho'}{\partial t} + \rho_0 \frac{\partial^2 \varphi}{\partial x^2} = 0. \tag{5.2.27}$$

Then, we perform the same operations with the Navier–Stokes equation for a sufficiently slow flow

$$\frac{\partial u}{\partial t} + \underbrace{u \frac{\partial u}{\partial x}}_{\sim 0} = -\frac{1}{\rho}\frac{\partial p}{\partial x}, \tag{5.2.28}$$

obtaining

$$\frac{\partial^2 \varphi}{\partial t \partial x} + \frac{1}{\rho_0}\frac{\partial p'}{\partial x} = 0. \tag{5.2.29}$$

From (5.2.29) we get

$$\frac{\partial \varphi}{\partial t} + \frac{p'}{\rho_0} = const, \tag{5.2.30}$$

and

$$\frac{\partial^2 \varphi}{\partial t^2} + \frac{1}{\rho_0} \frac{\partial p'}{\partial t} = 0. \tag{5.2.31}$$

The connection between the deviation in pressure p' and in density ρ' is

$$p' = \underbrace{\left(\frac{\partial p}{\partial \rho}\right)_S}_{c_s^2} \rho'. \tag{5.2.32}$$

Finally, combining (5.2.27) and (5.2.31) with (5.2.32), we find that

$$\frac{1}{c_s^2} \frac{\partial^2 \varphi}{\partial t^2} - \frac{\partial^2 \varphi}{\partial x^2} = 0. \tag{5.2.33}$$

Thus, we find out what quantity obeys the wave equation during the sound propagation: the hydrodynamic potential. The speed of its propagation is determined by (5.2.1) and now we understand how this function concerns the sound wave.

Also, it should be noted that, of course, if the hydrodynamic potential obeys the equation (5.2.33), then it is a function of a type $\varphi = \varphi(x - c_s t)$ and, consequently, the velocity of the medium (a gas or a liquid) has the same dependence $u = \partial \varphi / \partial x = u(x - c_s t)$, that is, the velocity can be described with (5.2.23) too.

Now let us examine solids. They cannot flow (usually, they are elastic; see the next section), and the whole consideration given above makes no sense here. Thus, we start from the beginning.

In solids, stress causes displacements of the elements: if the initial coordinate of a point (one may simply imagine a point marked on the solid by a pen) was x_1, then after the deformation caused by stress we would find this point at coordinate x_2, the corresponding displacement is $\varepsilon = x_2 - x_1$. We may consider displacements in all three directions; moreover, the deformation along one coordinate leads to deformations in other directions: when we stretch a rubber band, it narrows. The corresponding parameter is called the Poisson coefficient χ: the lateral compression to the longitudinal expansion ratio.

The dynamics of displacement along i-th coordinate follows the equation

$$\rho \frac{\partial^2 \varepsilon_i}{\partial t^2} = \frac{\partial \sigma_{ik}}{\partial x_k}, \tag{5.2.34}$$

which represents a variation of the Newton law formulated for an elementary volume of a solid; we used such relations for liquids too, with replacement $\partial \varepsilon_i / \partial t \to u_i$ (see the next Section 5.3). To be short, we use uniform designations of coordinates: $x = x_1, y = x_2, z = x_3$. Here and everywhere else, mute summation is used (if we see a double index in a single term, we must take a sum of all the values of this repeating index, for instance $a_{kk} = a_{11} + a_{22} + a_{33}$).

For solids, the stress tensor has a complicated form

$$\sigma_{ik} = \frac{E}{1+\chi} \left(\varepsilon'_{ik} + \frac{\chi}{1-2\chi} \varepsilon'_{ll} \delta_{ik} \right). \tag{5.2.35}$$

Here E is the Young modulus and ε'_{ik} is the deformation tensor:

$$\varepsilon'_{ik} = \frac{1}{2} \left(\frac{\partial \varepsilon_i}{\partial x_k} + \frac{\partial \varepsilon_k}{\partial x_i} + \frac{\partial \varepsilon_l}{\partial x_i} \frac{\partial \varepsilon_l}{\partial x_k} \right). \tag{5.2.36}$$

Considering deformations that depend only on the coordinate $x = x_1$ (that is, all other derivatives are zero), we obtain

$$\frac{\partial^2 \varepsilon_x}{\partial t^2} = \underbrace{\frac{E(1-\chi)}{\rho(1+\chi)(1-2\chi)}}_{c_{sl}^2} \frac{\partial^2 \varepsilon_x}{\partial x^2}, \tag{5.2.37}$$

$$\frac{\partial^2 \varepsilon_{y,z}}{\partial t^2} = \underbrace{\frac{E}{2\rho(1+\chi)}}_{c_{st}^2} \frac{\partial^2 \varepsilon_{y,z}}{\partial x^2}. \tag{5.2.38}$$

Thus, we see that there are two speed of sound in a solid: c_{sl} corresponds to the longitudinal wave, with displacements along the wave, and c_{st} corresponds to the transverse wave, with displacements athwart the direction of the wave propagation.

From the mechanical point of view, the equation (5.2.37) means about the same as the equation (5.2.33): the displacement $\varepsilon_x(x - c_{sl}t)$ and, therefore, the velocity is a similar function $u(x - c_{sl}t)$. In gases and liquids, as in solids, sound is the propagation of small disturbances. The distinction lays in another plane: for solids, we have a different relation for the speed of sound. The speed of sound in solid c_{sl} depends not on the function representing the equation of state (as in liquids), but on the elastic properties of the substance. Note that the equation of solid state is a thing-in-itself, however; not the worst way to obtain it is to combine that adiabatic derivative of pressure on density with the coefficient in the right-hand side of (5.3.37).

Note that equations (5.2.37), (5.2.38) can be rewritten with the compression modulus K which is related to the Young modulus as

$$K = \frac{E}{3(1-2\chi)}, \tag{5.2.39}$$

that is, the longitudinal acoustic velocity is

$$c_{sl} = \sqrt{\frac{3K(1-\chi)}{\rho(1+\chi)}}. \tag{5.2.40}$$

This representation can be more convenient, especially if we want to consider liquids too: for an incompressible fluid $\chi = 1/2$ and $c_{sl} = \sqrt{K/\rho}$.

The last interesting matter that can also be discussed goes something like 'In space, no one can hear you scream'.[5] The direct interpretation of this phrase is settled: in vacuum (in space) there is no substance, therefore, there is no medium to conduct a sound. However, it should be remembered that vacuum does not mean an absolute absence of medium, but only a low-pressure gas. Even in the deep space, a single particle may be found in volume of ~ 1 to, maybe, 10 liters (there exist various estimations of this parameter). Thus, vacuum is not empty indeed. Then we may look at the relation for the speed of sound (5.2.3): it contains no dependence on the density of the medium (in the frame of its correctness—for a perfect gas). Thus, we may insert in (5.2.3) the temperature of the low-pressure gas (we call it vacuum), which is around several kelvins (the famous value 2.7 K suits only for the relic radiation, particles may have higher temperature); thus, one may expect sound waves in space at speed ~ 10 m/s.

However, no one really heard any sounds in space. Why? Because to apply the whole hydrodynamic consideration, we have to consider spatial scales which are much larger than the

---

[5] The tagline of the movie Alien by Ridley Scott.

mean free path of the particle in the medium. In space, where no one hears you scream, the mean free path is huge: this value can be estimated in meters as $10^{-2}/p$, where p is measured in pascals, so for pressure $p \sim 10^{-20}$ Pa (which corresponds to one particle per liter at temperature ~ 3 K), we have mean free path of 10^{18} meters, which is around 100 light years. Thus, the spatial scale suitable for hydrodynamic consideration, i.e., to use such quantity as 'sound', must be very large. The complete phrase should be 'In space, no one can hear you scream, since we are too small', but, evidently, for the sake of brevity, authors chose the tagline as it sounds now.

5.2.4 Shock wave in a perfect gas

Now, let us return to the conditions of the adiabatic shock wave (5.2.5–7). As we stated above, to operate these equations we must know the caloric equation of state—the dependence $h(T)$. For any substance $C_p = (\partial h/\partial T)_p$, for a perfect gas, the enthalpy of which depends only on temperature, we may replace $\partial \to d$ (i.e., consider the whole derivative), and get for enthalpy

$$h = h_0 + C_p T, \tag{5.2.41}$$

where h_0 is some constant; we are not interested in it because, finally, we want to get the difference $(h_1 - h_2)$ for equation (5.2.7). Then, for a perfect gas

$$C_p - C_V = \tilde{R}, \; \mathrm{p} = \rho \tilde{R} T \tag{5.2.42}$$

thus, isobaric heat capacity can be represented through $\gamma = C_p/C_V$ as

$$C_p = \frac{\gamma \tilde{R}}{\gamma - 1} = \frac{\gamma}{\gamma - 1}\frac{p}{\rho T}; C_V = \frac{\tilde{R}}{\gamma - 1}. \tag{5.2.43}$$

With (5.2.41) and (5.2.43), we may also find that

$$h_1 - h_2 = C_p T_1 - C_p T_2 = \frac{\overbrace{\gamma \tilde{R} T_1}^{c_{s1}^2} - \overbrace{\gamma \tilde{R} T_2}^{c_{s2}^2}}{\gamma - 1} = \frac{c_{s1}^2 - c_{s2}^2}{\gamma - 1}, \tag{5.2.44}$$

that is, the enthalpy is defined, finally, by the speed of sound; c_{s1} is the speed of sound in the initial, non-disturbed gas, c_{s2} is the speed of sound behind the shock wave.
In this way, we have instead of (5.2.7) the relation

$$\frac{\gamma}{\gamma - 1}\frac{p_1}{\rho_1} + \frac{\tilde{u}_1^2}{2} = \frac{\gamma}{\gamma - 1}\frac{p_2}{\rho_2} + \frac{\tilde{u}_2^2}{2}. \tag{5.2.45}$$

From (5.2.5) and (5.2.6), we may get

$$\tilde{u}_2^2 - \tilde{u}_1^2 = \tilde{u}_2^2\left(1 - \frac{\rho_2^2}{\rho_1^2}\right) = (p_1 - p_2)\left(\frac{1}{\rho_1} + \frac{1}{\rho_2}\right), \tag{5.2.46}$$

which, accompanied by (5.2.45), gives a relation that connects $\delta p = p_2/p_1$ to $\delta \rho = \rho_2/\rho_1$:

$$\delta p = \frac{1 - \gamma + \delta\rho(\gamma + 1)}{1 + \gamma - \delta\rho(\gamma - 1)}, \tag{5.2.47}$$

$$\delta\rho = \frac{\gamma - 1 + \delta p\,(\gamma + 1)}{\gamma + 1 + \delta p\,(\gamma - 1)}. \tag{5.2.48}$$

These equations are called the Hugoniot adiabat; it describes the correlation between pressure and density in adiabatic shock waves and it significantly differs from the Poisson adiabat which may give from (5.2.8) the relation

$$\delta p = \delta\rho^{\gamma}. \tag{5.2.49}$$

Adiabats (5.2.47) and (5.2.49) are shown in Fig. 5.2.2.

Also, note that from (5.2.48) it follows that for $\delta p \to \infty$ the corresponding density tends to

$$\delta\rho_{max} = \frac{\gamma + 1}{\gamma - 1}, \tag{5.2.50}$$

which represents a limitation to compression of gas in a shock wave: contrary to the Poisson adiabat that can provide any degree of compression, the Hugoniot adiabat provides only compression of diatomic gas with $\gamma = 1.4$ by 6 times. That is a limit, and it can be easily explained from the thermodynamic point of view. The Poisson adiabat describes a reversible process of compression, when entropy of the initial state is equal to the entropy of the final state. The Hugoniot adiabat corresponds to an irreversible process, when the final value of entropy is higher than the initial one. The irreversible process is always less effective than the equilibrium one, thus the compression is lesser now. Indeed, the entropy of a perfect gas is

$$S = S_0 + C_V\,\ln T - \tilde{R}\ln\rho = S_0 + \frac{\tilde{R}}{\gamma - 1}\ln\frac{p}{\tilde{R}\rho} - \frac{\tilde{R}}{\gamma - 1}\ln\rho^{\gamma-1} = \underbrace{S_0 - \frac{\tilde{R}}{\gamma - 1}\ln\tilde{R}}_{S_0'} + \frac{\tilde{R}}{\gamma - 1}\ln\frac{p}{\rho^{\gamma}}. \tag{5.2.51}$$

Thus, the difference between the two values of entropy is

$$S_2 - S_1 = \frac{\tilde{R}}{\gamma - 1}\ln\left(\frac{\delta p}{\delta\rho^{\gamma}}\right) > 0, \tag{5.2.52}$$

because the Hugoniot adiabat gives larger values that the Poisson one; see the Fig. 5.2.2.

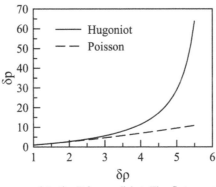

Fig. 5.2.2. The Hugoniot adiabat compared to the Poisson adiabat. The first one tends to infinity while $\delta\rho \to 6$ for diatomic gas.

Temperature can be found with the same method. From (5.2.7) with (5.2.41) and (5.2.46), for $\delta T = T_2/T_1$ we have

$$\delta T = \frac{\gamma + 1 + \delta\rho^{-1}(1-\gamma)}{\gamma + 1 + \delta\rho(1-\gamma)}. \tag{5.2.53}$$

Again, at $\delta\rho \to (\gamma + 1)/(\gamma - 1)$ temperature $\delta T \to \infty$.

Then we find the velocity of the shock wave. In the system (5.2.5–7), $\tilde{u}_1 = w$, thus, from (5.2.5), we get $\tilde{u}_2 = w/\delta\rho$, and we have from (5.2.44)

$$w = \sqrt{\frac{2\left(c_{s2}^2 - c_{s1}^2\right)}{(\gamma-1)\left(1-\delta\rho^{-2}\right)}} = c_{s2}\sqrt{\frac{2\left(1-\delta T^{-1}\right)}{(\gamma-1)\left(1-\delta\rho^{-2}\right)}}. \tag{5.2.54}$$

The ratio $\delta w = w/c_{s2}$ is shown in Fig. 5.2.3. As we see, our reasoning from the beginning of this section, the shock wave speed is roughly equal to the speed of sound in the disturbed zone, i.e., behind the shock wave; exactly, almost everywhere $w < 2c_{s2}$.

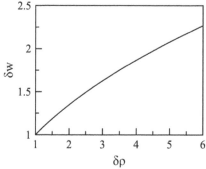

Fig. 5.2.3. The dependence of the shock wave speed (relative to the speed of sound behind the shock wave) on the ratio $\delta\rho = \rho_2/\rho_1$ in diatomic gas; the ratio w/c_{s1} can be obtained from these values by multiplying by $\sqrt{\delta T}$ from (5.2.52).

We may calculate the velocity u_2—the velocity of the medium behind the shock wave. From (5.2.5) we have

$$u_2 = w(1 - \delta\rho^{-1}). \tag{5.2.55}$$

Thus, from the obtained equations, we can see that at the limit $\delta\rho \to 1$: $\delta T \to 1$ (no density jump, no temperature jump), $w \to c_{s1} = c_{s2}$ (the 'shock wave' propagates at the speed of sound in the non-disturbed medium; both media are non-disturbed in this case, however), $u_2 \to 0$ (both gas behind the shock wave and the piston do not move). In sum, this case corresponds to the common sound wave spreading in a homogeneous medium without any effect on it. In other words, sound wave is the limiting case of the shock wave.

5.2.5 Shock wave of decompression

In Section 5.2.1, we discussed a shock wave caused by suppression, when the piston in Fig. 5.2.1 moves from left to right. Now, let us consider an opposite situation: initially, the piston abruptly (very fast) moves to the left, causing decompression in the medium. Can we expect a shock wave in this case?

From the consideration given above, the negative answer is clear. When the piston moves left, causing underpressure, molecules in the medium do not get exceeding energy; on the contrary, the medium will be cooled down under those conditions. Thus, there are no areas in the medium within the tube with special provisos to produce high-energy molecules.

The excellent monography (Zeldovich and Raizer 1967) contains the detailed discussion about impossibility of shock waves of decompression in 'regular' medium from various positions. Actually, for a very rare case of condensed medium, such special conditions can be formed. However, the condensed medium is a special case anyway.

5.2.6 Shock wave in a condensed medium

The system of equations (5.2.5–7) is suitable for every medium, not only a perfect gas, not only a gas at all. To obtain certain equations for the shock wave like in previous sections (the Hugoniot adiabat etc.), we must only know the equation of state.

The equation of liquid state, however, is much more difficult than the Clapeyron equation for a perfect gas. Generally, we have no certain relation yet, but we follow the formal way. The dependence of the pressure in the liquid can be represented as the series

$$p = \sum_{i=0}^{\infty} A_i(T)\rho^i. \tag{5.2.56}$$

This equation has almost general area of application, except the region in vicinity of critical point. Such representation does not oblige to anything; coefficients A_i can be obtained from experimental data, calculated by means of some theoretical methods, etc. Anyway, the attempt to use this equation cannot provide a convenient analytical result. However, we may simplify the expression (5.2.54) and represent it only with two terms:

$$p = p_0 + A\rho^n. \tag{5.2.57}$$

In (Zeldovich and Raizer 1967), a similar approach was discussed; to be exact, the caloric equation of state was considered where constant $A(S)$ depends on entropy, but the derivative at shock adiabat was taken by assumption $A = const$ indeed.

From (5.2.57), (5.2.1) and (5.2.20) we have the equation

$$c_s^2 = \gamma A n \rho^{n-1} \approx A n \rho^{n-1}, \tag{5.2.58}$$

since for liquids $\gamma \approx 1$, which can be used to establish the correlation between A and n.

Now take a close look at the problem of finding the adiabat correlation like the Hugoniot equation but for a liquid. For the condensed phase, we have the same equation (5.2.46) that was derived from common equations (5.2.5) and (5.2.6). Instead of (5.2.45), we have the most common and complicated equation yet

$$2(h_2 - h_1) = (p_2 - p_1)\left(\frac{1}{\rho_1} + \frac{1}{\rho_2}\right). \tag{5.2.59}$$

Our aim is to obtain an analogue of the Hugoniot adiabat for this equation. At this point, it is natural to get the proper kind of the function of enthalpy. Above, for a perfect gas, we use the Joule law: enthalpy is a function only on T. Here we apply the common knowledge that any thermodynamic potential, including enthalpy, can be represented as a function of two variables from triad (p, ρ, T). For our purposes, we choose p and ρ, so, at first glance, we may represent in general case

$$h(p, \rho) = \sum_{i,j} A_{i,j} p^i \rho^j. \tag{5.2.60}$$

Alas, as above, here we may admit that we cannot handle such a huge construction and so we must truncate it. Thus, the first way is to extract the single term that, as we hope, would allow us to obtain an acceptable result:

$$h(p,\rho) = h_0 + Ap^m\rho^n. \tag{5.2.61}$$

It is clear that for $m = -n = 1$ and for $A = C_p/\tilde{R}$, we have the correlation for a perfect gas.

Parameters in (5.2.61) can be obtained directly by approximation of experimental data. However, it should be understood that such a simple correlation would feature a great error margin almost inevitably. The initial moderate error margin in the function leads to a huge error in its derivatives, which determines other thermodynamic functions. Thus, in view of the inevitable error, we have to hold some minimal self-consistence of the function (5.2.61).

For these purposes, we take thermodynamic correlations into account. First, from the second law of thermodynamics, we have

$$\frac{1}{\rho} = \left(\frac{\partial h}{\partial p}\right)_S. \tag{5.2.62}$$

Then, the derivative in (5.2.61) can be represented as

$$\left(\frac{\partial h}{\partial p}\right)_S = \left(\frac{\partial h}{\partial p}\right)_\rho + \left(\frac{\partial \rho}{\partial p}\right)_S\left(\frac{\partial h}{\partial \rho}\right)_p = \left(\frac{\partial h}{\partial p}\right)_\rho + \frac{1}{c_s^2}\left(\frac{\partial h}{\partial \rho}\right)_p. \tag{5.2.63}$$

Combination with (5.5.61) and (5.5.10) gives the equation

$$\left(\frac{\partial h}{\partial p}\right)_\rho + \frac{1}{c_s^2}\left(\frac{\partial h}{\partial \rho}\right)_p = \frac{1}{\rho} \tag{5.2.64}$$

that connects two partial derivatives of enthalpy and can be used to find the coefficient A in (5.2.61)

$$Amp^{m-1}\rho^{n+1} + \frac{Anp^m\rho}{c_s^2} = 1. \tag{5.2.65}$$

This relation establishes the connection between the constant A and the speed of sound c_s.

Thus, for the representation (5.5.61), using the same designations as above $\delta p = p_2/p_1$ and $\delta\rho = \rho_2/\rho_1$, we get from (5.2.59) with (5.2.61) the shock adiabat for a liquid

$$2Ap_1^{m-1}\rho_1^{n+1}\left(\delta p^m - \delta\rho^n\right) = (\delta p - 1)\left(1 + \frac{1}{\delta\rho}\right). \tag{5.2.66}$$

However, the function (5.2.61) does not provide the convenient relation for enthalpy. For most liquids, dependences $h(p, \rho = const)$ and $h(p = cosnt, \rho)$ differ significantly.

The function $h(p)$ is very smooth. For water at $\rho = 990$ kg/m³, enthalpy is 200 kJ/kg at pressure of 1 MPa and 240 kJ/kg at pressure of 10 MPa, i.e., it varies by ~ 10% when the argument varies by an order of magnitude. In contrast, the function $h(\rho)$ is quite sharp: at $p = 1$ MPa, enthalpy is 47 kJ/kg at density of 1000 kg/m³ and 200 at density of 990 kg/m³, i.e., it varies by ~ 100% of its value when the argument varies by one percent.

Thus, it is much more convenient to represent the function $h(p, \rho)$ in a different manner:

$$h(p, \rho) = h_0 + A(1 + B \ln p)\rho^m. \tag{5.2.67}$$

In this equation and in such relations in general, the function of a form ln p—the logarithm of dimensioned quantity—demands explanations. Indeed, the result of the operation depends on the dimension that has been used in the argument of the function. Some would use pascals and get one results, others would use atmospheres and obtain a different value. Strictly, the correlations in the form of (5.2.67) should be avoided; if someone uses them, as we are, then one must provide the precise dimensions of the variables. Our correlation assumes that p is measured in pascals.

Similar to (5.2.65), from (5.2.67) we wave the additional relation for the speed of sound:

$$c_S = \sqrt{\frac{A(1+B\ln p)m\rho^m}{1-\dfrac{AB\rho^{m+1}}{p}}}. \qquad (5.2.68)$$

For correlation (5.2.67), we may get some quantitative results. First, let us estimate the parameters for expression (5.2.67). For water, to compare data at pressures $\sim 10^{5-6}$ Pa and densities $\sim 950 - 1000$ kg/m³: $A = 8.64 \cdot 10^{51}$ J/(kg(kg·m⁻³)¹⁵·⁵), $B = 0.01$, $m = -15.5$, $h_0 = -1.9 \cdot 10^4$ J/kg. The corresponding function is shown in Fig. 5.2.4. Of course, such huge values in the degree themselves (and, correspondingly, huge value of A) admit that such a simple function as (5.2.67) is not very convenient (to say softly) for precise calculations. However, we have no need for precise calculations: we want to obtain estimations.

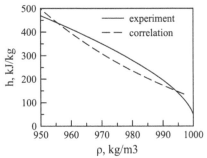

Fig. 5.2.4. Correlation (5.2.67) in comparison with experimental results for water at 10 atm.

With (5.2.67), one may obtain the expression for the shock adiabat in the form

$$A\rho_1^{m+1}\left(\delta\rho^m + \delta\rho^m \ln p_1 - B - 1\right) + AB\rho_1^{m+1}\delta\rho^m \ln \delta p = p_1\left(\delta p - 1\right)\left(1 + \delta\rho^{-1}\right). \qquad (5.2.69)$$

The expression (5.2.69) is valid in the range of approximation (5.2.67) which, for density, is shown in Fig. 5.2.4. As we see, this range is quite narrow; so, we may try to get another relation for the shock adiabat in a liquid.

Let us rewrite the basic equation as

$$2\underbrace{\left(h\left(p+\Delta p, \rho+\Delta\rho\right)-h\left(p,\rho\right)\right)}_{\Delta h} = \Delta p\left(\frac{1}{\rho}+\frac{1}{\rho+\Delta\rho}\right). \qquad (5.2.70)$$

i.e. we use the representation of enthalpy through the pressure p and density ρ again and show the differences in p and ρ.

Then, for the difference of enthalpy behind the shock wave and in front of it, we have

$$\Delta h = h\left(p+\Delta p, \rho+\Delta\rho\right)-h\left(p,\rho\right) = h_\rho\Delta p + h_p\Delta\rho + \frac{h_{pp}\Delta p^2}{2} + \frac{h_{\rho\rho}\Delta\rho^2}{2} + h_{p\rho}\Delta p\Delta\rho \quad (5.2.71)$$

where h with indexes means the corresponding derivation in the form

$$h_x = \left(\frac{\partial h}{\partial x}\right)_y, h_{xy} = \frac{\partial^2 h}{\partial x \partial y}, \text{ etc.} \tag{5.2.72}$$

and we still have the correlation (5.2.64)

$$h_\rho = \left(\rho^{-1} - h_p\right) c_s^2. \tag{5.2.73}$$

Correspondingly, we must expand the right-hand side of (5.2.70) to be in series up to the second order of deviations of pressure and density:

$$\Delta p \left(\frac{1}{\rho} + \frac{1}{\rho + \Delta \rho}\right) \approx \frac{2\Delta p}{\rho} - \frac{\Delta p \Delta \rho}{\rho^2}. \tag{5.2.74}$$

Thus, combining equations (5.2.71) and (5.2.74) into (5.2.70), we get the equation

$$\underbrace{\frac{h_{\rho\rho}}{2}}_{A} \Delta \rho^2 + \underbrace{\left(h_\rho + h_{p\rho}\Delta p + \frac{\Delta p}{2\rho^2}\right)}_{B} \Delta \rho + \underbrace{h_p \Delta p + \frac{h_{pp}\Delta p^2}{2} - \frac{\Delta p}{\rho}}_{C} = 0, \tag{5.2.75}$$

which has the obvious solution for $\Delta \rho$

$$\Delta \rho = \frac{-B \pm \sqrt{B^2 - 4AC}}{2A}. \tag{5.2.76}$$

Let us analyze the derivatives in (5.2.76). The dependences of enthalpy and the speed of sound are shown in Fig. 5.2.5 and 5.2.6.

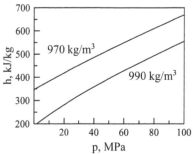

Fig. 5.2.5. Enthalpy of liquid water in the function of pressure for two values of density. Data from the NIST database.

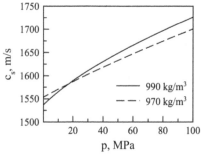

Fig. 5.2.6. Speed of sound in liquid water as the function of pressure for two values of density. Data from the NIST database.

First, we see that $h_p > 0$, $h_{pp} < 0$. As estimations, one may accept values $h_p \sim 10^{-3}$ J/kgPa (indeed, here the multiplier is greater than 1), $h_{pp} \sim 10^{-9}$ J/kgPa^2. Thus, with (5.2.71) we have, using the speed of sound $c_s \sim 10^3$ m/s, $h_\rho \sim -1$ J·m^3/kg^4; $h_{\rho\rho} \sim -10$ J·m^6/kg^7. Possibly, it is time to remind that the sign '$\sim$' means the multiplier of an order of unity. Thus, we see that $A < 0$, $B < 0$, $C > 0$, and the only option is to choose a negative sign before the square root in (5.2.76), otherwise the density difference will be negative.

The equation (5.2.76) represents the final answer—the adiabat for a weak shock. However, we want to simplify it further. For a weak wave, we may expect anyway that $\Delta\rho \sim \Delta p/c_s \sim 1$ kg/m^3, like for the isoentropic Poisson adiabat. Thus, even for $\Delta p \sim 1$ MPa, we have $\Delta\rho \sim 1$ kg/m^3. Thus, we may neglect all the terms that contain $\Delta\rho^2$ and $\Delta p\Delta\rho$, and get the simple equation for the adiabat

$$\Delta\rho = \frac{\Delta p}{c_s^2}\left(1 - \zeta\Delta p\right),\ \zeta = \frac{-h_{pp}}{2\left(h_p\rho - 1\right)}. \tag{5.2.77}$$

This is the simplest form of the adiabat of a weak shock wave. As it follows from the previous consideration, $\zeta \sim 10^{-10}$ Pa^{-1} (it is time to combine these factors together: $h_p\rho$ is lower by an order of several units, so we multiply the result by 2). The function (5.2.77) is represented in Fig. 5.2.7 schematically; of course, for calculations we may put $c_s = const = 1600$ m/s (see Fig. 5.2.6).

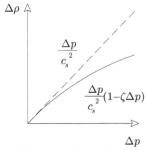

Fig. 5.2.7. The adiabat for a weak shock wave in liquid. Liquids are almost incompressible, so the solid line is very close to the dashed one indeed.

Relation (5.2.77) can be reformulated for our weak case in another manner. If

$$\zeta\Delta p \ll 1, \tag{5.2.78}$$

then we can rewrite (5.2.77) as an explicit dependence for $\Delta p(\Delta\rho)$:

$$\Delta p = \frac{c_s^2\Delta\rho}{1 - \zeta c_s^2\Delta\rho}. \tag{5.2.79}$$

Then we return to the Hugoniot equations (5.2.5), (5.2.6). From the first two equations we may obtain the relations for velocities on both sides relative to the shock front:

$$\tilde{u}_1^2 = \frac{\Delta p}{\rho_1\left(1 - \rho_1/\rho_2\right)} = \frac{\Delta p}{\left(\rho_1/\rho_2\right)\left(\rho_2 - \rho_1\right)} = \frac{\Delta p\delta\rho}{\Delta\rho}, \tag{5.2.80}$$

$$\tilde{u}_2^2 = \frac{\Delta p}{\rho_2\left(\rho_2/\rho_1 - 1\right)} = \frac{\Delta p}{\left(\rho_2/\rho_1\right)\left(\rho_2 - \rho_1\right)} = \frac{\Delta p\delta\rho^{-1}}{\Delta\rho}, \tag{5.2.81}$$

where we use the designations as in the rest of this chapter:

$$\Delta p = p_2 - p_1,\ \Delta\rho = \rho_2 - \rho_1,\ \delta\rho = \frac{\rho_2}{\rho_1}. \tag{5.2.82}$$

We may substitute (5.2.79) with (5.2.80) and (5.2.81) to get the explicit dependence of velocities on the deviation in density:

$$\tilde{u}_1^2 = \frac{c_s^2 \delta\rho}{1 - \zeta c_s^2 \Delta\rho}, \tag{5.2.83}$$

$$\tilde{u}_2^2 = \frac{c_s^2 \delta\rho^{-1}}{1 - \zeta c_s^2 \Delta\rho}. \tag{5.2.84}$$

Note that the equation (5.2.83) also defines the shock wave velocity if the shock front propagates in a steady liquid (see Section 5.2.1). Again, using the fact that the last term in the denominators is small, one may rewrite (5.2.83), (5.2.84) with replacing

$$\left(1 - \zeta c_s^2 \Delta\rho\right)^{-1} \approx 1 + \zeta c_s^2 \Delta\rho, \tag{5.2.85}$$

if this way seems more convenient.

It should be emphasized that the equations obtained above cannot be considered for a perfect gas because through all the derivation, we used the fact that the medium is almost incompressible and $\Delta\rho/\rho_1 \ll 1$, which defines the parameter ζ as it was derived above.

5.2.7 Cavitation in a droplet

Now we are ready to consider the final subject of this section: how shock waves may cause cavitation in a liquid. It may seem weird, but we can observe cavitation in a droplet that strikes a solid surface. Probably, this is the most amazing example of cavitation: a bubble arises inside a droplet that hits a surface.

As it follows from experiments and numerical simulations (we especially recommend (Kondo and Ando 2016)), when a droplet falls on a solid surface at high speed, its bottom deforms and produces a shock wave that propagates into the bulk of a droplet. When this wave reaches the upper side of the droplet, it reflects back downward, focusing on the upper part of the droplet and causes a rupture of the liquid (see Fig. 5.2.8). According to (Kondo and Ando 2016), for the initial speed

Fig. 5.2.8A. After the collision at high speed, the liquid at the bottom is stressed and shock front is formed inside the droplet.

Fig. 5.2.8B. The shock front propagates upward. In the bottom side, liquid jets are formed.

Fig. 5.2.8C. Once the shock wave reaches the top, it goes back to the bulk of the droplet as a reflected tension wave.

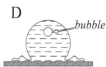

Fig. 5.2.8D. Focusing at some point on the upper side of the droplet, the reflected tension wave ruptures the liquid, and the cavitation bubble is formed.

of the droplet at 110 m/s (actually, in the corresponding experiment the droplet has been stricken by a metal slider) at the moment of rupture, the pressure inside a droplet is negative and is around –20 MPa.

Thus, 'cavitation is everywhere'. Even in an object that is, at first glance, so unlikely to cavitate—a drop of liquid—we may observe the origin of a cavern.

5.3 Rheology

5.3.1 Among models

Physics does not describe nature directly. It uses models, which must

- ✓ describe some properties of a real object,
- ✓ be simple in use.

With time, physics developed proper models for almost every object or process, so sometimes we do not even think that we are dealing with a physical model but not with a real object. We consider the motion of points (instead of bodies of finite sizes), assume that fluids are incompressible media or consider gases without interaction of particles, etc. However, in most cases, models provide reasonable results, freeing us from the detailed analysis of fundamentals of these models.

Sometimes, absolutely different models, indeed, describe the properties of the same physical nature. The stress response of the applied strain is a complicated reaction; however, we clearly differentiate two limiting cases:

- ✓ stress linearly depends on deformation; this approach corresponds to elastic media; most solids behave so;
- ✓ stress depends on the rate of deformation; we describe the fluid properties in this way.

After one-minute thinking, anyone would say that it is impossible to attribute any body to one of two opposite groups, sharing nothing in common between them. In other words, after one-minute thinking anyone would invent rheology—the science field that explores the common relationships between strains and stresses.

Preliminary, we may note some matters of the problem. From the technical point of view, the simplest way to obtain the general stress–strain correlation is to mix the dependencies of strain and the strain rate in some proportions. This is a productive recipe for obtaining the correlation for practical use.

The more complicated way is to analyze the physical foundations of limiting cases. On one side, we have the dependence on strain ε, on the opposite side—of its rate $\dot{\varepsilon}$. Thus, one may assume that the result depends on the time scale of the process τ: if this scale is short, then the rate ε/τ is high, and we must take it into account. For any substance—liquid or solid.

However, the time scale depends on the inner structure of the medium. In liquids, molecules move relative to each other more 'willingly' than in solids, where this motion can be possible only under special conditions. Thus, almost any influence, at almost any rate, in a fluid causes the relative motion of its parts, and the stress within it almost always must depend on the strain rate, not on the strain.

The exception is intriguing. When we describe a fluid with such property as viscosity, we imply motion and friction of macroscopic parts of the liquid—not separate molecules. Thus, strictly, one has to wait until the collective motion of particles begins; then we may call this motion the 'flow'. As long as the collective movement does not start, this is not the flow—this is the preliminary stage, when the fluid does not flow yet. Because it is immobile, one may consider a liquid at these time scales as solid.

In other words, at short time scales solids behave like fluids, but at very short time scales fluids behave like solids. Sic. However, skeptical minds who dislike any paradoxes may note that this is a tensioned combination: finally, at a very-very short time scale all molecules are steady; thus, in this way of explanation, we may consider the medium as a solid at zero temperature, which is kind of absurd. Well, explanation is a complicated exercise, and the edges of permissible are fuzzy...

5.3.2 The mechanics of continuous media

Continuous medium is a substance with properties (such as density, temperature, etc.), which can be properly defined and vary relatively weakly, so their spatial dependence is smooth.

Following the usual procedure, let us consider a selected 'particle' of the substance—the elementary volume $dV = dx_1 dx_2 dx_3$ of the continuous medium. As for any particle, the dynamics of volume obeys the Newton law:

$$\frac{d\vec{P}}{dt} = \vec{F}, \tag{5.3.1}$$

where $\vec{P} = m\vec{v}$ is the momentum of the 'particle'. In our case, the elementary volume has the mass $dm = \rho dV$.

Generally, in (5.3.1) $\vec{F}$ are all the forces that act on the volume, including forces of gravity, pressure, electrical field, etc. We are interested only in the inner forces—forces that arise due to stress in the very medium, so further we will neglect all other ones.

Let $\vec{\varepsilon}(x, y, z)$ be the deformation of the medium at the given point; these deformations condition the inner forces we are looking for—the stresses inside the medium. Note since our volume is elementary, then the forces must be differentially small too, that is, $\vec{F} \to d\vec{F}$.

Then, three components of the force dF_i (here i denotes direction: dF_1 corresponds to the force along the coordinate x_1) act on each of the cube's faces, giving the resulting force for each pair of the opposite faces; faces can be denoted by the index k, where k corresponds to the axis that is normal for this face. Now the resulting force that acts on all three pairs of opposite faces in the direction along the axis x_i is

$$dF_{i1} + dF_{i2} + dF_{i3} = \sum_{k=1}^{3} dF_{ik}, \tag{5.3.2}$$

Then, we introduce the stress σ_{ik}—the force F_{ik} acting on the given face multiplied by the area of the given face, i.e.,

$$dF_{ik} = d\sigma_{ik} dx_l dx_n, l \neq n; l, n \neq k. \tag{5.3.3}$$

Note that σ_{ik} represents a tensor, i.e., consists of nine components.

Finally, we get from (5.31) the equation for one projection of force and velocity

$$\frac{d\rho u_i}{dt} dx_1 dx_2 dx_3 = \sum_{k=1}^{3} d\sigma_{ik} dx_l dx_n, l \neq n; l, n \neq k. \tag{5.3.4}$$

When we divide (5.3.4) by the product $\prod_{k=1}^{3} dx_k$, from every term in the sum from the right-hand side of (5.3.4) we obtain the derivative $\partial\sigma_{ik}/\partial x_k$. Thus, using the silent sum rule,[6] we may rewrite (5.3.4) in a form

$$\frac{d\rho u_i}{dt} = \frac{\partial\sigma_{ik}}{\partial x_k}. \tag{5.3.5}$$

---

[6] In hydrodynamics, it is very convenient to use the summation rule $a_{ik}b_k = \sum_{k} a_{ik}b_k$, that is, we imply the mute summation on the repeated index again.

Keep in mind that we omit all other forces from here; these forces can be added for certain problems, but this case is not about them now.

Through the whole derivation of (5.3.5), we successfully avoid any considerations about the kind of the medium (liquid or solid), as well as about the certain physical nature of the inner stresses σ; we only mention in passing that stress must somehow depend on the deformation ε. Now it is time to finish our mathematical model.

Traditionally, all condensed media are divided in two types: elastic solids and fluent liquids. This does not mean that any medium can be necessarily attributed to one of these types, this means only a tradition.

For elastic solids, the common form of a stress tensor can be expressed directly with deformations ε (Landau and Lifshitz 1970). For small deformations, when one may neglect the quadratic term in the deformation tensor (the last term in (5.2.35)), we have:

$$\sigma_{ik} = G\left(\frac{\partial \varepsilon_i}{\partial x_k} + \frac{\partial \varepsilon_k}{\partial x_i} - \frac{2}{3}\frac{\partial \varepsilon_l}{\partial x_l}\delta_{ik} \right) + K\frac{\partial \varepsilon_l}{\partial x_l}\delta_{ik}. \tag{5.3.6}$$

where we use the Kronecker symbol: $\delta_{ik} = 1$ if $i = k$, $\delta_{ik} = 0$ if $i \neq k$. In the theory of elasticity, the parameter G is called the shear modulus, and K is the compression modulus; $K = \beta^{-1} = \rho(\partial p/\partial \rho)$.

For fluids, stress depends not only on deformations themselves, but on their rate—on the velocity $\vec{u} = \partial\vec{\varepsilon}/\partial t$; the common expression for the stress tensor is (Landau and Lifshitz 1987):

$$\sigma_{ik} = \eta\left(\frac{\partial u_i}{\partial x_k} + \frac{\partial u_k}{\partial x_i} - \frac{2}{3}\frac{\partial u_l}{\partial x_l}\delta_{ik} \right) + \zeta\frac{\partial u_l}{\partial x_l}\delta_{ik}. \tag{5.3.7}$$

In hydrodynamics, the parameter η is viscosity, and ζ is the second (or the volume) viscosity.

We will not discuss the similarity of expressions (5.3.6) and (5.3.7) now, but concentrate on more general issues.

5.3.3 Fluidity and elasticity: The traditional way

As we have seen above, the correlation between the stress tensor σ and the deformation ε may be of two kinds: stress may depend immediately on ε or on $\dot{\varepsilon} = \partial\varepsilon/\partial t$, i.e., on velocity. The first case corresponds to elastic media (i.e., for solids in their ideal representation), the second one corresponds to viscous liquids. In simple words, stress may depend either on the deformation of the media (solids), or on the deformation rate (liquids).

However, it is simple to point out the examples of media that do not correspond to any of the cases listed above. Plastic solids, regular solids at the temperature range close to the melting point, liquids close to the crystallization point and many other cases. Thus, in general case, one may conclude—it is the first thing that comes to mind—that stress depends both on ε and on $\dot{\varepsilon}$. The 'either–or' dilemma is ambiguous; there may exist a more complex situation for a certain medium.

There are many models to describe tension in plastic-elastic media. The usual way to construct the correlation $\sigma(\varepsilon, \dot{\varepsilon})$ is to combine elastic and plastic elements into certain schemes. The closest analogy is electrotechnics: two elements may be connected in parallel, in sequence or in a more complicated combination. In each case, we obtain that the total voltage or the total current is the sum of separate values, while the other quantity is the same for the total scheme: for instance, for the series connection, the total values $U_t = U_1 + U_2$, $I_t = I_1 = I_2$.

Below, we illustrate these ideas for a simplified 1D-case (West et al. 2003). We assume that for the elastic element

$$\sigma_e = C_e\varepsilon, \tag{5.3.8}$$

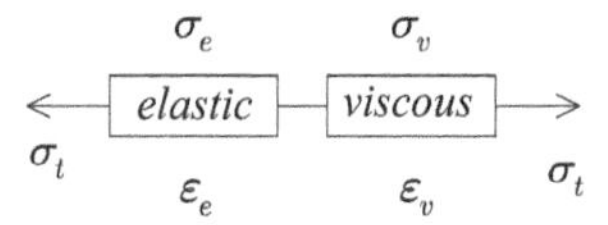

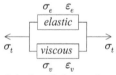

Fig. 5.3.1. The Maxwell elastic-viscous element. The total deformation is the sum of deformations of each separate element, the total stress is similar for each element and for the total.

while for the viscous one

$$\sigma_v = C_v \dot{\varepsilon}. \tag{5.3.9}$$

Now we will combine these elements.

The series connection is called the Maxwell model. Here (see Fig. 5.3.1) the net stress is equal to the tension of each element while the total strain is the sum of the strains. Thereby,

$$\sigma_t = \sigma_e = \sigma_v, \; \varepsilon_t = \varepsilon_e + \varepsilon_v. \tag{5.3.10}$$

Taking the time derivative from the second equation and using the first one, we obtain

$$\dot{\varepsilon}_t = \underbrace{\dot{\varepsilon}_e}_{\dot{\sigma}_e / C_e} + \underbrace{\dot{\varepsilon}_v}_{\sigma_v / C_v}, \tag{5.3.11}$$

thereby, we get the correlation between the tension and the deformation of the Maxwell's element in the form of a differential equation

$$\dot{\varepsilon}_t = \frac{\dot{\sigma}_t}{C_e} + \frac{\sigma_t}{C_v}. \tag{5.3.12}$$

The parallel connection—the Kelvin model, see Fig. 5.3.2—corresponds to the same strains and additive stress of elements:

$$\sigma_t = \sigma_e + \sigma_v, \; \varepsilon_t = \varepsilon_e = \varepsilon_v. \tag{5.3.13}$$

Fig. 5.3.2. The Kelvin element. Deformations of elastic and viscous elements are the same, while the total stress is the sum of stresses of both elements.

Therefore, we have directly with (5.3.8) and (5.3.9):

$$\sigma_t = C_e \varepsilon_t + C_v \dot{\varepsilon}_t. \tag{5.3.14}$$

And so on. One may combine the combinations, obtaining the results that would be consistent with experiments for the given media.

However, there exists a much more elegant method to get the description of the medium that is neither elastic, nor viscous.

5.3.4 The fractional derivative

The standard designation for the derivative of the function f of the order α is $f(\alpha)$, that is, $f^{(0)} = f(t)$, $f^{(1)} = df/dt$, etc. So, both correlations (5.3.6) and (5.3.7) can be written in the common form (West 2003)

$$\sigma_{ik} = A\left(\frac{\partial \varepsilon_i^{(\alpha)}}{\partial x_k} + \frac{\partial \varepsilon_k^{(\alpha)}}{\partial x_i} - \frac{2}{3}\frac{\partial \varepsilon_l^{(\alpha)}}{\partial x_l}\delta_{ik}\right) + B\frac{\partial \varepsilon_l^{(\alpha)}}{\partial x_l}\delta_{ik}. \tag{5.3.15}$$

Here $\varepsilon^{(\alpha)}$ is the time derivative of the order α. Thus, for an elastic solid $\alpha = 0$, $A = G$, $B = K$; for a viscous fluid $\alpha = 1$, $A = \eta$, $B = \zeta$.

We are interested in the common case—in a medium which is an intermediate between a solid and a fluid. From this point of view, we are interested in the medium for which the parameter α in (5.3.15) is between 0 and 1. Considering the meaning of the parameter α, we have to conclude that for a general medium, the stress tensor depends on the fractional order derivative of deformation. But how could it be?

In can be easy. Fractional order derivatives become more and more popular in mathematical physics (Samko 1993). Their initial sense, that made them applicable in physics, was exactly the interpolation of the derivation operator[7] for non-integer cases.

Today, there are many forms of fractional derivatives. Since this book is not devoted to this specific matter, we only mention two representations of fractional derivatives.

The classical form of the fractional operator is the Riemann–Liouville derivative

$$f_a^{(\alpha)}(x) = \frac{1}{\Gamma(1-\alpha)}\frac{d}{dx}\int_a^x \frac{f(t)\,dt}{(x-t)^\alpha}, \tag{5.3.16}$$

where a is an arbitrary parameter, $\Gamma(z) = \int_0^\infty x^{z-1}e^{-x}\,dx$ is the gamma-function.

The Marchaud derivative is more convenient for physical applications

$$f^{(\alpha)}(x) = \frac{\alpha}{\Gamma(1-\alpha)}\int_{-\infty}^x \frac{f(x)-f(t)}{(x-t)^{1+\alpha}}\,dt. \tag{5.3.17}$$

Generally, fractional derivatives open new horizons not only for rheology problems: reformulation of the Navier–Stokes equations with fractional derivatives (not only in the manner presented above, of course) creates a completely different description of turbulence. However, this is another story.

We return to rheology problems: how one can describe the elastic and plastic properties of media.

5.3.5 The solidity of a fluid

The fact that liquids have elastic properties is not surprising and has been known for quite a long time: in his work, Robert Hooke examined 'the phenomena of springy bodies whether solid or liquid' (Moyer 1977). Liquids resist compression, and this property does not distinguish them from solids. Liquids transfer small elastic oscillations, and, as we saw in previous sections, one may use both descriptions for the speed of sound in a liquid equally: (i) assuming the potential flow in it, (ii) assuming the elastic oscillations.

However, of course, fluids are not usually described as elastic media because of many reasons. First of all, the Young modulus for incompressible fluids in accordance with (5.2.38) at finite K and at Poisson ratio $\chi = 1/2$ is $E = 0$; this is not a surprise if we take into account that elasticity necessarily means compression (or tension). If we are not in a mood to discuss what came first—a chicken or an egg (the Young modulus E or the compression modulus K)—and consider the fact that the Hooke law is never applied for fluid, we may omit the question. Then, the stress tensor, as we

---

[7] To be exact, the integration operator, but the derivative follows from it immediately.

know, is expressed with velocities, not with deformations themselves. Thus, we have no practical reasons to use the elastic approach for liquids.

However, another matter is interesting. A liquid fills the jar completely, repeating its form. This fact may mean that liquids cannot resist the shear deformation absolutely, that is, any stress at the tangential direction to its surface causes the motion. In other words, one may expect that for liquids in (5.3.6) $G = 0$: indeed, we all see how fluids flow from any small influence. On the other hand, we never observe (we mean in domestic conditions) the flow at very short time scales. Some time must be spent for a fluid to be involved in a flow (Galileo and his principle of inertia can be mentioned here); within this time scale, this is not a fluid yet. Particularly, the absence of flow under the applied shear stress means non-zero G for the fluid.

This property of the fluid—finite shear modulus for liquids—was observed in experiments for periodic influence on a liquid: in (Kielczynski et al. 2001, Hutton and Phillips 1972) at frequency of $\sim 10^6$ Hz (epoxy resins, diethylhexyl phthalate), and in (Bazaron et al. 1972, 1978, 2015) at $\sim 10^5$ Hz (ethylene glycol, cyclohexanol, vaseline oil, etc.).

Thus, indeed, some fluids—very viscous fluids, we may note—demonstrate the properties of solids at short time scales.

5.3.6 The fluidity of a solid

Let us take a look at elasticity at the elementary—molecular—level.

The elasticity of a solid medium can be explained as follows. Atoms in a solid are placed in lattice sites, oscillating around the equilibrium positions (Fig. 5.3.3); this circumstance distinguishes solids from liquids, where atoms may travel more or less (rather less) freely. Then, an external perturbation occurs at the surface, so the atoms on the surface displace from their equilibrium positions. This offset of the surface particles causes the increase in the potential energy of interaction between them and their neighbors. The repulsive forces between particles arise: these forces tend to return the surface particles back to their initial places, and push their neighbors deeper into the medium.

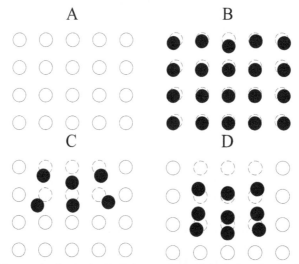

Fig. 5.3.3. The solid medium awaits the shock (A). If the magnitude of shock (i.e., the kinetic energy of the displaced particles, marked grey) is small, then the all the particles of the medium have time to offset slightly to their new positions (marked grey) and compensate the initial perturbation (B) by common forces. Thus, the solid shows its elastic properties: one particle leans on another, so the elastic wave works. But if the kinetic energy of the grey particles is high (C), then only their direct neighbors would feel the impact, other particles of the medium would not know that the penetration takes places. Grey atoms move into the medium, encountering resistance only from the particles that have been stricken immediately (D). Such a medium, despite being solid from the thermodynamic point of view, behaves like a fluid, if we consider its resistance properties.

The fate of the surface particles depends on their initial (right after the disturbance) kinetic energy. If the increased potential energy between them and their 'close neighbors' is sufficient (i.e., is at least higher than the kinetic energy by an order of magnitude) to stop them or noticeably drag them, then the process turns to the reverse phase: the displaced particles would start to get back to their equilibrium sites. Otherwise, surface atoms would continue to move deeper into the medium, continuing to push other atoms of the solid.

Meanwhile, the neighbors of the surface atoms, moving deeper into the medium, cause a local squeeze of the solid. New particles are involved in the process: now the neighbors of neighbors get an increased potential energy, and together all these particles try to prevent further penetration of the surface particles deeper into the medium.

Therefore, the deeper the atoms from the surface penetrate the medium, normally:

✓ the more and more particles come into action—the deeper are the layers of the medium drawn in this process; this does not mean that the initial atoms from the surface push all of them directly—atoms were pushed by their close neighbors;

✓ the total force from the compressed medium stops further penetration.

However, for this scheme to be realized, we need a certain time scale—not too short. To feel the response from the neighbors of neighbors (and of their neighbors, etc.), atoms in the medium must have enough time to displace, to gain the increased potential energy from their neighbors, and so on. Ideally, while the surface atoms shift at noticeable distance, other particles in the medium would have enough time to create the elastic wave in the solid: they would have time for short displacements, so the disturbance of potential energy keeps propagating through all the medium.

For very fast surface perturbations, this scheme is incorrect. If the surface particles are dislodged abruptly, the disturbance wave in the solid does not have time to propagate. The surface atoms, having a colossal kinetic energy, push away the atoms of the solid that stand in their way, but these spurned particles do not have time to involve other atoms of the medium—they do not have time to call the rest for help, so to say. Thus, such a process of propagation of atoms from the surface deeper into the medium concerns only them and those particles with which they interact directly; other particles remain immobile. Of course, we slightly exaggerate, and some additional particles of the medium would be involved in the process too, but the main character of the process of such a fast penetration is just like that. One may say that, since the collective effects in the solid are negligible, in this situation the solid behaves like a liquid.

From this consideration, it is clear what velocity is the scale for this 'anomalous' penetration. To ignore the solid and its elastic response, the initial disturbance must have a velocity that significantly exceeds the speed of sound in this solid.

Note that in the practical case, the surface atoms do not travel on their own: they are pushed by an external medium like a high-speed jet of a fluid. Thus, the type of the medium—liquid or solid—depends on the time scale: at the short ones, a solid is not solid but liquid. When a high-speed jet strikes a solid surface, it hits another liquid indeed.

Finally, we may remind that solids lose their elastic properties at very high pressures. This is a well-known fact which can be explained from a similar position: the elastic wave cannot propagate through a strongly strained medium, when atoms are noticeably displaced from their equilibrium positions. This process is out of our interest, unlike the jet strike that will be discussed further.

5.4 The effects of a fluid on a solid surface

5.4.1 Overview

Erosion is one of those beautiful things which prevents our technical devices from ever working as expected. Erosion, i.e., the surface wear due to material loss, does not allow a long work. Sometimes, the propeller screws on ships had to be replaced after only several weeks of work (Metter 1948), which, evidently, had not been included in the original cost estimate.

The fist qualitative explanations of the cavitation effect on the surface erosion were simple and a bit naïve. They thought that cavitation simply removes the products of corrosion on the construction elements, i.e., the initial reason for the cavitation damage was considered to be of chemical origin. Indeed, one may even find some confirmations for such an assumption. Erosion increases significantly with temperature (for instance (Metter 1948, Auret et al. 1993), which is not surprising from the common point of view: when temperature increases, all the elementary processes go faster. Thus, at first glance, we may refer to cavitation as a chemical phenomenon.

However, it quickly becomes clear that the rate of cavitation damage far exceeds any conceivable rate of corrosion, so other considerations were required.

The first physical mechanism of cavitation erosion was given in 1928 by S. Cook (Cook 1928). He reasonably assumed that cavitation somehow causes a hydraulic impact on the surface. This main statement of the theory is the basic part of modern theories; thus, this work survived the minor details.

Cook considered a collapse of a bubble that creates the fluid motion at velocity u. When this flow strikes the surface, its kinetic energy (per the volume unit) $\varepsilon_k = \rho u^2/2$ turns into its potential energy $\varepsilon_p = \beta p^2/2$, where $\beta = \rho^{-1}(\partial\rho/\partial p)_s$ assuming the adiabaticity of the process. Therefore, from the equality $\varepsilon_k = \varepsilon_p$ we once again obtain the water hammer equation for pressure:[8]

$$p = \rho u \sqrt{\left(\frac{\partial p}{\partial \rho}\right)_s} = \rho u c_s. \tag{5.4.1}$$

In accordance with (5.4.1), for a liquid at $\rho \sim 10^3$ kg/m^3 and $c_s \sim 10^3$ m/s we see that even for velocity of only ~ 10 m/s the pressure of the impact would be about 100 atm. However, Cook's theory could not explain the very fact that a collapsing bubble creates a fluid flow directed at the solid surface, since he considered only the spherical-symmetric collapse.

On the other hand, this detail is not crucial if we take into account the asymmetric bubble collapse near the solid wall (see Chapter 4), where the formation of a jet was examined. This jet has the velocity much greater than 10 m/s, and, in view of this phenomenon, we may conclude that, principally, the main aspect of the cavitation erosion has a good explanation: cavitation erosion is caused by water hammers from cavitating bubbles. However, of course, this simple answer is not full, and below we will discuss some possible additional factors, including the interaction between a jet and a surface.

So, it is time to consider the other side of the coin. Here and below we discuss how the fluid acts on the surface, that is, what are the processes in the liquid that cause effects on the solid. It is interesting, however, how the surface reacts to the fluid influence.

As for theoretical approaches, we may mention (Berniche et al. 2002), where the model of cavitation erosion is proposed. A series of successive impacts is considered: every cavitation impact produces a strain of a certain profile, and these strains superimpose; if a superimposed strain exceeds the rupture strain, we get the erosion of the surface.

Generally, this is a very difficult and, probably, even slightly ambiguous question. On the one hand, the certain mechanism of fracture of the surface is interesting at least from the following point of view: if we want to defend the surface from cavitation, what can we do—change the chemical properties of the material, or increase its hardness, or, otherwise, increase its plasticity, or change something else? On the other hand, the main effect of cavitation (see below) is too crude: these are impacts of a magnitude of $\sim 10^7$ Pa and higher (see below), so any modifications of the surface have

---

[8]   See also Section 5.1, where we derive this equation using another language, i.e., a slightly different physical description. The increase in potential energy occurred after the work that has been performed on a liquid.

a finite area of influence, so to say. Nevertheless, some methods to prevent cavitation erosion can be discussed (Toloian 2015):

✓ polish the surface to prevent the formation of bubbles on it; as we will see below, the hollow on the surface tends to grow;

✓ covering a metal surface with a soft coating such as plastic, if it is possible;

✓ cathodic protection: the protected surface is at negative potential—due to a connection either to an external electric power source, or to an anode made from a less electronegative material.

The last method is usually applied to prevent the corrosion of the surface; in case where the protecting potential is too high, hydrogen bubbles arise at the surface, but the role of these bubbles can be useful too (Toloian 2015).

5.4.2 Cavitation erosion: What we see...

Thus, as we see, cavitation is the source of erosion of solid surfaces. Possibly, by the way, this is the first reason why cavitation is studied: its damage resource is so high that it can destroy a metal surface in several hours.

Many works that investigate the cavitation erosion have a melancholic mood: they contain a lot of photos with piles of distorted metal with comments in a style of 'look what this beast (i.e., cavitation) does.'Our Section 1.4 from the first chapter is designed in almost the same fashion. However, in this section, we will discuss some presumptive reasons of the destructive forces of cavitation.

Following tradition, we begin with a photo (see Fig. 5.4.1). There we see a metal surface that looks like a mob of miniature elves hit it with small hammers (and at the same time dug a cave in it). The spatial scale of irregularities on the surface is less than 0.1 millimeter.

Of course, this landscape is caused by the effect of small jets produced in collapsing bubbles. That is the first reason that came to mind, but not the last one surely.

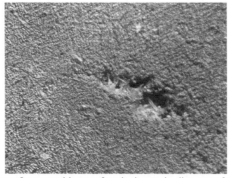

Fig. 5.4.1. The eroded titan surface after several hours of cavitation on it; diameter of the hole is approximately 1.5 mm. The surface is covered in traces of microjets. The photo is taken with an optical microscope. The total width is around 3 mm.

5.4.3. ...and what we don't

The collapse of a bubble may cause high temperature inside of it. This matter will be discussed further in details: in Chapter 7, we analyze a hypothesis that may predict (for some cases) the temperatures in collapsing bubbles higher than 10^3 K, and in Chapter 9, where the possibility of high temperatures is considered to explain the phenomenon of sonoluminescence.

Sonoluminescence is the light emission during dynamic processes in a two-phase system. It can be observed for a single oscillating bubble in a jar (the so-called single-bubble sonoluminescence) or in a cloud of bubbles (multi-bubble sonoluminescence). Thus, some theories predict that during the collapse phase high temperature causes light emission, and, indeed, observations demonstrate that the light emission occurs during the phase of bubble collapse.

On the other hand, we can arrange a cloud of bubbles near an acoustic waveguide (see Chapter 6) and observe sonoluminescence right there. Thus, in accordance with that concept, we have high-temperature bubbles on the metal surface of an acoustic waveguide. In such conditions, one may also expect that the metal surface contacting with a medium that hot (in some theories, the predicted values of temperature are around 10^4K and higher) would melt. Thus, it is interesting to get a view of the surface of the waveguide at a much better resolution.

The microphotos of the titanium waveguide are shown in Fig. 5.4.2A and 5.4.2B.

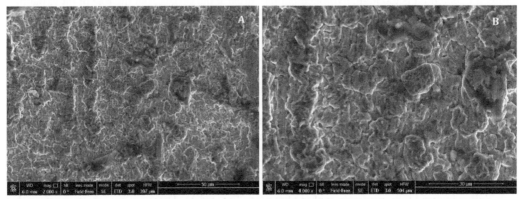

Fig. 5.4.2A,B. The titanium waveguide had suffered from cavitation and emitted sonoluminescence, so one may expect to find some melted zones here. The same part of the waveguide at different magnification is shown. The photos are taken with an electron microscope.

We cannot discover the traces of melting on this or on any such photos (Biryukov 2018). Of course, this means exactly what it seems: possibly, somewhere, one can find the evidence of the melting process. It is always difficult to prove the absence of something: thousands of evidences would be insufficient. The corresponding analysis must be complex: a theory, a lot of various experiments, etc. We will discuss such matters in Chapter 9.

Here we may conclude that we do not observe any evidence of the high-temperature effect of cavitation on the metal surface.

5.4.4 Mechanical sources of damage: Bubbles

Thus, if we cannot establish the thermal character of the erosion damage definitely, then we should consider some mechanical ones. One may divide the cavitation effects in two groups: macroscale effects and the microscale ones.

Microscale effects, generally, are caused by the bubble collapse. One can imagine the following processes:

✓ at first, the bubble collapse in a liquid leads to a pressure wave that affects the wall directly; this is a comparatively small influence on the surface;

✓ then, inside the bubble collapsing near the surface a jet can be formed; this jet has a perceptible impact on the surface;

✓ these two processes can be considered together (Dular et al. 2006): the bubble collapse in the bulk provokes a special bubble dynamics in vicinity of the surface, finally producing the high-speed jet.

Macroscale effects cause erosion due to formation of bubble clouds near the surface or due to formation of a cavitation vortex (Berchiche et al. 2002, Terwisga et al. 2009)—the large-scale structure arises during the developed cavitation. Almost always, these effects amplify the microscale effects, either due to collective effects in the bubble cloud, where pressure waves from separate bubbles may summarize and produce a heavier impact, or because of the prolonged influence of the cavitation flow caused by the vortex.

5.4.5 The effect of microjets

Let us estimate the parameters of bubble collapse in the bulk of liquid. The velocity scale of a 'calm collapse' of a symmetric bubble $u_c \sim \sqrt{\Delta p / \rho}$, this equation can be obtained even from the dimensions of physical quantities. For typical values $\Delta p \sim 10^3$ Pa, $\rho \sim 10^3$ kg/m^3, we have $u_c \sim 10$ m/s. A more precise relation for u_c also contains the multiplier $\sqrt{R_0^3 / R^3 - 1}$ (see Chapter 7), which tends to infinity when the final radius R is much smaller than the initial one R_0; but this is not truly a real case: since the gas in the bubble compresses, it will prevent the significant shrinking of the bubble (see Chapter 7), so usually R is smaller than R_0 by several times. Thus, while, in theory, one may expect any great velocities of the bubble collapse, in practice, the rate of this process is much slower. Also, it should remind the essential question to this mechanism: even if the bubble collapses at any rate, it is difficult to imagine how it transfers the velocity of the liquid to the surface.

However, there exists another process that may translate the energy of the fluid to the surface—microjets. In the previous chapter, we have seen that the bubble collapse near a solid wall produces a pit directed at the wall. Briefly, the collapse must be accompanied with the flow of the surrounding fluid, and in vicinity of the wall the velocity of the fluid is around zero because of impermeability. Initially, due to these circumstances, the external (relative to the wall) side of the bubble collapses much more rapidly than the part of the bubble near the wall. Then, the external side of the bubble produces a jet that strikes a wall at velocity that significantly exceeds u_c: in accordance with numerical simulations, the velocity of the jet is higher than u_c by an order of magnitude or even two: its value is of an order of $\sim 10^2$ m/s or even $\sim 10^3$ m/s. A fluid of such high kinetic energy hits a surface as a hammer and causes the mechanical damages. Precisely, microhammers cause microdamages.

Now let us take a look at the surface damaged by cavitation. In Fig. 5.4.3, we see that the surface is irregularly covered with traces of 'point impacts' visible as dark dots; note that this the same surface that was pictured in the Fig. 5.4.1 (but another part of it): an electron microscope provides a better resolution at similar magnification as obtained with an optical microscope. These traces of point impacts are comparatively rare: the main part of the surface does not contain them. Upon these circumstances, one may assume that among the common effects of cavitation, whatever they may be, some special events do occur.

In other words, we realize that damages may be of two types. The first one is the microdent— these dents can be observed on the main area of the solid surface. This type of damage creates the

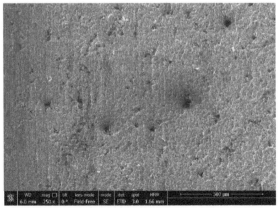

Fig. 5.4.3. The photo is taken with an electron microscope. Possibly, these microcraters can be explained by the effect of microjets created due to an asymmetric bubble collapse that took place in vicinity of the surface.

main landscape of the surface pictured in Fig. 5.4.3. It is logical to assume that this relief is caused by 'simple' impacts of microhammers on the metal surface.

However, there may be the second type of surface destruction. One can see microholes of diameter ~10 μmon the surface; these microholes are very rare, especially if we take into account the fact that the surface shown in Fig. 5.4.3 was created by dozens of hours of cavitation. That is, every square micrometer has been exposed to prolonged effects of cavitation, but we observe only a few occurrences of such a strong cavitation erosion effect on it.

These rare effects of cavitation, probably, can be explained based upon reasons discussed in section 5.3. If we presume that sometimes, with small probability, microjets reach anomalous velocity, of several kilometers per second, then we get another pattern of interaction between a jet and a surface. This is not a simple impact of a liquid 'microhammer' on a metal surface—at such fluid velocity, the properties of a metal surface do not correspond to solids.

As we saw in the previous section, solids behave as liquids on short temporal scales, their elasticity cannot manifest while disturbances spread too fast inside of them. Thus, a jet formed in a collapsing bubble, if it obtains an enormous velocity, may drill a hole in a metal surface. Because for this jet,this is not a solid surface anymore—it can be considered as a liquid one.

The physics of this process can be described by a simplified consideration based on the theory of cumulative jets (like in (Lavrent'ev and Shabat 1977)). Let us consider a jet of length l_j that moves at velocity u_j and penetrates a liquid target—sic—at velocity u_t (see Fig. 5.4.4). The second fluid—the target—also flows in this manner.

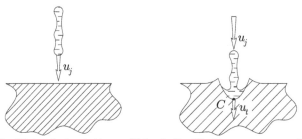

Fig. 5.4.4. Impact of the liquid jet on the target. At a very high velocity of the jet, one may assume that the target is not a solid but a liquid, so we have the contact of two fluids: the jet penetrates the liquid of a different density.

Inside the target, the hole moves on distance l_t at time τ; at the time τ, the whole jet penetrates the target at velocity $u_j - u_t$; thus

$$\tau = \frac{l_j}{u_j - u_t} = \frac{l_t}{u_t}.$$ (5.4.2)

At the contact point C, we may demand that normal components of the momentum flux density tensor are equal for both liquids, which can be written in a coordinate system where this point is immobile; we can write

$$p_j + \frac{\rho_j \left(u_j - u_t\right)^2}{2} = p_t + \frac{\rho_t u_t^2}{2},$$ (5.4.3)

For equal pressures $p_j = p_t$ in the contact point, we have

$$\rho_j \left(u_j - u_t\right)^2 = \rho_t u_t^2.$$ (5.4.4)

Combining (5.4.4) and (5.4.2), we get

$$l_t = l_j \sqrt{\frac{\rho_j}{\rho_t}}. \tag{5.4.5}$$

Thus, for the water jet $\rho_j = 10^3$ kg/m³ impacting the titanium surface with $\rho_t = 4500$ kg/m³, we see that the depth of penetration is approximately equal to a half of the length of the initial jet.

The approach considered above (a) looks strange, and (b) is simplified. One may be very surprised that we assume a solid piece of metal as a fluid; moreover, in (5.4.3) we use the approach of a perfect fluid. However, for velocities of an order of a few kilometers per second, this approach is quite reasonable: for such a high velocity, $\rho u^2 \sim 10^{10}$ Pa, which significantly exceeds the elastic stress in the metal.

The final equation (5.4.5), in its simplicity, has other interesting features: it does not depend on the properties of materials except for density and does not contain the velocity of the jet, that is, a jet with velocity of 10 km/s would drill the same hole as a jet of velocity of 5 km/s. The second issue follows from the 'threshold' character of the phenomenon: if we may consider two interacting objects as two fluids, which is possible for sufficiently giant velocities, then the mathematical description turns to hydrodynamics that excludes velocity from the solution. As for the properties, it is a more complicated matter. At least, for such high velocities, compressibility should be taken into account. On the other hand, looking at the structure of the correlation (5.4.4), we may suppose that compressibility factors must arise under the root, so the final difference between two correlations must be minor. The description presented above is very clear, and this is its excellence, not the precision of calculations made basing on it.

As for justification of this approach, we may refer to the monography (Lavrent'ev and Shabat 1977), where the comparison of the experimental results and of the theory is given. It turns out that even this simple model provides reasonable results for velocities of ~ 5 km/s.

The next question is where the jets obtain such a high speed, considering the fact that the theory (see Chapter 4) gives only velocity slightly higher than 100 m/s. On this argument, one may answer that these results were obtained from calculations (Plesset and Chapman 1971) for a spherical bubble, while calculations in (Voinov 1979) give that the velocity of the jet strictly depends on the shape of the initial bubble. Various deformations of the bubble's shape may decrease or increase the jet velocity, and some of them can provide values comparable with the velocity of cumulative jets. These are very rare cases, as we see from the photos (if these dots correspond to the jets indeed, of course, not to some other reason: the theory given above can explain these holes, but this fact does not mean that these holes are caused by the microjets for sure), but, as far as we can judge, they do take place.

5.4.6 Electrical discharges

This mechanism of cavitation effect is rarely discussed, but it should be.

There exists an idea (expressed first by Jakov Frenkel) that a double layer on a metal surface can produce electrical discharges on that surface. The ins of this theory will be considered in Chapter 8; here we only show an overview of the effect of such discharge on the surface.

If such discharges exist, they are weak-current objects. The voltage on the double layer is not high anyway, so we cannot assume high current on the surface. Thus, the heating effect will be small; however, above we observed (from another initial proposition) the absence of high-temperature traces on surface. These possible discharges are weak sparks (in the gas discharge terminology), so the only effect of them is of chemical nature: in plasma, even at its low parameters, many reactions proceed at a higher rate.

It can seem that these effects are negligible in comparison with liquid-hammer strikes at intensity of thousands of atmospheres. From this straight point of view, it is definitely so. However, such influence on the surface may noticeably change the mechanical properties of the surface (through changing the chemical ones), and we think that it is too early to dismiss these effects. Despite the circumstance that the fact of existence of such discharges needs proof.

5.4.7 Positive feedback

Probably the worst thing that happens with the cavitation erosion is the kind of feedback that leads to increase of the damage of the surface.

In Fig. 5.4.5, one can see the development of the cavitation erosion process. The titanium waveguide was placed into glycerol and produced ultrasonic cavitation (at frequency of ~ 20 kHz); in the figure, we present three photos corresponding to snapshots at intervals of around three minutes of cavitation. One can see that the initial hole grows rapidly, while the neighboring sites remain intact (on such a short time scale, of course; with time, the whole surface of the waveguide will be covered with the same potholes).

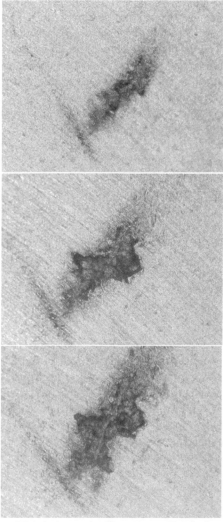

Fig. 5.4.5. Once cavitation has started, it is bad news, it will never stop. The titanium waveguide under the influence of cavitation erosion; the time gaps between these consequent images are around 3 minutes of cavitation. We can observe the growth of the initial hole; the approximate size of the cavern is ~ 1 millimeter.

Thus, the hollows arisen during the cavitation process tend to develop. On an inhomogeneous surface, cavitation is more intense→cavitation erosion is higher→inhomogeneity is more pronounced→cavitation is stronger, etc. In other words, the cavitation erosion has just begun, and it will continue (up to the limit corresponding to the significant destruction of the surface, see Chapter 1).

Therefore, the more uneven the surface is, the stronger the cavitation damage will be. From this point of view, on the contrary,the smoother the surface, the less the cavitation erosion.

Conclusion

In this chapter, we considered two problems that are of intrinsic interest: hydraulic shock and rheology. Both of them have a clear connection with cavitation; to be exact, cavitation, hydraulic shock and viscoelastic properties of liquids are tightly interwoven.

The first chain: hydraulic shock causes a tension wave—cavitation happens—a cavitating bubble produces a microjet at the surface—rheology must be taken into account to analyze that impact.

The second chain: a liquid drop falls on a solid surface—compression and rheology must be considered—a shock wave arises—a cavitating bubble is born inside the drop.

And so on. All separate parts must be examined before one may connect them together, and this chapter is an introduction to all these matters.

Hydraulic shock means, roughly, two things: one is evident, the other is not.

When a liquid jet strikes a solid obstacle, this is, probably, the first example of hydraulic shock that comes to mind. If the velocity of a fluid is u, the speed of sound in it is c_s and its density is ρ, then this liquid mass acts on a solid with pressure

$$p = \rho c_s u, \tag{5.C.1}$$

Another type of hydraulic shock happens in a pipe, when a valve closes suddenly. The liquid mass, as in the previous case, impacts the valve, but this is only the first phase of a complicated process, when the pressure near the valve changes periodically, fading over time. When the pressure drops enough, cavitation occurs in the pipe.

Rheology is the science field that investigates the properties of a continuous medium. We all know that solids are elastic while fluids are viscous, but this is only a simplified version of real things. Indeed, the properties of the medium depend even on time—on time scales on which the problem must be considered.

Usually, fluids do not resist the shear stress, but one may observe them at short time scales—under the pressure that alternates at 10^5–10^7 Hz.

Usually, solids are elastic, but under the impact of a high-speed jet they behave like liquids: if the jet is tiny and fast (with speed of around a few kilometers per second), then its blow has a completely different character than the strike of a metal hammer. Such a spray acts just like a cumulative jet: to it, the metal surface is only a liquid medium. The impact of this jet drills a hole in the surface of depth

$$l = l_{jet} \sqrt{\frac{\rho_{liquid}}{\rho_{surface}}}. \tag{5.C.2}$$

The impact of microjets is the main erosion mechanism during cavitation, but not the only one.

First, not every microjet has such an enormous velocity, so not every jet can drill a hole in the surface. The solid surface that has suffered from cavitation, mainly, looks rather rumpled than holey. Indeed, if the jet velocity is only 10 m/s, even in this case the pressure in accordance with (5.C.1) is ~10^7 Pa—not a bad strike.

Other—non-mechanical—effects of cavitation on a surface are chemical reactions (which are undisputed), thermal influence (which is questionable) and electrical discharges (which are exotic). Actually, many special investigations are required to clarify the role of each separate effect on the whole picture.

One thing is certain: cavitation erosion is a feedback-effect. Once cavitation perforates even a small hole, everything is over. This hollow will grow rapidly because it becomes a point of intense cavitation, so here the damage processes will intensify, and so on.

References

ASTM G32-16. *Standard Test Method for Cavitation Erosion Using Vibratory Apparatus*. Book of Standards. 03.02.

Auret, J. G., O. F. R. A. Damm, G. J. Wright and F. P. A. Robinson. 1993. Cavitation erosion of copper and aluminium in water at elevated temperature. *Tribology International*, 26: 421–9. https://doi.org/10.1016/0301-679X(93)90082-C.

Badmaev, B. B., T. S. Dembelova, D. N. Makarova and Ch. Zh. Gulgenov. 2017. Shear elasticity and strength of the liquid structure by an example of diethylene glycol. *Technical Physics*, 62: 14–7. https://doi.org/10.1134/S1063784217010042.

Bazaron, U. B., Deryagin, B. V. and K. T. Zandanova. 1972. Investigation of the shear elasticity of liquids at different shear angles. *Doklady Akademii Nauk SSSR*, 206: 1325–8 (in Russian).

Bazaron, U. B., B. V. Deryagin, O. R. Budaev and B. B. Badmaev. 1978. Determination of low frequency complex shear modulus of liquids according to the measurements of shear wave length. *Doklady Akademii Nauk SSSR*, 238: 50–53 (in Russian).

Berchiche, N., Franc, J. P. and J. M. Michel. 2002. A cavitation erosion model for ductile materials. *Journal of Fluids Engineering*, 124: 601–6. https://doi.org/10.1115/1.1486474.

Biryukov, D. A., Val'yano, G. E. and D. N. Gerasimov. 2018. Damage of an ultrasonic-waveguide surface during cavitation accompanied by sonoluminescence. *Journal of Surface Investigation: X-ray, Synchrotron and Neutron Techniques*. 2018, 12: 175–8. https://doi.org/10.1134/S1027451017060040.

Bombardieri, C., Traudt, T. and C. Manfletti. 2019. Experimental study of water hammer pressure surge. *Progress in Propulsion Physics*, 11: 555–570. https://doi.org/10.1051/eucass/201911565.

Bowden, F. P. and J. H. Brunton. 1961. The deformation of solids by liquid impact at supersonic speeds. *Proceedings of the Royal Society of London. Series A. Mathematical and Physical Sciences*, 263: 433–50. https://doi.org/10.1098/rspa.1961.0172.

Chandra, S. and C. T. Avedisian. 1991. On the collision of a droplet with a solid surface. *Proceedings of the Royal Society of London. Series A. Mathematical and Physical Sciences*, 432: 13–41. https://doi.org/10.1098/rspa.1991.0002.

Cook, S. S. 1928. Erosion by water-hammer. *Proceedings of the Royal Society of London. Series A. Mathematical and Physical Sciences,* 119:481–8. https://doi.org/10.1098/rspa.1928.0107.

Derjaguin, B. V., B. B. Badmaev, U. B. Bazaron, Kh. D. Lamazhapova and O. R. Budaev. 1995. Measurement of the Low-Frequency Shear Modulus of Polymeric Liquids, 29: 201–9. https://doi.org/10.1080/00319109508031637.

Dular, B., Stoffela, B. and B. Širokb. 2006. Development of a cavitation erosion model. *Wear*, 261: 642–655.

Dyre, J. C., and W. H. Wang. 2012. The instantaneous shear modulus in the shoving model. *The Journal of chemical physics*, 136: 224108-1–5. https://doi.org/10.1063/1.4724102.

Eisenberg, P. 1969. *Cavitation and impact erosion—concepts, correlations, controversies*. Philadelphia: American Society for Testing and Materials.

Engel, G. E. 1955. Waterdrop collisions with solid surfaces. *Journal of Research of the National Bureau of Standards*, 54: 281–98. https://doi.org/10.1615/AtomizSpr.v11.i2.40.

Field, J. E., Lesser, M. B., and J. P. Dear. 1985. Studies of two-dimensional liquid-wedge impact and their relevance to liquid-drop impact problems. *Proceedings of the Royal Society of London. Series A. Mathematical and Physical Sciences*, 401: 225–249. https://doi.org/10.1098/rspa.1985.0096.

Field, J. E., Dear, J. P., and J. E. Ogren. 1989. The effects of target compliance on liquid drop impact. *Journal of Applied Physics*, 65: 533–40. https://doi.org/10.1063/1.343136.

Ganz, S. 2012. *Cavitation: Causes, Effects, Mitigation and Application*. Hartford: RensselaerPolytechnic Institute.

Geng, J., X.-L. Yuan, D. Li and G.-S. Du. 2017. Simulation of cavitation induced by water hammer. *Journal of Hydrodynamics*, 29: 972–8. https://doi.org/10.1016/S1001-6058(16)60811-9.

Gorham, D. A., and J. E. Field. The failure of composite materials under high-velocity liquid impact. *Journal of Physics D: Applied Physics*, 9: 1529–41. https://doi.org/10.1088/0022-3727/9/10/018.

Granato, A. 1996. The shear modulus of liquids. *Le Journal de Physique IV*, 6: C8-1–9. https://doi.org/10.1051/jp4:1996801.

Haller, K. K., Y. Ventikos and D. Poulikakos. 2002. Computational study of high-speed liquid droplet impact. *Journal of Applied Physics*, 92: 2821–8. https://doi.org/10.1063/1.1495533.

Haller, K. K., D. Poulikakos, Y. Ventikos and P. Monkewitz. 2003. Shock wave formation in droplet impact on a rigid surface: lateral liquid motion and multiple wave structure in the contact line region. *Journal of Fluid Mechanics*, 490: 1–14. https://doi.org/10.1017/S0022112003005093.

Hess, S., Kröger, M. and W. G. Hoover. 1997. Shear modulus of fluids and solids. *Physica A*, 239: 449–66. https://doi.org/10.1016/S0378-4371(97)00045-9.

Heymann, F. J. 1968. On the shock wave velocity and impact pressure in high-speed liquid-solid impact. *ASME Journal of Basic Engineering*, 90: 400–402. https://doi.org/10.1115/1.3605114.

Heymann, F. J. 1969. High speed impact between a liquid drop and a solid surface. *Journal of Applied Physics*, 40: 5113–5122. https://doi.org/10.1063/1.1657361.

Huang, Y. C., Hammitt, F. G. and T. M. Mitchell. 1973. Note on shockwave velocity in highspeedliquidsolid impact. *Journal of Applied Physics*, 44: 1868–9. https://doi.org/10.1063/1.1662464.

Huang, F., J. Mi, D. Li and R. Wang. 2020. Impinging performance of high-pressure water jets emitting from different nozzle orifice shapes. *Geofluids*. https://doi.org/10.1155/2020/8831544.

Hutton, J. F. and M. C. Phillips. 1972. Transition in properties of Indium at 170 K. *Nature Physical Science*, 238: 141–2. https://doi.org/10.1038/physci238141a0.

Jansson, M. 2017. Water hammer induced cavitation - A numerical and experimental study. *SICFP2017*, 256–60. https://doi.org/10.3384/ecp17144256.

Johnson, W. and G. W. Vickers. 1973. Transient stress distribution caused by water-jet impact. *Journal Mechanical Engineering Science*, 15: 302–10. https://doi.org/10.1243/JMES_JOUR_1973_015_052_02.

Joliffe, K. H. 1966. The application of dislocation etching techniques to the study of liquid impacts. *Philosophical Transactions of the Royal Society of London. Series A, Mathematical and Physical Sciences*, 260: 101–20. https://doi.org/10.1098/rsta.1966.0035.

Kalkwijk, J. P. T. and C. Kranenburg. 1971. Cavitation in horizontal pipelines due to water hammer. *Communications on hydraulics*. https://doi.org/10.1061/JYCEAJ.0003106.

Kaspar, J., J. Bretschneider, S. Jacob, S. Bonß, B. Winderlich and B. Brenner. 2007. Microstructure, hardness and cavitation erosion behavior of Ti-6Al-4V laser nitride under different gas atmospheres. *Surface Engineering*, 23: 99–106. https://doi.org/10.1179/174329407X169430.

Kazama, T., K. Kumagai, Y. Osafune, Y. Narita and S. Ryu. 2015. Erosion of grooved surfaces by cavitating jet with hydraulic oil. *Journal of Flow Control, Measurement & Visualization*, 3: 41–50. http://dx.doi.org/10.4236/jfcmv.2015.32005.

Kempf, G. and E. Foerster. 1932. *Hydromechanische Probleme des Schiffsantriebs*. Berlin: Springer-Verlag. https://doi.org/10.1007/978-3-642-47554-2.

Kielczynski, P., Pajewski, W. and M. Szalewski. 2001. Determination of the complex shear modulus of viscoelastic liquids using cylindrical piezoceramic resonators. *201 IEEE Ultrasonic Symposium 323–326*. https://doi.org/10.1109/ULTSYM.2001.991634.

Kiyama, A., W. Pajewski, M. Szalewski. 2016. Effects of a water hammer and cavitation on jet formation in a test tube. *J. Fluid Mech.*, 787: 224–236. https://doi.org/10.1109/ULTSYM.2001.991634.

Kodura, A. 2016. An analysis of the impact of valve closure time on the course of water hammer. *Archives of Hydro-Engineering and Environmental Mechanics*, 63: 35–45. http://dx.doi.org/10.1515/heem-2016-0003.

Kondo, T. and K. Ando. 2016. One-way-coupling simulation of cavitation accompanied by high-speed droplet impact. Physics of Fluids, 28: 033303-1–12. https://doi.org/10.1063/1.4942894.

Krahl, D. and Weber, J. 2016. Visualization of cavitation and investigation of cavitation erosion in a valve. *10th International Fluid Power Conference*, 1: 333–48.

Kranenburg, C. 1972. *The effect of free gas on cavitation in pipelines induced by water hammer*. Proceedings of the First International Conference on Pressure Surges, BHRA, Canterbury, UK; 41–52.

Krumenacker, L., Fortes-Patella, R. and A. Archer. 2014. Numerical estimation of cavitation intensity. *IOP Conference Series: Earth and Environmental Science*, 22: 052014. https://doi.org/10.1088/1755-1315/22/5/052014.

Landau, L. D. and E. M. Lifshitz. 1987. *Fluid Mechanics*. Oxford: Pergamon Press.

Landau, L. D. and E. M. Lifshitz. 1970. *Theory of Elasticity*. Oxford: Pergamon Press.

Larock B. E., Jeppson, R. W. and G. Z. Watters. 2000. *Hydraulics of Pipeline Systems*. CRC Press.

Lavrent'ev, M. A. and B. V. Shabat. 1977. *Hydrodynamics problems and their mathematical models*. Moscow: Nauka (in Russian).

Lema, M., J. Steelant, F. P. Lopez and P. Rambaud. 2012. Experimntal characterization of the priming phase using a propellant line mock-up. *Space Propulsion*.

Lesser, M. B. 1983. The impact of compressible liquids. *Annual Review of Fluid Mechanics*, 15: 97–122. https://doi.org/10.1146/annurev.fl.15.010183.000525.

Madadnia, J., D. K. Shanmugam, T. Nguyen and J. Wang. 2008. A study of cavitation induced surface erosion in abrasive waterjet cutting systems. *Advanced Materials Research*, 53-54: 357–62. https://doi.org/10.4028/www.scientific.net/AMR.53-54.357.

Marinin, V. G. 1994. Effects of surface treatment on the cavitation erosion of high-chrome steel, zirconium, titanium and their alloys. *NEA*, 163–7.

Matikainen, V., K. Niemi, H. Koivuluoto and P. Vuoristo. 2014. Abrasion, erosion and cavitation erosion wear properties of thermally sprayed alumina based coatings. *Coatings*, 4: 18–36. https://doi.org/10.3390/coatings4010018.

Metter, I. 1948. Physical nature of cavitation and the mechanism of cavitation-damages. *Uspekhi Fizicheskikh Nauk*, 35: 53–79 (in Russian).

Mitelea, I., O. Oancă, I. Bordeaşu and C. M. Crăciunescu. 2016. Cavitation erosion of cermet-coated aluminium bronzes. *Materials*, 9: 204-1–11. https://doi.org/10.3390/ma9030204.

Moyer, A. E. 1977. Robert Hooke's ambiguous presentation of "Hooke's Law". *Isis*, 68: 266–75.

Muhammad, A. A. 2019. Hydraulic transient analysis in fluid pipeline: A Review. *Journal of Science Technology and Education*, 7: 291–9.

Naguib, N. W. M., D. Ulrike, S. Johannes and Z. G. Karl-Heinz. 2007. The effect of surface finish and cavitating liquid on the cavitation eroision of alumina and silicon carbide cceramics. *Ceramics – Sikikaty*, 51: 30–9.

Orlenko, L. P. 2008. Physics of the explosion and shock. Fizmatlit (in Russian).

Plesset, M. S. and R. B. Chapman. 1971. Collapse of an initially spherical vapour cavity in the neighbourhood of a solid boundary. *Journal of Fluid Mechanics*, 47: 283–90. https://doi.org/10.1017/S0022112071001058.

Polanco, J., M. S. Virk, U. N. Mughal, S. Victor, D. P. José, V. Antonio and A. Orlando. 2015. Encapsulated water hammer: theoretical/experimental study. *World Journal of Engineering and Technology*, 3: 290–5. http://dx.doi.org/10.4236/wjet.2015.33C043.

Puosi, F. and D. Leporini. 2012. Communication: Correlation of the instantaneous and the intermediate-time elasticity with the structural relaxation in glassforming systems. *The Journal of Chemical Physics*, 136: 041104-1–4. https://doi.org/10.1063/1.3681291.

Rein, M. 1993. Phenomena of liquid drop impact on solid and liquid surfaces. *Fluid Dynamics Research*, 12: 61–93. https://doi.org/10.1016/0169-5983(93)90106-K.

Reiner, M. 1958. *Rheology.* Springer-Verlag, 1958. https://doi.org/10.1007/978-3-662-43081-1_4.

Sadafi, M., Riasi, A. and S. A. Nourbakhsh. 2012. Cavitating flow during water hammer using a generalized interface vaporous cavitation model. *Journal of Fluids and Structures*, 34: 190–201. https://doi.org/10.1016/j.jfluidstructs.2012.05.014.

Samko, S. G. 1993. *Fractional Integrals and Derivatives: Theory and Applications.* CRC Press.

Skalak, R. and D. Feit. 1966. Impact on the surface of a compressible fluid. *Journal of Engineering for Industry*, 88: 325–31. https://doi.org/10.1115/1.3670955.

Terwisga, T. J. C., Z. Li, P. A. Fitzsimmons and E. J. Foeth. 2009. Cavitation Erosion—A review of physical mechanisms and erosion risk models. *Proceedings of the 7th International Symposium on Cavitation CAV2009, 41-1–13.*

Toloian, A. 2015. The cause of cavitation and prevention methods of this destructive phenomenon in pumps. *World Essays Journal*, 3: 141–9.

Voinov, O. V. 1979. Calculation of the parameters of the high-velocity jet generated by a collapsing bubble. *ZhurnalPrikladnoiMekhaniki i TekhnicheskoiFiziki*, 3: 94–9 (in Russian).

Walters, P. E. and R. A. Leishear. 2018. When the joukowsky equation does not predict maximum water hammer pressures. *Proceedings of the ASME 2018 Pressure Vessels and Piping Conference PVP2018*, 84050: 1–10. https://doi.org/10.1115/1.4044603.

West, B. J., M. Bologna and P. Grigolini. 2003. *Physics of Fractal Operators.* Springer. https://doi.org/10.1007/978-0-387-21746-8.

Xiaoujun, Z. 2002. *Effect of Surface Modification Processes on Cavitation Erosion Resistance.* Ph. D. Thesis, Curitiba: Universidade Federal Do Parana.

Zeldovich, Ya. B. and Yu. P. Raizer. 1967. *Physics of Shock Waves and High-Temperature Hydrodynamic Phenomena.* Academic Press.

Zijlstra, A. G. 2011. *Acoustic Surface Cavitation.* Ph.D. Thesis, University of Twente: Gildeprint Drukkerijen.

CHAPTER 6

Acoustic Cavitation

6.1 Acoustic waves

6.1.1 A wave of sound

First, let us agree on the terminology. Here and almost everywhere below, we will call an acoustic wave of any frequency as 'sound', despite the tradition in accordance with which it is assumed that the audible range is from 10–20 Hz to 20–22 kHz, and this range is called the 'sound', while acoustic waves of higher frequencies are called 'ultrasound' and lower frequencies are referred to as 'infrasound'. Only sometimes we will use term 'ultrasonic' for high-frequency oscillations, since, as it seems today, some special effects appear only at comparatively high frequencies.

Moreover, the audible range is a very subjective attribute. The upper level decreases with age, and, really, many middle-aged people do not hear sounds above ~ 16–17 kHz.[1] The low frequencies are perceived 'by the body', they are rather felt than heard. On the other hand, usually, we are dealing with much narrower sound ranges, even when we listen to music.

Actually, the sound that we hear is only the time-dependent pressure acting on the ear drum—nothing more. Every tone, every signal, every music can be represented as the dependence $p(t)$, which many of us, possibly, have seen on the screen of sound-editing software (see Fig. 6.1.1). It is even slightly sad that all the magic of music can be expressed with such a simple function, but this is true; to be exact, this is a half-truth because the signal of one of two channels is represented in that figure.

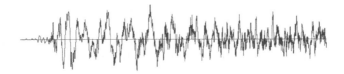

Fig. 6.1.1. This is how the beginning of 'White Room' by 'Cream' looks at the micro-sound-scope.[2]

Thus, finally, when we concern sound, we are interested in the propagation of pressure waves in the medium. Or, as we discussed above, sound is the wave of disturbances in the medium. We also have seen that in an acoustic wave the potential of velocity obeys the wave equation (see Chapter 4); for the one-dimensional case, this is

$$\frac{\partial^2 \varphi}{\partial t^2} - c_s^2 \frac{\partial^2 \varphi}{\partial x^2} = 0, \tag{6.1.1}$$

---

[1]   You may test your hearing with special tone generators, if you are not afraid of possible disappointing result.

[2]   Obtained with 'Sound Forge' software.

where the speed of sound (or the acoustic velocity as a more preferable term here) is defined as

$$c_S = \sqrt{\left(\frac{\partial p}{\partial \rho}\right)_S},$$ (6.1.2)

that is, determined by the adiabatic (isentropic) derivation of pressure on density. Also, since the deviation of pressure $\Delta p = p - p_0$, where p_0 is the constant initial pressure, is

$$\Delta p = -\rho \frac{\partial \varphi}{\partial t},$$ (6.1.3)

then we see that

$$\frac{\partial^2 p}{\partial t^2} - c_S^2 \frac{\partial^2 p}{\partial x^2} = 0,$$ (6.1.4)

which means that the pressure obeys the wave equation too: we may obtain (6.1.4) by differentiating (6.1.1) on time. Probably, one may omit all these procedures and simply state that if the acoustic wave is a wave (sic), then all the parameters change in this wave in accordance with the wave equation, so the equation for pressure is (6.1.4). Nevertheless, we may note one useful thing: the special correlation between the pressure deviation in a wave (6.1.3) and the velocity

$$\upsilon = \frac{\partial \varphi}{\partial x} = -\frac{1}{c_S} \frac{\partial \varphi}{\partial t},$$ (6.1.5)

which has a form

$$\Delta p = \rho c_S \upsilon.$$ (6.1.6)

This expression will be used below (see also Chapter 5). The relation (6.1.5)—its second equality—follows from the equation (6.1.1). From (6.1.6), it also follows that velocity is $\upsilon = \Delta p / \rho c_S$, that is, ρc_S can be referred to as the wave (or the acoustic) resistance.

The solution to the hyperbolic equation (6.1.4) is any function of variable $x \pm c_S t$; for the wave solution, we may get

$$p = p_0 + p_a \sin(kx \pm k c_S t),$$ (6.1.7)

where k is the constant defined by the boundary and initial conditions. Of course, we used the simplest form: here instead of sine there may be a cosine function, or both. It is more important that the periodic term in (6.1.7) can be of both signs: either positive or negative.

Cavitation takes place when pressure is sufficiently low. The investigation of this phenomenon in 'natural conditions' leads to the need to design channels of special forms, in which the pressure drop would correspond to the observed one. If we are interested in the nuances of cavitation in these exact conditions (for instance, in such type of pipeline), then such experiments are irreplaceable.

However, we may be interested in the fundamental features of cavitation: the dynamics of cavitating bubbles, pressure dynamics, changes in composition of the gas inside a cavitating bubble, and so on. If so, on one hand, we do need a special design of the experimental setup—it is sufficient that cavitation can be observed; on the other hand, in such an experiment we need many more controlled parameters; to explore cavitation, we want to change the size of bubbles, the rate of their growth, their number, etc.

Producing cavitation with an acoustic field of alternate pressure is a very convenient way which satisfies all the demands. Amplitude p_a can be easily adjusted to obtain any low value of pressure we need (in reasonable limits, of course). For the oscillating pressure at amplitude p_a and cyclic frequency ω at the point $x = 0$, that is

$$p(0) = p_0 + p_a \sin(\omega t),$$ (6.1.8)

we may get from the equation (6.1.4) the solution

$$p(x,t) = p_0 + \frac{p_a}{2}\left[\sin\left(\omega t - \frac{\omega x}{c_s}\right) + \sin\left(\omega t + \frac{\omega x}{c_s}\right)\right], \tag{6.1.9}$$

or, using $k = \omega/c_s$, it can be represented simply as

$$p = p_0 + p_a \sin(\omega t)\cos(kx). \tag{6.1.10}$$

In these equations, p_0 is a constant external pressure: hydrostatic pressure, atmospheric pressure, etc.

Now consider the distribution of pressure in the medium between two acoustic sources at the distance L (Fig. 6.1.2); each of them generates pressure p_a at cyclic frequency ω. For this case, the solution of (6.1.4) is

$$p = p_0 + p_a \sin(\omega t)\left[\cos(kx) + \frac{1-\cos(kL)}{\sin(kL)}\sin(kx)\right]. \tag{6.1.11}$$

Here we do not consider such matters as wave reflections; these problems will be treated in Section 6.1.3.

The distribution of relative pressure

$$\delta p = \frac{p - p_0}{p_a \sin(\omega t)} \tag{6.1.12}$$

from the relative coordinate $\delta x = kx$ is shown in Fig. 6.1.3 for $KL = 1$. We see that this function has a maximum in the middle of the region at $x = L/2$, the value of which is determined by the value of KL.

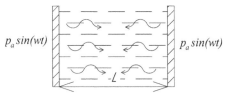

$p_a sin(wt)$ $p_a sin(wt)$

$-L-$

Fig. 6.1.2. Two flat acoustic sources create an acoustic field in a region of length L.

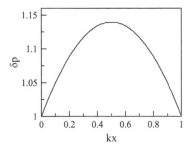

Fig. 6.1.3. Pressure distribution between two acoustic sources. Maximum depends on the function (6.1.11), see also Fig. 6.1.4.

Let us analyze which parameters are defining and which ones are definable. The wave number k depends on the frequency of the acoustic source and of the acoustic velocity. The first value is directly connected with the type of the acoustic generator; possibly, this is a fixed value for the given type of transducer (see following sections). The acoustic velocity is also a constant (here we neglect the dependence on temperature, etc.). Thus, k is almost given from above to the experimenter. But the length of the working part L can be adjusted—this is the size of the vessel or something similar. Pressure magnitude depends on the function

$$f(kL) = \frac{1 - \cos(kL)}{\sin(kL)}. \tag{6.1.13}$$

Apart of it is shown in Fig. 6.1.4: it is clear that the function reaches infinity at $L = \pi/k$ and changes the sign at this point. The corresponding maximum pressure is reached at $kx = kL/2$:

$$\delta p_{\max} = \cos(kL/2) + f(kL)\sin(kL/2). \tag{6.1.14}$$

Thus, to reach a higher pressure in the middle, we may adjust the length of the working section closer to the value

$$\frac{\pi c_s}{\omega} = \frac{\lambda}{2}, \tag{6.1.15}$$

where λ is the wavelength of the sound wave in the given liquid. Practically, of course, we may get close to the optimum, but never reach infinite pressure.

Finally, we must explain how we get the solutions of (6.1.4). A direct, complete and bothersome way is to use the variable separation method: represent the solution in a form

$$p(x,t) = p_x(x)p_t(t), \tag{6.1.16}$$

insert it into (6.1.4), obtaining two separate equations for p_x and p_t, use boundary conditions, and so on.

We may choose another way. If the solution can be represented in the form (6.1.16), then one may notice that each function must be a harmonic function: sine, cosine, or its combination; in this case, the second derivative gives the same function multiplied by the factor at the function argument, with the minus sign. That is, for the time dependence—dependence on variable ωt—we have after derivation—$\omega^2 p$, and for the spatial part—described by the argument kx—the corresponding function $-k^2 p$. Thus, from (6.1.4) we see that $k = \omega/c_s$; this is required to eliminate this expression at any p. Then, we see that the time part of the function p must correspond to (6.1.6), i.e., must have a form $p_t \sim \sin(\omega t)$. Then, the spatial part px should be represented as

$$p_x = A\cos(kx) + B\sin(kx), \tag{6.1.17}$$

and from the boundary conditions $p(0) = p(L) = p_0 + p_a \sin(\omega t)$, we get

$$A = p_a, B = \frac{1 - \cos(kL)}{\sin(kL)} p_a, \tag{6.1.18}$$

and, combining all the functions and all the constants, including P_0, we get (6.1.11).

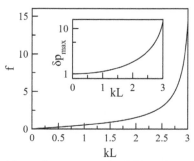

Fig. 6.1.4. Function (6.1.11) and the maximum pressure for different distances between two acoustic sources.

6.1.2 Origin of sound: Transducers

Because of its nature, sound may originate from any disturbance of the medium by the bodies placed in it, or by the surfaces surrounding this medium. From some point of view, we are lucky that our ears are not very sensible, otherwise we would live in a permanent chaos of surrounding sounds. If we could hear the sound of falling leaves, it may be poetical but not very practical, since we need useful information from our sensory organs. Meanwhile, generation of sound by such objects is quite problematic: see, for example (Prokhorov and Chashechkin 2011, Phillips et al. 2018) for the sound of falling drops.

For experiment on acoustic cavitation, we need sound sources of high power and, sometimes as much as possible, of a certain frequency. However, it is useful to understand, by the way, that even a monochromatic wave tends to dispart into waves of smaller wavelengths and, therefore, without any special implementations and contrivances the generator produces sound at various frequencies in the medium.

Let us consider the simplest sound generator—a guitar string. This device is fixed on both sides; consequently, displacements at points A and B (see Fig. 6.1.5) must be zero. Thereby, the number of half-waves must be integer n, that is, the length of the string is related to the wavelength as

$$l = \frac{\lambda n}{2} \tag{6.1.19}$$

Here $n = 1$ corresponds to the main (the lowest) frequency of vibration, other frequencies are overtones (harmonics). The vibration of the string cannot consist only of the main tone: it also contains harmonics, which appear with time, when the oscillation on the main frequency loses stability.

Let the string tuned, for example, to E (which corresponds to $n = 1$ in (6.1.19)). Overtones arising at $n = 2$ give E again (but an octave higher—this sound is emitted by the string which is twice as short), vibration related to $n = 3$ gives B,[3] and so on. The main harmonic, by the way, defines the harmony of the musical sound. When two sounds—two notes—contain similar overtones, we perceive them as a consonance, it is a pleasure for our ears to listen to such resembling sounds: these sounds sound similar, we are relaxed and calm. Otherwise, we would—to be exact, our ancestors, since today our ears are trained with different kinds of modern music from the early childhood—consider these two tones as dissonance.

The relation (6.1.19), observed from the musical point of view, explains

- ✓ why we divide the entire range into octaves: the first harmonics is of the doubled frequency; indeed, E1 and E2 are different tones but they are so similar that we name them similarly;

- ✓ what is a quint: this is the second harmonic which appears from the main tone; therefore, this is the main 'secondary' tone in the mood, fully connected in our musical imagination to the main tone of the mood—we used to consider it 'bound' to the main tone; that is why the quint is also called a 'dominant' in relation to the main tone—a tonic;

- ✓ what is a quart (subdominant): a tonic is the quint (shifted by an octave) for the quart.

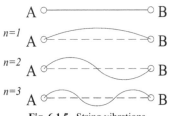

Fig. 6.1.5. String vibrations

---

[2] We use modern non-classic notations.

In simple words, the simple expression (6.1.19) explains why music teachers tormented you by the circle of fifths in music theory lessons, while you wanted immediately to learn how to play like Eddie Van Halen. It is also interesting that the consonance is the matter of habits as well: during fast playing, when the string has no time to lose stability and, therefore, to produce additional harmonics, we do not hear any overtones (they are absent), but still perceive the sounds like a 'consonance'—our musical imagination complements our hearing; dodecaphony, consequently, has some theoretical foundations too.

A string is almost an inexhaustible matter, but usually physicists use other sources of sound, especially in experiments on acoustic cavitation. But even in this case, we will see similar phenomena: in an elastic medium (a liquid in a jar, for example: as we discussed in Chapter 5, liquids have more elasticity than one may assume), vibration tends to break into smaller waves.

First, we will consider the creation of sound of high power. It cannot be a sound amplifier, like in a domestic acoustic system, because of the following reasons. To get (and control, which is also important) cavitation, we need to rule over the pressure inside a liquid. We cannot reach this effect by irradiating it with acoustic waves: even a loud sound does not produce significant pressure in the surrounding medium. To be exact, we may achieve a valuable pressure in some medium with an external acoustic source, the irradiation of which will be passed from this source to the investigated medium through air, but this method would be very ineffective. To obtain an acceptable result, the acoustic source must be fit snugly within the medium. In this way, generally, there may be two schemes (see Fig. 6.1.6) to connect the acoustical source with a liquid.

The first scheme implies that an ultrasonic transducer is immersed into water; this scheme is used for investigations of the multi-bubble cavitation; results are presented in Section 6.4.

The second scheme is used to obtain the single-bubble cavitation: to observe a single oscillating bubble. The last case, actually, uses two acoustic sources on the opposite sides of the vessel in an attempt to fix the bubble in its place stationary. This variant is much more complicated, and often, even if we manage to obtain a single bubble in the vessel, it moves anyway. On the other hand, the first expected consequence of the effective acoustic power supply in such a scheme may be the destruction of the vessel, of course: when the pressure impulse is high, the jar breaks; note that due to the character of such investigations, the jar must be made of a transparent substance, and glass is the first thing that comes to mind. Every scientific group which begins investigations of acoustic cavitation starts with a pile of broken glass. With experience, of course, they come to a less destructive way.

Now it is time to discuss what 'acoustic sources' are we talking about. Such elements—transducers—are supplied by electric power which transforms into acoustic power there; in the schemes presented in Fig. 6.1.6, the derived acoustic power does is not wasted.

Let us consider transducers more closely. Acoustic transducers are the devices which transform electrical signals into sound (or ultrasound). Below, we consider two basic types of them: piezoelectric transducers and the magnetostriction ones, the principles of which may be described from the common positions.

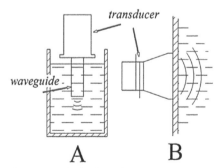

Fig. 6.1.6. Two schemes to use transducers in cavitation investigations. A—transducer supplied with a waveguide which is immediately immersed into liquid; B—transducer mounted on the wall of a vessel with a liquid.

As we know (see Appendix A), in a general case, thermodynamic equations (such as the first law) must be written with all kinds of forces acting on the body—not only the work of expansion. For a dielectric in an electrical field, we have

$$d\Phi = -SdT + Vdp - PdE, \tag{6.1.20}$$

where $P = \varepsilon_0(\varepsilon - 1)E$ is the polarization density of the dielectric with dielectric permeability ε; other designations are standard. Actually, this is an approach now, since the piezoelectrical effect takes place in crystals, where the expansion work represents a more complicated term because of their anisotropy.

Since the Gibbs potential Φ is the function of state, and $d\Phi$ is a full differential, then all cross derivatives must be equal, that is

$$\frac{\partial^2 \Phi}{\partial E \partial p} \equiv \left(\frac{\partial V}{\partial E}\right)_{p,T} = -\left(\frac{\partial P}{\partial p}\right)_{E,T} \equiv \frac{\partial^2 \Phi}{\partial p \partial E}. \tag{6.1.21}$$

Equation (6.1.21) has many meanings and contains many effects, so to say. At first, we see that the volume may change with the electric field strength—this is the so-called electrostriction effect; its analogue in a magnetic field will be considered below. Also, the right-had side of the equation (6.1.21) reflects some connection between the polarization density P and pressure p. If we have to deal with the dependence $P(p)$ in its direct, physical sense,[4] i.e., suppose the change in the polarization density due to the change in pressure, then this is the manifestation of a direct piezoelectric effect. The opposite case, for the dependence $p(P)$, when the variation in P leads to variation in pressure, corresponds to the reverse piezoelectric effect.

Also note that

✓ the change in pressure is connected with deformation;
✓ theoretically, all the effects listed above are bound in the expression (6.1.21), but in a special form: this relation contains the derivation $\partial P/\partial p$ at the constant electric field strength, which is important.

Thus, the expression (6.1.21) shows only a relation between electrostriction and the change in physical properties of the medium (change of permeability ε) with pressure, not between electrostriction and the piezoelectric effect as we usually understand it, when the variation in polarization density also includes the variation in the electric field strength. Consequently, the expression (6.1.21) gives

$$\left(\frac{\partial V}{\partial E}\right)_{p,T} = \underbrace{-E\varepsilon_0\left(\frac{\partial \varepsilon}{\partial p}\right)_{E,T}}_{-\chi} = E\chi, \tag{6.1.22}$$

where, just in case, we accentuate that in a non-linear medium there also may be $\varepsilon(E)$, but further we neglect the possible dependence $\chi(E)$. From (6.1.22), we see that the change in volume in electrostriction is a quadratic effect on the electric field:

$$\Delta V = \frac{\chi E^2}{2}. \tag{6.1.23}$$

---

[4] The term 'dependence' has different meaning in mathematics and physics. In mathematics, any function of a kind of $z = f(x,y)$ is a dependence. Physics, however, operates also with casual relationships. One may say that the acceleration of a body depends on the force acting on it, and on its mass (the second Newton law $a = F/m$), but it would be strange to declare that the mass depends on the force and acceleration as $m = F/a$.

In its turn, the piezoelectric effect is linear on E: if we consider the common form of the derivative $\partial P/\partial p$ with variable E and a connection between pressure and deformation in the simplest linearized form, then we have $E \sim p \sim \Delta V$. In some expositions, the difference in dependencies $\Delta V \sim E$ or $\Delta V \sim E^2$ distinguishes the reverse piezoelectric effect (the first case) from electrostriction (the second one).

Generally, all effects (electrostriction and both piezoelectric effects) are widely used, so widely that they are mixed up in their explanation. Various 'piezo-' devices are the wide range of apparatus which use the fact that in some media, pressure (or deformation) causes electricity and vice versa. The fact that pressure may generate an electrical signal is used in various types of detectors (microphone is the first of them); here we are interested in a reverse case: how to generate a pressure with an electrical impulse.

Piezo transducers are usually made of ceramics (of special sort—piezo ceramics) or crystals of ferro electrics which are placed between metal electrodes. From some point of view, this is a condenser with a special dielectric inside it. The common scheme of a transducer is shown in Fig. 6.1.7. Electricity is conducted to the piezoceramic discs which produces striction in response to an electric signal; other elements are designed to form an acoustic signal (reflector and the first part of irradiator) and to change it (the second part of irradiator may have various forms, made of different materials, etc.).

Piezoelectric transducers are very suitable for many applications, but they have one serious disadvantage: they cannot produce high power. Usually, the power of piezo transducers is $1 - 10$ W, rarely $\sim 10^2$ W, while sometimes we need kilowatts. If so, another type of transducers must be considered.

Now let us return to the equation of kind of (6.1.20). Symmetrically with the previous consideration, for magnetics in a magnetic field the corresponding equation has a form

$$d\Phi = -SdT + Vdp - MdH, \tag{6.1.24}$$

with magnetization M and magnetic field strength H; therefore, by analogy

$$\frac{\partial^2 \Phi}{\partial H \partial p} \equiv \left(\frac{\partial V}{\partial H}\right)_{p,T} = -\left(\frac{\partial M}{\partial p}\right)_{H,T} \equiv \frac{\partial^2 \Phi}{\partial p \partial H}. \tag{6.1.25}$$

From this equation, we see that the volume may depend on magnetization or on the magnetic field strength. It may be so for ferro magnetics; iron and nickel are the simplest examples. Applying the magnetic field, we may cause a change in the volume of a body (a plate or a rod, usually). This change is very small—the relative extension, as a rule, $\Delta l/l \sim 10^{-5}$. An instance of the scheme of a magnetostriction transducer is shown in Fig. 6.1.8. Indeed, this is a generalization of the simplest rod scheme; alternate current with adjusted frequency is used, usually with additional constant magnetic field (created by direct current or even by a permanent magnet). The frequency of vibration of the transducer coincides with the frequency of the exciting current; the most effective regime of work is achieved when the excitation frequency coincides with the natural frequency of the magnetostrictor, i.e. at resonance. Out of resonance, the magnetostrictor works with significant irregularity—with beats that may cause damage of the whole construction.

two discs of piezoceramics

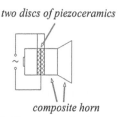

composite horn

Fig. 6.1.7. The scheme of a piezo transducer.

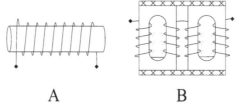

Fig. 6.1.8. The idea (A) and its realization (B) of a magnetostriction transducer. We need to place a ferromagnetic into the magnetic field, which is created by electrical current (direction is shown with arrows). Due to striction (change in sizes) in the magnetic field, this element causes the deviation in pressure on its surface.

In our (and many other) experimental setup, magnetostrictoris mounted to the ultrasonic waveguide (see Fig. 6.1.6)—the intermediate between the transducer and the investigated medium. Simply, the waveguide is a solid metal rod, the function of which is to transfer the vibration of magnetostrictor to the liquid. By changing the geometry of this rod, one may change the parameters of ultrasound irradiation using the same magnetostrictor.

6.1.3 Sound reflection and focusing

Sound is a wave—the oscillations of particles of a medium; therefore, we may apply all the general results concerning wave reflection to sound. Many results presented below have direct associations with the problems of light propagation. From one point of view, it may be sufficient to refer to the well-known results from optics. On the other hand, in order to understand the internal mechanisms of acoustic laws, it is useful to obtain the results 'from the beginning'.

Below, we will consider the propagation of sound waves which are predicted by the wave equation for pressure (6.1.4). As we discussed above, the solution of the wave equation (6.1.4) is the harmonic function; in this section, we represent it with the help of an imaginary exponent, i.e., in the form $\sim e^{ikx}$.

Let us consider the interaction of a sound wave with a plane obstacle (Fig. 6.1.9) in the case of normal incidence of wave. At the boundary of two media, we have three waves:

- ✓ incident wave; $p_{in} = p_{in}^0 e^{i(k_1 x - \omega t)}$;
- ✓ reflected wave; $p_r = p_r^0 e^{-i(k_1 x + \omega t)}$;
- ✓ passed wave; $p_p = p_p^0 e^{i(k_2 x - \omega t)}$.

Thus, here we consider only oscillating pressures; the constant background pressure is put to zero.

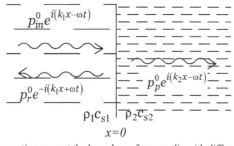

Fig. 6.1.9. Acoustic waves at the boundary of two media with different properties.

We immediately take into account that the reflected wave propagates in the reverse direction; also, the number of waves depends on the acoustic velocity, so it is different in two media. At the boundary, the pressures must be equal, so

$$p_{in}^0 + p_r^0 = p_p^0. \tag{6.1.26}$$

The velocity of oscillating particles in the waves must satisfy the condition

$$\upsilon_{in} - \upsilon_r = \upsilon_p, \tag{6.1.27}$$

so, using the expression (6.1.6), we have

$$\frac{p_{in}^0}{\rho_1 c_{s1}} - \frac{p_r^0}{\rho_1 c_{s1}} = \frac{p_p^0}{\rho_2 c_{s2}}. \tag{6.1.28}$$

Introducing the reflection coefficient[5]

$$R = \frac{p_r^0}{p_{in}^0} \tag{6.1.29}$$

and the transmittance coefficient

$$T = \frac{p_p^0}{p_{in}^0}, \tag{6.1.30}$$

we obtain for these quantities from (6.1.27) and (6.1.28)

$$R = \frac{\rho_2 c_{s2} - \rho_1 c_{s1}}{\rho_1 c_{s1} + \rho_2 c_{s2}}, \; T = \frac{2\rho_2 c_{s2}}{\rho_1 c_{s1} + \rho_2 c_{s2}}. \tag{6.1.31}$$

Thus, if the acoustic resistance is the same $\rho_1 c_{s1} = \rho_2 c_{s2}$, then the pressure wave passes through the boundary without any changes: $R = 0$, $T = 1$. The reverse case will be observed when the second medium is much denser than the first one: $\rho_2 c_{s2} \gg \rho_1 c_{s1}$; for example, it may be a gas and a solid, that is why walls have sound-isolating properties.

Now let us consider a case when the direction of the incident wave has the angle θ with axis x (see Fig. 6.1.10), that is, the pressure waves become

$$p_{in} = p_{in}^0 e^{i\left(k_1 x \cos\theta_{in} + k_1 y \sin\theta_{in} - \omega t\right)}, \tag{6.1.32}$$

$$p_r = p_r^0 e^{-i\left(k_1 x \cos\theta_r - k_1 y \sin\theta_r + \omega t\right)}, \tag{6.1.33}$$

$$p_p = p_p^0 e^{i\left(k_2 x \cos\theta_p + ik_2 \sin\theta_p - \omega t\right)}. \tag{6.1.34}$$

At first, we may derive the laws of reflection and transmittance of the wave. The boundary distorts the propagation along the axis x, not along the axis y. Thus, there must be an equality for y-components in phases of the incoming wave (6.1.32) and of the reflected wave (6.1.33), and also for the incoming wave and the passed wave (6.1.34). From these conditions, we have

$$\theta_{in} = \theta_r, \tag{6.1.35}$$

$$k_1 \sin\theta_{in} = k_2 \sin\theta_p. \tag{6.1.36}$$

---

[5] Such coefficients can also be defined through wave energies; if so, they will be different. This variant is preferred, but here we discuss only the basic principles, so we may be satisfied with the terms used in the text.

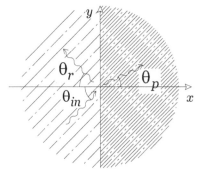

Fig. 6.1.10. The reflection and refraction of a wave incidenting at an arbitrary angle.

The last equation can be rewritten in a more familiar form

$$\frac{\sin\theta_{in}}{c_{s1}} = \frac{\sin\theta_p}{c_{s2}}. \tag{6.1.37}$$

These equations represent the Snell law[6] of wave refraction; it has a well-known analogue in optics.

Then one may obtain the parameters of reflection and transmittance from the conditions like (6.1.26) and (6.1.27). To be exact, the condition (6.1.26) stays the same because pressure is a scalar quantity and, therefore, cannot react to such circumstance as the angle of incidence. The condition (6.1.27) transforms: the velocities are now directed at some angles to the boundary, while the condition (6.1.27) concerns only the normal component of velocities. Thereby, instead of (6.1.27) we have

$$\upsilon_{in}\cos\theta_{in} - \upsilon_r\cos\theta_r = \upsilon_p\cos\theta_p. \tag{6.1.38}$$

Here we did not use the condition (6.1.35). Any velocity in (6.1.38) is connected to the corresponding pressure as $\upsilon = p/\rho c_s$.

Thus, we may note that in order to obtain the answer—values for the reflection coefficient R and for the transmittance coefficient T—one may simply replace the corresponding acoustic velocities in (6.1.31) as $c_s \to c_s/\cos\theta$, so we get, using (6.1.35):

$$R = \frac{\rho_2 c_{s2}/\cos\theta_p - \rho_1 c_{s1}/\cos\theta_{in}}{\rho_1 c_{s1}/\cos\theta_{in} + \rho_2 c_{s2}/\cos\theta_p}, \tag{6.1.39}$$

$$T = \frac{2\rho_2 c_{s2}/\cos\theta_p}{\rho_1 c_{s1}/\cos\theta_{in} + \rho_2 c_{s2}/\cos\theta_p}. \tag{6.1.40}$$

Of course, these expressions transform into (6.1.31) for $\theta_{in} = \theta_r = 0$.

Sound reflection can be used to concentrate the acoustic pressure. This is a very applicable scientific field for many practical things: from listening devices to amplification systems. The roots of such devices go back centuries: various architecture constructions allowed to listen what happens in guest rooms or even far away from a castle; such structures (rooms with vaulted walls) were called in ancient Russia as 'hearings'—thanks to them, the inhabitants of the castle knew what was happening far outside. Later, sound mirrors ('acoustic radars') were used, most famous of them were built at Denge in Kent—huge concrete constructions created from 1928 to 1935.

---

[6] Note that the full name of the main author of this law is Willebrord Snellius; Snell is the shortened form.

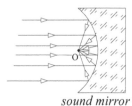

sound mirror

Fig. 6.1.11. A sound mirror—the surface that reflects parallel waves toward a single point O, where the sound wave concentrates—increasing its magnitude.

The principle of a sound mirror is illustrated in Fig. 6.1.11. Sound is mirrored from the paraboloid surface so that the parallel sound waves are concentrated at the selected point, where a microphone (or something else) can be set.

Actually, sound may provide much more information that one can imagine. The old popular question of Lipman Bers 'can one hear the shape of a drum'—which originally concerns the frequency spectrum of a membrane and the unambiguity of this spectrum to the shape of the sound source—got an unexpected answer in present days. Special apps allow to determine the shape of a key based on the sounds produced when a key is inserted into a keyhole.

For our purposes, we are interested in sound focusing to obtain a more intensive pressure wave, where the dynamic processes of a growing/collapsing cavitating bubble may have significant magnitude. In experiments, such an effect can be reached only with a special shape of the vessel; note that usually vessel geometry is not calculated with rigid theory (like the shape of lenses and mirrors of a telescope), but is chosen 'by eye', especially taking into account that the vessel is selected from the given line-up of jars.

On the contrary, in general, acoustic focusing is a very delicate subject. Acoustic focusing devices may be constructed after complicated calculations using special materials (Håkansson et al. 2005) or made as an array of ordinary soda cans where the 'extraordinary' focusing has been observed (Maznev et al. 2015). The last choice is not a kind of a 'scientific joke' or something similar: the application of non-delicate methods to obtain very delicate results is a usual matter for acoustic science. Reverberation (sound effects caused by reflections in rooms) strictly depends on subtle details. The legend reads that during the renovation of the Mariinsky theatre in 1970s, a pile of broken glass was found under the orchestra pit; that pile was thrown away as garbage, and as a result, the acoustic properties of the hall were significantly decreased. From the technical point of view, such results are explainable: broken glass added high frequencies, which strongly influences the overtones.

6.2 Experimental issues

6.2.1 Cavitation of a single bubble

The experimental setup to explore single bubble cavitation is shown in Fig. 6.2.1. With variations, this scheme is used in many works.

To produce a stable (or quasi-stable) acoustic field, two ultrasonic piezoceramic transducers with adjustable frequency are used; these transducers are set on the external side of the walls. For a flat configuration, as it follows from the consideration in the previous chapter, for the best result the frequency must satisfy the condition $KL = \pi$, where L is the distance between two transducers, that is, the frequency would be

$$\nu = \frac{c_s}{2L}, \tag{6.2.1}$$

which gives for water where the speed of sound is ~ 1500 m/s and for our jar of 6.5 cm in diameter the value of frequency is ~ 11 kHz. In practice, this is not the case because of the complex geometry of the cavity and the transducers, from the point of view of acoustics. The acoustic modes in a cavity

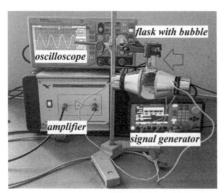

Fig. 6.2.1. Experimental setup.

area much more complicated matter (Russell 2009) than a simple condition (6.2.1); for the first mode in the flat geometry, they can be established either by the sophisticated calculations or empirically. In short, there exists a number of resonance frequencies for standing waves in a cavity, which can be calculated semi-analytically. We did not concentrate here on the calculations, since, first, our geometry is not spherical, and, second, the boundary conditions provided by the transducers are very confusing too. Finally, choosing the empirical way, we found that for that vessel the best frequency is from ~ 20 kHz to ~ 30 kHz.

The vessel is made of glass (organic glass is another option), and this elastic medium generates pressure waves from the whole surface of the vessel—not as two dot sound sources, as it would be if the point transducers were mounted inside a jar. With a spherical form of the vessel and transducers of finite size, one may try to concentrate sound waves in the center of the vessel, that is, to produce an alternate acoustic field of a sort of $p = p_0 \sin \omega t$. If successful, one may expect that a single acoustic bubble would appear in the center. If so, we would observe the 'pure' cavitation: the bubble appears at the negative stage of the external pressure, and we may treat the oscillations of the 'natural' bubble, the composition of which corresponds exactly to bubbles appearing in free conditions.

Unfortunately, often this strict approach does not provide a suitable result. The bubble does not appear where we expected—in the center of the vessel, so we inject a bubble seed into the vessel with a needle. In such experiments, the cavitating bubble contains more of the gaseous phase—we mean not the vapor of the corresponding liquid—than in natural conditions. Such a bubble is larger and the pressure inside it is higher than in a 'natural' bubble, which appears only because of the rapture of a liquid. On one hand, this bubble has too many distinctions from the natural object we are interested in. On the other hand, the dynamics of bubble oscillations can be investigated there more conveniently, since, again, the bubble is larger and can be observed and controlled more easily.

There exists a compromise. One may not place a seed in a fluid—the nucleus of a ready-to-use bubble. Fluid can be aerated with the same needle of an empty syringe. The added gas will be dissolved evenly throughout the jar; thus, the probability of formation of a cavitating bubble will be increased. In this case, the bubble can be considered 'natural', despite the evident external help.

To register bubble oscillations, we used a high-speed Phantom VEO-E 310L camera, which can record a video at up to 10^6 frames per second.

6.2.2 Cavitation of many bubbles

This is a more regular type of cavitation—cavitation that we saw in the first two chapters of this book. In this case, we want to observe a large number of cavitating bubbles; thus, we do not need to focus acoustic waves, ignoring the fact that the acoustic field in the vessel is an important factor for the process inside it anyway. So, if we have no desire to design the shape of a vessel specially, we may simply place an ultrasonic waveguide into a fluid, turn on the power and watch a couple of bubbles form under the waveguide.

For our own purposes, we are interested in multibubble cavitation for three reasons:

✓ investigation of sonoluminescence;

✓ investigation of the surface damage during cavitation;

✓ investigation of the multi-bubble cavitation itself.

Besides these factors, of course, someone may have interest in other details of the cavitation process, such as wave formation on a free fluid surface, etc. But we concentrate on these three items.

The experimental setup is shown in Fig. 6.2.2; it is based on the standard IL10-2.0 ultrasound generator: power output up to 2 kW. In short, here we use a single magnetostriction transducer (working frequency ~ 20 kHz), equipped with a waveguide which is immersed into the fluid.

The main set of elements of this setup are the parts which produce ultrasound directly: an ultrasonic generator, an ultrasonic transducer (magnetostrictor) and an ultrasonic waveguide. In theory, the magnetostrictor produces ultrasound of the same frequency; in practice, this frequency is slightly 'floating', so sometimes it must be re-adjusted. Working outside of the resonant frequency is a dangerous mode for the magnetostrictor that can lead to damaging the device. Unfortunately, this mode is potentially interesting for investigating special patterns of sonoluminescence, when the cavitation cloud is gone from the waveguide directly, forming the cavitation region at a distance from the waveguide. In this mode, thin jets spread from the waveguide to the bulk of the fluid. Also, in this mode we may easily observe the patterns of special form on the waveguide surface. It is always a dilemma: to take a risk or not, and only the experimenter personally can make the decision.[7]

For the common reader, it could be interesting how the proper frequency of the ultrasonic generator is set. The answer is—by ear. When the frequency is properly adjusted, the waveguide placed in a fluid emits a low, uniform hissing sound, in contrast to the high uneven whistling sound of an untuned generator. Indeed, to treat acoustic cavitation some special skills in acoustics are required.

The part of the setup which contains the magnetostriction transducer demands intensive cooling due to a large heat release there. Heating influences the work of the transducer and may even destroy

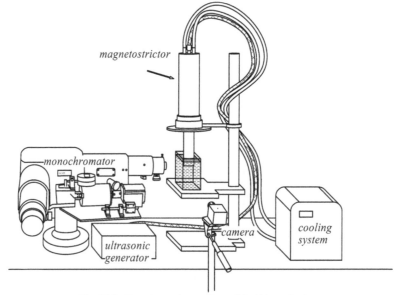

Fig. 6.2.2. Experimental setup for multibubble cavitation.

[7]   We risked, obtained interesting results (see chapter 9) and destroyed both of our ultrasound generators. Science requires sacrifice.

it; thus, a cooling contour is absolutely necessary. Possibly, this is an auxiliary element from the point of view of the creation of ultrasonic oscillations, but this is an important thing for operability of the whole setup.

The waveguide is a special element. From a certain point of view, it is the basic element of the scheme since it implements the direct effect on the investigated medium. We used titanium wave guides (with addition of 4% of aluminum) to analyze the influence of conductivity of its surface on the light emission; during cavitation, we covered them with dielectric paints, but did not achieve proper results: under the ultrasonic impact, the paint layer peeled off quickly, and we cannot present here any certain results about that matter. Possibly, sputtering may provide a better result.

Actually, the vessel is also an ambivalent element of the setup. Initially, we used glass vessels, which were quickly destroyed by the intensive impact of the ultrasonic waveguide. When we got tired of collecting the pieces of broken glass, we switched to plastic vessels; all the results presented below were obtained with them. Plastic does not get destroyed as easily as glass because it dampens the shock through vibrations of its surface. These vibrations can be felt directly by touch: under ultrasonic vibrations, the vessel becomes very slippery, as if we are touching a smooth oily surface. It would be interesting to investigate the behavior of such a construction in a fluid: how effective would an ultrasonic impact be from reducing of the hydraulic resistance.

The investigated media were: glycerol, water, polypropylene, saline, silicon oil, propylene glycol. Since our main interest was sonoluminescence, and glycerol produces the brightest glow, if this attributive can be applied to a light which can be seen only in a complete darkness, almost all valuable results were obtained with glycerol. Particularly, the destruction of the waveguide surface presented in Chapter 5 also was observed in glycerol.

The diagnostics depends on the investigated processes. To analyze the consequences of ultrasonic cavitation on the surface, an electron microscope was used (see photos in Chapter 5). The sad side of such treatment is the total destruction of the surface (to prepare samples for the microscope). Optical treatments are more sparing and allow to conduct repeating researches, analyzing the process of surface destruction in time.

Temperature diagnostics was provided by a thermocouple placed in the liquid at a distance ~ 1 cm under the waveguide. Surely, the main principle of thermocouple stowage is broken here: thermocouple wires must be laid along isotherms. This requirement must be fulfilled to exclude heat leaks along the wires: the thermocouple junction must be in thermal equilibrium with the surrounding medium, it must get the heat only from the medium and release the heat only to the medium—only in these conditions we can be sure that the thermocouple obtains the temperature of the medium. However, for our case we did not recognize any isotherms, while the temperature measured in experiments can be used only as reference data and is not of its own interest. Thus, we can submit before some uncertainties of the temperature measurement.

Note that somewhere the temperature measurements promised interesting results, but deceived us. When working with glycerol for a long time, glycerol turns into the modification known as the 'yellow glycerol'. In this type of liquid, we saw a strange effect: under the ultrasound influence, temperature of the liquid begins to grow, then stabilizes at ~ 30°C for time period of ~ 1 minute, and then continues to grow. When we replaced yellow glycerol with the regular (transparent) one, this effect was gone. So, we still don't know what it was: we never published this result before and mention it here only because of the free form of the given section.

Since our aim was the investigation of light emission during cavitation, the equipment to register the visual part is a very important component of the setup.

To record a video of the luminescence, we had to use a very sensitive camera AVDERT-9346V. This very camera was also used to obtain many other pictures shown in this chapter. To capture a video at high frequency, with many of frames per second (fps), of an order of $\sim 10^5$ fps, we used the Phantom VEO-E 310L camera.

To measure the spectrum we, unfortunately, cannot use standard spectrometers because of their insufficient sensitivity. We made a special spectrometer, the construction of which was based on the

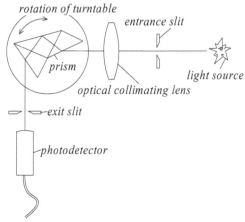

Fig. 6.2.3. The scheme of the monochromator. The input light falls on the prism which expands the light into a spectrum. The collimator cuts a certain wavelength, and the detector mounted at the end of the outer tube registers the monochromatic light. The wavelength selection is carried out by the rotation of the turntable. Optical schemes in tubes are shown approximately.

old Soviet prism monochromator UM-2. Monochromator is the device which represents the light as a spectrum; we placed the Hamamatsu H8711-300 photodetector in the image plane. A micromotor rotates the position of the prism and, therefore, the photodetector is irradiated with the light of different wavelengths: the wavelength, therefore, is explicitly connected with the prism position. To measure the intensity of the given wavelength (that is, at the given position of the prism), we must wait at least several seconds, until the photodetector collects the signal of sufficient intensity, then change the position of the prism, wait again, and so on. Thus, the measurement of the whole spectrum lasts many minutes.

The problem is that the intensity of sonoluminescence may change in time due to many reasons. The liquid is heating, changing its properties; consequently, the intensity of sonoluminescence, generally, diminishes. This means that we may measure a lesser intensity on the wavelength λ_1 than on the wavelength λ_2, only because we measured on λ_1 later than on λ_2, not because the intensity on the wavelength λ_1 is indeed lower. To correct the measured signal to the common level of intensity of sonoluminescence, we measured this common level with the second Hamamatsu photodetect or S2281-01 and recalculated the measured values cooperatively.

The diaphragms of the monochromator were open to a width ~1 mm—to reach high irradiation at the photodetector. As a result, of course, the resolution of our spectrometer was poor—this the reverse side of the high sensitivity of the apparatus.

Thus, we have two schemes to induce acoustic cavitation. In the following sections, we consider the main properties of two types of cavitation: when a single bubble appears and oscillates or for the cavitation flow consisting of a large number of bubbles.

6.3 Cavitation of a single bubble

6.3.1 Bubble oscillations: Overview

Oscillations of a single acoustic bubble consist of two terms:

- ✓ pulsations driven by an external force;
- ✓ natural vibrations of the bubble.

The first term, we assume, is clear: alternating external pressure tends to change the volume of the bubble, and the frequency of such oscillations tends to coincide with the frequency of that external force. Does it mean that the bubble will vibrate precisely at that frequency? Surely not, since the bubble has its own opinion concerning variations of its volume.

Let us imagine a bubble in absence of an external force. Deviation of its volume leads to the change in the pressure inside it. The certain law for such change depends on the conditions inside and around a bubble; considering the perfect gas inside a bubble, for the adiabatic case there must be

$$pV^\gamma = const,$$ (6.3.1)

while for isothermal conditions

$$pV = const.$$ (6.3.2)

The variant to choose depends on conditions. In the next chapter, we will argue heatedly about this matter. The key point is that when the bubble grows, the pressure inside it decreases, and vice versa.

Thus, when a sudden external disturbance—we mean exactly disturbance, not a constant or variable force—shrinks the bubble, the pressure in it increases, tending to return the bubble to its initial state. Further, there may be two variants: (a) when the bubble reaches its initial size, the process will stop completely (static case), (b) the growing (restoring its normal volume) bubble skips the initial stage by inertia, gets larger volume than its starting value, so the pressure inside it drops below the normal value and the bubble will shrink again, and so on (dynamic case). The real case is the dynamic case—only dissipation can stop such pulsations with time.

Now we estimate the frequency of these natural pulsations of the bubble. All the equations and formulas will be given in the next chapter, here we use an old and semi-reliable method—analysis of dimensions of the key variables. A brief description of this method:

✓ establish key variables for this problem;

✓ write the dimensions of those variables with 'primal' dimensions: units of mass, time, length, temperature, and electrical current (in a common case);

✓ write the dimension of the variable of our interest;

✓ make sure that there may be a single combination of the key variables: multiplication of these variables at certain exponents gives the answer for the problem.

The finest element of this approach is to establish those 'key variables'. If we omit some important parameter, all the results can be thrown away. The next nuance is that the answer may contain some constants which are ignored in this method. In addition, the problem may have two parameters of the same dimension, which cannot be taken into account by this method fundamentally.

Indeed, one may exaggerate in this manner further, but usually – usually! – the analysis of dimensions works well; in Chapter 4, we have seen it. Let us apply this method to bubble oscillations.

Here we may select three key variables: the bubble radius R, pressure p, and the density of the liquid ρ; the liquid holds the bubble back: to grow, the bubble has to overcome the resistance from the liquid. In other words, we assume that deviation in bubble radius is about R, and deviation in pressure is about p.

Now let us establish dimensions of these variables:

$$[R] = m; [p] = \frac{N}{m^2} = \frac{kg}{m \cdot s^2}; [\rho] = \frac{kg}{m^3}.$$ (6.3.3)

We want to find the oscillation frequency v, the dimension of which is $[v] = s^{-1}$. Thus, from (6.3.3) we see that pressure and density must be used only as their ratio because both of them contain mass, which is foreign for v. This ratio contains a unit of time (as s^2), while the third—still unused, parameter R does not; thus, the root must be taken from the ratio p/ρ, and the exponent at R follows from here. Finally, we have the estimation of frequency

$$v \sim \frac{1}{R}\sqrt{\frac{p}{\rho}}.$$ (6.3.4)

In Chapter 7, we obtain the complete expression for oscillation frequency. The main difference is the ratio of the maximum radius to the initial one; as we mentioned above, that is an unavoidable point. Nevertheless, the relation (6.3.4) is quite suitable for estimations.

Normally, the pressure is $p \sim 10^5$ Pa and $\rho \sim 10^3$ kg/m^3. Thus, from (6.3.4) we have corresponding frequencies: for the millimeter bubble $v \sim 10^4$ Hz, for the bubble of radius $R \sim 10$ μm there must be $v \sim 10^6$ Hz, and so on.

Thereby, in the presence of an external force, the bubble will tend to do the same exercise: oscillate with its own frequency. If the bubble is sufficiently large, with the size in millimeters, then the frequency of its proper pulsations is roughly equal to the frequency of the external force which is about $\sim 10^4$ Hz, as we discussed in previous sections. This is a more complicated case.

An interesting case which is simpler is oscillations of small bubbles, the natural frequency of which is higher by an order (or even more) than the frequency of the external acoustic field. In this case, we have two characteristic values for the time period—caused by the external ultrasound field and conditioned by natural oscillations of the bubble. Note that this circumstance is suitable because, indeed, we are more interested in natural oscillations than in the ultrasonically induced ones: finally, in such experiments, we want to explore the natural cavitation anyway.

Let us try to imagine bubble dynamics in an external field. When the outer pressure is low, the bubble grows, but the pressure inside it decreases like in (6.3.1) or (6.3.2), or something intermediate. Thus, the bubble collapse will start earlier than the outer ultrasonic force changes its sign. Then, during the collapse stage, when the pressure inside the bubble increases, we would observe a competition between two processes: ultrasonic force tends to shrink the bubble, but the increasing pressure inside it tends to prevent this. This pulsation must take place with comparatively small amplitude: every attempt to squeeze the bubble meets an immediate reaction from the compressed medium inside it.

Thus, for a very small bubble, of the radius of ~ 10 micrometers or less, the dynamics of the bubble radius would have a form shown in Fig. 6.3.1.

It would be interesting to compare such dependence with experimental data, but it is possible only for a bubble so small that would oscillate at such a frequency that observing it would be accompanied with certain problems. For a larger bubble, we have another (unexpected for some) problem.

Fig. 6.3.1. The common time dependence of the bubble radius for small bubbles, for which the frequency of natural oscillations is higher than the frequency of the ultrasound field.

6.3.2 Bubble oscillations: Certain experimental view

In this section, we present the experimental results—evidences of bubble oscillations.

The common view is presented in Fig. 6.3.2. Here a tiny bubble appeared in the center of the vessel, its size is ~ 0.5 mm—average, because this bubble oscillates.

This bubble looks unpresentable: on one hand, one may want to explore its oscillations and, in common, scrutinize it closely. On the other hand, this desire may look excessive. Indeed, what we hope to see? The bubble is a bubble. A simple spherical bubble, the radius of which oscillates with time.

The unexpected answer to these doubts is presented in Fig. 6.3.3, where bubble oscillations at the frequency of the external acoustic field at ~ 26 kHz are presented. These are frames from a

Fig. 6.3.2. A single bubble in a jar irradiated by ultrasound with two piezo transducers.

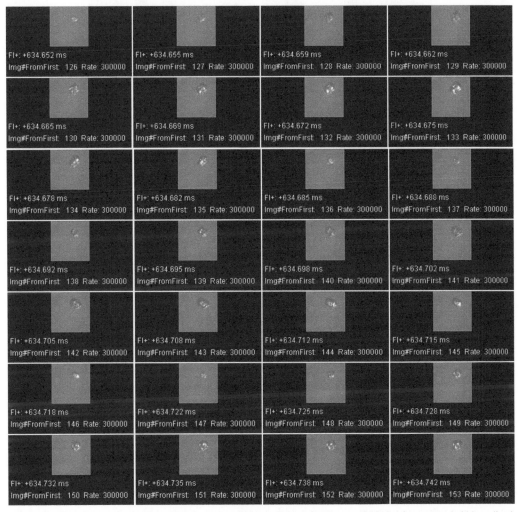

Fig. 6.3.3. Evolution of the cavitating bubble in the ultrasonic field at frequency ~ 26 kHz; the average bubble radius is ~ 0.5 mm. Captured at $3 \cdot 10^5$ fps, each frame has a size of 64×64 pixels.

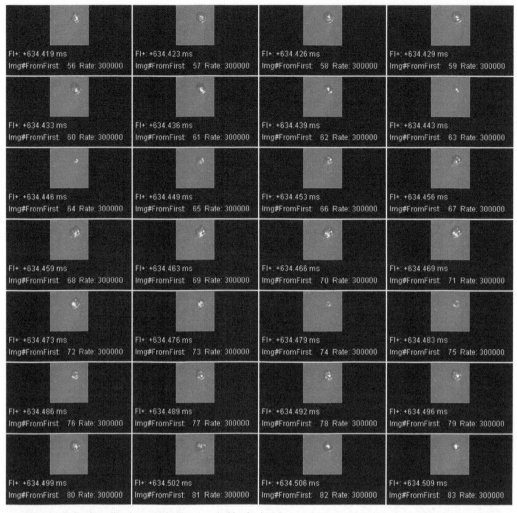

Fig. 6.3.4. During vibrations, the bubble may hold spherical form much better than it was presented in Fig. 6.3.3.

video captured at a frame rate of $3 \cdot 10^5$ fps—this is very fast, so, the frames are very small—only 64 by 64 pixels in size. Nevertheless, we may distinguish a very important thing there: the bubble is non-spherical indeed.

To be exact, we can see that the smallest bubble is always (at least, in our experiments) spherical: at the collapse phase, it obtains the form of a ball. But at the growing stage, the bubble loses stability of its shape, becoming very unstructured (see frames number 132—134 in Fig. 6.3.3, for example). Actually, it looks like a product of explosion, or something like that. Then, at the collapse phase, the bubble returns to its spherical shape, but not for long: at the next expansion phase, the process repeats itself. Of course, these losses of stability can be strong (as in Fig. 6.3.3), but can also be weak: in Fig. 6.3.4., we present the series when the bubble more or less keeps its sphericity (for two of three expanding phases shown in Fig. 6.3.3, the bubble approximately keeps the spherical shape).

Note that the time period of these pulsations is ~ 0.04 ms, which can be determined as time gaps between two consecutive frames with a bubble of minimum radius. This value correlates with the frequency of the ultrasound generator (26 kHz) and cannot be established with better accuracy, since we cannot exactly fix the minimum size of the bubble—it is most likely that the bubble captured on those frames is not minimal, so it is excessive to peer onto decimal digits at captions on those frames.

Fig. 6.3.5. During oscillations, a bubble of irregular shape may 'lose' its part, but with time, this part will be attracted to the main bubble, so the unity will be preserved.

Now we consider the impressive phenomenon. While oscillating, in its expanding phase, the bubble sometimes looks like it is ready to 'explode', and, really, this happens from time to time: the bubble divides in two (or even more) parts. However, this state does not last long: the parts join back into a whole bubble (see Fig. 6.3.5); they attract to each other like opposite charges.

Of course, here, on such small images, it is difficult to follow the details of the process. In the next section, we explore multi-bubble cavitation in an ultrasonic field and present larger pictures, which lead to the discussion of the Bjerknes effect.

We clearly see that bubble oscillations lead to non-sphericity of the bubble's shape at the growth stage.

6.3.3 Oscillations of bubble's shape

In section 6.3.1, we discussed the oscillations of volume—only the dependence $R(t)$, now we see that the shape of the bubble oscillates too. The bubble may not be shaped like a spherical ball, it can be a much more complicated thing.

Generally, how does a bubble know that it must be a ball? At equilibrium, a bubble takes the spherical form because of equal pressure at each point on its surface. Thereby, the tension must be uniform along the surface too; thus, we have the necessary conditions for sphericity. However, these conditions can be violated in two ways.

The fist way comes from the macroscopic ununiform forces on the bubble surface: it is possible that during oscillations in real conditions, the bubble lays in the non-uniform external pressure field. The second way is caused by the instabilities on the bubble surface, i.e., at the liquid–vapor interface, which was considered in Chapter 4. In practice, the second way looks more realistic for small bubbles of less than 1 mm in size, but the first one should not be forgotten too. In common, we may conclude that deviation from sphericity is caused by the Rayleigh–Taylor instability (see Chapter 4) on the bubble surface: that is the possible reason why the bubble is not spherical during the expansion phase (when a liquid moves away from a vapor), but is spherical during the collapse phase (when a liquid moves towards a vapor).

Now let us take a look at the problem from another, thermodynamic angle. A different formulation of problem: we transfer energy to the bubble; what part of this energy would be spent on work against the external pressure and what energy would be transferred to the change of the surface area? Indeed, the total work calculated for a vapor bubble is

$$A = \int_{V_1(t_1)}^{V_2(t_2)} p dV - \int_{F_1(t_1)}^{F_2(t_2)} \sigma dF, \tag{6.3.5}$$

where $V(t)$ and $F(t)$ are the volume and the surface area of the bubble at the corresponding moments of time. For a spherical bubble, the second term is negligible in comparison with the first one, simply if

$$\frac{2\sigma}{R} \ll p, \tag{6.3.6}$$

that is, if the Laplace pressure jump conditioned by the surface tension σ is much lower than the pressure p. However, the surface area of a bubble may significantly exceed $4\pi R^2$, if the bubble is non-spherical and R is only its average radius; therefore, the condition (6.3.6) does not state necessarily that the 'surface term' in (6.3.6) is much smaller than the 'volume term'.

Let us find the surface area of a bubble with 'strong waves'—the bubble with a very complicated surface shape. The complete answer must be based on the 3D analysis in spherical coordinates (r, θ, φ); our dependencies of interest are deviations of the function $r(\theta, \varphi)$, describing the radial coordinate of the bubble wall, from the average value R.

The surface area of a non-spherical bubble is

$$F = \int_0^{2\pi} \int_0^{\pi} \sqrt{r^2 + \left(\frac{\partial r}{\partial \theta}\right)^2 + \frac{1}{\sin^2 \theta}\left(\frac{\partial r}{\partial \varphi}\right)^2} \, r \sin\theta d\theta d\varphi \tag{6.3.7}$$

For small deviations of sphericity: $r = R + \xi$, with $|\xi| \ll R$, this expression can be simplified. For this case, using the approximation of a kind of $\sqrt{1+\varepsilon} \approx 1+\varepsilon/2$ for $\varepsilon \ll 1$, we get

$$F = \int_0^{2\pi} \int_0^{\pi} \left(1 + \frac{1}{2r^2}\left(\frac{\partial \xi}{\partial \theta}\right)^2 + \frac{1}{2r^2 \sin^2 \theta}\left(\frac{\partial \xi}{\partial \varphi}\right)^2\right) r^2 \sin\theta d\theta d\varphi. \tag{6.3.9}$$

It is important to understand, however, that the deviation of sphericity is defined not only by the amplitude of disturbance ξ, but also by the spatial scale of disturbance. For instance, let us consider a sphere of radius R with a developed surface obeying the harmonic law: $r = R + \delta\cos k\theta \cos k\varphi$. We have then

$$F = \int_0^{2\pi} \int_0^{\pi} \sqrt{r^2 + \delta^2 k^2 \left(\sin^2 k\theta \cos^2 k\varphi + \cos^2 k\theta\right)} \, r \sin\theta d\theta d\varphi. \tag{6.3.10}$$

As one may expect *apriori*, for a strongly developed surface—for short-wave distortion when $k \to \infty$—the main part of the integral (6.3.10) is determined by the second term under the root at any value of amplitude of disturbance δ, even if $\delta \ll R$, in this case:

$$F \approx \delta k R \int_0^{2\pi} \int_0^{\pi} \sqrt{1 - \sin^2 k\theta \sin^2 k\varphi} \, \sin\theta d\theta d\varphi. \tag{6.3.11}$$

For a large value of k, i.e., for fast oscillations of the integrand function, we may put the value of square root to the mean value, using $\overline{\sin^2\theta} = 1/2 : \sqrt{1 - \sin^2 k\theta \sin^2 k\varphi} \approx \sqrt{3}/2$, and

$$F \approx \sqrt{12}\pi\delta kR, \qquad (6.3.12)$$

with $k >> R/\delta$ as the condition for correctness of (6.3.12); due to the last inequality, the surface of the 'golf ball' (6.3.12) is larger than the sphere area $4\pi R^2$.

Note the 'golf ball' is a quite real thing for a vapor bubble. In Fig. 6.3.6, we may see the texture on the bubble arising on the needle, which is sometimes used in experiments (see Section 6.2).

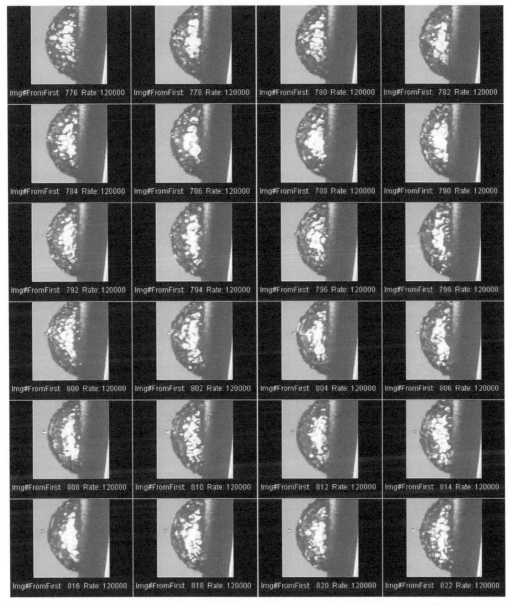

Fig. 6.3.6. A bubble may have the shape of a 'golf ball' and 'spit' with a smaller bubble. Here the bubble appeared in the ultrasonic field on the needle of 1 mm in diameter.

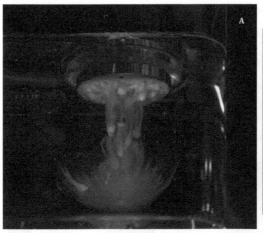 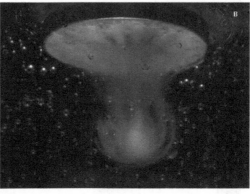

Fig. 6.4.1. Cavitation jets. *A*—common view, *B*—a detailed picture where separate bubbles are recognizable. Two bubbles in the center will be merged with each other, see Section 6.4.3.

6.4 Multibubble cavitation

6.4.1 Common description

In this section, we discuss the results obtained on the experimental setup described in Section 6.2.2. Multibubble cavitation is surprisingly an easier object to investigate, despite the fact that the phenomenon itself is more complex than the acoustic oscillation of a single bubble.

When the ultrasonic generator, shown in Fig. 6.2.2, is on, the waveguide begins to impact the liquid (glycerol for all the pictures presented below). Displacements of the waveguide are very small (much less than one millimeter); thus, we cannot see them by eye. However, we easily see the result of these impacts on the liquid. The cavitation flow appears below the waveguide; this flow is caused, obviously, by pressure acting on the liquid from the waveguide. Since, however, this pressure is alternating, the flow has a non-stationary character. When the waveguide goes up, creating underpressure at the waveguide, the fluids tend to flow up, but, of course, it has no sufficient place there. Thereby, the flow has an oscillatory character: noticeable movement down, then a weak attempt to return back up, then the next stage of down flow, etc. As a result, we see the cavitation jet spreading from the waveguide toward the bulk of the liquid, see Fig. 6.4.1.

In this mode, we may also observe sonoluminescence—the light emission of mysterious nature from the cavitation area; we will consider this phenomenon in Chapter 9; there is no intrinsic light emission on these photos—all the bright regions are the reflections of external light. Note that the 'smooth' picture given above can be observed when the frequency of an ultrasonic generator is well adjusted, that is, corresponds to the resonance frequency. In this case, a magnetostriction transducer moves easily, and the process has one characteristic time scale—the frequency of ultrasonic oscillations. In another case, when the frequency of an ultrasonic generator is tuned slightly out of resonance frequency (now we are talking about the case when the frequency of the ultrasonic generator is set to 20520 Hz instead of the 'correct value' of 20500 Hz, for example), we may observe another mode, under the risk of destroying all the equipment. When the formation of the strong cavitation jet is obstructed, we may clearly see the extraordinary structures formed on the waveguide.

6.4.2 Foam on a waveguide

As we mentioned above, under the ultrasonic vibrations of the waveguide the fluid flow tries to repeat these oscillations, but with desperate success: the flow back toward the waveguide is restricted by the waveguide itself. Due to these attempts to leave the near-surface area, very small

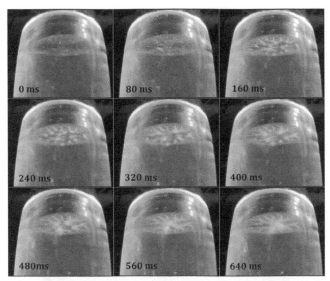

Fig. 6.4.2. The foam on the waveguide. When such tiny jets spread into the bulk of the liquid, they are sometimes called 'streamers'; on the waveguide, they form a 'fractal'. One may see that this 'net' appears on various parts of the waveguide surface (80 ms) and grows, increasing its brightness, from the peripheral area to the center. The huge 'foam bubble' formed in the center at 640 ms will be pushed into the fluid further.

bubbles stay near the waveguide, forming mysterious structures on it (see Fig. 6.4.2). Actually, the trace of such structures can also be seen in Fig. 6.4.1A (above the jet, one may see the fragmented foam on the waveguide surface), but it is represented crisper in Fig. 6.4.2.

Such foam structures were observed often; sometimes they are called 'acoustic streamers' (Pelikasis et al. 2004)—by analogy with the spark discharge (the first stage of spark discharge—a narrow channel with weak conductivity and weak glow—is called a 'streamer' in gas discharge physics). On the other hand, such term is more suitable for the light-emitting filiform object spreading from the waveguide to the fluid (Biryukov et al. 2020); anyway, the nature of glowing streamers is the same—these are thin jets of foam.

Again, for the terminology matters: sometimes, these net-like objects were classified as fractals (Skokov et al. 2007). It is difficult to establish firmly whether these figures are indeed fractals—self-similar structures of the same form at any scale[8]—but, actually, some properties of fractals can be definitely found. On a stationary picture, these structures look like tree branches (see the frame corresponding to 560 ms, for example), as if they are growing from the center of the waveguide to the periphery, in the opposite direction, if someone would analyze the frame corresponding to 240 ms. However, the full video explains that the formation of that 'fractal form' starts simultaneously everywhere, throughout the whole waveguide surface (see the frame for 80 ms).

Note that these structures pulsate. In Fig. 6.4.3, these pulsations are shown; these frames are taken from the video at 3200 frames per second which captures ultrasound cavitation at frequency ~ 20 kHz. Thus, here we see at least seven times less frames than it should be; for the same reason, by the way, it seemed in old movies that the carriage wheels rotate in reverse.[9] Fortunately, insufficient time resolution is not a great problem for us: we see the variation in brightness of the foam clearly: the period of foam pulsations is 10 ms, as we see. On second glance, it is a slightly strange value,

---

[8]   Of course, any real object cannot be self-similar at any scale due to the molecular structure reasoning: on a certain spatial scale, the molecular structure of an object prevails. We may call such objects as fractals only at the scales of intermediate range.

[9]   If a wheel makes 9/10 of a full rotation during the time between two frames, it would seem for us that it made 1/10 of a turn in the opposite direction.

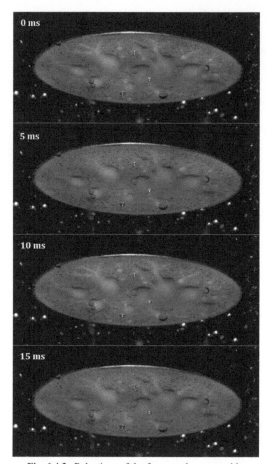

Fig. 6.4.3. Pulsations of the foam on the waveguide.

since one may try to relate the foam pulsations to the ultrasonic oscillations (it looks logical at least), but the measured value is much higher.

It is an interesting matter, indeed, like that carriage wheel. Does the measured period of pulsations relates to the real time period of foam oscillations, or is this an artifact caused by multiple frequencies of ultrasonic pulsations and the frame-per-second parameter of the camera? The time period of the ultrasonic waveguide is ~ (1/20500) ~ 0.04878 ms; in the presented mode of video fixation, the time gap between two frames is ~ (1/3200) ~ 0.3125 ms (we give so many digits and the 'approximately equal' sign since we are not sure how the exactly the number of frames per second differs from the value of 3200 fps set on the camera). These time values are related as 32:205—there are 205 full oscillations of a bubble between 32 frames fixed on the camera. Of course, we must keep in mind that all the values given above are approximate, but anyway, we should conclude that the measured period of 10 ms is the artifact of the measurement procedure. On each 32nd frame, we see the bubble at the peak of its form (so to say); indeed, we see the bubble that made 205 oscillations to this time. On each 16th frame, we see the bubble at its minimum; actually, it made 102 full oscillations and a half oscillation more. The measurement at a high frequency, at a much higher fps-parameter of the camera, shows that the frequency of the foam oscillation is ~ 0.05 ms indeed (Biryukov et al. 2020); here we do not present the frames fixed at 390 000 fps because they are not very photogenic.[10] By the way, all this situation, where we 'clearly see' the time period of

---

[10] Precisely, only a well-trained eye may detect something on frames that small with such a low resolution (high frequency is achieved at such a cost). Computer processing is required for such frame series, which was done in (Biryukov et al. 2020).

10 ms for oscillations, is a good illustration for those who suppose that experiments provide solid, unambiguous facts that cannot be interpreted obliquely.

Now let us return to the common properties of foam. In (Mayer and Varaksina 2007), it was reported that these structures leave the corresponding seal on the surface—the foil would have exactly the same form as the structures. We never observed such an effect, and this phrase means what it sounds: we reasonably agree that this effect may take place. At the same time, we observed the destruction of the surface precisely at the locations of intensive cavitation (see Chapter 5); thereby, we do not wonder if these 'trees' leave a seal after them on the surface.

The nature of such self-similar structures is more intriguing. We have seen that these objects arise initially at different locations, and then grow. Considering the fact that these structures consist of small bubbles, their growth means the accession of new bubbles to the given structure; actually, we even saw in Fig. 6.4.2 how this process occurs. Thus, it seems that bubbles must attract each other to fabricate such a structure.

Starting from another point, we may come to the same conclusion. Standing on the absolutely formal position, we may declare that usually[11] the formation of a self-similar structure demands some interaction of its parts. The analogy is the growth of a crystal: the new part of an object 'wants' to set a place in this structure since it is somehow coupled with other parts of that structure.

What sort of interaction could it be? The first phenomenon that comes to mind when we discuss the interaction of particles is 'electricity'. This is much more natural as one may suppose, if we take into account the origin of the word 'electricity': it appeared in attempts to explain the attraction properties of amber (see Chapter 8). On the other hand, today under 'electricity' we mean a certain phenomenon connected with electric and magnetic fields. It is very doubtful that bubbles obtain opposite electrical charges (if they get charges of the same sign, they will repulse each other, not attract) or gain some magnetic properties.

Possibly, bubbles indeed may interact one with another, but if so, we may observe it explicitly, on a large scale, so to say. If it is so, some mechanical—non-electrical—explanation may possibly exist.

6.4.3 The Bjerknes effect: Observation

Sometimes, bubbles behave very strangely, like they have been magnetized. The whole history is presented in Fig. 6.4.4.

Many bubbles appeared in the vicinity of the ultrasonic waveguide (see Fig. 6.4.1B). We follow two of them. Initially, those bubbles behave independently: they moved separately, each according to its own business. But when they get closer, we observe a very strange thing: they were attracted to each other, like charged particles or something similar. They abruptly changed the direction of moving: they moved to meet each other, collided, and merged.

The presumption that these bubbles carried electrical charges of opposite signs looks fantastical a bit; even if a bubble can take some charge (see Chapter 8), that charge would be rather of the same sign, not the opposite.

However, we may see that the bubbles oscillate. It is not well seen in Fig. 6.4.4: during oscillations, bubbles are 'blurred' on the frame, so we chose frames where the bubbles look more 'sharp'. But indeed, such oscillations do take place.

One may assume that these oscillations are the key to understanding this phenomenon. Oscillation of a bubble generates the variation in the pressure around it. This pressure acts on the neighboring bubble, and conversely, the deviation in pressure caused by that second bubble affects the bubble number one. As a result, we may expect that these bubbles can attract each other

---

[11] 'Usually' is a beautiful scientific word that must be forbidden, strictly speaking. This word means a declaration of a common elusive principle instead of solid facts directly concerning the subject of discussion. But, so far this word is legal, we use it: for this very case, the formation of fractal structures is not clear as of today, if we do not consider a percolation process.

somehow: not because the bubbles acquired some 'attractive' properties themselves, but because they may interact by means of the surrounding medium.

Surely, the explanation given above is yet incomplete: we have to ascertain some details of such type of interaction. This matter will be finished in the next chapter devoted to the theoretical description of oscillating bubbles. Here we may only assure that 'the theorem is true,'[12] and this phenomenon is called the 'Bjerknes effect'. The detailed explanation of bubble interaction, as well as their interaction with a solid surface (why these small bubbles attached to the waveguide surface?) will be discussed in the next chapter.

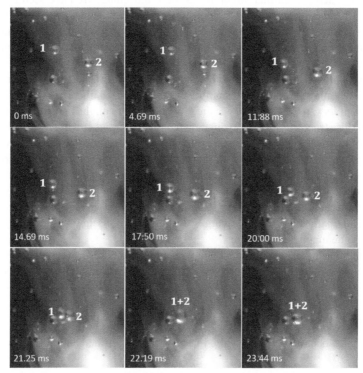

Fig. 6.4.4. Two oscillating bubbles attracted to each other. Frames from the video at 3200 fps (frames per second). Frames from that video are also presented in Fig. 6.4.3: the foam pulsates above these dancing bubbles. The first frame indeed is a part from the picture presented in Fig. 6.4.1B, that can be seen in Fig. 6.4.3.

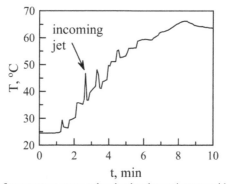

Fig. 6.4.5. The time dependence of temperature measured under the ultrasonic waveguide at the distance of ~ 5 mm from it. The peaks correspond to the instants when hot jets came from the waveguide.

---

[12] Augustin–Luis Cauchy could not explain the theorem to his apprentice, a young earl. In desperation, Cauchy claimed: "Theorem is true, I swear on my honor", and the earl said: "Oh, why didn't you say it right away? Of course, I trust such a respected man."

6.4.4 Temperature of a liquid

The temperature of a liquid is a more important problem than it may seem at first glance. The details will be discussed in Chapter 9; here we only present the experimental data collected by the thermocouple placed under the waveguide (see scheme 6.2.2).

The thermogram is presented in Fig. 6.4.5. We see that temperature increases noticeably, at the average rate of ~ 5°C per minute. But the irregular character of this dependence is more interesting.

The reason behind this irregularity is the heated mass of fluid in the jets coming down from the waveguide. In Fig. 6.4.5, one may see that the jets induced from the waveguide have increased the temperature: the maximums on the graph correspond to the events when the hot fluid mass arrived to the thermocouple. On the other hand, we see that these maximums are not huge—of about ~ 10°C.

Thus, we may conclude that the temperature of the fluid increased under ultrasonic impacts, but not dramatically. The observed temperature rise can be explained by the friction processes on the waveguide without additional hypotheses. Let us keep this fact in mind for Chapter 9.

6.5 Acoustic chemistry

6.5.1 Pressure and temperature

It is an interesting fact that new chemical compounds can be found in the medium after acoustic cavitation. Investigation of such processes is a separate scientific line; the corresponding science area is called 'acoustic chemistry' or 'sonochemistry'. There exist several reviews and monographs (Luche 1998, Mason and Lorimer 2002, Ashokkumar et al. 2007, Kentish and Ashokkumar 2010), where this matter is discussed in detail—adjusted for the complexity of the problem. Indeed, many processes occurring under the influence of an intensive acoustic field are mysterious, and in this section we present only a brief introduction to the problem.

We start at the very beginning. Chemical reactions are processes of association or dissociation of atoms and molecules into new compounds. In a common case, if a reaction goes in one direction, then it may go in the opposite direction too; for example, if molecules A and B associate into molecule C, then molecule C can dissociate into A and B correspondingly.

A reaction of synthesis of two elements into a third one can be written in the form

$$\nu_A A + \nu_B B \rightleftarrows \nu_C C, \tag{6.5.1}$$

meaning that ν_A (stoichiometric coefficient) molecules of A associating with ν_B molecules of B produce ν_C molecules of C, can be represented in a more lapidary form

$$\sum_{i=1}^{3} \nu_i X_i = 0, \tag{6.5.2}$$

where stoichiometric factors are taken with the sign '+' for initial elements and with the sign '−' for the products. Any reaction can be written in the form (6.5.2) by replacing $3 \rightarrow N$, where N is the number of reactive components (both starting elements and the products of a reaction).

From the point of view of thermodynamics, chemical equilibrium conforms to the rule

$$\sum_{i=1}^{N} \nu_i \mu_i = 0, \tag{6.5.3}$$

where μ_i is the chemical potential of components (see Appendix A). For mixture of ideal gases, μ_i can be represented in the form

$$\mu_i = kT \ln p_i + \mu_{0i}(T), \tag{6.5.4}$$

where p_i is the partial pressure of i-th component:

$$p_i = c_i p, \tag{6.5.5}$$

where $c_i = n_i / \sum_{i=1}^{N} n_i$ is the concentration of i-th component, p is the total pressure of the entire gas mixture.

With (6.5.4), we have from (6.5.3):

$$\sum_{i=1}^{N} v_i kT \ln c_i + \sum_{i=1}^{N} v_i kT \ln p + \sum_{i=1}^{N} v_i \mu_{oi}(T) = 0, \tag{6.5.6}$$

and moving the terms with p and T to the right-hand side and elevating the expression,

$$\prod_{i=1}^{N} c_i^{v_i} = \frac{\exp\left(-\sum_{i=1}^{N} v_i \mu_{oi}(T)/kT\right)}{p^{-\sum_{i=1}^{N} v_i}} = K(p,T). \tag{6.5.7}$$

The function $K(p, T)$ is called the equilibrium constant. As we see from (6.5.7), the concentration of a reagent (to be exact, the distribution of the reagent concentrations) depends on pressure and temperature of the system. Above, we saw that cavitation processes are accompanied by significant changes in pressure and, in some cases, in temperature.

That is the first and the simplest reason, why acoustic cavitation may affect the chemical composition of the medium. Note that (6.5.7) is obtained only for perfect gases, but the same qualitative inferences are true for any media, including liquids. As it will be discussed in Chapter 7, the pressure in a liquid increases greatly when a bubble collapses within it; therefore, one should expect the variation in the component composition of the liquid, not only inside the bubble.

6.5.2 Heterogeneous catalysis

Above, we discussed two kinds of chemical reactions: in a bubble and in the bulk of a liquid. However, in (Luche 2002) a third type of chemical reactions is considered: reactions at the interface. In this way, the 'vapor–liquid' boundary is considered as the surface of heterogeneous catalysis.

Catalysis is a common term to depict any way of accelerating a chemical reaction. One of the possible ways is to place a surface in the gas volume; usually, this surface is solid, but in our case the surface represents an interface (see Fig. 6.5.1).

To react, molecules A and B must collide in a gas. This is a comparatively rare case, because one has to deal with 3D chaotic motion of molecules. Another case happens with absorbed molecules A' and B': these molecules are temporally captured by the surface (in our case, by the interface); they cannot penetrate deeply into the condensed medium and diffuse along the 2D surface layer between

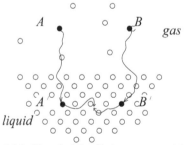

Fig. 6.5.1. The principle of heterogeneous catalysis.

the molecules of a liquid. The probability to meet one another for such 'trapped' molecules is much higher is this case, that is, the rate of the corresponding chemical reaction is increased.

Of course, the scheme works better when a surface contains some 'real' catalysts K—the intermediates in a reaction

$$v_A A + v_B B + K \rightarrow v_C C + K \qquad (6.5.8)$$

which directly accelerate the reaction. But, even in absence of them, the very 'developed' interfacial surface (because of acoustic cavitation, there is a lot of surface inside a liquid) plays a conspicuous role in the chemical processes.

6.5.3 Free radicals

Many reactions occur because new molecules arise under the effect of the acoustic field—free radicals. The simplest example is hydroxyl OH^- in water, appearing after dissociation of a molecule of water

$$H_2O \rightarrow H^+ + OH^- \qquad (6.5.9)$$

Free radicals are very chemically active; thus, they initiate reactions of new types in comparison with the absence of acoustic irradiation.

The formation of free radicals and their importance for chemical reactions are undoubted and described in all the books cited above. However, one problem remains: where do free radicals come from? Why, for example, water dissociates under the acoustic effect?

These and similar questions are unresolved completely up to date. The next two chapters, partially, give some answers but, indeed, the complete physics of a cavitating bubble remains undiscovered.

6.5.4 Plasma

Thus, as we just saw, a cavitating medium contains charged particles; in other words, it contains plasma.

In general, reactions between charged particles are completely different from their analogues among neutral molecules. A good example is the ArH^+ ion: neutral atoms of argon and hydrogen do not react at all, but the interaction between Ar and H^+ may produce a stable ion (Sode et al. 2013, Mitchell et al. 2015, Adboulanziz et al. 2018).

The role of plasma in sonochemical reactions depends on the type of plasma, i.e., on conditions in cavitating bubbles; this is an open question yet, despite the semi-common opinion that a collapsing bubble shrinks the gas so strongly that enormous (up to 10^4 K or even higher) temperatures can be reached and gaseous plasma is naturally produced. Many scientists believe in this mechanism; however, from the theoretical point of view, there may be various mechanisms to torch up plasma in a gaseous bubble in absence of an external electric field, and the possibility of any of them is impugned from some scientists. Here we enter an open area.

The simplest alternative theory of bubble plasma is based on the molecular-level consideration of the cavitation processes. One may divide the cavitation processes in two groups:

(1) the growth of small bubbles which were already in a liquid: small bubbles of air or undissolved gases, etc.; this problem was considered in the main part of this chapter;

(2) the rapture of a liquid caused by negative pressure because of acting of stretching forces on the liquid; this process was discussed in Chapter 3.

In the last case, intuitively, it is difficult to expect that molecules would hold their neutrality under such extremal conditions, especially long, complex molecules like alcohols;even such a simple molecule as H_2O may be torn apart. The most popular analogue of a mechanically induced electrization is the famous experiments of the radiation upon the peeling of a tape. Or one may mention the electric field observed at rock destruction (Li et al. 2015). And so on: sugar splitting, domestic triboelectricity which can be easily observed when you touch a metal door handle and get an electric strike, etc. The first explanation of electrization of a liquid during cavitation was proposed by Jakov Frenkel (Frenkel 1940, 1955).

Thus, if we only assume that after the rapture of a liquid mass different parts of molecules turned up on different sides of the forming cavity, then all other steps to the plasma emerging in a bubble are almost obvious, at first glance, because: (*a*) a bubble born this way is not empty—it would be filled with a vapor evaporated from the wall, (*b*) the charged walls of a cavity produce an electric field, (*c*) this electric field produces plasma from the vapor because of electrical discharges. However, all three steps are not as clear as they may look. Because of their importance, we critically discuss all items amply.

As for point *a*, the liquid must evaporate into a cavern. The main factor that determines the evaporation process is that there must be molecules (or atoms) at the liquid surface that may overcome the bounding energy from other molecules. There are problems with the last point in some liquids, for instance, in glycerol which evaporates very slowly. Thus, the bubble may be almost empty.

One may think that item *b* is undisputable: surely, if walls contain charged particles, then these particles produce an electric field. But in a common case, different sides of a cavity would contain equal amounts of charged particles of both signs; thereby, the total charge of the wall is close to zero. One should expect only local electric field in places where the uniformly charged molecules are concentrated. The analogy with triboelectricity is not so direct because the contact of heterogeneous substances leads to definite signs of charges appearing on each substance. During the rapture of a liquid, it is difficult to expect such certainty for identical substances on both sides of the gap.

The last item *c* implies special conditions too. Briefly, an electrical discharge is the process of electron breeding in an electric field: an electron gains high kinetic energy from the electric field, then strikes an atom, and one more electron appears because of ionization. Then two electrons gain high kinetic energy in the electric field, strike two atoms, etc. This pretty scheme demands one important condition: the spatial scale of the medium must significantly exceed the mean free path *l* of an electron in this medium, which is questionable for small bubbles: the mean free path of an electron in a comparatively weak electric field *E* is (Raizer 2001)

$$lp \sim 3 \cdot 10^{-2} \text{ cm·torr at } E/p \sim 4 \div 50 \text{ V/(cm·torr)} \tag{6.5.10}$$

For atmospheric pressure $p = 760$ torr, we get $l \sim 4 \cdot 10^{-5}$ cm; this a short scale for the bubble of ~ 10 μm but at lower pressures or in smaller bubbles, electrons may fly through the gas without any collisions.

Of course, the objections of the electrization theory may be protested themselves. For instance, Frenkel defended the point *b* with fluctuations: because of random variations, the charges would not be distributed on opposite walls 50 on 50.

A more detailed analysis of hypothesis of electrization of a liquid during cavitation will be discussed in a special place—in Chapter 8. Here we may only answer an evident question: do experiments give certain answers about the plasma properties? The short answer is 'no'; but the explanation demands at least a brief introduction to plasma diagnostics, which will be done in Chapter 8.

Anyway, the rapture of a liquid may give interesting effects to a cavitation problem. If the charged particles appear during cavitation, they may play a significant role in bubble dynamics, which must be taken into account in the future.

6.5.5 *Sonochemical efficiency*

Everywhere in this chapter we had to deal with a single bubble. Also, in this section, we discussed chemical reactions near a single bubble. But, of course, the whole medium takes part in the process: the whole bulk gains and dissipates energy from the external acoustic source. This dissipated energy can be estimated with sonochemical reactions (Koda et al. 2003) by determining the concentration of the hydroxyl radical. To measure it, various reactions with OH can be used.

For instance, one may use the reaction

$$Fe^{2+} + OH \rightarrow Fe^{3+} + OH^- \tag{6.5.11}$$

which undergoes in the so-called Flicke solution, which consists of $Fe(NH_4)_2 (SO_4)_2 6H_2O$ (1 mmol/dm³), H_2SO_4 (0.4 mmol/dm³) and NaCl. The F^{3+} output can be measured by a spectroscope—the line on 304 nm. Then, calibrating the setup—collating the dissipated energy measured directly with a calorimeter and the amount of F^{3+} ions—we obtain the instrument which allows to determine the energy dissipated in the vessel by measuring the number of Fe^{3+} ions.

In practice, this method is quite precise to measure the energy dissipated in a vessel. The 'mechanical' method can be poorly suitable because this energy depends on the shape of the vessel, due to the inhomogeneity of the acoustic field.

Conclusion

Acoustic cavitation is a very simple way to obtain and explore bubbles without heating the liquid. This can be achieved, generally, in two ways: we may get a single bubble in the center of a vessel or produce a whole cavitation cloud under an ultrasonic waveguide.

We may highlight the following details of the process:

- ✓ in an external ultrasonic field of frequency v, a bubble oscillates at the basic frequency v—perhaps this is an obvious result; oscillations at a higher frequency cannot be resolved with sufficient quality to observe additional details;
- ✓ during oscillations, bubbles lose the sphericity of their shape, generally; sometimes, their shape may be close to a sphere, but, strictly, at the expansion phase a bubble does not have a spherical form, even if its radius is ~ 0.5 mm;
- ✓ for larger bubbles, of radius ~ 1 mm, regular oscillations of their shape can be observed directly; such pulsations can be easily described with periodic functions on the surface;
- ✓ during the collapse phase, the shape of a bubble is almost spherical;
- ✓ the likely mechanism of shape oscillations is the Rayleigh–Taylor instability;
- ✓ the deviation from sphericity may be so strong that it can lead to bubble division: a small part of a bubble splits away from the main mass of a bubble;
- ✓ however, with time, very quickly, that chipped part returns back to the parental bubble;
- ✓ the attraction of bubbles can also be seen during multi-bubble cavitation; this is the so-called Bjerknes effect—bubbles attract to each other because they interact with the pressure field created by a neighboring bubble; the complete explanation will be given in the next chapter;
- ✓ during multibubble cavitation, complicated bubble structures can be formed near an ultrasonic waveguide.

In the next chapter, we consider the theoretical description of a cavitating bubble 'purified' in such way.

References

Abdoulanziz, A., F. Colboc, D. A. Little, Y. Moulane, J. Zs. Mezei, E. Roueff, J. Tennyson, I. F. Schneider and V. Laporta. 2018. Theoretical study of ArH+ dissociative recombination and electron-impact vibrational excitation. *Monthly Notices of the Royal Astronomical Society*, 479: 2415–20. https://doi.org/10.1093/mnras/sty1549.

Aoki, R., N. Thanh-Vinh, K. Noda, T. Takahata, K. Matsumoto and I. Shimoyama. 2015. Sound focusing in liquid using a varifocal acoustic mirror. 2015. *28th IEEE International Conference on Micro Electro Mechanical Systems (MEMS)*, 925–7. https://doi.org/10.1109/MEMSYS.2015.7051111.

Ashokkumar, M., J. Lee, S. Kentish and F. Grieser. 2007. Bubbles in an acoustic field: An overview. *Ultrasonics Sonochemistry*, 14: 470–5. http://doi.org/10.1016/j.ultsonch.2006.09.016.

Biryukov, D. A., D. N. Gerasimov and E. I. Yurin. 2020. The cavitation process in proximity to an ultrasonic waveguide. *Journal of Physics: Conference Series*, 1683: 022016-1-5. https://doi.org/10.1088/1742-6596/1683/2/022016.

Buckingham, M., B. Berknout and S. Glegg. 1992. Imaging the ocean with ambient noise. *Nature*, 356: 327–9. https://doi.org/10.1038/356327a0.

Crum, L. A. 1975. Bjerknes forces on bubbles in a stationary sound field. *The Journal of the Acoustical Society of America*, 57: 1363–70. https://doi.org/10.1121/1.380614.

Doinikov, A. A. and S. T. Zavtrak. 1995. On the mutual interaction of two gas bubbles in a sound field. *Physics of Fluids*, 7: 1923–30. https://doi.org/10.1063/1.868506.

Frenkel, J. 1940. Electrical phenomena associated with cavitation caused by ultrasonic vibrations in a liquid. *Zhurnal Physicheskoi Khimii*, 14:305 (in Russian).

Frenkel, J. 1955. *Kinetic theory of liquids*. Dover Publications.

Gaitan, D. F., L. A. Crum, C. C. Church, and R. A. Roy. 1992. Sonoluminescence and bubble dynamics for a single, stable, cavitation bubble. *The Journal of the Acoustical Society of America*, 91: 3166–3183. https://doi.org/10.1121/1.402855.

Håkansson, A., Cervera, F. and J. Sánchez-Dehesa. 2005. Sound focusing by flat acoustic lenses without negative refraction. *Applied Physics Letters*, 86: 054102-1–3. https://doi.org/10.1063/1.1852719.

Isakovich, M. A. 1973. *General Acoustics* (Obshaya Acustica). Nauka:Moskow (in Russian).

Ishimaru, L. and R. Hyoudou. 2002. Sound focusing technology using parametric effect with beat signal. *Proceedings. 11th IEEE International Workshop on Robot and Human Interactive Communication*, 277–81. https://doi.org/10.1109/ROMAN.2002.1045635.

Karpuk, M. M., D. A. Kostyuk, Y. A. Kuzavko and V. G. Shavrov. 2004. Ultrasonic piezoceramic transducers with a magnetoacoustic layer. *Technical Physics Letters*, 30: 1005–8. https://doi.org/10.1134/1.1846841.

Kentish, S. and M. Ashokkumar. 2011. The physical and chemical effects of ultrasound. *In*: Feng, H., G. Barbosa-Canovas and J. Weiss (eds.). *Ultrasound Technologies for Food and Bioprocessing. Food Engineering Series*. Springer: New York. https://doi.org/10.1007/978-1-4419-7472-3_1.

Kikuchi, Y. 1969. *Ultrasonic transducers*. Corona Publishing Company:Tokyo.

Koda, S., T. Kimura, T. Kondo and H. Mitome. 2003. A standard method to calibrate sonochemical efficiency of an individual reaction system. *Ultrasonics Sonochemistry*, 10: 149–56. https://doi.org/10.1016/s1350-4177(03)00084-1.

Leighton, T. G., Walton, A. J. and M. J. W. Pickworth. 1990. Primary Bjerknes forces. *European Journal of Physics*, 11: 47–50. https://doi.org/10.1088/0143-0807/11/1/009.

Li, Z., Wang, E. and M. He. 2015. Laboratory studies of electric current generated during fracture of coal and rock in rock burst coal mine. *Journal of mining*, 2015. https://doi.org/10.1155/2015/235636.

Luche, J.-L. 1998. *Synthetic Organic Sonochemistry*. Springer US. https://doi.org/10.1007/978-1-4899-1910-6.

Mason, T. J. and J. P. Lorimer. 2002. *Applied Sonochemistry: Uses of Power Ultrasound in Chemistry and Processing*. Wiley-VCH.

Mason, W. P. (ed.). 1964. *Physical Acoustics: Principle and Methods*. Academic Press.

Mayer, V. V. and E. I. Varaksina. 2007. Elastic waves physics in scholar research (Fizika uprugih voln v uchebnih issledovaniyah). Fizmatlit: Moscow (in Russian).

Maznev, A. A., G. Gu, S. Y. Sun, J. Xu, Y. Shen, N. Fang and S. Y. Zhang. 2015. Extraordinary focusing of sound above a soda can array without time reversal. *New Journal of Physics*, 17: 042001. https://doi.org/10.1088/1367-2630/17/4/042001.

Mettin, R., I. Akhatov, U. Parlitz, C.-D. Ohl and W. Lauterborn. 1997. Bjerknes forces between small cavitation bubbles in a strong acoustic field. *Physical Review E*, 56: 2924–31.https://doi.org/10.1103/PhysRevE.56.2924.

Mitchell, J. B. A., O. Novotny, J. L. LeGarrec, A. Florescu-Mitchell, C. Rebrion-Rowe, A. V. Stolyarov, M. S. Child, A. Svendsen, M. A. El Ghazaly and L. H. Andersen. 2005. Dissociative recombination of rare gas hydride ions: II. ArH+. *Journal of Physics B: Atomic, Molecular and Optical Physics*, 38: L175.https://doi.org/10.1088/0953-4075/38/10/L07.

Moussatov, A., Granger, Ch. and B. Dubus. 2003. Cone-like bubble formation in ultrasonic cavitation field. *Ultrasonics Sonochemistry*, 10: 191–5. https://doi.org/10.1016/s1350-4177(02)00152-9.

Oguz, H. N. and A. Prosperetti. 1990. A generalization of the impulse and virial theorems with an application to bubble oscillations. *Journal of Fluid Mechanics*, 218: 143–62. https://doi.org/10.1017/S0022112090000957.

Ohl, C.-D., A. Philipp and W. Lauterborn. 1995. Cavitation bubble collapse studied at 20 million frames per second. *Annalen Der Physik*, 507: 26–34.https://doi.org/10.1002/andp.19955070104.

Ohl, C.-D., T. Kurz, R. Geisler, O. Lindau and W. Lauterborn. 1999. Bubble dynamics, shock waves and sonoluminescence. *Royal Society*, 357: 269–94. https://doi.org/10.1098/rsta.1999.0327.

Pelekasis, N., A. Gaki, A. Doinikov and J. A. Tsamopoulos. 2004. Secondary Bjerknes forces between two bubbles and the phenomenon of acoustic streamers. *Journal of Fluid Mechanics*, 500: 313–47. https://doi.org/10.1017/s0022112003007365.

Phillips, S., A. Agarwal and P. Jordan. 2018. The sound produced by a dripping tap is driven by resonant oscillations of an entrapped air bubble. *Scientific Reports*, 8: 9515. https://doi.org/10.1038/s41598-018-27913-0.

Prokhorov, V. E. and Y. D. Chashechkin. 2011. Sound generation as a drop falls on a water surface. *Acoustical Physics*, 57: 807–18. https://doi.org/10.1134/S1063771011050137.

Raizer, Yu. P. 2001. *Gas discharge physics*. Springer-Verlag: Berlin.

Russell, D. A. 2010. Basketballs as spherical acoustic cavities. *American Journal of Physics*, 78: 549–54. https://doi.org/10.1119/1.3290176.

Skokov, V. N., A. V. Reshetnikov, A. V. Vinogradov and V. P. Koverda. 2007. Fluctuation dynamics and 1/f spectra characterizing the acoustic cavitation of liquids. *Acoustical Physics*, 53: 136–40. https://doi.org/10.1134/S1063771007020042.

Sode, M., T. Schwarz-Selinger and W. Jacob. 2013. Ion chemistry in H2-Ar low temperature plasmas. *Journal of Applied Physics*, 114: 063302. https://doi.org/10.1063/1.4817526.

Spadoni, A. and Ch. Daraio. 2010. Generation and control of sound bullets with a nonlinear acoustic lens. *Proceedings of the National Academy of Sciences*, 107: 7230–4. https://doi.org/10.1073/pnas.1001514107.

Wang, Ch., F. Jiang, S. Shao, T. Yu and C. Guo. 2020. Acoustic Properties of 316L stainless steel hollow sphere composites fabricated by pressure casting. *Metals*, 10: 1047. https://doi.org/10.3390/met10081047.

Yon, S., M. Tanter and M. Fink. 2003. Sound focusing in rooms: The time-reversal approach. *The Journal of the Acoustical Society of America*, 113: 1533–43. https://doi.org/10.1121/1.1543587.

Zhang, Y., Y. Zhang and Sh. Li. 2016. The secondary Bjerknes force between two gas bubbles under dual-frequency acoustic excitation. *Ultrasonic Sonochemistry*, 29: 129–45. http://dx.doi.org/10.1016/j.ultsonch.2015.08.022.

CHAPTER 7

Dynamics of a Cavitating Bubble

7.1 The common problem of dynamics of a cavern in a liquid

7.1.1 Dynamics of a void in a liquid

In the previous chapter, we described the experiments of acoustic cavitation which demonstrate many interesting features. It is a natural interest to affirm these results with a more or less strict theory, and, of course, many attempts were made to date. In the present chapter, we will follow this approach too, but first we have to discuss some basic physical principles that lie beneath the macroscopic hydrodynamical description as a whole and the special approximations that are used in this certain problem.

Our aim is to describe the evolution of a cavitating bubble under the external ultrasonic pressure and, in more general terms, to describe the processes inside a vapor bubble occurring at non-equilibrium initial conditions. For example, we want to understand the evolution of the cavitation cavern formed when the external pressure drops momentarily. The common difficulties for such a problem are small spatial scales (a cavitating bubble may have a size of micrometers or even less) and strong non-equilibrium conditions caused by the short temporal scale of the process (the collapsing time may be of microseconds or even less). These, slightly extreme, conditions complicate the mathematical formulation of the physical problem.

While we describe a bubble in a liquid, we want to use the hydrodynamical approach for both phases. In the most general case, we need to consider the dynamics of two phases: liquid and gas, and to express the connection between them through the condition on the interfacial surface. It may seem that we can solve this problem directly, without any approach, using the standard computational codes, i.e., with CFD modeling. However, this is not the case because of two main reasons at least:

1) We have to know the relation between the parameters of the liquid phase and of the vapor one at the interface. In this part, we have to be sure that we take into account all the main processes on the surface (such as evaporation, condensation, deformation of the surface shape, etc.). This is a relatively complicated matter; generally, the formulation of the boundary conditions leads us to an uncharted territory. Therefore, CFD meets an unresolved problem: even if one thinks that hydrodynamic equations are the absolute law, we cannot connect two equations—for the liquid and for the vapor—between them.

2) Hydrodynamic equations are not some God-given laws in the ultimate form. They have many restrictions: as concerning the correlation for the tensor of viscous stress, as for the limiting of the continuous-media description. For many problems concerning the dynamics of a cavitating bubble, we go far away from the hydrodynamic description because the spatial scales are too small. For many problems of the cavitating bubble, the kinetic description looks preferable. Let us consider the last statement closely.

Hydrodynamics, that is, the mathematical description based on Navier–Stokes equations, is correct only for continuous media, in cases when the free path length l is much smaller than the spatial scale L of the system:

$$l/L \ll 1. \tag{7.1.1}$$

The mean free path can be estimated by an order as

$$l \sim \frac{10^{-2}}{p \, [Pa]}, [m] \tag{7.1.2}$$

Thus, the scale of a gaseous cavern, for instance the radius of the bubble, must be not too small. For atmospheric pressures $p \sim 10^5$ Pa, we have $l \sim 10^{-7}$ m; thus, the diameter of the bubble must be at least ~ 1 μm if we want to analyze its dynamics with hydrodynamic equations. However, a bubble may be much smaller than this value. Thereby, if we want to describe the vapor dynamics, we may need the kinetic approach (with the Boltzmann equation or its analogues). This is not the last issue.

The short spatial scale means that the energy transfer between molecules is insufficiently effective, i.e., the equilibrium distribution functions on molecules' velocities (in Maxwellian form) is not established. This situation concerns not only the dynamics of the vapor phase, but also the thermodynamic description of the vapor.

Then, to calculate properties of the vapor in a bubble, or even simply to describe the bubble dynamics, we have to know the correlation between the parameters of the vapor phase at least, i.e., we need the dependence $p = f(\rho, T)$—the equation of state. In the simplest case, when we want to use the Clapeyron equation, we meet two main difficulties: (*a*) we must be sure that the equation of state even exists and (*b*) have sufficient fundamentals to use the perfect gas equation for it.

The reasoning (*a*) may look strange. How the equation of the state cannot exist? In a common case, especially in strong non-equilibrium conditions, we have to use the distribution function for gas molecules $f(\vec{x}, \vec{v}, t)$. With this function, one may construct thermodynamic functions—gas density n, pressure p, temperature T of a kind

$$n = \int f d\vec{v}, p_{x,y,z} = 2\int m v_{x,y,z}^2 f d\vec{v} d\vec{x}, \frac{3kT}{2} = \int \frac{m v^2}{2} f d\vec{v} d\vec{x} \tag{7.1.3}$$

For an equilibrium case, the Maxwellian $f(\vec{x}, \vec{v}, t) \sim \exp\left(-\frac{m v^2}{2kT}\right)$ with correlations (7.1.3) provides the Clapeyron equation $p = nkT$. However, in a non-equilibrium system, the distribution function may depend not on the total velocity $v^2 = v_x^2 + v_y^2 + v_z^2$, but on the separate components of velocity: $f(\vec{x}, v_x, v_y, v_z, t)$; thus, $p_x \neq p_y \neq p_z$ – the simplest example of such a case is a gas near the evaporation surface. Moreover, in this case we have no universal scale T for the kinetic energy; the parameter T defined in accordance with (7.1.1) does not represent a unique measure for the distribution functions in such a media. Sometimes, instead of T parameters T_x, T_y, T_z are used (each of them describes the distribution function on the corresponding velocity); besides, the fact that temperature is a scalar value and, therefore, $\vec{T}$ is some kind of nonsense. Therefore, the construction $p = f(n, T)$ does not exist in such a system—Q.E.D.

Now, let us assume that thermal equilibrium is established. This fact does not mean that the perfect gas equation suits the given problem. Usually, for a low or a moderate density ρ (like at the atmospheric pressure and at room temperature), we may use the Clapeyron equation; this is an obvious approach that found applications in almost all the works about the cavitation void. But a perfect gas is only a limiting case: in general, the state equation can be represented as the expansion of the compressibility factor

$$z = \frac{p}{\rho \tilde{R} T} = 1 + \sum_{j=1}^{\infty} B_j(T) \rho^j. \tag{7.1.4}$$

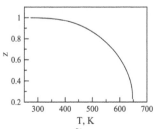

Fig. 7.1.1. Dependence of the compressibility factor $z = p/\rho\tilde{R}T$ on temperature for saturated water vapor: thermodynamic parameters correspond to saturation line. The gas may be considered perfect (i.e., satisfy the Clapeyron equation) for $z = 1$. As one can see, this approach is correct only for temperatures up to $\sim 400-450$ K. At higher temperatures, the saturated vapor does not obey the Clapeyron equation—more complicated equations of state are required. Data taken from the NIST database.

Equation (7.1.4) predicts the Clapeyron law, i.e., unity on its right-hand side, only for $\rho \to 0$. Figure 7.1.1 illustrates (7.1.4) for saturated water vapor.

Summing up the reasons listed above: we have no strict mathematical apparatus to describe the evolution of a cavitation cavern. Complicated calculations based on the full hydrodynamic description may be easily crashed because of the (forgotten) physical restrictions of the initial mathematical model; for example, if they use the Clapeyron equation as the equation of state. Then, we have to be humble and use a simplified mathematical description of the problem. In this, simplified, way the description of the void is reduced to the investigation of a spherical bubble that obeys the so-called Rayleigh equation (considered below). If so, we want to examine the limitations of the spherical approximation of the bubble form. Actually, the mathematical model of a cavitating cavern is a very complicated problem.

Last but not least: the description of bubble dynamics may demand some unexpected features such as thermal or non-thermal radiation, triboelectricity, electrical discharges, etc. Such exotic matters will be discussed in the next two chapters.

7.1.2 Hydrodynamic description

Let us briefly analyze the full hydrodynamical implementation. For the liquid phase, the Navier–Stokes must be considered. For an incompressible fluid, in the continuous equation

$$\frac{\partial \rho}{\partial t} + \frac{\partial \rho u_k}{\partial x_k} = 0 \tag{7.1.5}$$

one may put $\rho = const$ and, therefore, get instead of (7.1.5) the equation

$$\frac{\partial u_k}{\partial x_k} = 0; \tag{7.1.6}$$

So, the Navier-Stokes equation is

$$\frac{\partial u_i}{\partial t} + u_j\frac{\partial u_i}{\partial x_j} = -\frac{1}{\rho}\frac{\partial p}{\partial x_i} + \nu\frac{\partial^2 u_i}{\partial x_k \partial x_k} + g_i, \tag{7.1.7}$$

or, a vector form

$$\frac{\partial \vec{u}}{\partial t} + \vec{u}\nabla\vec{u} = -\frac{1}{\rho}\nabla p + \nu\Delta\vec{u} + \vec{g}, \tag{7.1.8}$$

where all the notions are standard (see Appendix B), $\nu = \mu/\rho$; all the properties are supposed to be constant. In a usual case, the gravity term is negligible.

The main proviso for the incompressibility approximation is the sufficiently small Mach number

$$M = \frac{u}{c_s} < 0.3. \tag{7.1.9}$$

With (7.1.9), the change in density $\delta\rho$ at the adiabatic deceleration would be less than 5%,[1] which is within the experimental error margin. At first glance, for liquids where $c_s \sim 10^3$ m/s it is hard to observe velocities of a liquid contradicting with (7.1.9). However, condition (7.1.9) is not a common case, and for the extreme but realistic conditions this inequality may fail even for such a simple system as a droplet facing a solid surface.

A similar description can be formulated for the vapor phase; however, in the gas phase we have more reasons to consider the compressibility of a medium. In some cases, the speed of the bubble collapse is comparable with the speed of sound c_s or even higher; therefore, the inequality (7.1.9) is incorrect and we have to consider a vapor as a compressible fluid.

For the vapor phase, the description via the incompressible approach has additional restrictions, because density may vary with temperature: for a vapor, the adiabatic (or isobaric) coefficient $\rho{-}1(\partial\rho/\partial T)_{s,p}$ is much higher than for a liquid. It is useful to remember this fact: sometimes, a vapor is considered as an incompressible liquid even for heat transfer problems.

Compressible fluids (and gases) demand a much more complicated description. Additional terms arise in the Navier-Stokes equation: the viscous stress has a form (Landau and Lifshitz 1987)

$$\tau_{ij} = \mu\left(\frac{\partial u_i}{\partial x_j} + \frac{\partial u_j}{\partial x_i} - \frac{2}{3}\delta_{ij}\frac{\partial u_k}{\partial x_k}\right) + \zeta\delta_{ij}\frac{\partial u_k}{\partial x_k}, \tag{7.1.10}$$

where δ_{ij} is the Kronecker symbol: $\delta_{ij} = 1$ only if $i = j$. For an incompressible fluid due to (7.1.6) the correlation (7.1.10) leads to (7.1.7) with the term $\partial\tau_{ij}/\partial x_j$. But for the case $\partial u_k/\partial x_k \neq 0$, we face not only the complication of equation caused by increasing of the number of terms: in this case, we have to determine the second viscosity ζ, the term that describes the connection between the viscous stress and the volume change. There are different ideas about ζ; for instance, $\zeta = 0$, $\zeta = 2\mu/3$, etc., see reviews (Graves and Argrow 1999, Sharma and Kumar 2019). In this book that is not devoted to theoretical fundamentals of hydrodynamics, we cannot discuss this specific topic. We may only conclude that the mathematical description of a compressible liquid is a complicated matter.

We do not discuss here the problems concerning the turbulence of a flow; this is the common hard place of hydrodynamics.

Then, the cavitation process can be accompanied with heating and energy transfer; thus, a corresponding equation is required. In the simplest form, this equation can be formulated as

$$\rho c_p\left(\frac{\partial T}{\partial t} + u_k\frac{\partial T}{\partial x_k}\right) = \lambda\frac{\partial^2 T}{\partial x_k \partial x_k}, \tag{7.1.11}$$

where we neglect the heat source.

Let us calculate the variables in equations. In each phase, both in a liquid and in a vapor, we need to know six quantities: three components of velocity, pressure, density and temperature. The last three terms can be connected by the equation of state: $p = f(\rho,T)$; taking into account this fact, we see that there are six equations in this problem, as it must be: three Navier-Stokes equations (for each of three projections), the continuity equation, the heat transfer equation and the equation of state. It must be this way because above we presented a more or less general description of the hydrodynamical problem; there is no problem from this side.

---

[1] This follows from the Bernoulli equation for the adiabatic drag: $u^2/2 = \Delta p/\rho$ with $\Delta p = c_s^2\Delta\rho$; thus, we see that $\delta\rho = \Delta\rho/\rho = u^2/2c_s^2$.

Consequently, we have the complete mathematical problem: hydrodynamical equations can be written for each phase—both for the vapor and the liquid; then, we need the boundary conditions for these equations. The external boundary conditions are a usual point, the conditions at the vapor–liquid interface are much more interesting.

7.1.3 Boundary conditions

Thus, we have to connect both phases with additional relations at the interface. In a common case, there may be four types of boundary conditions:

for velocities,

for pressure,

for mass flux,

for energy flux.

These conditions will be considered in detail in Section 7.3. Here we mention the basic principles.

A flux of any quantity (mass, momentum, energy) may have a singularity at the surface due to the source of the corresponding quantity on this surface. We denote the difference in fluxes as $[J] = J_v - J_l$ (the difference between the flux in a liquid and in a vapor), and the production of a quantity per unit of area as Q; the dimension of the flux is $\dfrac{[\text{quantity}]}{\text{m}^2\text{s}}$, under the quantity there may be kg, W, etc. In this notation,

$$[J] = \dot{Q}. \qquad (7.1.12)$$

The most important jump on the surface occurs with the mass and the energy fluxes due to evaporation or condensation. Suppose that due to evaporation, the mass $\dot{m}_s$ of vapor leaves the area unit per unit of time. If so, then for the mass flux

$$[J_m] = \dot{m}_s, \qquad (7.1.13)$$

while for the heat flux

$$[q] = \dot{m}_s h_{\text{lg}}, \qquad (7.1.14)$$

where h_{lg} is the enthalpy of the 'liquid–vapor' phase transition.

For equilibrium, i.e., for Maxwellian distribution functions, both for molecules from the evaporated surface and for molecules in the bulk of the gas, the mass flux can be expressed as

$$\dot{m}_s = \beta \underbrace{\frac{p_g}{\sqrt{2\pi m k T_g}}}_{condensation} - \alpha \underbrace{\frac{p_{sat}}{\sqrt{2\pi m k T_l}}}_{evaporation}. \qquad (7.1.15)$$

Formula (7.1.15), called the Hertz-Knudsen equation, means that the fluxes on the surface correspond to Maxwellian distribution functions with appropriate coefficients: β for condensation (the fraction of particles arrived from the gas and attached to the surface) and α for evaporation (the meaning is more complicated, indeed (Gerasimov and Yurin 2015)). Also, the flux of the evaporated molecules corresponds to the equilibrium parameters defined for the temperature of the surface, i.e., p_{sat} is the saturation pressure of a vapor at temperature T_l.

At first glance, equations (7.1.13–15) provide the full description. However, the mass flux $\dot{m}_s$— the difference between the fluxes of evaporation and condensation—can be difficult to determine. In a common case, the conditions at the evaporated surface are strongly non-equilibrium (Gerasimov and Yurin 2018); thus, the simple equation (7.1.15) is far from the final answer. In a phenomenological consideration, this fact leads to dependence of coefficients α and β on the parameters p_g, T_g and T_l.

In addition, we may formulate the expression for the pressure difference in contacting phases. The jump in pressure at the interface is determined by the Laplace term at least:

$$p_\upsilon - \tau_{rr\upsilon} - p_l + \tau_{rrl} = \frac{2\sigma}{R}. \tag{7.1.16}$$

Here $\tau_{rr\upsilon}$ and τ_{rrl} are the radial components of the viscous stresstensor (7.1.10) for the gaseous phase and for the liquid one correspondingly. Usually, the gas is assumed to be motionless; thereby, $\tau_{rr\upsilon} = 0$. Condition (7.1.16) will be applied for the problem in the next section.

7.1.4 Numerical simulation of a bubble in a liquid

There may be two 'direct' ways to apply hydrodynamical equations to the evolution of a bubble in a liquid:

(A) to solve Navier-Stokes equations directly for both phases with proper boundary conditions at the liquid–vapor interface, which must be found itself;

(B) to consider a 'mixture' of vapor and liquid: at a given point (i.e., in the given elementary volume), we have the fraction κ of a liquid and the fraction $(1 - \kappa)$ of a vapor. The volume fraction is determined by the equation

$$\frac{\partial \kappa}{\partial t} + \frac{\partial (\kappa u_i)}{\partial x_i} = \frac{\dot{m}}{\rho_l}, \tag{7.1.17}$$

where the mass flux can be found as the net $\dot{m} = \dot{m}_c - \dot{m}_e$ of the evaporation $\dot{m}_e$ and condensation $\dot{m}_c$ rates; for instance, in (Ghahramani 2019) the following equations were used:

$$\dot{m}_c = C_c \kappa (1 - \kappa) \frac{3\rho_l \rho_g}{\rho_m R_B} \sqrt{\frac{2}{3\rho_l |p - p_t|}} \max (p - p_t, 0), \tag{7.1.18}$$

$$\dot{m}_e = C_e \kappa (1 + \kappa' - \kappa) \frac{3\rho_l \rho_g}{\rho_m R_B} \sqrt{\frac{2}{3\rho_l |p - p_t|}} \min (p - p_t, 0) \tag{7.1.19}$$

with parameters

$$\kappa' = \frac{V_{nuc}}{1 + V_{nuc}}, \quad V_{nuc} = \frac{\pi n_0 d_{nuc}^3}{6}, \tag{7.1.20}$$

$$R_B = \sqrt[3]{\frac{3}{4\pi n_0} \frac{1 + \kappa' - \kappa}{\kappa}}, \tag{7.1.21}$$

where n_0 is the number of nuclei per a unit of volume and d_{nuc} is the nucleation site diameter. Therefore, κ' has the meaning of a volume fraction of nuclei and R_B is the generic radius.

With κ, any parameter of a vapor–liquid mixture is supposed to be an additive quantity: for example, density (which is evident)

$$\rho_m = \kappa \rho_l + (1 - \kappa) \rho_g; \tag{7.1.22}$$

viscosity (which is a more questionable relation)

$$\mu_m = \kappa \mu_l + (1 - \kappa) \mu_g. \tag{7.1.23}$$

Moreover, even if both the vapor phase and the liquid one are incompressible separately, that is, for each phase one would have the equation (7.1.6), for a mixture we would have the equation

$$\frac{\partial u_k}{\partial x_k} = \left(\frac{1}{\rho_l} - \frac{1}{\rho_g} \right) \dot{m}. \tag{7.1.24}$$

The relation (7.1.24) means that due to the change in the volume fraction of a liquid (i.e., for the non-zero mass flux $\dot{m}$), the density of the mixture varies.

Then, one should solve the Navier–Stokes equations for the vapor–liquid mixture with its 'effective' properties. This way B (named also VOF—Volume Of Fluid method) was used, among other works, in (Ghahramani et al. 2019, Lechner et al. 2017, Minsier 2010, Sarkar 2019, Zahedi et al. 2014). This method turned out to be very productive, especially for such complicated problems as the bubble collapse near a wall: such calculations demonstrate the formation of microjets inside a bubble, as it was discussed in previous chapters.

Another method that can be used for the numerical simulation of a bubble in a liquid is the MDS—molecular dynamics simulation, where an ensemble of separate particles (atoms or molecules) is considered. This is a more or less direct way to analyze the problem, despite the difficulties concerning the modeling of molecules. However, to describe a bubble surrounded by a liquid, one needs a lot of particles: at least $\sim 10^{5-6}$; this circumstance demands great computing resources. It is possible that the calculation problems are responsible for the lack of MDS results for bubble dynamics, while the benefits of MDS are clear: this method allows the investigation without restrictions of the continuous medium approach (see the beginning of this chapter) that can be essential for extreme modes of the bubble collapse. In future, we hope that the number of these works will increase; to date, the number of corresponding works is small: (Matsumoto et al. 2000, Schanz et al. 2012).

However, besides the direct solution of the Navier–Stokes equation, there also exists another way to consider the dynamics of a bubble in a liquid. This way leads not only to tremendous computer calculations, but to clear analytical results too.

7.2 The Rayleigh equation

7.2.1 Derivation from a simple consideration

The Rayleigh equation represents the dependence of the vapor bubble radius on time. This is the simplest model of a cavitation cavern, which nevertheless provides good qualitative and even quantitative results.

The first method of derivation of the Rayleigh equation follows from the simple physical assumption: a growing spherical bubble makes work on the surrounding liquid mass, pushing it away from the bubble; of course, these reasons suit the reverse situation—for a collapsing bubble (with opposite signs of terms). Thus, in this way, we neglect any thermal effects, considering only the mechanical aspects.

Let us put the pressure in a liquid at the bubble wall as p_b, the pressure in a liquid at infinity is p_∞. In the simplest case, which is normally used, one may suppose that the pressure in a liquid in vicinity of the bubble wall is lower than the pressure inside a bubble p_{in} on $2\sigma/R$, i.e.,

$$p_b = p_{in} - \frac{2\sigma}{R}. \tag{7.2.1}$$

This is a stationary correlation; the full form will be discussed further: here we use, as it was titled, only primitive ideas.

The substance inside the bubble consists of, in a general case, the vapor phase of the surrounding liquid and external gases (air, for the simplest example). The difference between these substances is important: the amount of external gases is presumably a constant (in some assumptions, indeed, this a questionable matter), while the mass of the vapor phase varies due to evaporation/condensation processes. Thus, in a common case, the pressure inside the bubble is

$$p_{in} = p_v + p_g. \tag{7.2.2}$$

The terms in (7.2.2) will be discussed in the next section. In the considered framework, we would not flesh out the exact form of pressure in the liquid at infinity p_∞, we provide the corresponding correlation in the next chapter. Here it's enough for us to take into account the pressure difference

$$\Delta p = p_b - p_\infty = p_{in} - p_\infty - \frac{2\sigma}{R}. \tag{7.2.3}$$

If $\Delta p > 0$, one should expect that the bubble tends to grow, otherwise the bubble tends to collapse.

The work of the growing bubble $\Delta p dV$ spends on the kinetic energy dE_{kin} of the surrounding liquid: the walls of the expanding bubble push the liquid mass; it is logical to assume that the flow would hold a spherical symmetry. The distribution of velocity in the liquid can be found from the continuous equation, where in spherical coordinates only the radial component of velocity exists:

$$\frac{1}{r^2}\frac{\partial}{\partial r}\left(r^2 u_r\right) = 0. \tag{7.2.4}$$

Integrating and taking into account that on the bubble wall—at $r = R$—the velocity $u_r = \dot{R}$, we have

$$u_r = \frac{R^2 \dot{R}}{r^2}. \tag{7.2.5}$$

Here the dot above the variable denotes the time derivative.
Thus, the kinetic energy of the whole mass of liquid is

$$E_{kin} = \int_R^\infty \rho \frac{u_r^2}{2} dV = \int_R^\infty \rho \frac{R^4 \dot{R}^2}{2r^4} 4\pi r^2 dr = 2\pi\rho R^3 \dot{R}^2. \tag{7.2.6}$$

Following the line mentioned above, we may conclude that

$$\Delta p \frac{dV}{dt} = \frac{dE_{kin}}{dt}; \tag{7.2.7}$$

or consider a stationary deviation of parameters, i.e., without differentials dt in (7.2.7). Because we consider a spherical bubble, $V = \frac{4}{3}\pi R^3$, and we obtain

$$4\pi R^2 \dot{R}\Delta p = 2\pi\rho\left(3R^2\dot{R}^3 + 2R^3\dot{R}\ddot{R}\right) \tag{7.2.8}$$

$$\frac{3}{2}\dot{R}^2 + R\ddot{R} = \frac{\Delta p}{\rho}. \tag{7.2.9}$$

Indeed, from some point of view, the Rayleigh equation is ambiguous, a more useful equation follows from the equation (7.2.7) directly:

$$4\pi R^2 \Delta p dR = 2\pi\rho d\left(R^3 \dot{R}^2\right). \tag{7.2.10}$$

Those readers who prefer clarity before completeness may jump immediately to Section 7.4.2. But we will use the standard long way: we obtained the Rayleigh equation (7.2.9), then, we have to solve it with various methods. Therefore, despite the fact that many useful results follow from (7.2.10), we will spend some time to discuss the analytical solutions only for (7.2.9) in the beginning of Section 7.4 and numerical calculations in Section (7.4.5).

7.2.2 Derivation from hydrodynamics

It is a usual situation, when the 'clear' physical consideration raises some possible questions from the critically acclaimed mind. How strict is the derivation of the Rayleigh equation? Does this equation follow from the fundamental equations?

Let us consider the Navier–Stokes equation for a spherically symmetric flow of incompressible fluid—the fluid around a spherical bubble:

$$\frac{\partial u_r}{\partial t} + u_r \frac{\partial u_r}{\partial r} = -\frac{1}{\rho}\frac{\partial p}{\partial r} + \nu\left(\frac{\partial^2 u_r}{\partial r^2} + \frac{2}{r}\frac{\partial u_r}{\partial r} - \frac{2u_r}{r^2}\right). \tag{7.2.11}$$

Because of incompressibility, we have (7.2.5). Using (7.2.5) for u_r, one may see, first, that the bracket in (7.2.11) vanishes. Then, integrating (7.2.11) with (7.2.5) with respect to r from the bubble radius R to infinity, we obtain

$$\underbrace{\left(R^2\ddot{R} + 2R\dot{R}^2\right)\int_R^\infty \frac{dr}{r^2}}_{R\ddot{R}+2\dot{R}^2} - \underbrace{2R^4\dot{R}^2\int_R^\infty \frac{dr}{r^5}}_{\dot{R}^2/2} = \underbrace{-\frac{1}{\rho}\int_R^\infty \frac{\partial p}{\partial r}dr}_{(p_b - p_\infty)/\rho}. \tag{7.2.12}$$

Thus, we obtain the same equation as (7.2.9) again. This is not a surprise because of the nature of the Navier-Stokes equation—the conservation law of momentum. Note a nuance: with this derivation, we miss the stage that corresponds to (7.2.7) and leads directly to (7.2.10).

Within the hydrodynamics approach, we may also clarify the pressure difference (7.2.3): in a common case, the Laplace pressure jump on the bubble surface is determined by the full relation (7.1.16):

$$\left(-p + \tau_{rr}\right)_{liquid} - \left(-p + \tau_{rr}\right)_{bubble} = \frac{2\sigma}{R}. \tag{7.2.13}$$

Only for stationary media (7.2.13) turns into (7.2.3). One may assume that the gaseous phase rests; this is an assumption, but, as we may judge, a reasonable assumption. Thus, the second bracket on the left side of (7.2.13) transforms into $-p_{in}$. However, the fluid flows, so the corresponding terms for the incompressible fluid are

$$\left(-p + \tau_{rr}\right)_{liquid} = -p_b + 2\mu\frac{\partial u_r}{\partial r}\bigg|_{r=R} = -p_b - 4\mu\frac{\dot{R}}{R}, \tag{7.2.14}$$

$$p_{in} = p_b + \frac{2\sigma}{R} - \frac{4\mu\dot{R}}{R}, \tag{7.2.15}$$

gives us instead of (7.2.3)

$$\Delta p = p_b - p_\infty = p_{in} - p_\infty - \frac{2\sigma}{R} - \frac{4\mu\dot{R}}{R}. \tag{7.2.16}$$

The relation (7.2.16) gives the full answer for the incompressible fluid.

It is useful to compare the order of the last term in (7.2.16) with typical values of the rest ones. During bubble oscillations in water, the velocity may be $\dot{R} \sim 1$ m/s for a bubble of radius $R \sim 10$ μm; so, we have with $\mu \sim 10^{-3}$ Pa·s for $4\mu\dot{R}/R \sim 10^{2\pm3}$ Pa. This term is much lower than the surface tension term for the same bubble radius: $2\sigma/R \sim 10^4$ Pa. Therefore, generally, the viscous effects do not play a big role during bubble oscillations. It should be remembered for calculations: the main purpose of the Rayleigh equation is to give us an adequate qualitative picture, so the excess terms must be neglected. Of course, neglecting must be circumspect: for low pressures, for very viscous fluids (like glycerol), the last term on the right-hand side of (7.2.16) must be saved.

7.2.3 Modifications of the Rayleigh equation

Above we presented the derivations of the classic form of the Rayleigh equation. Since 1917, when Rayleigh used this equation to analyze the collapse of a spherical cavity (Rayleigh 1917), many other forms of his equation were obtained and applied.

Implying consideration of acoustic cavitation, we begin from the work of Noltingk and Nepppiras (Noltingk and Neppiras 1950). To describe the dynamics of a spherical cavity at an alternating acoustic field, they used the periodic function for the pressure in a liquid:

$$p_\infty = p_a - p_0 \sin \omega t. \tag{7.2.17}$$

Also, they assume isothermal conditions for the gas inside a bubble, which lead to the condition $p_g V_b = const$ for a bubble of volume $V_b = 4\pi R^3/3$, so if at the origin the bubble radius was R_0 at the pressure $p_a + 2\sigma/R_0$, then at any time the pressure inside it is

$$p_g = \left(p_a + \frac{2\sigma}{R_0} \right) \frac{R_0^3}{R^3}. \tag{7.2.18}$$

In more modern works, the oscillations of a cavity are assumed to be adiabatic; this approach lead to the replacement $3 \to 3\gamma$ in (7.2.17) (γ is the ratio of the isobaric and isochoric specific heats; see the next section). With numerical calculations, Noltingk and Neppiras found solutions $R(t)$ for two cases: for isothermal gas and for the void, i.e., for the case $p_g = 0$. They also obtained some approximate analytical solutions; we will discuss similar constructions in Section 7.4.

Keller and Miksis analyzed a more complicated equation than (7.2.9) (Keller and Miksis 1980). They considered a general case of a compressible fluid, when the viscous stress is expressed by (7.1.7) with $\zeta = 0$; consequently, one may obtain instead of (7.2.15)

$$p_{in} = p_b + \frac{2\sigma}{R} - \frac{4\mu}{3} \left(\frac{\partial u_r}{\partial r} - \frac{u_r}{r} \right) \Bigg|_{r=R}. \tag{7.2.19}$$

For an incompressible fluid, there exists the condition $\dfrac{u_r}{r} = -\dfrac{1}{2} \dfrac{\partial u_r}{\partial r}$, following from (7.2.4), and we go back from (7.2.19) to (7.2.15) again. But for a compressible fluid, more complicated mathematics must be applied. Using the velocity potential φ, so that

$$u_r = \frac{\partial \varphi}{\partial r}, \tag{7.2.20}$$

one can rewrite the continuity equation as

$$\frac{\partial \rho}{\partial t} + \frac{\partial \varphi}{\partial r} \frac{\partial \rho}{\partial r} + \rho \underbrace{\frac{1}{r^2} \frac{\partial}{\partial r} \left(r^2 \frac{\partial \varphi}{\partial r} \right)}_{\Delta \varphi} = 0 \tag{7.2.21}$$

and the Navier–Stokes equation in the form of

$$\frac{\partial^2 \varphi}{\partial r \partial t} + \frac{\partial \varphi}{\partial r} \frac{\partial^2 \varphi}{\partial r^2} = -\frac{1}{\rho} \frac{\partial p}{\partial r} + \frac{4\nu}{3} \Delta \varphi. \tag{7.2.22}$$

Then one may consider a 'weak compressibility': put $\rho = const$ in (7.2.22) and consider the wave equation for φ

$$\frac{1}{c_s^2} \frac{\partial^2 \varphi}{\partial t^2} = \Delta \varphi, \tag{7.2.23}$$

where c_s is the speed of sound in the fluid. Integrating (7.2.22) with respect to r from R to infinity (as in previous 7.2), differentiating the obtained equation on time, finally we have at $r = R$

$$\frac{\partial \varphi}{\partial t} + \frac{1}{2}\left(\frac{\partial \varphi}{\partial r}\right)^2 = -\frac{1}{\rho}\left(p_{in} - p_\infty - \frac{4\nu}{R}\frac{\partial \varphi}{\partial r} - \frac{2\sigma}{R}\right). \tag{7.2.24}$$

With (7.2.23), the equation (7.2.24) gives (after applying the derivative to it one more time)

$$\ddot{R}\left[4\nu - R\left(\dot{R} - c_s\right)\right] = \frac{1}{2}\dot{R}^3 + \left(\dot{R} + c_s\right)\frac{\Delta p_{in\infty}}{\rho} - c_s\left(\frac{3}{2}\dot{R}^2 + 4\nu\frac{\dot{R}}{R} + \frac{2\sigma}{\rho R}\right) +$$
$$+ R\dot{R}\Delta\dot{p}_{in\infty} + \left(\dot{R} + c_s\right)\dot{\Phi}(t) \tag{7.2.25}$$

where $\Delta p_{in\infty} = p_{in} - p_\infty$, and $\Phi(0,t)$ is the velocity potential of the sound field acting on the bubble; in (Keller and Miksis 1980), $\dot{\Phi} = (p_0/\rho)\sin(t + R/c_s)$ was used.

Thus, the equation (7.2.25) generalizes the Rayleigh equation for a bubble in an acoustic field for a (weakly) compressible fluid.

We mention the generalization from the work (Sinkevich et al. 2012). In this work, the consideration took into account the evaporating mass flux $\dot{m}_s$ from the bubble wall; the corresponding equation has the form

$$\rho\left(R\ddot{R} + \frac{3}{2}\dot{R}^2\right) = p_{in} - p_\infty - \frac{2\sigma}{R} - 4\nu\frac{\dot{R}}{R} + \ddot{m}_s R - \frac{\dot{m}_s^2}{2\rho}. \tag{7.2.26}$$

Evaporation may play a significant role for bubble dynamics, if it exists. In the next section, we discuss the matters concerning determination of special conditions for a cavitating bubble—conditions inside a bubble and on its walls.

7.3 The boundary conditions for a bubble in a liquid

7.3.1 Pressure and temperature inside a bubble

From the previous consideration, we see that the mathematical formulation for the problem of a cavitating bubble may include the Navier–Stokes equation or the Rayleigh equation as the essential approximation for a more or less common case. But for any description, we need a correlation for the pressures appearing in (7.2.15): the pressure inside a bubble p_{in} and a pressure in a liquid far away from the bubble p_∞. For many practical cases, the outer pressure can be considered as constant: $p_\infty = const$. As for the pressure inside a bubble, one may imagine two situations:

1) This pressure is a constant, so with $p_{in} = const$ we have a very comfortable condition $\Delta p = const$. This may be applicable, for instance, for a large bubble (the surface term is negligible) filled with pure vapor at constant temperature.

2) In a common case, p_{in} may obey the intrinsiclaw, i.e., it must be calculated on its own. The simplest case: the pressure of the gaseous phase inside a bubble may depend on the temperature of this phase or on the temperature on the surrounding liquid.

The second way is much harder and, unfortunately, is more realistic. Let us explore this way step-by-step.

In a general case, the bubble is filled not only with vapor, but also with a gas. Here we distinct the vapor (product of evaporation of the surrounding liquid) and the extraneous gas (such as air). Thereby, the pressure inside a bubble consists of two parts: the vapor pressure p_v and the gas pressure p_g:

$$p_{in} = p_v + p_g \tag{7.3.1}$$

The simplest way to construct this correlation is to put the vapor pressure inside the bubble as the saturation pressure corresponding to the temperature of the liquid:

$$p_v = p_s(T_l), \tag{7.3.2}$$

where T_l is the temperature of the liquid. In other words, we assume that at any moment of time we have the total thermodynamic equilibrium. In this approach, we suppose that the bubble dynamic process is sufficiently *slow* in comparison with the rate of establishing of the thermodynamic equilibrium. Thus, the equation (7.3.2) implies isothermal conditions at the bubble wall. Thus, in the case of saturation one may hope that the vapor pressure, determined by constant temperature, remains constant.

To formulate the expression for the pressure of a non-condensing gas, we have to know what process occurs inside the gaseous phase in the bubble. Assuming the processes inside the cavitating bubble are *fast*, we may state the absence of heat transfer between the gas and the liquid wall of the bubble. In such conditions, we may use the Poisson equation for adiabatic process in a gas

$$p_g V^\gamma = const, \tag{7.3.3}$$

where $V = 4\pi R^3/3$ is the volume of the bubble and $\gamma = C_p/C_V$ is the adiabatic exponent.

During the adiabatic process, the temperature within the collapsing bubble increases; the corresponding equation follows from (7.3.3):

$$p_g^{1-\gamma} T_g^\gamma = const \text{ or } T_g V^{\gamma-1} = const, \tag{7.3.4}$$

that is, the temperature inside a bubble is connected with its radius as

$$T_g R^{3\gamma-3} = const. \tag{7.3.5}$$

For example, if a bubble filled with air ($\gamma = 1.4$) has been decreased by half, the temperature inside it would be increased by 2.3 times.

As we emphasized above, we have a controversy in such a physical picture depicted above. If we want to consider both components of the gas, the characteristic time of the process must be

$$\tau_e \ll \tau \ll \tau_h, \tag{7.3.6}$$

where τ_e is the time to establish the equilibrium pressure of the vapor phase (due to evaporation and condensation processes on the bubble wall), τ_h is the characteristic time scale of heat exchange.

To defend this approach, one may declare that the processes of evaporation and condensation are very fast, i.e., the time scale τ_e is very small. Usually, evaporation (or condensation) fluxes have values of the order of kilograms per square meter per second (Gerasimov and Yurin 2018) or even much more. Even for a huge bubble of radius ~ 1 mm at pressure ~ 1 atm (i.e., for the vapor mass $\sim 10^{-9}$ kg), the time scale τ_e of the process is less than ~ 1 ms. It may seem that this fact means that the pressure instantly 'adjusts' the bubble wall temperature (i.e., the temperature of the surrounding liquid), but actually the time scale of the bubble collapse can be much smaller than this value (see Section 7.4). Thus, from this point of view, the adiabatic approach can be correct for the bubble collapse, but not in any conditions.

The main problem lays on another plane: usually, we have to consider both gaseous components—vapor and extraneous gas. Can we use equations (7.3.2) and (7.3.3) simultaneously, combining them into (7.3.1)? If so, what is the temperature of the gas mixture inside a bubble? In accordance with (7.3.1), the vapor parameters correspond to the temperature of the surrounding liquid, while the extraneous gas must obey the correlation (7.3.5). Since masses of particles (of vapor and gas) are close to each other, one may reasonably expect that the temperatures of vapor and gas would be similar: the energy transfer between particles is effective, and all components of the gaseous mixture have the same temperature.

Thus, simple answer of these questions is no, we cannot use (7.3.2) and (7.3.3) combined. However, one may suppose that the liquid surrounding the bubble has the same temperature following the temperature of the gas inside a bubble, that is, in (7.3.2) we must use the temperature of gas. Indeed, there must be phase equilibrium between the vapor and a tiny layer of liquid at the interface; is it possible this tiny layer can be heated with such a small amount of heat that the overall process can be assumed adiabatic? This line looks correct especially for a weak evaporation. On the other hand, of course, the evaporation process may be weak and, if so, its heat leaks can be negligible. However, if the evaporation is weak, then one may also neglect the term p_v in (7.3.1), and eliminate the problem completely.

Thus, our discourse meets a dead end. It is time to consider the whole problem directly, without lengthy reasonings.

From the point of view of thermodynamics, a bubble is an open system (due to vapor). For such a system, the first law states the following:

$$dQ = dU + p_{in}dV - \mu dm. \tag{7.3.7}$$

Here p_{in} is the net pressure (7.3.1) of a gas of a vapor, $m \equiv m_v$—the mass of the vapor inside a bubble. For the positive change $dm > 0$ at $dV < 0$, it is required that $dQ < 0$: some heat must withdraw from the bubble to produce the vapor mass dm (or vice versa with opposite signs for condensation; however, this case is more interesting: for evaporation, it is clear that the energy goes from the bubble, but not the energy allotted because of condensation that goes to the bubble). The corresponding term can be written as—$h_{lg}dm$, assuming sufficient equilibrium state; the work function for evaporation will be discussed in the next section.

Then, based on presumption of a perfect gas in the bubble (yes, we have to do so: we find the generalization of the (7.3.5) which is obtained for the perfect gas itself), the term

$$dU = \underbrace{\left(\frac{\partial U}{\partial T}\right)_{V,m}}_{C_V} dT + \underbrace{\left(\frac{\partial U}{\partial V}\right)_{T,m}}_{0} dV + \underbrace{\left(\frac{\partial U}{\partial m}\right)_{V,T}}_{\tilde{u}} dm . \tag{7.3.8}$$

The second term vanishes because of the Joule law. Note that

$$\tilde{u} = \left(\frac{\partial U}{\partial m}\right)_{T,V} \neq \left(\frac{\partial U}{\partial m}\right)_{S,V} = \mu, \tag{7.3.9}$$

see Appendix A.

Then, we have instead of (7.3.7)

$$\left(\mu - \tilde{u} - h_{lg}\right)dm = C_V dT + p_{in}dV . \tag{7.3.10}$$

Then, because of the Clapeyron equation, we can rewrite (7.3.10) in the form

$$\frac{\mu - \tilde{u} - h_{lg}}{C_V} dm = dT + \underbrace{\frac{\nu R_G}{C_V}}_{\gamma-1} \frac{T}{V} dV, \tag{7.3.11}$$

since from the Mayer equation $\nu R_G = C_p - C_V$. Finally, dividing by dV, we obtain the desired equation. It should be emphasized that differentials in (7.3.11) correspond to the process of variation of the bubble size: growth or collapse. This is not an adiabatic process: some heat leaks form the bubble to evaporate the liquid on the bubble wall. Correspondingly, the derivatives $\partial T/\partial V$ and $\partial m/\partial V$ used below mean the deviation of T or m on V in the same process:

$$\underbrace{\frac{\mu - \tilde{u} - h_{lg}}{C_V}}_{f} \frac{\partial m}{\partial V} = \frac{\partial T}{\partial V} + \left(\gamma - 1\right)\frac{T}{V} . \tag{7.3.12}$$

Another subtle trick: here the derivative

$$\frac{\partial m}{\partial V} = \frac{\partial (\rho V)}{\partial V} = \rho + V \frac{\partial \rho}{\partial V}, \tag{7.3.13}$$

but not density ρ. Derivative $\frac{\partial m}{\partial V}$ is a function of the process; for the absence of evaporation, $\frac{\partial m}{\partial V} = 0$. For evaporation during the bubble collapse, when $dm > 0$ and $dV < 0$, this derivative is negative.

The equation (7.3.12) contains the term f, which describes all the processes connected with the vapor mass change and, in a general case, can be a function on T and V. In the simplest case, one may put $f = const$, and the solution of (7.3.12) gives us the modified Poisson law for a bubble in the following:

$$TV^{\gamma-1} = C + \frac{f}{\gamma} V^{\gamma}. \tag{7.3.14}$$

The constant C from (7.3.14) corresponds to the initial parameters (denoted by index '0' below), so (7.3.15) can be represented as

$$TV^{\gamma-1} = T_0 V_0^{\gamma-1} + \frac{f}{\gamma} \left(V^{\gamma} - V_0^{\gamma} \right). \tag{7.3.15}$$

Let us sum up and repeat the assumptions for (7.3.16):

(1) the gas is perfect;

(2) parameter f is different for evaporation and condensation;

(3) parameter f is supposed to be constant, but, actually, it varies with thermodynamic parameters such as temperature and volume (or pressure).

Let us estimate a new term in (7.3.15). To calculate f, we will use only the latent heat of evaporation, i.e., to compare the terms $TV^{\gamma-1}$ and fV^{γ}/γ, we will compare $\frac{T}{V}$ and $\frac{h_{lg}}{C_V} \frac{\partial m}{\partial V}$, since $\gamma \sim 1$. Using specific volume $\upsilon = V/m$ and specific isochoric heat capacity $\tilde{c}_V = C_V/m$, one may find that these terms would be comparable if

$$\frac{\partial m}{\partial V} \sim -\frac{T \tilde{c}_V}{\tilde{\upsilon} h_{lg}}, \tag{7.3.16}$$

This relation estimates the rate of evaporation required to distort the plain adiabatic scheme of the bubble collapse. For gases, typical values are $\tilde{\upsilon} \sim 1$ m³/kg, $\tilde{c}_V \sim 10^3$ J/(kg·K), and for $h_{lg} \sim 10^6$ J/kg and for temperature $T \sim 10^2$ K, we get $\left| \frac{\partial m}{\partial V} \right| \sim 10^{-1}$ kg/m³. Possibly, it is not ambiguous to explain the meaning of this derivative again: during the compressing by volume dV, the evaporation process leads to increase in the vapor (therefore, in the gaseous) mass by dm. Thus, the obtained value means that when the bubble volume decreases by 1 m³ (sic! only for estimations, and then, why cannot we really consider huge bubbles?), the vapor mass in this bubble increases by 0.1 kg. Considering the fact that any cubic meter of vapor weighs about 1 kg, the obtained value does not look unreal. Actually, with this value we even stay within the frame of the perfect gas approach.

However, this means that the influence of evaporation on bubble dynamics can be significant. The leaks of heat may be not negligible; thus, equations (7.3.3–7.3.5) are beyond their applicability. If so, let us develop the approach following from (7.3.10) to obtain more appropriate forms of the modified Poisson equation.

This increase of the vapor mass in a bubble during its collapse is caused, finally, by the rise of temperature inside a collapsing bubble. Therefore, we may rewrite the key derivative

$$\frac{\partial m}{\partial V} = \underbrace{\frac{dm}{dT}}_{m_T} \frac{\partial T}{\partial V} \tag{7.3.17}$$

and use this relation in (7.3.12), obtaining

$$\frac{\partial T}{\partial V}\underbrace{\left[1 - \frac{\left(\mu - \tilde{u} - h_{\lg}\right)}{C_V} m_T\right]}_{\omega} + \left(\gamma - 1\right)\frac{T}{V} = 0. \tag{7.3.18}$$

Supposing $\omega = const$ (which also demands the condition $m^T = const$), we get the 'corrected' adiabatic law:

$$TV^{\frac{\gamma-1}{\omega}} = const. \tag{7.3.19}$$

The choice between (7.3.15) and (7.3.20) depends on what condition is filled with more physical sense: $\dfrac{\partial m}{\partial V} = const$ (leads to (7.3.15)) or $m_T = const$ (leads to (7.3.19)). We prefer (7.3.19): besides other reasons, this condition is more laconic. It is a joke, of course, but many physicists believe that 'correct' laws are shorter.

Equation (7.3.19) may also be interpreted as a corresponding equation for a polytropic process with index $1 + \dfrac{\gamma-1}{\omega}$. One may even construct the heat capacity of such a process:

$$C_n = C_V \frac{n-\gamma}{n-1} = C_V \frac{\omega-1}{\gamma-1}. \tag{7.3.20}$$

Finally, one may note that it is possible to obtain these results directly from (7.3.10), substituting there $dm = m_T dT$, introducing a new heat capacity, etc.

How evaporation from the liquid wall of the bubble affects the temperature inside it? Qualitatively, the answer follows directly from (7.3.10). External forces do a work on the bubble (term $p_{in}dV$ with negative sign at dV: bubble collapses); this work is spent on (a) increasing the internal energy (term $C_V dT$), and (b) the heat leaks to the liquid for evaporation (term $h_{\lg}dm$). Thus, the more the heat leaks, the lesser the temperature increases. In the limiting case, it follows from the equation (7.3.10) that temperature can remain constant: all the work undergoing on the bubble will be spent on evaporation of the surrounding liquid. The same thing follows from the obtained equations; for example, in (7.3.19) the exponent at V is lower than the one for a regular adiabatic process ($\gamma - 1$), because $\omega > 1$.

With equations (7.3.15) and (7.3.19), one may find the temperature in a collapsing bubble for each value of volume. Then, with the Clapeyron equation, it is possible to find the total pressure inside a bubble:

$$p_{in} = \underbrace{\left(\frac{m_g}{\mu_g} + \frac{m_v}{\mu_v}\right)}_{v}\frac{R_G T}{V}, \tag{7.3.21}$$

where μ_g and μ_v are the molecular weights of gas and of vapor, m_g is (a constant) mass of the gas in a bubble, m_v is (a variable) mass of the vapor, R_G is a universal gas constant.

The method presented in this section looks preferable in comparison with a line where pressures of vapor and gas are being calculated separately, like in (7.3.2–4). On the other hand, of course, relations (7.3.15) and (7.3.20) are only approximations. We have little reason for $\dfrac{\partial m}{\partial V} = const$, the stipulation $\dfrac{dm}{dT} = const$ also is only an approach.

However, the relation (7.3.20) helps to understand the real dynamics of the temperature inside a collapsing bubble; this formula is quite suitable for qualitative consideration of the issue. Equations (7.3.15) and (7.3.30) contain the variation in vapor mass (in (7.3.16) through f, in (7.3.20) through ω), and one should know these parameters to calculate temperature through these relations. To calculate vapor mass, one must know the mass flux; for example, the change in vapor mass at time Δt is

$$\Delta m = \int_0^{\Delta t} \dot{m}_s S dt. \tag{7.3.22}$$

In its turn, the mass flux is not a determining parameter, it is a function of the process. Moreover, it is possible that this calculation of the evaporation flux in a collapsing bubble is the most complicated part of the problem.

7.3.2 Mass flux at the bubble surface

During the bubble collapse, the temperature inside a bubble increases. The heat flux from the hot vapor falls onto the liquid at the bubble wall, provoking evaporation. The additional gaseous mass significantly influences the process: pressure increases, preventing the collapse, while temperature decreases, in accordance with (7.3.20). In this section, we discuss evaporation in such conditions. In the beginning, we consider a more common problem.

Let us consider the boundary of two phases (see Fig. 7.3.1). Usually, the heat flux is continuous, but not in this case: some heat will be spent to transform the substance at the interface into a new phase. For certainty, we will consider the case when phase 2 turns into phase 1 after heating: one needs to spend heat to transform phase 1 into phase 1; for instance, phase 1 is a liquid and phase 2 is a solid, or phase 1 is a vapor while phase 2 is a liquid.

Denote the heat flux falling from phase 1 onto the selected volume as q_1; the heat flux from that volume into the bulk of phase 2 is q_2. Phase 2 in the volume $dSdx$ will be fully transformed into phase 1 during the time dt; the total heat $(q_1 - q_2)dSdt$ has gone into the phase transition of mass dm at the latent heat of phase transition h_{ph}.[2]

$$\left(q_1 - q_2\right)dSdt = h_{ph}\underbrace{\rho dSdx}_{dm}, \tag{7.3.23}$$

ρ is the density of phase 2—the phase that undergoes the phase transition.

From this equation, one may obtain two useful correlations. The first one defines the heat flux jump at the phase interface:

$$q_1 - q_2 = h_{ph}\rho \underbrace{\frac{dx}{dt}}_{\upsilon}, \tag{7.3.24}$$

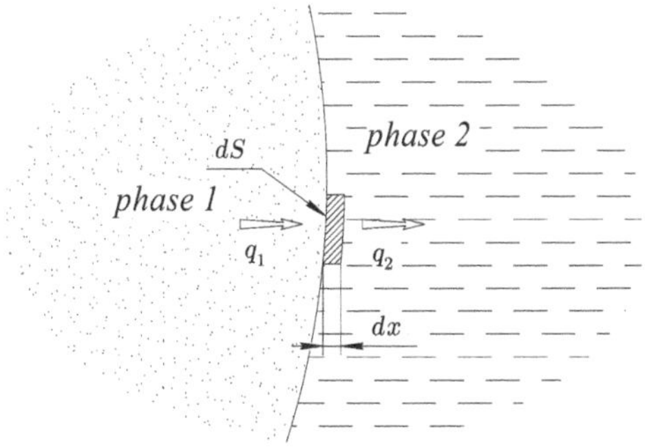

Figure 7.3.1. The heat flux jump at the phase transition interface.

---

[2]   We denote this value differently in comparison with h_{lg} for a generalization.

which connects this jump with the speed of the interface boundary υ. The problem of heat transfer with such boundary conditions is usually referred to as the Stephan's problem.

The next relation establishes the flux of mass conversion $\dot{m}_s = \dfrac{dm}{dSdt}$ at the interface:

$$\dot{m}_s = \frac{q_1 - q_2}{h_{ph}}. \tag{7.3.25}$$

This correlation can be used to determine the evaporation flux on the bubble wall. Supposing that all the heat proceeded from the vapor phase (phase 1) was disbursed for the phase transition, i.e., the heat flux into phase 2 $q_2 = 0$, we have the correlation that connects the mass flux $\dot{m}_s$ and the heat flux from the vapor qs at the interface. For water at atmospheric pressure, $h_{ph} = h_{lg} \approx 2.3 \cdot 10^6$ J/kg; thus,

$$\dot{m}_s \left[\frac{kg}{m^2 s} \right] \approx 4.3 \cdot 10^{-7} q_s \left[\frac{W}{m^2} \right]. \tag{7.3.26}$$

Equation (7.3.26) connects the heat and the mass fluxes. To be exact, it implies that the heat flux is a determining parameter, while the mass flux is the function of it. At first glance, this interpretation perfectly corresponds to the real physical picture: heat flux causes the phase transition—evaporation, in particular.

However, one needs to have good understanding of the evaporation process. Solids do not turn into liquid until their temperature reaches the temperature of phase transition. In contrast to this, a liquid may emit molecules into the vapor phase at any temperature. To leave a surface, a molecule (or an atom) must have a sufficient kinetic energy that exceeds its bounding energy. This is possible at any temperature: particles in a liquid are distributed on their energies, and one particle in billions will always be capable of this, having enough energy for this. Leave a cup of water in a room, and you will be convinced. In some cases, for a saturated vapor, the evaporation flux will be balanced out by the condensation one, but this is a rare circumstance.

Another important thing: the bounding energy that must be exceled by the evaporating molecule varies widely because of the energy distribution of particles in a liquid. So, the value h_{lg} that corresponds to the equilibrium phase transition may be inadequate for another energy distribution of liquid molecules, just like distribution on kinetic energy or distribution on potential energy. Indeed, this problem is very complicated; some theoretical aspects and their discussion can be found in (Gerasimov and Yurin 2018). Equation (7.3.24) must be verified in non-equilibrium conditions of the liquid surface; likely, the proportional factor will be different.

Then, any molecule from the vapor phase striking the liquid surface may kick out a molecule from it—that depends on the initial energy of the molecule in a liquid; the net results depend on three distribution functions:

(a) distribution function on kinetic energy of gas molecules striking the liquid;

(b) distribution function on kinetic energy of molecules in the liquid;

(c) distribution function on potential energy of molecules in the liquid.

Thus, it is hard to expect that the total flux of evaporated molecules would depend only on two integral parameters: the total energy flux from the gas (circumscribes the point a) and the latent heat of evaporation (depicts points b and c). In such a complex object as a bluntly collapsing bubble, the mean parameters for the relation of a kind of (7.3.26), experimentally obtained for a quasi-equilibrium cases, may be far from truth.

To date, there are no final recipes for the function $\dot{m}_s$. At least, it should be remembered that the correlation (7.3.25) may give a large error in a general, non-equilibrium case.

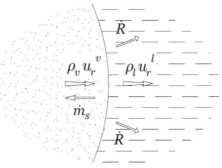

Fig. 7.3.2. Mass fluxes on the evaporation surface; the direction of velocities follows the velocity in a liquid. The mass flux $\dot{m}_s$ is defined in the laboratory system of coordinates.

7.3.3 Flows in a bubble and in a liquid

Equation (7.1.12) can be used to the find the mass fluxes from a vapor and from a liquid. Mass cannot appear at the surface (of course, we will not consider the pairing effect for this problem); thereby, the mass fluxes at the interface are equal. We only have to take into account the movement of the interface, i.e., of the bubble wall (see also Fig. 7.3.2). Therefore, the normal, i.e. radial, components of velocities in the laboratory coordinate system

$$\underbrace{\rho_\upsilon\left(u_r^\upsilon - \dot{R}\right)}_{vapor} = -\dot{m}_s = \underbrace{\rho_l\left(u_r^l - \dot{R}\right)}_{liquid}. \tag{7.3.27}$$

It follows from (7.3.27) that the radial component of velocity of the liquid is not equal to the velocity of the bubble growth R, as it was assumed during the derivation of the Rayleigh equation; actually, it is

$$u_r^l = \dot{R} - \frac{\dot{m}_s}{\rho_l}. \tag{7.3.28}$$

This correction of the Rayleigh equation was made in (Sinkevich et al. 2012). The additional term has clear physical sense: to provide the mass flux on the surface (because of evaporation), there must be an additional 'mass supply' from the liquid to the interface.

However, estimations of this additional term give that even for a mass flux at the surface $\dot{m}_s \sim 10^2$ kg/m²s (this is a huge, theoretically calculated flux of pure evaporation, without any condensation) the second term of equation (7.3.28) gives additional velocity of only ~ 0.1 m/s. This is a negligible correction in comparison with usual velocities of bubble dynamics ~ 1 m/s, especially considering the uncertainties in the calculation of mass fluxes.

The more interesting results follow from (7.3.27) for the gaseous phase. Usually, when simulating bubble dynamics, we neglect the gas flow inside it. For example, one may conclude it from the continuous equation: assuming spherical symmetry, one may restrict the consideration only with the radial velocity, and from the equation

$$\frac{1}{r^2}\frac{d\left(r^2 u_r^\upsilon\right)}{dr} = 0 \tag{7.3.29}$$

we get the single solution restricted at origin: $u_r^\upsilon = 0$.

However, if so, then we have from (7.3.27) the relation for the velocity of bubble growth

$$\dot{R} = \frac{\dot{m}_s}{\rho_\upsilon}. \tag{7.3.30}$$

What does it mean? How can we obtain a certain correlation for the bubble growth velocity, which does not contain any other parameters of the process, like the pressure difference in the bubble and in the liquid? From the physical intuition, one may assume that R must depend only on the external parameters, but not be only a function of evaporation rate. Furthermore, in a common case the bubble contains both vapor and gas or even gas alone; in such cases, the relation (7.3.30) gives quite a strange result.

The answer leads to understanding that, indeed, the velocity of vapor is non-zero. If $u_r^v \neq 0$, everything falls into place: $\dot{R}$ and $\dot{m}_s$ are defined independently. It could seem that the continuous equation is broken, but, in fact, everything is fine. At first, it should not be forgotten that the proviso (7.1.9) is not the only condition for $\rho = const$. The vapor in the bubble is compressible, we may even say well compressible, regardless of its velocity. Because of evaporation, the vapor mass inside a bubble changes; therefore, density varies too. For a variable density, the continuous equation has no longer a form of (7.1.6), that is, $div\overline{u} = 0$. Besides, there may be other components of velocity except u_r^v.

The difference in the tangential velocity between the vapor and the liquid at the interface is usually set to zero because of the sticking condition, which is suitable for viscous fluids. Of course, a skeptical mind may question this thesis: is a vapor 'sufficiently viscous' to stick a liquid? Cannot there be a slip? In a general case, yes, especially for strong gradients of temperatures along the normal for the surface. Anything can happen in a general case, but for bubble dynamics these circumstances look negligible.

Finally, we may discuss the pressure inside a bubble. For the full hydrodynamics approach, this problem is related to the velocity field, etc. Let us simplify the problem and consider the Rayleigh problem: a growing or a collapsing bubble inside a liquid. Can we assume $p(x, y, z) = const$—the pressure to be equal at any point inside a bubble?

As we know, pressure equalizes at the speed of sound c_s. Consequently, the time scale to establish a uniform pressure in a bubble of radius R is R/c_s; for the speed of sound $c_s \sim 10^2$ m/s in a bubble of radius $R \sim 10^{-5}$ m, this is $\sim 10^{-7}$ s. Thus, the consideration of such a bubble will be correct for time scales $\gg 10^{-7}$. It is important to understand it while analyzing the calculations in the next section.

7.4 Analytical solutions of the Rayleigh equation

7.4.1 Simple scaling

The Rayleigh equation is too complicated to obtain its exact solution in a common case for any function $\Delta p(R,R,t)$. Then one may follow one of two ways: (1) find partial solutions for special particular cases of Δp, or (2) find special forms of Δp which correspond to simple solutions.

The second way may possibly look somewhat provocative. However, today the role of analytical solutions has been changed. Earlier, in the pre-computing era, the formulae obtained from the differential equation were the single way to obtain any quantitative information about this equation. Huge complicated expressions, usually in a form of series on special functions, allowed to get numerical results. Today, every engineer has numerical instruments that allow him to obtain the demanded quantitative information directly from the differential equation. Indeed, in many cases the numerical solution of differential equation is more convenient than summing huge terms of a poorly-convergent series. But the role of analytical solutions was never restricted by the utilitarian role of a calculator. Probably, the main role of an analytical solution is to get the clear picture of the considered phenomenon which lets us understand the quantitative issues of the physical nature. Thus, here we will use the second way. Let us take a closer look at the Rayleigh equation

$$\frac{3}{2}\dot{R}^2 + R\ddot{R} = \frac{\Delta p}{\rho}.$$

(7.4.1)

Here we have a single constant—the density of a liquid ρ; also, one may see the arbitrary pressure function Δp and a very complicated differential operator in the left-hand side of the equation. Suppose that some prodigy mathematician somewhere will obtain the solution $R(t)$ for the Rayleigh equation (we mean, for arbitrary $\Delta p(R, \dot{R}, t)$) as a series on special functions; we do not mean the variant when some special function $F(t)$ would be directly defined as the solution of the Rayleigh equation for a special kind of the pressure, computed numerically and given in such form in reference tables. Even in this eventual case, to clear the behavior of the bubble's radius one would have some simple limit for this solution; for example, the main terms of polynomial expansion of the obtained solution. Following this way, one should try to find the solution in the form of

$$R = At^n, \tag{7.4.2}$$

with constants A and n that must be defined from the expansion of $F(t)$ in corresponding series.

However, we may choose another way: to find these constants directly from the equation (7.4.1). Inserting (7.4.2) into the differential operator, we have

$$\frac{3}{2} A^2 n^2 t^{2(n-1)} + A^2 n(n-1) t^{2(n-1)}. \tag{7.4.3}$$

Thus, we see that both terms are proportional to $t^{2(n-1)}$; this was predictable based on the reasons of dimensions of these terms. If the function (7.4.2) provides correct solutions (in the main value at least), then the right-hand side of (7.4.2) must be of the form $\sim t^m$ (in the main value at least); comparison would give us the condition $2(n-1) = m$ to find the value of n.

The simplest situation when we have the demanded form for $\Delta p \sim t^m$ is the expanding of a gaseous bubble (without vapor, see Section 7.3. for details) due to enormous pressure inside it. This inner pressure is much higher than other terms in Δp, like the pressure of the liquid, surface tension, etc. In this case, if the gas inside a bubble obeys the Clapeyron's equation

$$p_{in}V = \nu R_G T \tag{7.4.4}$$

and the number of moles $\nu = const$ (the mass of gas does not change, i.e. no processes of evaporation or condensation occur on the bubble walls), then we have for the adiabatic conditions the Poisson equation (see previous section)

$$p_{in}V^\gamma = const. \tag{7.4.5}$$

For a spherical bubble $V = \frac{4}{3}\pi R^3$, we get

$$p_{in} = \frac{C}{R^{3\gamma}}, \tag{7.4.6}$$

where the constant $C = p_0 R_0^{3\gamma}$ determined by the values of the pressure and the radius at the initial moment of time.

Finally, inserting (7.4.6) into (7.4.1) with the equation (7.4.2) for radius, we see that the right side of (7.4.1) is $\sim t^{-3\gamma n}$. Thereby, $2(n-1) = -3\gamma n$, and

$$n = \frac{2}{2+3\gamma}. \tag{7.4.7}$$

Thus, the approximate solution has the scaling form $\sim t^n$ with n from (7.4.7). But it should be noted that according to this law we have the initial radius $R = 0$ at $t = 0$; so, this solution cannot be applied for primary stages of bubble growth: application range of the correlation $R = At^n$ corresponds to time values for $At^n >> R(0)$.

7.4.2 Exact analytical solution for adiabatic conditions (autonomous equation)

It is possible to obtain the exact solution of the Rayleigh equation for the function $\Delta p(R)$, i.e., when the Rayleigh equation is autonomous (does not contain an explicit dependence on time).

Let us take a closer look on the whole problem of bubble growth and collapse in adiabatic conditions. This is a usual approximation for theoretical investigation of a cavitating bubble used in many works (Flynn 1975, Noltingk and Neppiras 1950); further, we will discuss this assumption closer.

In Section 7.3, we have established that these conditions are quite appropriate. Briefly, assuming adiabaticity means presumption of a very fast process of cavitation: a bubble grows too rapidly and heat exchange does not take off any significant amount of heat during the bubble growth cycle. Thereby, the temperature of a bubble changes in accordance with the Poisson law for an adiabatic process if we assume a perfect gas inside a bubble which is, in truth, the strictest point of the consideration.

For adiabatic conditions, the pressure inside the bubble obeys the power-law correlation (7.4.6). Besides it, we have to take into account non-zero pressure of the surrounding medium. Here we will use the simplest case: the outer pressure in a liquid at infinity is $p_\infty = const$.

Consequently, the Rayleigh equation for the considered problem has the form

$$\frac{3}{2}\dot{R}^2 + R\ddot{R} = \frac{\dfrac{C}{R^{3\gamma}} - p_\infty}{\rho}. \tag{7.4.8}$$

Using symbol U for growth rate $\dot{R}$, we may rewrite this equation as the system

$$\dot{R} = U,$$

$$\dot{U} = \frac{C}{\rho R^{3\gamma+1}} - \frac{p_\infty}{\rho R} - \frac{3}{2}\frac{U^2}{R}. \tag{7.4.9}$$

At first, one may see a stationary point in this system: for $U = 0$ (zero growth rate) the bubble radius $R_s = \left(\dfrac{C}{p_\infty}\right)^{\frac{1}{3\gamma}}$ also corresponds to zero acceleration, i.e., $\dot{U} = 0$. This means that a stationary value of the bubble radius can be expected: a bubble of radius R_s would exist forever if this stationary state is stable. To test the stability of this point, we must use standard operations: find the linearized matrix for the system (7.4.9) and find its proper values at this point.

The linearized matrix of the system (7.4.9) has the form

$$\begin{bmatrix} \left(\partial\dot{R}/\partial R\right) & \left(\partial\dot{R}/\partial U\right) \\ \left(\partial\dot{U}/\partial R\right) & \left(\partial\dot{U}/\partial U\right) \end{bmatrix}, \tag{7.4.10}$$

correspondingly, the proper values for the linearized matrix can be found from the equation

$$\begin{vmatrix} 0-\lambda & 1 \\ -\dfrac{C(3\gamma+1)}{\rho R^{3\gamma+2}} + \dfrac{p_\infty}{\rho R^2} + \dfrac{3U^2}{2R^2} & \dfrac{3U}{R}-\lambda \end{vmatrix} = 0, \tag{7.4.11}$$

where values (R, U) correspond to the research point $(R_s, 0)$. Thereby, we obtain the simplest quadratic equation from (7.4.11)

$$\lambda^2 = -3\gamma\frac{p_\infty}{\rho R_s^2}. \tag{7.4.12}$$

The equation has two complex conjugate roots $\lambda_{1,2} = \pm i \left(3\gamma \dfrac{p_\infty}{\rho R_s^2} \right)^{1/2}$. This type of a stationary point is called a 'center'; this is not a stable point: values of R and U change periodically in vicinity of this point. Because of absence of other stationary points, we may conclude that the bubble radius (and, therefore, its growth rate) oscillates.

The physical picture of these oscillations is clear: due to overpressure, the bubble radius begins to grow. During the expansion, the pressure inside the bubble drops, and external pressure causes the bubble to collapse. Since we suppose no dissipation processes (we assume a reversible adiabatic process inside a bubble), these oscillations are eternal.

From the mathematical point of view, this means that the system (7.4.9) has an integral: some function on R and U remains constant. Geometrically, this function represents a closed line on the (U, R)–plane; this curve corresponds to the differential equation

$$\frac{dU}{dR} = \frac{C}{\rho U R^{3\gamma+1}} - \frac{p_\infty}{\rho U R} - \frac{3}{2}\frac{U}{R}. \tag{7.4.13}$$

At first, we see that this equation gives a symmetrical solution on U: equation (7.4.14) can be rewritten for U^2:

$$\frac{dU^2}{dR} = \frac{2C}{\rho R^{3\gamma+1}} - \frac{2p_\infty}{\rho R} - \frac{3U^2}{R}. \tag{7.4.14}$$

Equation (7.4.14), in its turn, is an ordinary linear heterogeneous differential equation of the first order of a kind $y' + 3y/x = f(x)$. Integrating this simple equation, we obtain

$$U^2 = \frac{A}{R^3} + \frac{2C}{3\rho(1-\gamma)R^{3\gamma}} - \frac{2p_\infty}{3\rho}. \tag{7.4.15}$$

Here the constant A should be taken from the initial value. If at the time of origin we have zero rate of bubble growth, then we must get from (7.4.15) $U = 0$ for $R = R_0$, thus

$$A = \frac{2p_\infty R_0^3}{3\rho} + \frac{2C}{3\rho(\gamma-1)R_0^{3(\gamma-1)}}. \tag{7.4.16}$$

Also, using the coefficients $B = \dfrac{2C}{3\rho(\gamma-1)}$ and $D = \dfrac{2p_\infty}{3\rho}$, we may conclude based upon (7.4.15) that during bubble oscillations its radius R is correlated with its growth rate $U = \dot{R}$ as

$$\frac{A}{R^3} - \frac{B}{R^{3\gamma}} - U^2 = D. \tag{7.4.17}$$

The equation (7.4.17) is the basic correlation for bubble dynamics in adiabatic conditions. We may establish the main parameters of a cavitating bubble from (7.4.17). First, let us find the maximum rate of bubble growth. Finding the extremum of the function $U(R)$

$$U = \pm \sqrt{\frac{A}{R^3} - \frac{B}{R^{3\gamma}} - D} \tag{7.4.18}$$

(where signs '+' and '–' correspond to the expanding and collapsing phases correspondingly), we get the value R_U where $\partial U/\partial R = 0$:

$$R_U = \left(\frac{\gamma B}{A} \right)^{\frac{1}{3(\gamma-1)}}. \tag{7.4.19}$$

In its turn, the maximum rate of growth (and collapse too, of course) $U_{max}(R_U)$ is

$$U_{max} = \sqrt{\left(\frac{\gamma B}{A^\gamma}\right)^{\frac{1}{1-\gamma}}\left(1-\gamma^{-1}\right)-D}.$$

(7.4.20)

The next problem is to find the analytical expression for the minimum and maximum radiuses R_{max} of a cavitating bubble. There may be two cases: (A) at the beginning, the external pressure is lower than the pressure inside the bubble, so the bubble expands and the initial radius R_0 is the minimum radius; (B) at $t = 0$, the external pressure is higher than the inner pressure, bubble collapses and R_0 is the maximum radius.

Let consider the case A (the bubble grows). This problem is native to the problem of cavitation, when vapor caverns expand due to a pressure drop. For this, we want to find the largest radius R_{max}; this value corresponds to zero growth rate (at the end of the expansion), i.e., it should be taken from the non-linear algebraic equation

$$\frac{A}{R^3}-\frac{B}{R^{3\gamma}}-D=0.$$

(7.4.21)

This equation has two roots: one of them is R_0, another one is the radius R_{max} of interest. Noting that since usually the second term is significantly smaller than the first one, we may find out the rough estimation of the radius

$$R_{max} = \left(\frac{A}{D}\right)^{\frac{1}{3}}.$$

(7.4.22)

However, this estimation is too rough. We will clarify it in the following way. Representing (7.4.21) in the form

$$\frac{1}{R^3}\left(A-\frac{B}{R^{3(\gamma-1)}}\right)-D=0,$$

(7.4.23)

we insert (7.4.22) instead of R into hooks. This iterative procedure (consisting of a single iteration) gives a better estimation

$$R_{max} = \left(\frac{A-B(D/A)^{\gamma-1}}{D}\right)^{\frac{1}{3}}.$$

(7.4.24)

The case B (the bubble collapses) is even easier. The minimal radius R_{min} must be obtained with the same equation (7.4.21). Assuming a significant collapse, i.e., a sufficiently small radius $R_{min} \ll R_0$, we may neglect the constant D in (7.4.21). Therefore, we have

$$R_{min} \approx \left(\frac{B}{A}\right)^{\frac{1}{3(\gamma-1)}}.$$

(7.4.25)

Note that, strictly, we need $R_{min}^3 \ll R_0^3$ which is easier to observe in a real situation.

Finally, we want to find the time period of the total 'growth–collapse' cycle. If the bubble expands, this quantity may be found as

$$\tau = 2\int_{R_0}^{R_{max}}\frac{dR}{U(R)},$$

(7.4.26)

where $U(R)$ must be taken from (7.4.18); factor '2' takes into account both phases: expanding and collapsing. For a collapsing bubble, we have a similar expression by replacing $R_0 \rightarrow R_{min}$, $R_{max} \rightarrow R_0$.

The expression (7.4.26) is very complicated and its integrating can be provided numerically, but it is always interesting and useful to have an approximate solution. At last, our aim here is not to obtain the precise correlation but rather get a reasonable estimation of basic parameters. Thus, we spent some time on the analytical evolution of (7.4.26).

First, note that usually the main contribution in (7.4.26) is conditioned by the stage of growth where the rate U is small (because $\tau \sim 1/U$), and, taking into account the dependence $U(R)$, where radius R is large. Then we take a closer look on the dependence (7.4.18) $U \sim \sqrt{f(R)}$. For the function $F(R)$, we may notice that the term $\sim R^{-3}$ is much larger than $R^{-3\gamma}$ for large R since $\gamma > 1$. Neglecting the constant D, we have the approximate correlation for the growth rate at a late stage of bubble dynamics:

$$U \approx \sqrt{\frac{A}{R^3}}. \tag{7.4.27}$$

Using (7.4.27), we get the period from (7.4.26), supposing $R_{max} \gg R_0$

$$\tau \approx \frac{4}{5}\sqrt{\frac{R_{max}^5}{A}}. \tag{7.4.28}$$

We should say that the expression (7.4.28) with its factor '4/5' is too ambitious. The dependence (7.4.27) is an approximation only for a part of the bubble trajectory; at the beginning of growth, this expression predicts an enormous growth rate and, as a result, it underestimates the contribution of the early stage to a common period. Thus, a more correct statement would be: the total time of the 'growth–collapse' cycle is

$$\tau = \chi\sqrt{\frac{R_{max}^5}{A}}, \tag{7.4.29}$$

where the parameter $\chi \sim 1$, that is, the equation (7.4.29) should be written as $\tau \sim \sqrt{R_{max}^5 / A}$. Indeed, the expression (7.2.29) claims to be no more than an estimate by an order of magnitude, and its only excuse is its simplicity. This correlation implies an extended phase of slow growth of a large bubble. In general, one should expect such situation for a great pressure difference Δp between the inner pressure at the starting point in time and the external pressure, i.e., for $p_0 \gg p_\infty$. If so, we may neglect the first term in the equation for A (7.4.16), writing it in a simplified form for (7.4.27) and (7.4.29) as

$$A \approx \frac{2p_0 R_0^3}{3\rho(\gamma - 1)}. \tag{7.4.30}$$

Now we have an exact solution for a cavitating bubble in adiabatic conditions as many analytical estimations of the key parameters. Then we need to 'feel' the physical quantities of the cavitation process.

7.4.3 Dynamics of a cavitating bubble: An example

To illustrate bubble dynamics, we consider a vapor bubble of initial radius $R(0) = R_0 = 10$ μm in water at pressure $p(0) = p_0 = 10^5$ Pa and $T(0) = 300$ K. Let the external pressure drop down suddenly to $p_\infty = 10^3$ Pa; we are interested in changes in the main parameters.

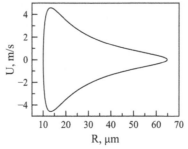

Fig. 7.4.1. Phase portrait of a cavitating bubble.

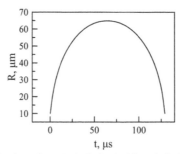

Fig. 7.4.2. The dependence R(t) obtained with analytical calculations.

The main result is the phase portrait in coordinates (R, U), calculated in accordance with (7.4.17), and shown in Fig. 7.4.1; from here we see the limit values: the bubble grows up to 65 μm, and the maximum growth rate is 4.6 m/s. As it was discussed above, the portrait represents a closed curve without a beginning and an end.

On the other hand, such a graph as Fig. 7.4.2 is not a usual interpretation; the dependence $R(t)$ is more convenient. We may obtain this graph with the first equation of (7.4.9), rewritten as

$$dt = \frac{dR}{U}, \tag{7.4.31}$$

(the same method as for the period (7.4.26)) where U is the function of R corresponding to (7.4.17). Thus, from (7.4.18) we have the dependence $t(R)$

$$t(R) = \int_{R_0}^{R} \frac{dR}{U(R)} = \sum_{i=0}^{N} \frac{\Delta R}{U(R_i)}, \tag{7.4.32}$$

if we use the simplest representation of integral with $i = 0$ corresponding to R_0 (radius at the moment $t = 0$) and $i = N$ corresponding to R (radius at the moment t). The step ΔR must be reasonably small;

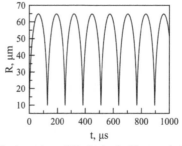

Fig. 7.4.3. The dependence R(t) obtained with numerical calculations.

of course, one may use a more complicated and more precise method to take the integral. The correlation between R and t obtained with (7.4.19) is shown in Fig. 7.4.2.

The analytical expression (7.4.20) gives the maximum growth rate U_{max} = 4.6 m/s at the bubble radius R_U = 13.3 μm. The largest radius calculated with (7.4.23) is R_{max} = 65.1 μm. Numerical calculations directly from the equation (7.4.32) give R_{max} = 64.8 μm. Thus, during oscillations in such conditions (initial inner pressure 10^5 Pa, outer pressure 10^3 Pa), the bubble grows by 6.5 times: the life of such a bubble in a liquid consists of the consequence of cycles presented in Fig. 7.4.2; the results on numerical simulations are presented in Fig. 7.4.3. As one may see, the time periods of oscillations coincide.

The rate of the bubble expanding (and collapsing) is shown in Fig. 7.4.4; as we see, acceleration is enormous at the early stage of bubble growth. Note that the acceleration $\dot{R} = U$ can be represented directly from (7.4.9) by replacing U^2 from (7.4.17):

$$\ddot{R} = \frac{C}{\rho R^{3\gamma+1}} + \frac{3B}{2R^{3\gamma+1}} - \frac{3A}{2R^4}. \tag{7.4.33}$$

Changes in temperature and pressure are the more interesting results concerning the physics of bubble dynamics. Both of these parameters can be calculated with Poisson equations for an adiabatic process:

$$p(t) = p_0 \left(\frac{R_0}{R(t)} \right)^{3\gamma}, \ T(t) = T_0 \left(\frac{R_0}{R(t)} \right)^{3\gamma-3}. \tag{7.4.34}$$

The results are shown in Fig. 7.4.5 and 7.4.6. As for pressure, one may depict that at the final stage of the expansion the pressure inside a bubble drops even lower than the external pressure 10^3 Pa: the bubble expands by inertia. The results for temperature are more interesting: as we know, the

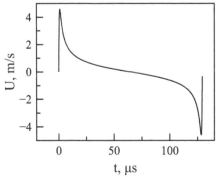

Fig. 7.4.4. The dependence of the bubble growth rate during a single oscillation period.

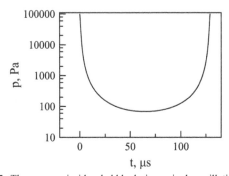

Fig. 7.4.5. The pressure inside a bubble during a single oscillation period.

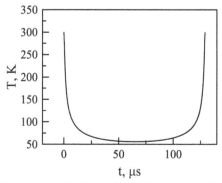

Fig. 7.4.6. The temperature inside a bubble during a single oscillation period.

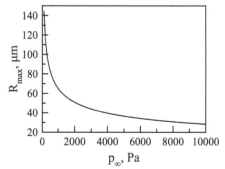

Fig. 7.4.7. The dependence of the maximum radius on the external pressure.

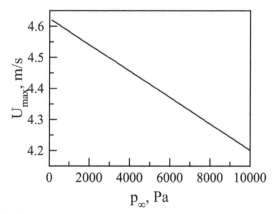

Fig. 7.4.8. The dependence of the maximum growth rate on the external pressure.

adiabatic expansion is a very effective cooling process and, as we see, the bubble expansion is not an exception. The obtained temperatures are extremely low; so, we cannot expect that the vapor mass inside a bubble remains constant.

Finally, we illustrate the dependence of key parameters–maximum radius R_{max}, maximum growth rate U_{max}, τ—on the pressure in the liquid. One may see that various parameters depend on the external pressure differently: the dependence $U_{max}(p_\infty)$ is almost linear unlike the function $R_{max}(p_\infty)$.

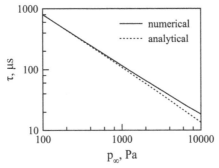

Fig. 7.4.9. The dependence of the oscillation period on the external pressure.

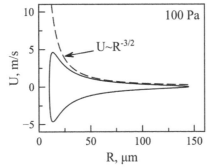

Fig. 7.4.10. Phase portrait for the external pressure 100 Pa.

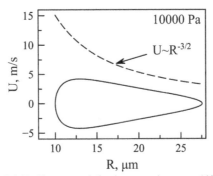

Fig. 7.4.11. Phase portrait for the external pressure 10^4 Pa.

The dependence of the oscillation period on the external pressure is shown in Fig. 7.4.9. Here 'numerical' means that the value is calculated directly from the integral (7.4.32), while the 'analytical' solution is the result of (7.4.29) at $\chi = 1.5$ (chosen for better agreement). It is easy to understand why the correlation (7.4.29) is adequate for $p_\infty = 100$ Pa, and deviates seriously for $p_\infty = 10^4$ Pa: in Fig. 7.4.10 and 7.4.11, the corresponding phase portraits are shown. For 100 Pa, the figure contains an easily recognizable part $U \sim R^{-3/2}$ (used to obtain (7.4.29)), while for 10^4 Pa it does not. It should be rather surprising that the correlation (7.4.29) gives an adequate estimation at all.

7.4.4 The generalized solution for arbitrary Δp(R)

Of course, the solution presented in previous sections is not universal. One can imagine an infinite number of various situations: different laws of the inner pressure $p(R)$, the Laplace jump at the bubble wall $2\sigma/R$, etc. Thus, we need a more common correlation $U(R)$ in a more scholastic form.

The solution presented above can be generalized; the common expression can be obtained in the same way. To not overwhelm the result, we present only the final correlation.

So, for the Rayleigh equation written in the form

$$\frac{3}{2}\dot{R}^2 + R\ddot{R} = \frac{\Delta p(R)}{\rho} \tag{7.4.35}$$

in initial conditions

$$R(0) = R_0, \ \dot{R}(0) = 0, \tag{7.4.36}$$

there exists the integral for $U = \dot{R}$

$$U^2 = \frac{2}{\rho R^3} \int_{R_0}^{R} R^2 \Delta p(R)dR. \tag{7.4.37}$$

The particular case for $\Delta p(R) = p_0 \left(\dfrac{R_0}{R}\right)^{3\gamma} - p_\infty$ was considered in Sections 7.4.3 and 7.4.4. Here we present the common solution that actually was presented high above—in Section 7.2. The correlation (7.4.37) directly represents the integral of the equation (7.2.9), which was derived during the derivation of the Rayleigh equation.

Finally, it should be noted that expression (7.4.37) is the common solution of the problem, even in a general case when we have $\Delta p(R,t)$; for instance, the explicit dependence on time is crucial for the problem of acoustic cavitation under an external sound field, where $p_\infty \sim \sin(\omega t)$. The equation (7.4.37) is correct anyway, because of its origin.

Of course, the solution in the integral form is too complicated to find closed expressions in a general case of $\Delta p(R,\dot{R},t)$, because to integrate such a function we cannot put $\dot{R} = const$ and $t = const$ in (7.4.37). Even in the simplest case for periodically changing pressure, the procedure to obtain a solution from (7.4.37) is very difficult because this integral contains dependence $R(t)$ in an implicit form (or, to be more precise, $t(R)$). However, in some cases one may use approximations of (7.4.37) assuming that the pressure is almost constant.

Indeed, if one cycle of bubble oscillations takes ~ 10 µs, then the external pressure varying with frequencies $\sim 10^4$ Hz or slower (i.e., with periods of ~ 100 µs or longer) can be approximately considered as a constant. Since the oscillation period depends on a magnitude of pressure, the correctness of the discussed approximation demands a complex analysis for each particular case. However, in general, the correlation (7.4.37) would be tried for acoustic cavitation: assuming $t = const$ in the term of acoustic pressure $p_a\sin(\omega t)$, one may use correlations obtained in this section.

7.4.5 Isothermal growth of a bubble

Why does a gas increase its temperature during compression? The first law of thermodynamics reads that the work δA done on the body is expended on changing the internal energy dU and leaks out as heat δQ:

$$\delta A = dU + \delta Q; \tag{7.4.38}$$

note that here δA and δQ have opposite signs compared to the 'usual' formulation when δA means the work of gas and δQ is the heat conducted to the system (see Appendix A).

In an adiabatic process, the energy is 'locked' in the system: $\delta Q = 0$ and all the work is spent to increase the internal energy of a system, i.e., the temperature of gas in our case. That is the reason why temperature grows in an adiabatic process.

However, the absence of heat exchange is not a unique situation. Probably, heat leaks from the system so effectively that all the energy gained from the work would go out due to heat exchange, i.e., $\delta A = \delta Q$ and $dU = 0$, that is, $T = const$. This is an isothermal case; actually, it needs special conditions, first of them is the low rate of the process (see Section 7.3): usually a system cannot lose heat intensively.

To be absolutely strict, the real physical conditions for a cavitating bubble are neither adiabatic nor isothermal. A gas (vapor) inside a bubble increases its internal energy because of the work of compression forces and loses its energy giving heat to the surrounding liquid.

In this section, we look closer at the condition $T = const$ inside a bubble; this approach is used from time to time in some works. This condition means that temperature at any point of gas remains constant; the isothermal condition contains requirement of two fast heat exchange processes: heat leaks from a bubble to a liquid and energy redistributes inside the bubble itself. Because heat transfers through a gas much slower than through a liquid, it is logical to assume that processes inside a bubble limit the total rate of the process. Only when the heat transfer is 'completed', the temperature of the system goes to equilibrium, and the problem satisfies the isothermal conditions.

On the contrary, adiabatic conditions take place for fast processes, when a system (a bubble in our case) 'has no time' for sufficient heat transfer. Heat exchange is too slow in comparison with the rate of warming (due to compression), so the temperature obeys the adiabatic law.

At first glance, for bubble oscillations one can expect that adiabatic conditions are fulfilled by default. Indeed, a typical time period of oscillations is ~ 10 μs or even $\sim 10^2$ μs (see Fig. 7.4.9); these are very short times—at first glance.

However, let us estimate the heat transfer of a cavitating bubble closely. Ignoring the phase transitions on the bubble wall, the slowest process that causes temperature equalization is the conductive heat transfer; any convection would speed up the process. The character time for molecular heat transfer τ is correlated with the spatial scale L as

$$\tau \sim \frac{L^2}{a},\tag{7.4.39}$$

where a is the thermal diffusivity. For gases $a \sim 10^{-5}$ m²/s; thus, for a bubble of radius (taken as the spatial scale) $\sim 10^{-5}$ m, we get the time scale $\tau \sim 10^{-5}$ s, i.e., the same value as the bubble oscillation period. Moreover, for instance, the full period for low pressures (Fig. 7.4.9) is much longer than this value. If so, adiabatic conditions for the bubble are questionable. Note that extremely low temperatures during the expansion phase are caused only by the assumption of the adiabatic character of the process.

Thus, the consideration of the isothermal conditions is not ambiguous at least. Let us return to the common form of the phase portrait (7.4.37). For an isothermal case, the pressure inside a bubble obeys the law $pR^3 = const$ (which follows from the Clapeyron equation for $T = const$); thereby, the pressure difference is

$$\Delta p = \frac{C}{R^3} - p_\infty.\tag{7.4.40}$$

Here $C = p_0 R_0^3$, unlike the adiabatic analogue from (7.4.6).
After integration, we have

$$U^2 = \frac{2C}{\rho R^3} \ln\left(\frac{R}{R_0}\right) - \frac{2p_\infty}{3\rho} + \frac{2p_\infty}{3\rho}\left(\frac{R_0}{R}\right)^3.\tag{7.4.41}$$

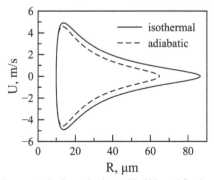

Fig. 7.4.12. Phase portraits for an isothermal bubble and for the adiabatic one.

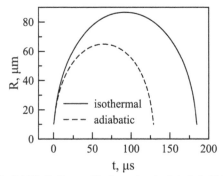

Fig. 7.4.13. Radiuses of isothermal and adiabatic bubbles.

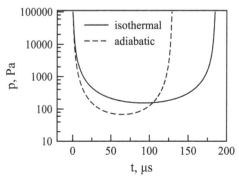

Fig. 7.4.14. Pressures inside isothermal and adiabatic bubbles.

Let us consider the example from the previous section, with $R_0 = 10$ μm, $p_0 = 10^5$ Pa, $p_\infty = 10^3$ Pa. The difference between two types of external conditions (to be exact, for internal conditions) is shown in Fig. 7.4.12.

In addition, in Fig. 7.4.13 we present the dependence of the bubble radius on time; again, one can see not qualitative but only the quantitative difference for adiabatic and isothermal conditions. For example, the maximum bubble radius differs by ~ 30%, the time period—about 40%.

The maximum rate of bubble growth also does not significantly differ: as we see from Fig. 7.4.12, for isothermal conditions the rate is higher by ~ 20%. The same conclusion follows from the dependence $p(t)$ for inner pressure (Fig. 7.4.14). Based on the close values of R, we may conclude that bubble volumes have close values too.

Thereby, we see that the condition $T = const$ gives close results for the adiabatic case except one important variable: the temperature itself. Adiabatic conditions lead to strange results for temperature, when the temperature drops down to cryogenic values while, for close dynamic characteristics such as R, $\dot{R}$, we do not observe any variation of T for the isothermal approach. This is a bad circumstance because we have a case when problem solving has strict dependence on the chosen theoretical model.

Because of the discrepancy of $\sim$ 20–40% in dynamic variables, we may conclude that an experiment has no good hints to resolve the problem: which approximation is correct—adiabatic or isothermal? Note, however, that analytical estimations (see above) give a more certain answer: a long period for isothermal oscillations does not contradict with the assumption that has been made, while the time period of adiabatic cycle is controversial with the initial suggestion about fast oscillations. The good news, at least, is that we have close results for both cases; it would be more interesting if we obtain short time scales at adiabatic assumption and long periods at the isothermal one.

We present the formula for key parameters of the cycle of an isothermal bubble.

The maximum radius of the growing bubble can be obtained by the 'single-iteration method' used above. We have

$$R_{max} \approx R_0 \left(\frac{p_0}{p_\infty} \ln \left(\frac{3(p_0 + p_\infty)}{p_\infty} \right) \right)^{\frac{1}{3}}. \tag{7.4.42}$$

For the considered example, this correlation gives 83 μm instead of 87 μm from numerical simulations: not the worst result for such a simple expression. Note that in our case $p_\infty \ll p_0$, that can be taken into account directly in (7.4.42).

At the radius

$$R_U = R_0 \exp \left(\frac{p_0 - p_\infty}{3p_0} \right) \tag{7.4.43}$$

(14 μm for our instance) the bubble has its maximum growth rate

$$U_{max} = \sqrt{\frac{2p_0}{3\rho} \left(\frac{R_0}{R_U} \right)^3 - \frac{2p_\infty}{3\rho}}. \tag{7.4.44}$$

The last expression gives 5 m/s; this value is very close to the growth rate of the adiabatic bubble (4.6 m/s).

The period may be found (to be exact, can be estimated) with the same method as in Section 7.4.3. Neglecting the constant in the function $U^2(R)$, we have

$$\tau \approx 2 \int_{R_0}^{R\,max} \frac{R^{3/2}\, dR}{\sqrt{Q\left(\ln R / R_0 + p_\infty / 3p_0 \right)}}, \tag{7.4.45}$$

where $Q = \dfrac{2p_0 R_0^3}{\rho}$. Then, note that all this approximation is more or less correct only for $p_\infty \ll p_0$ (see above); thereby, one may neglect this term in (7.4.45). Since the logarithm varies slowly, we assume that this term is a constant. Finally, we obtain the estimate for $R_{max} \gg R_0$

$$\tau \approx \chi \sqrt{\frac{R_{max}^5}{Q \ln \left(R_{max} / R_0 \right)}}. \tag{7.4.46}$$

Here we replace the multiplier again; $\chi \sim 1$. For the example considered in this section, using $\chi = 1.5$ as before, we obtain $\tau \approx 153$ μs from (7.4.46) while exact calculations give 185 μs. Notice once more that the correlation (7.4.46) only gives an estimation of the parameter τ.

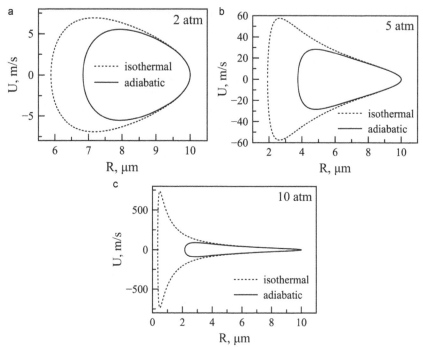

Fig. 7.4.15. Phase portraits for a collapsing bubble at different external pressures.

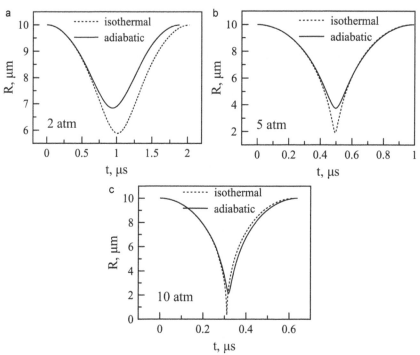

Fig. 7.4.16. Bubble radius of the collapsing bubble at different external pressures.

7.4.6 The bubble collapse

All previous considerations concern the case of pure cavitation: at the moment $t = 0$, the external pressure p_∞ is lower (much lower) than the inner pressure p_0 inside a bubble, so the bubble tends to grow. After a certain time, the expanding phase drags and the collapsing stage begins.

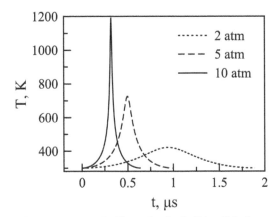

Fig. 7.4.17. Temperature inside a collapsing bubble; adiabatic approach.

Here we consider a 'reverse' problem: let us discus the bubble collapse, i.e., the case when $p_\infty > p_0$. We will consider both assumptions: adiabatic approximation and the isothermal one. Using correlations (7.4.15) and (7.4.44), we obtain phase portraits of corresponding systems; dependencies $R(t)$ can be obtained in the same way as described above.

We present the calculations for the initial bubble radius 10 μm and the inner pressure 10^5 Pa. The outer pressure varies: we consider pressures 2, 5 and 10 atmospheres.

First, we see that functions $R(t)$ are the same as in Fig. 7.4.2, 7.4.13 but 'transposed': in this case, for high external pressures, the points at maximum radius correspond to $R = R_0$. Another coincidence with previous results: we may notice that both the bubble radius R and the oscillation period τ are very close for isothermal and adiabatic conditions.

Then we emphasize important quantitative and qualitative differences. The quantitative distinction is the rate of bubble collapse: we may see velocities of ~ 10 m/s or even of ~ 10^2 m/s, exceeding the speed of sound. We did not obtain such huge values in previous calculations for a cavitating bubble.

Periods of such oscillations are about 1 μs (change from ~ 2 μs at 2 atm to ~ 0.6 μs at 10 atm). This is a crucial qualitative difference in comparison with a 'pure cavitation case' considered above: at such high frequencies, the heat transfer does not have time to happen. Thus, the adiabatic condition is an undisputable approximation for bubble oscillations at high external pressure.

We may also calculate temperature dynamics in an adiabatic case, the result is shown in Fig. 7.4.17. For external pressure 10 atm, the temperature peak reaches 1200 K; at oscillations for pressures ~ 1 atm, temperature remains at room values.

7.4.7 The pressure around a collapsing bubble of vapor

These results were obtained by Rayleigh himself in his pioneering work (Rayleigh 1917).

Let us consider the constant pressure difference $\Delta p = p_{in} - p_\infty$; as it was discussed in Section 7.3, this case may suit a vapor bubble surrounded by a liquid at constant temperature, i.e., $p_{in} = p_s(T)$. Then, for $\Delta p < 0$ one should expect the total bubble collapse: because the pressure inside a bubble remains the same, independent from the bubble radius, finally there must be $R \to 0$.

To avoid misunderstandings, further we replace $\Delta p \to -\Delta p$, that is, further everywhere in this section $\Delta p = p_\infty - p_{in} > 0$.

For the case of constant pressure, from (7.4.37) we have

$$U^2 = \frac{2\Delta p}{3\rho}\left(\frac{R_0^3}{R^3} - 1\right), \tag{7.4.47}$$

and for collapse $U < 0$, i.e.,

$$U = -\sqrt{\frac{2\Delta p}{3\rho}\left(\frac{R_0^3}{R^3}-1\right)}. \tag{7.4.48}$$

From (7.4.48) we have for the time of bubble collapse t_c, corresponding to $R = 0$

$$\int_0^{t_c} dt = \int_0^{R_0}\left[\frac{2\Delta p}{3\rho}\left(\frac{R_0^3}{R^3}-1\right)\right]^{-1/2} dR. \tag{7.4.49}$$

Introducing a new variable $z = R/R_0$, we get

$$t_c = R_0\sqrt{\frac{3\rho}{2\Delta p}}\underbrace{\int_0^1 \frac{z^{3/2}dz}{\sqrt{1-z^3}}}_{\dfrac{\sqrt{3}\Gamma(2/3)\Gamma(5/6)}{2\sqrt{\pi}}} = R_0\sqrt{\frac{9\rho}{8\pi\Delta p}}\Gamma\left(\frac{2}{3}\right)\Gamma\left(\frac{5}{6}\right) \approx 0.915 R_0\sqrt{\frac{\rho}{\Delta p}}. \tag{7.4.50}$$

For instance, a bubble of radius $R = 10$ μm in a liquid with density $\rho = 10^3$ kg/m³ completely collapses under pressure $\Delta p = 10^5$ Pa at time ~ 1 μs.

The pressure field in a liquid can be found with a simplified Navier–Stokes equation

$$\frac{\partial u}{\partial t} + u\frac{\partial u}{\partial r} = -\frac{1}{\rho}\frac{\partial p}{\partial r}, \tag{7.4.51}$$

that is, neglecting the viscous terms. For an incompressible fluid we have (7.2.5), and

$$\frac{\partial u}{\partial t} = \frac{\dot{U}R^2}{r^2} + \frac{2U^2 R}{r^2}, \quad \frac{\partial u}{\partial r} = -2U\frac{R^2}{r^3}. \tag{7.4.52}$$

For $\dot{U}$ we have from (7.4.6):

$$\dot{U} = -\frac{\Delta p}{\rho}\frac{R_0^3}{R^4}. \tag{7.4.53}$$

Thereby, from (7.4.51) with (7.4.52) and (7.4.53)

$$\frac{\partial p}{\partial r} = \Delta p\frac{R_0^3}{R^2}\frac{1}{r^2} - \frac{4\Delta p R}{3}\left(\frac{R_0^3}{R^3}-1\right)\frac{1}{r^2} + \frac{4\Delta p}{3}\left(\frac{R_0^3}{R^3}-1\right)\frac{R^4}{r^5}. \tag{7.4.54}$$

Taking this integral from $r = \infty$, where $p = p_\infty$, we get

$$p(r,t) = p_\infty + \frac{\beta_1(t)}{r} - \frac{\beta_2(t)}{r^4}, \tag{7.4.55}$$

where

$$\beta_1(t) = \frac{\Delta p R_0^3}{3R(t)^2} - \frac{4\Delta p R(t)}{3}, \quad \beta_2 = \frac{\Delta p R^4(t)}{3}\left(\frac{R_0^3}{R^3(t)}-1\right). \tag{7.4.56}$$

Of course, at any moment of time $p(R,t) = p_\infty - \Delta p$, as it must be. At the initial moment of time $\beta_2 = 0$ and

$$p(r,0) = p_\infty - \Delta p \frac{R_0}{r}. \tag{7.4.57}$$

As follows from (7.4.55), the pressure may reach a maximum

$$p_m = p_\infty + \frac{3}{4} \frac{\beta_1}{r_m} \tag{7.4.58}$$

at the point $r = r_m$

$$r_m = \sqrt[3]{\frac{4\beta_2}{\beta_1}}, \text{ or } \frac{r_m}{R} = \sqrt[3]{\frac{4\left(R_0^3 - R^3\right)}{R_0^3 - 4R^3}}. \tag{7.4.59}$$

Until the bubble is sufficiently large, i.e., the denominator under the root in (7.4.58) is negative, values r_m have no physical sense (monotonic dependence (7.4.58) illustrates this fact). The first instance, when a maximum of pressure really exists and coincides with $r_m = \infty$, corresponds to the bubble radius $R = R_0/\sqrt[3]{4}$. During the collapse, when $R \to 0$, r_m decreases and tends to zero as $r_m = \sqrt[3]{4}\,R$.

With respect to (7.4.59), the relation (7.4.58) may be expressed as

$$p_m = p_\infty + \Delta p \frac{R_0^3 - R^3}{4R^3} \sqrt[3]{\frac{R_0^3 - 4R^3}{4\left(R_0^3 - R^3\right)}}. \tag{7.4.60}$$

From (7.4.60), one may see that $p_m \to \infty$ at $R \to 0$ as

$$p_m \approx p_\infty + \Delta p \frac{R_0^3}{4^{4/3} R^3}. \tag{7.4.61}$$

In simple words, hydraulic shock appears in vicinity of a collapsing bubble. Due to $p_m \to \infty$, there may be very interesting processes near a collapsing bubble of such type. We will return to this matter in Chapter 9.

7.4.8 The effect of surface tension

Among various terms for pressure in the Rayleigh equation (or in its integral representation), surface tension is a nuance. With its term, the pressure difference is

$$\Delta p = p_{in} - \frac{2\sigma}{R} - p_\infty. \tag{7.4.62}$$

First, we want to estimate the second term on the right side of (7.4.62). For water at room temperature, the surface tension is ~ 70 mN/m. Consequently, the bubble radius must be about ~ 1 μm to make this term of an order of external pressure $p_\infty \sim 1$ atm. Based on this estimation, we may conclude that the surface tension term does not play a role during the expansion phase, but may be important for the bubble collapse.

Indeed, the equation (7.4.62) gives little more in addition to previous consideration. The equation (7.4.37) can be used with (7.4.62); for instance, for an adiabatic case we have

$$\Delta p = \frac{p_0 R_0^{3\gamma}}{R^{3\gamma}} - \frac{2\sigma}{R} - p_\infty, \tag{7.4.63}$$

and with (7.4.37), we have

$$U^2 = \frac{\alpha_1}{R^3} - \frac{\alpha_2}{R} - \frac{\alpha_3}{R^{3\gamma}} - \alpha_4, \tag{7.4.64}$$

where

$$\alpha_1 = \frac{2p_0 R_0^3}{3\rho(\gamma-1)} + \frac{2p_\infty R_0^3}{3\rho} + \frac{2\sigma R_0^2}{\rho}, \ \alpha_2 = \frac{2\sigma}{\rho}, \ \alpha_3 = \frac{2p_0 R_0^{3\gamma}}{3\rho(\gamma-1)}, \alpha_4 = \frac{2p_\infty}{3\rho}. \tag{7.4.65}$$

Of course, these coefficients differ from their analogues obtained above by one term in α_1 and α_2 itself.

7.4.9 The pressure around an oscillating bubble of gas

In this section, we combine all the results from previous sections and find the expression for the pressure around a collapsing bubble filled with gas. As it was shown above, for this case surface tension must be taken into account, that is, the connection between the growth (or collapse) rate U and the bubble radius R is defined by (7.4.64) for the adiabatic case. This is the single difference in comparison with the derivation given in the previous section: instead of (7.4.53), we have

$$\dot{U} = -\frac{3\alpha_1}{2R^4} + \frac{\alpha_2}{2R^2} + \frac{\alpha_3(3\gamma+1)}{2R^{3\gamma+1}}. \tag{7.4.66}$$

Then, (7.4.66) must be used in (7.4.52) and, finally, in (7.4.51) to integrate it. As a result, we obtain the relation of the same structure as (7.4.55):

$$p(r,t) = p_\infty + \frac{\xi_1}{r} - \frac{\xi_2}{r^4}, \tag{7.4.67}$$

with

$$\xi_1 = \frac{\rho\alpha_1}{2R^2} - \frac{3\rho\alpha_2}{2} + \frac{\rho\alpha_3(3\gamma-4)}{2R^{3\gamma-1}} - 2\rho\alpha_4 R, \tag{7.4.68}$$

$$\xi_2 = \frac{\rho\alpha_1 R}{2} - \frac{\rho\alpha_2 R^3}{2} - \frac{\rho\alpha_3 R^{4-3\gamma}}{2} - \frac{\rho\alpha_4 R^4}{2}. \tag{7.4.69}$$

For $r = R$, we see that the boundary condition is fulfilled:

$$p(R,t) = p_\infty - \frac{\rho\alpha_2}{2R} + \frac{3(\gamma-1)\rho\alpha_3}{2R^{3\gamma}} - \frac{3\rho\alpha_4}{2} = \frac{p_0 R_0^{3\gamma}}{R^{3\gamma}} - \frac{2\sigma}{R}. \tag{7.4.70}$$

This dependence also has a maximum at the point

$$r_m = \sqrt[3]{\frac{4\xi_2}{\xi_1}}, \tag{7.4.71}$$

but this maximum is not as sharp as the one for the 'classic' case $\Delta p = const$ considered above, where $R \to 0$ monotonically. For the case considered in this section, pressure oscillates as well as the bubble radius. Note that for a very small bubble, we have

$$\xi_1 \approx \frac{\rho\alpha_3(3\gamma-4)}{2R^{3\gamma-1}}, \ \xi_2 \approx -\frac{\rho\alpha_3 R^{4-3\gamma}}{2}, \tag{7.4.72}$$

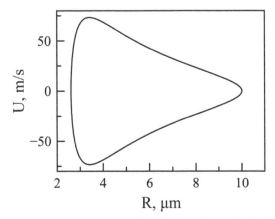

Fig. 7.4.18. Phase portrait for (7.4.64): adiabatic collapse, surface tension is taken into account.

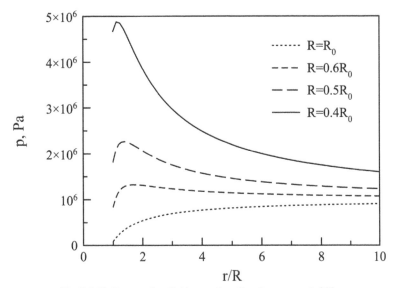

Fig. 7.4.19. Pressure in a fluid around a collapsing gaseous bubble.

and the dependence $p(r)$ is monotonic. It is possible, however, that a bubble will not collapse to a size this small—that depends on the pressure difference.

Further, we illustrate the bubble dynamics of a bubble of initial radius $R_0 = 10$ μm at initial pressure $p = 10^5$ Pa; the external pressure $p_\infty = 10^6$ Pa. The phase portrait for this system is shown in Fig. 7.4.18; it is very similar to the one in Fig. 7.4.15 (despite the fact that the adiabatic curve cannot be seen there well): this means that the surface tension, considered here, does not play an important role.

Distributions of pressure in the surrounding liquid are shown in Fig. 7.4.19 for different bubble radiuses. We did not display there the monotonic curve corresponding to the minimal bubble size of $0.26\,R_0$ (see Fig. 7.4.18). We see that dependences $p(r)$ are non-monotonic for intermediate values of bubble radius.

We may note that these results can also be applied for a bubble containing either a gas or a vapor by replacing γ in all expressions with $(\gamma - 1)/\omega$, as it follows from (7.3.20).

7.4.10 The Bjerknes effect: Explanation

The Bjerknes effect is the attraction of gaseous bubbles in a liquid, which was described and illustrated in Chapter 6, where we show how two pulsating bubbles attracted to each other. Actually, this may not be necessarily an attraction, as we presented there—it can also be a repulsion; the result depends on the oscillation phases. In this section, we give a common theoretical explanation of the Bjerknes effects.

The name of Bjerknes (father and son, who discovered the effect and explained it in the XIX century) now is connected with several effects. Usually, the primary and secondary Bjerknes forces are mentioned (Mettin et al. 1997, Zhang et al. 2016, Crum 1975, Leighton et al. 1990), but the explanation of these effects also leads to such matters as interaction of pulsating bubbles with solid walls and free surfaces (Birkhoff and Zarantonello 1957, Lavrent'ev and Shabat 1977).

The primary Bjerknes force—commonly, an analogue of the Archimed force—can be explained as follows. Let a small air bubble have a volume V and be located in a field of pressure p; we may consider it as a solid particle if we neglect the deviation of its shape; this is, in fact, a tiny point of many such theoretical models in general. Thus, the force acting on this bubble can be found as

$$F = -\oint p\,dS = -\int \nabla p\,dV \approx -V\nabla p. \tag{7.4.73}$$

The last equality is correct if the pressure gradient can be assumed constant, which is correct for sufficiently small bubbles. For the true Archimed force, the situation is simpler, since for that case $\nabla p = -\rho g$.

To describe the second Bjerknes force, we consider two bubbles of radiuses $R_1(t)$ and $R_2(t)$ at a distance L (see Fig. 7.4.1). As we discussed above, for each bubble the velocity distribution around it is

$$u = \dot{R}(t)\frac{R^2}{r^2}. \tag{7.4.74}$$

and, consequently, the pressure gradient of the pressure field generated by this very bubble is

$$\frac{\partial p}{\partial r} = -\rho\frac{\partial u}{\partial t} = -\frac{\rho}{r^2}\frac{d}{dt}\dot{R}R^2, \tag{7.4.75}$$

where ρ is the density of a liquid; we considered all these equations in previous sections in details.

Thus, to calculate the force acting on bubble number 2 from bubble number 1, we must substitute the pressure gradient from (7.4.75) for the equation for the primary Bjerknes force (7.4.73); we get

$$F = -V_2\frac{\partial p_1}{\partial r} = V_2\frac{\rho}{L^2}\frac{d}{dt}\dot{R}_1 R_1^2, \tag{7.4.76}$$

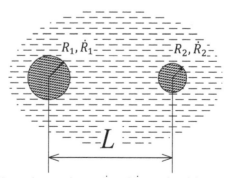

Fig. 7.4.20. Two oscillating bubbles; at the same instant, $\dot{R}_1$ and $\dot{R}_2$ may be of the same time, or of the opposite one: as a result, they would attract or repel each other.

which can be represented through the volume of the first particle $V_1 = \dfrac{4\pi R_1^3}{3}$ as

$$F = \frac{\rho}{4\pi L^2} V_2 \frac{d^2 V_1}{dt^2}. \tag{7.4.77}$$

Deriving this equation, we, indeed, neglect the deviation of the pressure gradient at the size of the bubble (of the second bubble): we calculate $\partial p/\partial r$ at the average distance L. This means that we assume that radiuses of bubbles are much smaller than the distance between them.

Now consider oscillating bubbles, i.e., when both volumes depend on time according to some periodic law with period τ. The average force in this case is

$$\overline{F} = \frac{1}{\tau} \int_0^\tau F dt = \frac{1}{\tau} \underbrace{[\dot{V}_1(\tau) V_2(\tau) - \dot{V}_1(0) V_2(0)]}_{0} - \frac{\rho}{4\pi L^2} \overline{\dot{V}_1 \dot{V}_2}. \tag{7.4.78}$$

The first term vanishes because the period τ, obviously, corresponds to $\dot{V}_1(0) = \dot{V}_1(\tau)$, $V_1(0) = V_2(\tau)$.

Thus, the average force depends on the correlation of the deviation of bubble volumes. If the volumes change synchronously, that is, when $\dot{V}_1 > 0$, $\dot{V}_2 > 0$ and vice versa, and, therefore, $\dot{V}_1 \dot{V}_2 > 0$, then the average force has a negative sign, bubbles are attracted to each other. Otherwise, when the phases of bubble oscillations are opposite, $\dot{V}_1 \dot{V}_2 < 0$ (when the first bubble grows, the second one diminishes), bubbles repulse. In an external ultrasound field, neighboring bubbles oscillate synchronously; thus, we observed their attractions.

The same result can also be obtained with using the language of hydrodynamic potentials (Birkhoff and Zarantonello 1957, Lavrent'ev and Shabat 1977) (see Chapter 4), which define the velocities with the relation

$$\vec{u} = \nabla \varphi. \tag{7.4.79}$$

In brief, the method of potentials is an analogue of electrostatics, where the electric field can be described by the sum of potentials corresponding to the sources of electricity (charges or dipoles). By analogy, the velocity field in a liquid around two bubbles can be represented as a sum of potentials of a kind $\sim qr^{-1}$ (here the parameter q is an analogue of the electric charge, as we see), which gives (7.4.1) and, from there, the value of the parameter q. Consideration of these potentials also leads to attraction for bubbles oscillating in the same phase, that is, with the same signs of parameters q_1 and q_2.

Moreover, a similar description gives results concerning the interaction of oscillating bubbles with walls and a free surface. Like in electrostatics, we may describe a field of a point source near a plane with certain boundary conditions on it with the second 'imaginary' point source.

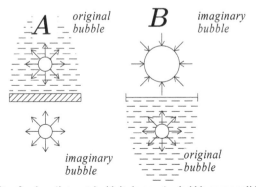

Fig. 7.4.21. A bubble in vicinity of a plane 'interacts' with its image. A—bubble near a solid surface interacts with the same 'imaginary' bubble, which oscillates in the same phase. B—bubble at the free surface interacts with the 'image', which oscillates in an antiphase.

For a plane with the condition $\varphi = const$ on it corresponding to a stationary free surface (along a free surface, velocity obeys the equation $u^2 = u_n^2 + u_\tau^2 = (\partial\varphi/\partial n)^2 + (\partial\varphi/\partial\tau)^2 = const$ because pressure is a constant there, therefore for a stationary surface $u = 0$ and $\varphi = const$), the field may be described with the second 'imaginary charge' of the opposite sign, that is, with a bubble oscillating in an antiphase with the first bubble; see Fig. 7.4.21.

For a plane with the condition $\partial\varphi/\partial n = 0$ corresponding to a solid wall (where the normal velocity $u_n = \partial\varphi/\partial n$ is zero because of the condition of impermeability), the second 'charge' has the same sign (see Fig. 7.4.21), which means that both bubbles oscillate in the same phase.

As we have seen above, bubbles that oscillate synchronously (i.e., have the same charges q) are attracted to each other; this condition spreads on the case of a real bubble and the imaginary one. Thus, due to Bjerknes forces an oscillating bubble would be attracted to a solid surface and would repel from the free surface of a fluid. On the other hand, the external flow may strongly distort such a simple behavior pattern.

One may think that the full explanation of the Bjerlkes effect is given above. This is not the case:

- ✓ the condition $R << L$ was used above; if not, the problem would have special features up to the qualitative difference (Doinikov and Zavtrak 1995);
- ✓ we did not consider nonlinear effects (coupling of bubble oscillations), which also give some additional nuances (Mettin et al. 1997) or quite a different result (Oquz and Prosperetti 1990);
- ✓ one may consider differently under the complicated form of an external force (Zhang et al. 2016);
- ✓ and many other issues which, generally, can be taken into account.

Meanwhile, we see that the theory presented in this section affirms and explains the experimental results presented in Chapter 6. Thus, despite the fact that many tiny features can be considered, as in many problems, surely, the simple approach explains the majority of the experimental data. It should not be forgotten when we consider complicated models.

7.4.11 Addition: Notes about stable points on a plane

Above, in Section 7.4.2 we briefly stated that the point of a 'center' type is a non-stable point. The discussion about stability of various types of stationary points was ambiguous among the explanation of the bubble physics, but this question is an interesting topic with some nuances. We feel that this matter needs a more detailed explanation, and here we provide it.

An autonomous system of two non-linear differential equations (i.e., equations that do not contain an explicit dependence on time) can be written in a common form

$$\dot{X} = F_X(X,Y)$$
$$\dot{Y} = F_Y(X,Y) \tag{7.4.80}$$

This system may have a stationary point where $\dot{X} = 0$, $\dot{Y} = 0$; this point satisfies the system of algebraic equations

$$F_X(X_0,Y_0) = 0$$
$$F_Y(X_0,Y_0) = 0 \tag{7.4.81}$$

The stability of the stationary point (X_0, Y_0) means that any disturbance from that point tends to decay, i.e., any deviation from this point vanishes. Mathematically, if we consider an initial point $(X_0 + u_x, Y_0 + u_y)$ in vicinity of the point (X_0, Y_0), then $u_x(t) \to 0$, $u_y(t) \to 0$ for a stable point (X_0, Y_0). One may investigate the stability of this point in the following way.

Since X_0, $Y_0 = const$, we have

$$\dot{u}_x = F_X \left(X_0 + u_x, Y_0 + u_y \right)$$
$$\dot{u}_y = F_Y \left(X_0 + u_x, Y_0 + u_y \right) \tag{7.4.82}$$

Assuming small deviations from the point (X_0, Y_0), one may expand functions in (7.4.82) into series, obtaining

$$\dot{u}_x = F_{XX} u_x + F_{XY} u_y$$
$$\dot{u}_y = F_{YX} u_x + F_{YY} u_y \tag{7.4.83}$$

Here $F_{XY} = \partial F_X / \partial Y$, etc.

One may try to find a solution of the system of linear differential equations (7.4.75) in the form

$$u_x = U_x e^{\lambda t}, u_y = U_y e^{\lambda t}, \tag{7.4.84}$$

and magnitudes must follow from the system of equations

$$U_x \lambda = F_{XX} U_x + F_{XY} U_y$$
$$U_y \lambda = F_{YX} U_x + F_{YY} U_y \tag{7.4.85}$$

which has nontrivial solutions only if the determinant of this system

$$\begin{vmatrix} F_{XX} - \lambda & F_{XY} \\ F_{YX} & F_{YY} - \lambda \end{vmatrix} = 0. \tag{7.4.86}$$

The corresponding quadratic equation, in a general case, has two roots (eigenvalues λ_1 and λ_2); thus, we finally have the general solution of (7.4.83):

$$u_x = U_{x1} e^{\lambda_1 t} + U_{x2} e^{\lambda_2 t}, u_y = U_{y1} e^{\lambda_1 t} + U_{y2} e^{\lambda_2 t}. \tag{7.4.87}$$

If both λ_1 and λ_2 are negative, the point (X_0, Y_0), called a 'stable node', is stable, just like in the case when $\lambda_{1,2} = a \pm ib$ with $a < 0$ (called a 'stable spiral point'). Any initial deviation from (X_0, Y_0) will fade in time.

Points with $\lambda_1, \lambda_2 > 0$ ('unstable node') or with $\lambda_{1,2} = a \pm ib$ at $a < 0$ ('unstable spiral point') are unstable (as follows from their names): any divergence from these points will grow with time.

The case $\lambda_1 < 0$ and $\lambda_2 < 0$ corresponds to an unstable point (a 'saddle'): deviation will increase along the components (U_{x2}, U_{y2}).

The most interesting case is a 'center' or an 'elliptic point': when $\lambda_{1,2} 0 = \pm ib$. For such eigenvalues, we see that initial deviations from the point (X_0, Y_0) would neither fade nor increase. Thus, we may think that the phase portrait in vicinity of an elliptic point looks like in Fig. 7.4.22.

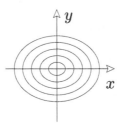

Fig. 7.4.22. An elliptic point.

For the purpose of stability analysis, one may conclude that this point is neither stable nor unstable; sometimes this is called a 'neutral stability'.

However, this case is more complicated. In accordance with the Grobman–Hartman theorem, linearization is correct only in absence of zero real part of eigenvalues of (7.4.86), otherwise the result is unpredictable: this may be a point with 'neutral stability' in its vicinity, but it may be something different.

An excellent example was given in (Tabor 1989) after Bender and Orszag. The system

$$\dot{X} = -Y + X\left(X^2 + Y^2\right)$$
$$\dot{Y} = X + Y\left(X^2 + Y^2\right) \tag{7.4.88}$$

has an elliptic point $(0, 0)$. However, this prediction based on the linear theory is wrong. Multiplying the first equation by X and the second equation by Y, summing up the equations we get

$$X\dot{X} + Y\dot{Y} = \left(X^2 + Y^2\right)^2. \tag{7.4.89}$$

Introducing the variable $R^2 = X^2 + Y^2$, we have the equation

$$\frac{1}{2}\frac{d\left(R^2\right)}{dt} = R^4, \tag{7.4.90}$$

and indeed

$$R^2(t) = \frac{R^2(0)}{1 - 2R^2(0)t}. \tag{7.4.91}$$

In other words, $R^2(t) \to \infty$ as $t \to 1/2R^2(0)$. This is not an elliptic point, this is an unstable spiral point.

One must be careful if the matrix has an eigenvector with zero real part.

7.5 Numerical solutions

7.5.1 *Non-autonomous systems*

In theory, the analytic integral obtained in the previous section can be applied for any case, but this method does not bring a real bargain for any case in comparison with a direct numerical solution of the Rayleigh equation.

One of such overcomplicated problems is cavitation under an external ultrasonic field. In such case, the external pressure contains the term

$$p_u = P_u \sin \omega t, \tag{7.5.1}$$

describing the acoustic pressure.

Acoustic cavitation was described in previous section from the experimental point of view. The theoretical analysis of this process is usually provided with the Rayleigh equation, see works (Gaitan et al. 1992, Prosperetty et al. 1984, Lauterborn 1976). Many of these papers concern the sonoluminescence phenomenon—the light emission from a cavitating bubble. One of the most popular explanations of this problem is the thermal hypothesis: during cavitation, a bubble collapses so intensively that temperature inside it increases to enormous values—thousands, tens of thousands of kelvins or even more. Thus, the heated gas emits thermal radiation, which is registered as sonoluminescence.

Again, analytical calculations based on the expression (7.4.28) with pressure explicitly depending on time in the form of (7.5.1) are too complicated. Numerical simulations can give quantitative results much easily, and we will use them in the next section. But we can explain qualitative issues of the cavitation process right here.

Let us begin our analysis from the expansion phase; thus, the pressure difference is written in the form

$$\Delta p = p_{in} + p_u \sin \omega t - p_\infty \qquad (7.5.2)$$

Here p_v is the vapor pressure inside a bubble, p_∞ is the external pressure except for the ultrasonic term described with (7.5.1); p_∞ includes the hydraulic pressure of the liquid column (usually negligible, because in laboratory experiments cavitation is examined on depths not deeper than $\sim$ 10 cm, which gives pressure of $\sim$ 1000 Pa) and the atmospheric pressure above the free surface of a liquid (usually about $\sim 10^5$ Pa). The frequency of 'ultrasound' exceeds 22 kHz, but usually in experiments this value may be inside the 'acoustic' range $\sim$ 20 kHz or less. The magnitude of acoustic pressure of an interest is great—of an order of $\sim$ 1 atmosphere at least.

At the beginning, the ultrasound pressure stretches a liquid: $p_u > 0$, and the bubble will grow unstoppably. The inner pressure p_v decreases, but the cause of the expansion is not only the difference $(p_v - p_\infty)$, as in problems from Section 7.4; the real cause of expansion is the stretching tension p_u. The expansion phase cannot stop until the acoustic pressure lets it do so. The expansion lasts until time $\sim \pi/2\omega$; after that time, when the acoustic pressure begins to diminish, other variants are possible.

It is interesting to analyze the temporal characteristics of ultrasonic cavitation. Frequency $\sim$ 20 kHz corresponds to the period of acoustic oscillation $\sim$ 50 μs. This value corresponds to proper frequencies of bubble oscillations during cavitation at low external pressures and exceeds the periods of oscillations when the external pressure is compared with the starting inner pressure (see Section 7.4). Thus, one should observe proper oscillations caused by the change in the internal pressure in a bubble at the last stage of a single ultrasonic period, when $p_u < 0$, i.e., the external ultrasonic pressure tends to collapse the bubble.

Finally, it may be predicted that oscillation under conditions (7.5.2) consists of three phases:

(1) intensive growth when the ultrasound pressure stretches a bubble;

(2) monotonic collapse after the term $p_u - p_\infty$ changes its sign to '−'; this does not happen immediately after this term becomes negative, of course, due to inertia;

(3) oscillations caused by deviation of pressure inside a bubble because of its volume change.

To provide numerical simulation of the Rayleigh equation, we have to assume a theoretical model of function $p_v(R)$: it can be the adiabatic law $p_v \sim R^{-3\gamma}$, or the isothermal one $p_v \sim R^{-3}$ or something else. An intermediate way may possibly look preferable: the isothermal model for slow phases and adiabatic description for fast intensive processes. However, such a mixed model looks overcomplicated: as we have seen above, the difference between two opposite models is not crucial. Unless we do not try to obtain the exact data for a certain problem but rather want to understand the general physical picture of the phenomenon, there should be a united correlation. In the following section, we will use the adiabatic model, expecting fast short-period oscillations of a bubble.

In this section, we use the Raylight-Plesset equation in the Prospretti formulation (Prospretti 1984)

$$\left(1 - \frac{\dot{R}}{c_s}\right) R\ddot{R} + \frac{3}{2}\dot{R}^2 \left(1 - \frac{\dot{R}}{3c_s}\right) = \left(1 + \frac{\dot{R}}{c_s}\right)\frac{1}{\rho_l}\left[p_B - p_u\left(t + \frac{R}{c_s}\right) - p_\infty\right] + \frac{R}{\rho_l c_s}\frac{dp_B}{dt}. \quad (7.5.3)$$

where $p_B = p_{in} - \frac{2\sigma}{R} - 4\mu\frac{\dot{R}}{R}$.

This form takes compressibility into account, which we can see due to the presence of the sound speed c_s and pressure derivation. We chose this form of equation because it was used in, possibly, one of the most widely known works (Gaitan et al. 1992). We also took the parameters of the numerical experiment from it. We will provide details of numerical modeling in the next subsection, but now we will discuss what interesting things can happen under the influence of an ultrasonic field which can cause the abovementioned temperature increase.

7.5.2 Oscillations in an ultrasonic field

At first glance, it seems that nothing special happens under the influence of an external pressure field. Let us say we want to find an amplitude that creates the temperature inside a bubble about thousands K—ten times of the initial one, for example. Then with (7.3.3) we have

$$p^{1-\gamma}T^\gamma = p_0^{1-\gamma}T_0^\gamma. \tag{7.5.4}$$

It means that for an air bubble ($\gamma = 1.4$), for such growth, the external pressure should increase by $\sim 3\,000$ times; for temperature growth by 5 times, it should increase by 300 times. However, in fact, such an increase can be achieved using the amplitude of ultrasonic vibrations of only half an atmosphere (if the right vibration frequency is used).

Previously, we found the dependence of the oscillation period (7.4.29). It is obvious that the unison work (with a frequency corresponding to the oscillation period) of the external pressure (7.5.2) will increase the oscillation amplitude. The estimation mentioned above correlates with a widely known resonant frequency (Lauterborn 1976)

$$\nu_0 = \frac{1}{2\pi R_0 \sqrt{\rho}}\left[3\gamma p_0 - \frac{2\sigma}{R_0} - \frac{4\mu^2}{\rho R_0^2}\right]^{1/2}. \tag{7.5.5}$$

Let us carry out a numerical experiment: place an air bubble with a radius of 21 μm in water at room temperature and atmospheric pressure. We will create an ultrasonic field with a given amplitude (0.4, 0.5 and 0.6 bar). We will vary the vibration frequency (from 0.1 to 1 of ν_0). The resonant frequency for such conditions is about 160 kHz.

Before showing any results, we should note that the simulation was carried out for the adiabatic case in absence of water vapor. We are missing out on a lot of the details discussed above and mentioned below. Let us keep in mind that many of them will prevent such temperature rise (to thousands of kelvins).

The maximum radius observed during the discussed numerical experiment is presented (after transition to the stationary mode) in Fig. 7.5.1.

As we can see from Fig. 7.5.1, there are sharp peaks on the maximum radius—frequency dependency. We can see a similar picture (but turned upside-down) for the minimum radius dependency.

The highest peak is near ν_0—the first harmonic; the second highest one is near $0.5\nu_0$, etc. Also, one may see the ultraharmonic near $0.75\,\nu_0$ appearing with 0.6 bar of driving pressure. There

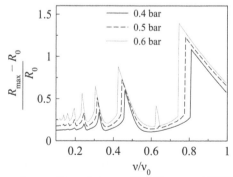

Fig. 7.5.1. The dependence of the maximum radius on frequency ($R_0 = 21$ μm, $\nu_0 = 160$ kHz) for different pressure amplitudes (0.4, 0.5 and 0.6 bar).

are many more subharmonics and ultraharmonics appearing with an increasing amplitude; also, the maxima increase significantly at relatively low frequencies (left side of the graph). But, in general, a similar picture will be observed for other parameters (amplitude and initial radius).

Bubble dynamics is obvious near the harmonics: during a single period of ultrasound oscillation, a bubble performs m oscillations when frequency is about $1/m$. For example, near frequencies $v = v_0/2$, two maximums are expected—first one caused by the natural bubble frequency and the second one caused by the same and the ultrasonic one (the maximum would be higher). The dynamics of the bubble radius for frequencies $v/v_0 = (0.8, 1/2, 1/3, 1/4)$ and amplitude 0.6 bar are presented in Fig. 7.5.2.

Unfortunately, such a fair result (Fig. 7.5.2) is observed not for all frequencies. In the subsection below, we will look at an example with a higher amplitude and a significantly lower frequency, and get possibly the most famous and outstanding image of bubble dynamics.

Going back to the maximum radius dependency (Fig.7.5.1), in the adiabatic approximation, the observed peaks mean temperature maximums (and minimums) (Fig. 7.5.3).

As was promised, with an amplitude of only a half of the atmosphere, we got the temperature of a thousand of kelvins (Fig. 7.5.2).

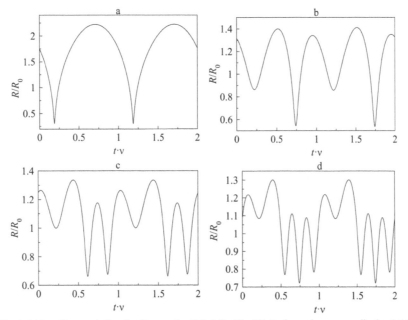

Fig. 7.5.2. The bubble radius evolution for frequencies 0.8, 1/2, 1/3, 1/4 (a, b, c, d correspondingly; 0.6 bar pressure amplitude).

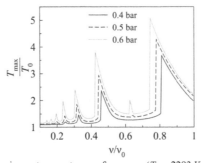

Fig. 7.5.3. The dependence of the maximum temperature on frequency ($T_0 = 2293$ K, $v_0 = 160$ kHz) for different pressure amplitudes (0.4, 0.5 and 0.6 bar).

The resonant frequencies caused a significant temperature rise, and now we see what is around and can discuss the numerical integration of the equation (7.5.3).

7.5.3 Numerical solution technique

We aimed to solve the equation (7.5.3) numerically. For solving this differential equation of the second order, the acceleration $\ddot{R}(t) = a(t)$ is expressed from the equation (7.5.3). Introducing the speed $U = \dot{R}$, we get the system

$$\begin{cases} \dot{U} = a \\ \dot{R} = U \end{cases}. \tag{7.5.6}$$

Integrating both equations from time moment t to $(t + h)$, where h is the integration time step,

$$\begin{cases} \displaystyle\int_{t}^{t+h} dU = \int_{t}^{t+h} a\,dt \\ \displaystyle\int_{t}^{t+h} dR = \int_{t}^{t+h} U\,dt \end{cases}. \tag{7.5.7}$$

Finally, integrating with the trapezoidal rule, we get

$$\begin{cases} U(t+h) = U(t) + h\dfrac{a(t+h)+a(t)}{2} \\ R(t+h) = R(t) + h\dfrac{U(t+h)+U(t)}{2} \end{cases}. \tag{7.5.8}$$

In other words, we used an improved Euler method for numerical integration. There is one detail that significantly complicates the solution of the system: to calculate the acceleration at time t, it is necessary to know the radius and speed at the same step, which means an implicit system. We use the Gauss–Seidel method for solving this system.

Now it's time to present the expression for $a(t)$:

$$a(t) = \frac{-\dfrac{3}{2}U^2\left(1-\dfrac{U}{3c_s}\right) + \dfrac{1}{\rho_l}\left[p_B - p_u\left(t+\dfrac{R}{c_s}\right) - p_\infty\right]\left(1+\dfrac{U}{c_s}\right)}{\left(1-\dfrac{U}{c_s}\right)R + \dfrac{4\mu}{\rho_l c_s}} +$$

$$+ \frac{\dfrac{R}{\rho_l c_s}\left[\dfrac{dp_{in}}{dt} + \dfrac{2\sigma U}{R^2} + \dfrac{4\mu U^2}{R^2}\right]}{\left(1-\dfrac{U}{c_s}\right)R + \dfrac{4\mu}{\rho_l c_s}}. \tag{7.5.9}$$

As one can see, we need to set the function for the inner pressure. The simplest way is to use pressure in the form of (7.2.8) with the mentioned replacement $3 \to 3\gamma$. For a more detailed research, we should complete the system with the energy conservation equation. Finally, we need the initial conditions. The most common are

$$U(0) = 0,$$

$$R(0) = R_0, \tag{7.5.10}$$

$$p_{in}(0) = p_0 = p_\infty + \frac{2\sigma}{R_0}.$$

With such initial conditions, in absence of ultrasonic exposure, the bubble will be at rest.

In fact, solving the equation (7.5.3) numerically in a dimensional form is not a good idea. While speeds can be of an order of hundreds of meters per second, the radius can be of an order of a micron. Such a scatter of values included in the same equation can lead to losses in computational accuracy. Not to mention the fact that the dimensionless view allows you to interpret the results clearer and generalize them.

The use of such a simple, albeit implicit integration method (the improved Euler method) allows to use a variable time step. It is useful for such a stiff equation. Moreover, it gives us an opportunity to catch the extremums. It is important because at such a moment the bubble reaches extreme conditions.

We suppose that at time t^*, we have an extremum; with the basic time step h, we will simply step over it. Our task is to find the correct time step $h^* = t^* - t$. Since this is an extremum, the speed $U(t + h^*)$ is equal to zero. Then we can use the vacated equation of system (7.5.8) to find the step instead of velocity.

$$\begin{cases} 0 = U(t) + h^* \dfrac{a(t + h^*) + a(t)}{2} \\ R(t + h) = R(t) + h^* \dfrac{0 + U(t)}{2} \end{cases}. \tag{7.5.11}$$

Obviously, to start 'catching the extremum' one should know that it exists. This fact is easily determined with the following condition:

$$U(t)U(t + h) < 0; \to t^* \in (t, t + h). \tag{7.5.12}$$

This technique (described above) is completely optional and like the improved Euler's method, it is quite stable. However, the mentioned technique allows to get a more accurate result.

We know what to expect from solving the equation (7.5.3) and how to solve it. Let us move on to a more detailed consideration of the result.

7.5.4 The simulation

In this subsection, we present the results of numerical simulations recalculated from the paper (Gaitan et al. 1992) for an air bubble oscillation (with the initial radius of 21 μm) in an acoustic field with 21 kHz frequency. This frequency corresponds to $\sim 0.14 \cdot v_0$. The magnitude of pressure is 1.2 bar.

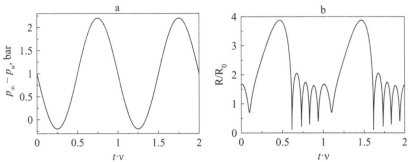

Fig. 7.5.4. The dependence of the external pressure and R/R_0 obtained with numerical simulations.

Over time, the system comes to a stationary state; usually, several dozens of periods are required. Fig. 7.5.4 shows dependences of external pressure and bubble radius during two periods.

We have a cyclic curve—this is a stationary state. During one period, the bubble collapses to a tenth of its initial size and expands by almost four times. In the context of adiabatic conditions, it leads to phenomenal values of the inner temperature and pressure (Fig. 7.5.5).

We got what we wanted—the resonant frequency of the ultrasonic field collapses the bubble so hard that temperature exceeds 3 000 K and pressure exceeds 10 000 atm. If we look at the speed of the bubble wall (Fig. 7.5.6), we can see that during the collapse the speed is around half of the Mach number (in relation to the speed of sound in a gas).

The Mach number in a gas significantly differs from the one in a liquid due to significant gas temperature change (the temperature of liquid and hence its sound speed are constant).

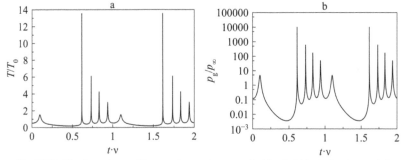

Fig. 7.5.5. The dependence of temperature and pressure inside a bubble during two periods

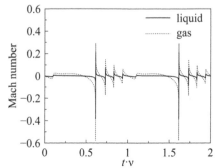

Fig. 7.5.6. The dependence Mach number obtained with numerical simulations.

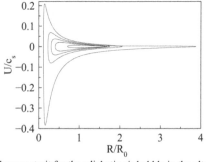

Fig. 7.5.7. Phase portrait for the adiabatic air bubble in the ultrasound field.

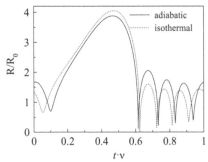

Fig.7.5.8. Comparison of the bubble radius for adiabatic and isothermal processes.

However, despite the fact that we did not take into account the compressibility of the gas and indeed did not solve the equations of motion for the gas at all, such a high velocity of the bubble wall is not a problem. The fact is that such a high speed is achieved near the initial radius R_0—with negligible pressure difference (in relation to the maximum pressure). As for the compressibility of the liquid, we took it into account in the equation (7.5.3).

Finally, we cannot deny ourselves the pleasure of constructing a phase portrait of such a magnificent system (Fig.7.5.7).

This result correlates with the one described in subsection 7.4.3. The phase portrait (Fig.7.5.7) looks like a collection of portraits for different constant Δp (some are effective).

7.5.5 Temperature variation in a collapsing bubble

The first thing that we would like to know is how significantly does the temperature variation change bubble dynamics? We already discussed that in subsection 7.4.6. But now we are dealing with a motion forced by an ultrasound field. To check that, we provide the same numerical simulation described above, but in the equation for the inner pressure, instead of the adiabatic exponent, we put the polytropic one with the value of unity (isothermal process).

Again, we get the closest curves for the radius—even closer than in subsection 7.4.6. It is caused by the fact that the same external pressure for both processes smoothens out differences.

Thus, we have seen that the temperature change inside a bubble is great, and heat from the center of the bubble cannot get in time to leak off to the liquid. We are dealing with a very complicated situation described in the previous section and specialized by the ultrasonic conditions.

Since temperature changes significantly during the collapse phase, one may expect many special problems caused by heating. At first, one may expect an additional vapor mass due to evaporation from the bubble walls; this matter was discussed above.

Conclusion

The full hydrodynamical description of a bubble in a liquid is very complex. Usually, a simplified approach is used, and the Rayleigh equation is the main implement for analyzing bubble dynamics. This equation represents the energy conservation during the bubble growth accompanied with pushing of a liquid mass:

$$\frac{3}{2}\dot{R}^2 + R\ddot{R} = \frac{\Delta p}{\rho}. \tag{7.C.1}$$

However, for many important practical cases, when the pressure difference depends only on the bubble radius, this equation is not as convenient as its integral: using for the growth rate $\dot{R} = U$

$$U^2 = \frac{2}{\rho R^3} \int_{R_0}^{R} R^2 \Delta p\, dR. \tag{7.C.2}$$

Here, for the considered case,

$$\Delta p = p_{in} - \frac{2\sigma}{R} - p_\infty, \tag{7.C.3}$$

where p_{in} is the pressure inside a bubble, p_∞ is the pressure in a liquid far away from the bubble; the second term describes surface tension. In a general case, p_{in} depends on the bubble radius.

For a typical cavitation problem, when the outer pressure suddenly drops, both (7.C.1) and (7.C.2) predict bubble oscillations: at the starting phase the bubble grows, the pressure inside it decreases, even lower than the outer pressure, the bubble slips this stage on inertia; the negative acceleration cannot stop the expansion immediately. Later, when time has passed and the pressure inside the bubble becomes very low, the expansion phase is replaced by the collapsing one.

With the equation (7.C.2), we can find the analytical estimations for the main parameters of bubble dynamics if we know the dependence $\Delta p(R)$, i.e., the dependence $p_{in}(R)$, for the given process. However, this is a big problem because we have two forks on the way.

The first fork: what is inside a bubble—a vapor, the mass of which changes due to processes of evaporation or condensation, or a constant amount of gas? A careful answer would be 'both', but this answer doubles the problem, while even the presence of a vapor is sufficient to over complicate the problem alone: the processes at the interface are strongly non-equilibrium, and a detailed kinetic analysis is required.

The second fork: is the process of bubble oscillation adiabatic or isothermal? Does heat leak from the bubble, or does it have no sufficient time to do so? When the vapor volume shrinks abruptly, molecules inside get an additional energy from the occluded walls and their energy of chaotic motion (i.e., temperature) tends to increase. Can it really increase or their additional energy is transferred to the surrounding liquid? Note that the second fork is connected with the first one: heat leaking from the bubble causes eventual evaporation.

Of course, the careful answer to the question can be: neither adiabatic nor isothermal, but this way does not help to construct any productive description. Really, a crucial part of the analysis for many applications is the choice between these two theoretical models: assumption about adiabatic conditions for the vapor in a bubble or allowance of the isothermal ones.

On the other side, the vapor bubble is located in the liquid, the temperature of which can be assumed to be the constant value T: for a such tiny thing as a bubble, the surrounding liquid is a good thermostat. Thus, there may be two limiting cases:

(A) the temperature inside the bubble cannot get in time to reach the temperature of the liquid during the compression process; this case completely corresponds to the term 'abruptly' and implies adiabatic conditions;

(B) the full process of heat transfer can be completed, all excessive heat leaks from the bubble to the surrounding liquid; this case means that the vapor will always be at the temperature of the liquid around it—the process is isothermal.

To figure out what process is closer to reality, we must compare the time scale of heat exchange, both within the bubble and from the bubble into the liquid, and the time scale of bubble oscillations.

The time scale for bubble oscillations, of course, is the period. For example, if a bubble of the inceptive radius R_0 at the initial pressure p_0 expands into a liquid of density ρ because the external pressure is low: $p_\infty \ll p_0$, then for adiabatic conditions the oscillation period can be estimated as

$$\tau \sim \sqrt{\frac{\rho(\gamma-1)R_{max}^5}{p_0 R_0^3}} \tag{7.C.4}$$

(γ is the adiabatic exponent); for isothermal conditions

$$\tau \sim \sqrt{\frac{\rho R_{max}^5}{p_0 R_0^3 \ln\left(R_{max} / R_0\right)}}.$$ (7.C.5)

Note that both expressions give values of the same order as a rule because even for $R_{max} \gg R_0$, there is $(\gamma - 1) \sim \ln(R_{max}/R_0)$. For such conditions, the comparison of calculated time periods with the typical time scale of heat exchange shows that the adiabatic approach is questionable, at least.

However, adiabatic conditions suit a collapsing bubble much better: when the outer pressure is high, the time period of oscillations is short.

Then, we have to return to the fork number one. In a case when the bubble contains not only gas but also vapor, we may use the 'corrected' Poisson law for the temperature of the gaseous phase in a bubble:

$$TV^{\frac{\gamma-1}{\omega}} = const,$$ (7.C.6)

where the parameter ω describes the heat and mass exchange due to evaporation. The equation (7.C.6) is a simple approximation that generalizes the adiabatic law for a pure gas.

An interesting thing that should be understood is that inertia plays an important role for bubble oscillations. During the bubble expansion into the liquid at a low pressure p_∞, the pressure inside the bubble drops to values $p_{in} < p_\infty$. During the bubble collapse under a high liquid pressure p_∞, the pressure inside the bubble increases to values $p_{in} > p_\infty$. Depending on the conditions inside the bubble, the highest value of pressure in the 'bubble–liquid' system may be reached in the liquid in vicinity of the bubble.

References

Ashokkumar, M. 2007. *Sonochemistry*. Kirk-Othmer Encyclopedia of Chemical Technology. https://doi.org/10.1002/04712 38961.1915141519211912.a01.pub2.

Birkhoff, G. and E. H. Zarantonello. 1957. *Jets, wakes, and cavities*. New York: Academic Press.

Bjerknes, V. F. K. 1906. *Fields of Force*. The Columbia University Press, The Macmillan company: London.

Crum, L. A. 1975. Bjerknes forces on bubbles in a stationary sound field. *The Journal of the Acoustical Society of America*, 57: 1363–70. https://doi.org/10.1121/1.380614.

Doinikov, A. A. and S. T. Zavtrak. 1995. On the mutual interaction of two gas bubbles in a sound field. *Physics of Fluids*, 7: 1923–30. https://doi.org/10.1063/1.868506.

Flynn, H. G. 1975. Cavitation dynamics. I. A mathematical formulation. *The Journal of the Acoustical Society of America*, 57: 1379–96. https://doi.org/10.1121/1.380624.

Gaitan, D. F., L. A. Crum, C. C. Church and R. A. Roy. 1992. Sonoluminescence and bubble dynamics for a single, stable, cavitation bubble. *The Journal of the Acoustical Society of America*, 91: 3166–83. https://doi.org/10.1121/1.402855.

Gerasimov, D. N. and E. I. Yurin. 2015. Parameters determining kinetic processes on an evaporation surface. *High Temperature*, 53: 502–8. https://doi.org/10.1134/S0018151X15040112.

Gerasimov, D. N. and E. I. Yurin. 2018. *Kinetics of Evaporation*. Springer Nature. https://doi.org/10.1007/978-3-319-96304-4.

Ghahramani, E., Arabnejad, M. H. and R. E. Bensow. 2019. A comparative study between numerical methods in simulation of cavitating bubbles. *International Journal of Multiphase Flow*, 111: 339–59. https://doi.org/10.1016/j.ijmultiphaseflow.2018.10.010.

Graves, R. E. and B. M. Argrow. 1999. Bulk viscosity: past to present. *Journal of Thermophysics and Heat Transfer*, 13: 337–42. https://doi.org/10.2514/2.6443.

Keller, J. B., and M. Miksis. 1980. Bubble oscillations of large amplitude. *The Journal of the Acoustical Society of America* 68:628–33. https://doi.org/10.1121/1.384720

Kentish, S. and M. Ashokkumar. 2010. *The physical and Chemical Effects of Ultrasound* in: Ultrasound technologies for Food and Bioprocessing, Feng H., G. V. Barbosa-Cánovas and J. Weiss. (Eds.). Springer. https://doi.org/10.1007/978-1-4419-7472-3_1.

Koda, S., T. Kimura, T. Kondo and H. Mitome. 2003. A standard method to calibrate sonochemical efficiency of an individual reaction system. *Ultrasonics Sonochemistry*, 10: 149–56. https://doi.org/10.1016/s1350-4177(03)00084-1.

Landau, L. D. and E. M. Lifshitz. 1987. *Fluid Mechanics*. Oxford: Pergamon Press.

Lauterborn, W. 1976. Numerical investigation of nonlinear oscillations of gas bubbles in liquids. *The Journal of the Acoustical Society of America*, 59: 283–93. https://doi.org/10.1121/1.380884.

Lavrent'ev, M. A. and B. V. Shabat. 1977. *Hydrodynamics problems and their mathematical models.* Moscow: Nauka (in Russian).

Lechner, C., M. Koch, W. Lauterborn and R. Mettin. 2017. Pressure and tension waves from bubble collapse near a solid boundary: A numerical approach. *The Journal of the Acoustical Society of America*, 142: 3649. https://doi.org/10.1121/1.5017619.

Leighton, T. G., Walton, A. J. and M. J. W. Pickworth. 1990. Primary Bjerknes forces. *European Journal of Physics*, 11: 47–50. https://doi.org/10.1088/0143-0807/11/1/009.

Luche, J.-L. 1998. Synthetic Organic Sonochemistry. Springer Science + Business media, LLC. https://doi.org/10.1007/978-1-4899-1910-6.

Mason, T.J. and Lorimer, J.P. 2002. *Applied Sonochemistry*. Wiley-VCH Verlag GmbH & Co. KGaA.

Matsumoto, M., K. Miyamoto, K. Ohguchi and T. Kinjo. 2000. Molecular dynamics simulation of a collapsing bubble. *Progress of Theoretical Physics Supplement*, 138: 728–9. https://doi.org/10.1143/PTPS.138.728.

Mettin, R., I. Akhatov, U. Parlitz, C.-D. Ohl and W. Lauterborn. 1997. Bjerknes forces between small cavitation bubbles in a strong acoustic field. *Physical Review E*, 56: 2924–31. https://doi.org/10.1103/PhysRevE.56.2924.

Minsier, V. 2010. *Numerical simulation of cavitation-induced bubble dynamics near a solid surface*. Universite catholique de Louvain.

Noltingk, B. E. and E. A. Neppiras. 1950. Cavitation produced by ultrasonics. *Proceedings of the Physical Society*, 63: 674–85. https://doi.org/10.1088/0370-1301/63/9/305.

Oquz, H. N. and A. Prosperetti. 1990. A generalization of the impulse and virial theorems with an application to bubble oscillations. *Journal of Fluid Mechanics*, 218: 143–62. https://doi.org/10.1017/S0022112090000957.

Prosperetti, A. 1984. Bubble phenomena in sound fields: part one. *Ultrasonics* 22: 69–77. https://doi.org/10.1016/0041-624X(84)90024-6.

Rayleigh Lord. 1917. On the pressure developed in a liquid during the collapse of a spherical cavity. *Philosophical Magazine* 34: 94–98.

Sarkar, P. 2019. *Simulation of cavitation erosion by a coupled CFD-FEM approach*. Université Grenoble Alpes.

Schanz, D., B. Metten, T. Kurz and W. Lauterborn. 2012. Molecular dynamics simulations of cavitation bubble collapse and sonoluminescence. *New Journal of Physics*, 14: 113019. https://doi.org/10.1088/1367-2630/14/11/113019.

Sharma, B. and R. Kumar. 2019. Estimation of bulk viscosity of dilute gases using a nonequilibrium molecular dynamics approach. *Physical Review E*, 100:013309. https://doi.org/10.1103/PhysRevE.100.013309.

Sinkevich, O. A., Glazkov, V. V. and A. N. Kireeva. 2012. Generalized Rayleigh–Lamb Equation. *High Temperature*, 50: 517–26. https://doi.org/10.1134/S0018151X12030194.

Tabor, M. 1989. *Chaos and Integrability in Nonlinear Dynamics*. New York: Wiley.

Zahedi, P., R. Saleh, R. Moreno-Atanasio and K. Yousefi. 2014. Influence of fluid properties on bubble formation, detachment, rising and collapse; Investigation using volume of fluid method. *Korean Journal of Chemical Engineering*, 31: 1349–61. https://doi.org/10.1007/s11814-014-0063-x.

Zhang, Y., Zhang, Y. and Li Sh. 2016. The secondary Bjerknes force between two gas bubbles under dual-frequency acoustic excitation. *Ultrasonics Sonochemistry*, 29: 129–45. http://dx.doi.org/10.1016/j.ultsonch.2015.08.022.

CHAPTER 8

Electrization of Liquids

8.1 Triboelectricity

8.1.1 Feel like an ancient Greek

Electrization due to friction (in modern terms, triboelectricity, or contact electrification) was discovered, or rather, observed, by Thales of Miletus, the 'father of philosophy'. A piece of amber rubbered by fur acquired the property to attract light objects; it is interesting that the term 'electricity' originates from the Greek word 'amber'—ηλεκτρον (pronounced similar to electron, by the way, that is, from the terminological point of view, an electron is a small piece of amber; of course, today we know that an electron is a small piece of everything).

Surely, Thales, who lived in around ~ 600 BC, could not explain the physical nature of triboelectricity. Many reasons for this circumstance can be pointed out, we name two:

✓ the nature of electricity was absolutely unknown at that time,

✓ triboelectricity turned out to be an insidious thing: even now, when the nature of electricity is well-known to us (as we think), we cannot explain triboelectricity in details.

As usual, poor understanding of the phenomenon does not prevent us from using it in our life. The most famous example is the photocopying Xerox devices; we dwell on a more scientific case—the discovery of the finite electron charge in the Millikan experiments. This experimental setup is now almost a staple of an educational physical laboratory, so we discuss it slightly thoroughly.

The scheme of such experiments is shown in Fig. 8.1.1. The method, in its rough explanation, is to measure the electric field strength that is required to hold a charged droplet in place: two forces act on the droplet—gravity mg and the electrical force qE; at equilibrium, we may find the charge of the particle from here:

$$q = \frac{mg}{E}. \qquad (8.1.1)$$

Millikan established that the charge q quantizes—its value is Ne, where e is the elementary charge, N is an integer number that shows how many elementary charges are located on the droplet.

From a more accurate theoretical background, some corrections to the relation (8.1.1) have been made; these are not crucial for our purposes. Another question is more interesting: from which source do those electrons appear? The experiment goes as follows: we spray a liquid (an oil) into the gap, apply the electric field from the electric power source, regulating E to obtain its value corresponding to equilibrium, until the droplet is suspended in a gas, and then look carefully inside the gap with microscope, trying to estimate the falling speed of the droplet to determine its size (and, consequently, its mass), when the electric power supply is off.

Usually, the size of the droplet (the droplet that can be suspended, other ones would leave a gap) is less than 1 μm, which corresponds to the mass of about ~ 10^{-15} kg. If a droplet has only a single elementary charge in it, then the electric field applied to the plates is ~ 10^5 V/m—this is a

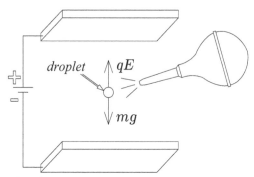

Fig. 8.1.1. The modern interpretation of the Millikan scheme. Electrical field is generated by two large parallel plates (the gap between them is about several millimeters); we presume that the droplet has a negative charge. Droplets come from a spray from a medical ball syringe; usually various oils are used.

rather small electric filed; it corresponds to the voltage of $\sim 10^2$ V applied to the gap of ~ 1 mm. This field is far below the critical values of electric field strength to provoke an electrical discharge here—of orders of $\sim 10^{6-7}$ V/m. This means that free electrons cannot be formed in the gas,[1] and, moreover, cannot be emitted from the negative electrode; for this, the electric field strength must be even higher.

Thus, we have a question—where do the electrons deposited on droplets come from? Millikan himself used an X-ray generator—its ionizing radiation produced additional charges; variation in droplet charge helped him to determine the mass of a droplet; actually, a serious battle took place around the Millikan's results, see (Franklin 1997). However, today in such experiments no additional ionizers are used; thus, droplets receive their electrons 'from nowhere'.

To make the problem more interesting, we may note that, actually, the charged particles setting on droplets are not necessarily electrons—that may also be positive ions. Inside the setup in our laboratory next door, the probability to obtain a positive or a negative charge for a droplet is approximately 50/50.

The usual explanation reads that charged particles—ions and electron—appear because of friction, i.e., when the oil is being sprayed from a pulverizer. It may be definitely so because other factors look less likely. Indeed, let us assume that somehow electrons are emitting from the cathode, for instance, because of the UV irradiation acting on it; despite the fact that no additional UV irradiation is used in such a scheme, the external light usually is very bright to see those tiny droplets; thereby, one may assume that the UV component of normal light is enhanced too. But anyway, if the gas is filled with electrons from the cathode, then droplets must always get a negative charge, while we see the controversial situation of equal probability for both signs of charge. Another hypothesis may state the cause lies in the external ionizing irradiation—from cosmic rays, or from the neighboring laboratory (which studies ionizing radiation, for example). In this case, charges would appear in a bulk of the gas, and, certainly, we would get charges of both signs there; so a droplet would be able to accumulate either positive or the negative ones. On the other hand, the intensity of the external ionizing radiation is negligible, as a rule.

So, electrization due to friction seems to be a likely candidate for the role of the source of charged particles in the setup from Fig. 8.1.1. In addition to other circumstantial evidences, we may point one more: the problem looks very uncertain, the 50/50 probability is a good marker that we deal with one of the most mysterious things—triboelectricity, a phenomenon the nature of which remains unclear for 2500 years.

---

[1] Electrical discharge in a gas is the process of formation of a conducting channel in a gas. The key feature of this process is the electron avalanche: an electron gets high kinetic energy in the electric field, hits an atom, a second electron appears, both electrons get high energy, hit two atoms, etc.

From some point of view, it is even funny that such a simple phenomenon—electrization due to friction—has no final solution today, when we seriously discuss the matters concerning the origin of our Universe. Up to this date, modern reviews are being published with titles like 'Long standing and unresolved issues in triboelectric charging' (Lacks and Shinbro 2019). Triboelectricity gave a taste of its quality as a tricky process, and many of its features keep getting explained in the same manner as several centuries ago.

For instance, when two pieces rub against each other, which piece obtains a positive charge, and which piece a negative one? For the pair 'Teflon – gold', all three answers may be correct (Teflon charges positively, negatively, or takes no charge at all) (Galembeck et al. 2014). We should emphasize that we are talking about experimental results.

Thus, having such a rich assortment of answers even to such simple, qualitative questions, it is no surprise that we betake ourselves to 'proven methods'. Any science (except mathematics) can be as empirical as we desire; thus, the attempt to lay out all materials in a row with respect to their property to obtain a certain electrical charge during their mutual friction seems no the worst idea. This is similar to the reactivity series in chemistry or something; for triboelectricity, the first progression was obtained by J.C. Wilcke in 1757; one of the modern versions of it (Galembeck et al. 2014) is from plus to minus

Glass → Mica → Polyamide → Wool → Aluminum → Paper → Steel → Wood → Amber → Polymethyl methacrylate → Cooper → Silver → Gold → Polyethylene terephthalate → Polycarbonate → Polystyrene → Polyethylene → Polypropylene → Polyvinyl chloride → Polytetrafluoroethylene

In simple words, when someone rubs a piece of glass on wool, the glass piece obtains a positive charge and the wool—a negative one, but when wool is rubbed on some 'poly-' substance, it gets a positive charge, while the other thing—a negative one. Attempts in constructing triboelectric series continues up to date (Zou 2019); on the other hand, this method can be criticized as a whole (Lacks and Shinbro 2019).

The matter that is directly connected with the triboelectric series (or with a possibility to construct such a structure in principle) is what particles transfer from one rubbed body to another: electrons, ions or both, or more complicated things like vacancies (see below)?

For the pair of two insulators, the fine answer, in a common style of triboelectricity, can be found in the monography by W.R. Harper (Harper 1967): "Montgomery would say that the carriers of charge are *always* electrons and Loeb that they are *generally* electrons: Henry feels that the question is still an open one. I am of the opinion that a definite answer can now be given which is that the carriers are *never* electrons—when the material being charged is strictly an insulator."

For the pair 'metal—dielectric (semiconductor)', the answer is more certain (Xu et al. 2018): the carriers are electrons; however, history taught us to expect surprises everywhere, when we deal with triboelectricity.

In common, there are many theoretical models that predict the charge transfer from one contacted body to another; they can explain some contact effects that take place in a static case, but the main problem is how to incorporate friction in such models. Meanwhile, at the molecular level, the friction of two surfaces looks like a war of the worlds. Bodies are never contacted as 'a flat surface with a flat surface': the unevenness on each surface takes the main impact. Thus, in the simplest case, friction causes a local increase in temperature,[2] which gives charge carriers (we are cautious with terminology here) additional chaotic energy to overcome the potential barrier and, therefore, to jump onto another body.

---

[2] Here we use the terms 'temperature' and 'chaotic energy' synonymously and write about caution when choosing terminology; of course, this is a bad combination: in absence of equilibrium, such a term as 'temperature' means nothing.

In a more complicated situation, that can be imagined-friction (macrolevel phenomenon) leads to devastations on the microlevel scale. Atoms are torn away from their parental body, and locally the electrical neutrality can be broken because of this very circumstance: new conditions for old charge carriers (typically from this body) occur, but also new charge carriers may appear in the contact region. For complex materials like 'poly-' things, the atom detachment can be even more dramatic, since many combinations of atoms may appear, and various parts of the chemical formula of this substance may find themselves on the opposite body.

Now, we stop the discussion of the complexity of triboelectricity for a while. Certain matters must be explored; for example, what is the difference between a conductor and an insulator. We have to start from the definitions from a very sophisticated science—quantum mechanics.

8.1.2 The quantum mechanics formalism

From some point of view,[3] studying quantum mechanics must be the basic element of modern education. However, today we have the same situation as decades ago: quantum mechanics is studied as an exotic part of general science.

Here we do not discuss the 'physics' of quantum mechanics—it is a matter so confusing that we do not dare to do it. Here we introduce the technical parts of it. Actually, such a point of view exists and that it is enough.

In accordance with quantum representation, the location of a particle is described with the wave function $\Psi(r)$, where r are the coordinates of the particle. A particle may be detected in r with probability density

$$w(r) = \Psi(r)\Psi^*(r) = |\Psi(r)|^2, \qquad (8.1.2)$$

where $\Psi^*(r)$ is the complex conjugate of the function. Because of the normalizing condition for the probability density, we have

$$\int |\Psi|^2 \, dr = 1. \qquad (8.1.3)$$

In general case, we should consider not a single particle, but the whole system; if so, wave function $\Psi(r)$ implies the function of the whole set of coordinates.

The Schrodinger equation describes the evolution of the wave function; for a single particle of mass m it reads

$$i\hbar \frac{\partial \Psi}{\partial t} = \hat{H}\Psi, \qquad (8.1.4)$$

where $\hat{H}$ is the Hamilton operator—an operator that corresponds to the energy of the system (in the given case, to the energy of a single particle):

$$\hat{H} = -\frac{\hbar^2}{2m}\Delta + U, \qquad (8.1.5)$$

where the potential energy of the particle U, and $\Delta = \dfrac{\partial^2}{\partial x^2} + \dfrac{\partial^2}{\partial y^2} + \dfrac{\partial^2}{\partial z^2}$ is the Laplace operator; so, the full Schrodinger equation is

$$i\hbar \frac{\partial \Psi}{\partial t} = -\frac{\hbar^2}{2m}\Delta \Psi + U\Psi. \qquad (8.1.6)$$

---

[3]   For example, Charles Kittel explained all the thermal science—the science field which always represents a domain of the classical approach, with exceptions for systems at low temperatures—in quantum language (Kittel 2000).

Thus, to know the evolution of the wave function, we must solve the full Schrodinger equation (8.1.6). Another way is to consider the state with the definite energy E, i.e., to consider the stationary Schrodinger equation

$$\hat{H}\psi(r) = E\psi(r). \tag{8.1.7}$$

Note that the wave function of the stationary equation depends on time: according to (8.1.6), we see that

$$\Psi(t,r) = \psi(r)e^{-iEt/\hbar}; \tag{8.1.8}$$

note that we may follow the reverse way: to decompose $\Psi(t, r) = \tau(t)\psi(r)$, insert it into the Schrodinger equation and obtain the representation (8.1.8) by solving the ordinary differential equation for $\tau(t)$.

For example, for a free particle, that is, in absence of any potential energy $U = 0$, so the total energy E is only the kinetic energy of the particle, we have from (8.1.7)

$$-\frac{\hbar^2}{2m}\Delta\psi = E\psi. \tag{8.1.9}$$

This equation has a solution in a common form

$$\psi(\vec{r}) = Ce^{i\vec{k}\vec{r}}, \tag{8.1.10}$$

where the wave vector $\vec{k}$ satisfies the condition following from (8.1.9)

$$\frac{\hbar^2\left(k_x^2 + k_y^2 + k_z^2\right)}{2m} = E, \text{ or } k^2 = \frac{2mE}{\hbar^2}. \tag{8.1.11}$$

Introducing the momentum of the particle p, so that the kinetic energy $E = p^2/2m$, we obtain that

$$k = \frac{p}{\hbar}, \tag{8.1.12}$$

and the total wave function (8.1.8) of the free particle is

$$\Psi = C\exp\left[-\frac{i}{\hbar}\left(Et - \vec{p}\vec{r}\right)\right]. \tag{8.1.13}$$

According to quantum mechanics, particles may be of two kinds,[4]

✓ bosons—any number of such particles can be found in the same state;

✓ fermions—only a single particle can be found in the given state.

All particles, aside from coordinates and momentums, are also characterized by spin—the intrinsic angular momentum. This is a special parameter which cannot be interpreted as the 'angular momentum of a particle spinning around its own axis' (Landau and Lifshitz 1977), despite the fact that, for the sake of simplicity, approximately half of non-specialists in quantum mechanics represent the spin exactly like that. Bosons have integer spin; fermions have half-integer spin.

Actually, this is enough material for the first time. Below we consider some special quantum problems, which can be used in this and in the next chapter.

---

[4] Theoretically, we may consider any number of particles in the given state.

8.1.3 Electrons in a condensed medium

Further, we will stay within the one-electron approach, that is, consider the wave function for only a single electron. In a solid—a medium with periodical structure—this electron is in a periodic field, that is, the potential energy U in the stationary Schrodinger equation (8.1.7) (written for the single electron) is a periodic function, i.e.,

$$U(\vec{r}) = U(\vec{r} + \vec{a}),\tag{8.1.14}$$

where $\vec{a}$ is the period of the lattice.

Then, looking at the Schrodinger equation, we see that its solution must have the same form as (8.1.14), that is, the wave function has a property

$$\psi(\vec{r} + \vec{a}) = \psi(\vec{r}) e^{i\vec{k}\vec{r}}.\tag{8.1.15}$$

Looking from slightly different angle, one may note that if the wave function is represented in the form

$$\psi(\vec{r}) = f(\vec{r}) e^{i\vec{k}\vec{r}},\tag{8.1.16}$$

then the function $f(\vec{r})$ has a periodic property, as it follows from (8.1.15)

$$f(\vec{r} + \vec{a}) = f(\vec{r}).\tag{8.1.17}$$

The statements (8.1.15) or the combined formulation of (8.1.16), (8.1.17) are the Bloch theorem; the last formulation is a more canonical form. We will not prove it here—this is a standard element of any university course on the physics of solids, for example (Blakemore 2013).

Let us take a look at (8.1.16) formally. In this equation, in a common case, the wave vector $\vec{k}$ can be real or imaginary. If this quantity is real, then the electron can travel in a solid without attenuation (to be exact, its wave function can). Such a state is called the 'allowed state'. On the contrary, if the wave number is imaginary, then the wave function would contain the factor e^{cr}, which inevitably leads to the infinite increase of ψ in some direction (parameter c here can be positive or negative—it does not matter, the corresponding direction can be chosen anyway). Of course, because of the requirement (8.1.3), such a state cannot exist throughout the whole spatial scale of the solid—this is called a 'forbidden state'. However, such states may exist locally, in vicinity of point $\vec{r}_0$, where the wave function can be represented as

$$\psi(\vec{r}) \sim \exp\left(-ik|\vec{r} - \vec{r}_0|\right),\tag{8.1.18}$$

where k is imaginary, that is, the wave function has a form $\sim e^{-cr}$—monotonically attenuated function concentrated around the center at the point $\vec{r}_0$ which corresponds, in accordance with the traditional representation, to defects of lattice or, which is more interesting, to the surface of solids (Tamm 1932).

For free electrons in a solid, the wave function must be simply of the form

$$\psi(\vec{r}) \sim e^{i\vec{k}\vec{r}},\tag{8.1.19}$$

which follows from (8.1.16) at $f = 1$ and coincides with the common form of the wave function for a free particle (8.1.10).

Thereby, from the condition of periodicity, we see that the wave vector $\vec{k}$ must satisfy the expression

$$e^{i\vec{k}\vec{a}} = 1,\tag{8.1.20}$$

which gives certain set values for $\vec{k}$; for a one-dimensional case

$$k = \frac{2\pi j}{a};$$

(8.1.21)

here j is an integer number; for a three-dimensional case

$$\vec{k} = \frac{2\pi j_x}{a_x}\hat{x} + \frac{2\pi j_y}{a_y}\hat{y} + \frac{2\pi j_z}{a_z}\hat{z}.$$

(8.1.22)

where $\hat{x}, \hat{y}, \hat{z}$ are the corresponding orts.

Thus, in accordance with (8.1.22), the phase space–space of the values of $\vec{k}$—is divided into elementary cells of volume

$$\left(\frac{2\pi}{a_x a_y a_z}\right)^3 = \frac{8\pi^3}{V}.$$

(8.1.23)

For the free electrons, there are no restrictions on energy; thus, energy may give any value, except for limitations connected with the discrete spectrum of k (8.1.21): since the energy is correlated with the wave vector k by (8.1.11), the energy of an electron is a lattice of period a

$$E = \frac{\hbar^2 k^2}{2m} = \frac{\hbar^2}{2m}\left(\frac{2\pi j}{a}\right)^2.$$

(8.1.24)

From (8.1.22), we see that the spectrum of E is discrete too.

Let us calculate the number of electrons N that can be placed in the phase space with maximum value of the wave number k_F (the sense of it and the meaning of the notation given below) and, correspondingly, the maximum value of energy

$$\varepsilon_F = \frac{\hbar^2 k_F^2}{2m}.$$

(8.1.25)

It follows from (8.1.24), (8.1.25) that all possible values of energy are located inside a sphere of radius k_F. Thus, we may calculate the number of electrons in the following way: divide the total volume of this sphere by the number of elementary cells of volume (8.1.23). However, we must take into account that any point of the phase—any value of $\vec{k}$—corresponds to two electrons with different spins. Thus, we have

$$N = \frac{2\int\limits_0^{k_F} 4\pi k^2 dk}{8\pi^3/V} = \frac{Vk_F^3}{3\pi^2}.$$

(8.1.26)

that is, the electron density $n = N/L^3$ is connected with the maximum value of the wave number k_F by the correlation

$$k_F = \left(3\pi^2 n\right)^{1/3}.$$

(8.1.27)

In other words, two electrons cannot be found in the same state; thus, any additional electron must be placed 'over' the previous one in k-space, and so on. The larger the number of electrons, the higher the maximum value k_F can be reached.

The corresponding energy—the maximum energy of electrons—is called the Fermi energy; from (8.1.25) and (8.1.27)

$$\varepsilon_F = \frac{\hbar^2}{2m}\left(3\pi^2 n\right)^{1/3}.$$

(8.1.28)

The model described above has some benefits (simplicity is the first of them) and can be applied to metals—solids (or sometimes liquids)—which may conduct electric current through electron flows. Meanwhile, we need much more from the theory: at least, we expect an explanation on which substance can be considered as metal and which cannot. Thus, we need a more developed approach, where the potential energy of electron cannot be neglected.

Such a theory can be constructed with the assumption of almost free electrons (Postnikov 1978). Suppose that the potential energy U in the Schrodinger equation is very small, so the wave function, approximately, stays the same as for free electrons, i.e., has a form of (8.1.10). Then we will consider a one-dimensional Schrodinger equation for a stationary state (8.1.7) with potential energy in the form of

$$U = U_0 e^{iqx}, q = \frac{2\pi}{a},$$

(8.1.29)

where, as it was assumed above, $|U_0|^2 = U_0 U_0^*$ is a small value. The wave function is a periodic function;therefore, we may represent it as a linear combination

$$\psi = \psi_1 e^{ikx} + \psi_2 e^{i(k-q)x}.$$

(8.1.30)

Using (8.1.29) and (8.1.30) for (8.1.7), we have, after some rearrangements

$$\left(\underbrace{\frac{\hbar^2 k^2}{2m}}_{E_k}\psi_1 - E\psi_1 + U_0\psi_2\right)e^{ikx} + \left(\underbrace{\frac{\hbar^2(k-q)^2}{2m}}_{E_{k-q}}\psi_2 - E\psi_2 + U_0^*\psi_1\right)e^{i(k-q)x} = 0.$$

(8.1.31)

Here we used the fact that potential energy is a real quantity, that is, from the condition $U = U^*$ we have $U_0 e^{iqx} = U_0^* e^{-iqx}$ and, finally,

$$U_0^* = U_0 e^{i2qx}.$$

(8.1.32)

Equation (8.1.31) must be satisfied for

✓ any x,

✓ non-zero amplitudes ψ_1 and ψ_2.

Because of the first condition, we must demand that expressions in both brackets in (8.1.31) are zero separately; thus, we have a system of equations

$$E_k\psi_1 - E\psi_1 + U_0\psi_2 = 0,$$

(8.1.33)

$$U_0^*\psi_1 + E_{k-q}\psi_2 - E\psi_2 = 0.$$

(8.1.34)

From the second condition, to obtain non-trivial solutions for ψ_1, ψ_2 from the system (8.1.33), (8.1.34), we must have zero determinant of this system; correspondingly, we obtain the energy of electron

$$E(k) = \frac{1}{2}\left[E_k + E_{k-q} \pm \sqrt{\left(E_k - E_{k-q}\right)^2 + 4|U|^2}\right].$$

(8.1.35)

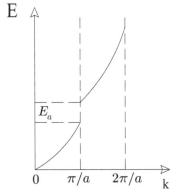

Fig. 8.1.2. Electron energy as a function of wave vector.

If $U = 0$, then from (8.1.35) $E(k) = E_k$ or $E(k) = E_{k-q}$—two independent values. But for $|U| \neq 0$, we have a completely different case: now $E(k)$ looks like Fig. 8.1.2.

For $k = \pi/a$, there is $E_k = E_{k-q} = E_a$, and the electron energy has two different values:

$$E_1 = E_a \pm |U|. \tag{8.1.36}$$

Thus, electron energy cannot obtain values in the range $(E_a - |U|, E_a + |U|)$—this is the forbidden range of energies, or, considering diagrams like (8.1.2), the forbidden zone.

Without a doubt, this theory is very simplifying. But it explains the main issue of electrons in a solid: their energy spectrum contains allowed zones divided by forbidden gaps. For real substances, the structure of such zones is complicated; meanwhile, the main principle is the same. Considering two zones—the valency zone and the conductivity one—we may classify the substances by the occupation of these zones: dielectrics are solids which have only an empty, fully occupied zone, while metal is a substance which has partially occupied zones.

In the next chapter, we will return to these matters; in the next sections, we will consider the exodus of electrons from the surface.

8.1.4 Electrons beyond a condensed medium

From the general point of view, triboelectricity is the charge transfer from one body to another. The simplest case, to be exact the most convenient for theoretical description, is the electron transfer; note that this way is quite considerable (see above and read below).

Usually, charge transfer is described as direct travelling of electron from the given medium to the neighboring one. Such ways will be discussed in further sections; here we observe another mechanism: electron leaves the surface and, possibly (which does not interest us now), may then enter another surface.

To leave the surface, electrons have to overcome the work function: at a solid surface, electron is located in the potential pit, that is, electrons must get additional energy to compensate that energy difference. This additional energy can be gathered from various sources, and each way to escape is named after the given energy source.

The first and obvious way to run is called 'thermoemission'. It is a direct way out: an electron has a sufficient kinetic energy that exceeds its bond energy. There is one potential problem in this way: the potential energy that holds an electron inside a body (at the surface) is of a few electron-volts, while one electron-volts corresponds to temperature

$$1 \text{ eV} = 1.6 \cdot 10^{-19} \text{ J} = 1.38 \cdot 10^{-23} \cdot T \text{ [K]}, \text{ i.e., } 1 \text{ eV} \approx 11\,600 \text{ K} \tag{8.1.37}$$

Yes, indeed: energy can be measured in kelvins, while temperature can be measured in electron-volts; this is a common thing in plasma physics.

Thus, one may think that to overcome the bond energy an electron must have temperature of tens of thousands of kelvins; if we assume that electrons are in equilibrium with the medium (i.e., with heavy particles), then we have to conclude that the temperature of the solid must be $\sim 10^4$ K. Of course, this is not so: the kinetic energy of the given electron and the temperature of the electron gas in the medium are different things.

As other particles, electrons are distributed on their energy. Temperature is only the measure of dispersion of kinetic energy,[5] and T is the typical value of kinetic energy, not the general value for any electron in this medium. The given electron may have any energy—as very high, as very low. The higher the temperature, the higher the common level of kinetic energy of electrons, i.e., the higher the probability of high kinetic energy for the given electron. Nothing more. Between electrons, like in any other collective, there is always a parvenu, which has much more energy than its neighbors. This electron may overcome the energy barrier, despite the fact that temperature is 0.1 eV while the barrier is 5 eV. The temperature increase also increases the amount of such parvenues; we may also notice that if temperature would be of an order of the work function, then the flux of them would be enormous: it would mean that an ordinary electron may leave the surface.

Another way to get sufficient energy to escape is photo-emission. Of course, we all know this mechanism: it is also called a photoeffect. According to the quantum theory, an electromagnetic wave is not an electromagnetic wave, or it is better to say, it is not only an electromagnetic wave (it is even better to say that some matters should not be discussed in detail because any clarification only brings more confusion).[6] An electromagnetic field consists of photons—particles that have no mass but have energy ε connected with its frequency v:

$$\varepsilon = hv, \tag{8.1.38}$$

where $h = 6.626 \cdot 10^{-34}$ J·s is the Planck constant. When a photon is absorbed by a bound electron, and the energy of this photon exceeds the work function A, this electron breaks loose and runs away; if the process takes place at the atomic scale (electron leaves a separate atom), this is called the 'inner photoeffect', if an electron leaves the surface, that is called the 'outer photoeffect'.

Ion-electron emission is a more complicated thing. Reading this term, one may imagine how an ion hits the surface with great kinetic energy, an electron obtains this energy and escapes from the surface. Actually, this is not a usual scheme since to knock out an electron from the surface, the ion must have the kinetic energy of about several electron-volts; for instance, the velocity of an atom of iron (atomic mass m is 10^{-25} kg) that has kinetic energy E of 3 eV, that is,

$$v = \sqrt{\frac{2E}{m}} \approx 3 \text{ km/s}. \tag{8.1.39}$$

It is difficult to organize such fast ions, but not impossible, of course. Nevertheless, ions may contribute to the electron escape, but in a different way.

When a positive ion is placed near the surface, it pulls out an electron from it, and this pair recombines in vicinity of this surface. The recombination energy, which is equal to the energy of ionization of a separate atom, significantly exceeds the work function of the surface (in other words, the ionization energy of the surface). Thus, in vicinity of the surface the energy excels, and this energy is greater than the work function, and the surface obtains that energy. Thereby, additional electron leaves the surface following the electron that recombined with that ion.

The process described above, by the way, also explains the fact that work function is not a constant indeed. The work function depends on conditions in the surrounding medium: if the surface

---

[5]   Note that, actually, in an equilibrium system temperature is the measure of dispersion of any sort of energy, including the potential one; see our book (Gerasimov and Yurin 2018).

[6]   For instance, try to understand what a photon is—a single particle of the electromagnetic field; an old but good entry to the problem is (Klyshko 1994).

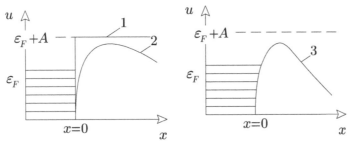

Fig. 8.1.3. Electron emission at an electric field. *1*— a theoretical (rectangular) shape of the potential barrier: electrons must have additional energy A (work function) to overcome this barrier. *2*—potential barrier at a moderate electric filed: the height of the barrier is slightly lower, and an electron may overcome this barrier more easily. *3*—potential barrier at a strong electric field ($\sim 10^8$ V/m). The key difference from case *2* is not the lower height, but its narrowness: now an electron may go through the barrier directly, without having to overcome it.

is placed in plasma, positive ions from the outer space will try to pull out electrons from this surface, decreasing the work function value.

The next way out from surface arises for an electron at the external electric field. The influence of the electric field can be twofold. The first way: the electric field slightly diminishes the height of the potential barrier—in the presence of an electric field, the work function A becomes smaller (Fig. 8.1.3), and the probability to overcome such a barrier is higher.

The second way is based especially on quantum mechanics. At a very strong electric field[7] $\sim 10^8$ V/m the barrier becomes very thin, and an electron has another way out: not over the barrier, but through it. This is the quantum tunneling effect: an electron, regardless of its energy, has a non-zero probability to jump to the other side. The probability of such a process can be described, approximately, as

$$w = \exp\left[-\frac{2}{\hbar} \int\limits_{0}^{b} \sqrt{2m\left(U\left(x\right)-\varepsilon\right)}dx \right], \tag{8.1.40}$$

where ε is the electron energy, $U(x)$ is the potential energy at the point x. Thus, the probability to 'break through the barrier' depends on the width of the barrier b that corresponds to the given energy ε. The narrower the barrier, the higher the probability. If the width of the barrier becomes ~ 10 angstrom, which is reachable at the electric field of $\sim 10^8$ V/m, then the electron flow becomes conspicuous. This process is called the auto-electron emission.

8.1.5 Contact phenomena

One may propose a theory of triboelectricity as a process of electrons leaving the surface under an external influence and the consecutive settlement of these electrons on the opposing body. For instance, let us consider the influence of the local temperature increase on the electron exodus. We all know that rubbing results in temperature growth, but, indeed, we do not know how much the local temperature grows. The macroscopic energy of friction, actually, is applied on local spatial scales; due to inhomogeneities of surfaces, the applied force acts on very small objects, see Fig. 8.1.4. These objects—surface protrusions—will be destroyed, and, scientifically speaking, the energy of their inelastic deformation will be turned into heat. We feel that rubbing bodies are heated, but we sense the macroscopic warming, when the heat that was locally excreted spreads on the macroscale—throughout the whole body or a large part of it. However, when we divide the total heat by the volume of small protrusions, where this heat was given off initially, we get a

---

[7] For comparison: the breakdown electric field in gases is smaller on about two orders—for instance, the breakdown field of air at atmospheric pressure is ~ 30 kV/cm.

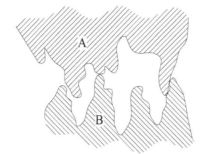

Fig. 8.1.4. This is how it looks on microscale; the characteristic size of inhomogeneities is up to 10 μm and can be even smaller. Now two surfaces are brought to a contact. Then let us imagine what will happen if these surfaces will move one along the other.

significant overheating. Thus, thermoemission can be tried for the role of the source of electrons during intensive friction.

Such models can be considered, especially taking into account the circumstance that there are no sufficient triboelectricity theories yet. In any case, the scheme when an electron transfers from one medium to another through an intermediate (air, in most cases) is not the only way that may be treated. The mechanisms which presume a direct transfer of an electron from one body to another must be considered too.

When two media which have free electrons are put into contact, then an electron in the given body must not overcome the work function A of the surface of that body. The question is: the work function of which body is higher? If $A_1 < A_2$, electrons will leave the medium #1; the medium #2 therefore will be charged negatively while the medium #1 would obtain a positive charge (since electrons left it). The increased electron density in medium #2 will stop the process sooner or later, when the emission currents will be balanced. Besides, there exist another way to interrupt the electron transfer: by establishing the outer contact potential difference, called the Volta potential,

$$\Delta\varphi_V = \frac{A_2 - A_1}{e} \tag{8.1.41}$$

which locks the current.

When two different metals are in a tight contact because of different Fermi energies in them—suppose for certainty $\varepsilon_{F1} > \varepsilon_{F2}$—electrons from medium #1 will diffuse into medium #2 because their kinetic energy, generally, is higher. This current will be locked by the internal contact potential difference—the Galvani potential

$$\Delta\varphi_G = \frac{\varepsilon_{F1} - \varepsilon_{F2}}{e}. \tag{8.1.42}$$

Of course, these mechanisms do not explain triboelectricity on their own, since both of them do not contain such an important issue as friction which is the driving force of triboelectricity. These ways of electron transfer are rather the additional elements to the whole puzzle.

Finally, it should be mentioned again that the nature of charge carriers transferred during triboelectricity, strictly, is not established firmly. Possibly, these are electrons, but they may be ions. The behavior of ions in vicinity of a solid surface deserves a separate section.

8.2 Double layer

8.2.1 *The charge carriers in a gas inside a liquid*

Of course, in a common practical case only electrons and ions may be charge carriers. Someone may possibly mention positrons or something from this area of physics, but here we deal with our domestic world. Thus, our choices are very restricted.

Everything is clear about electrons until we do not consider their quantum properties. As for ions, we have two quite different situations for a gaseous phase and for a liquid one.

An ionized gas can be termed as plasma. Indeed, plasma physicists use various definitions of plasma, the two most popular are

✓ plasma is an ionized quasi-neutral gas (i.e., a gas which consists of neutral particles, ions and of electrons, and the total charge of the system is zero);

✓ plasma is a medium, the properties of which are strictly determined by the interaction of charged particles from this medium with the intrinsic and the external electric field.

From the first definition, an electron beam or an electrolyte is not a plasma: the first object is not neutral, the second one is not a gas. On the other hand, the second definition has uncertain limits of application: the air that you breathe contains several electron–ion pairs per cubic centimeter; of course, such a system is not a plasma, but what about the spark discharge, where the ionization degree—the ratio of charged particles to the total number of particles—can be about 10^{-10}? For this reason, and also because the physical description of an ionized gas and of an ionized liquid differs significantly, below we will use the term 'plasma' only for an ionized gas.

Thus, we can say that in plasma the origin of charged particles is ionization by electrons or by photons; other sources of ionization are negligible. The most important way is ionization by electrons:[8] electrons get high energy from an external source (external electric field), hit atoms and produce additional electrons. As a result of a single impact, a single-ionized atom—a single-charged positive ion—appears. To get additional charge, i.e., to lose one more electron, the single-charged ion must wait until the next impact, hoping that this impact would bring exactly ionization, not recombination—association with that impacting electron. Therefore, concentration of multi-charged ions is not high only if the temperature of the whole medium is not too high (i.e., does not noticeably exceed the value of $\sim 10^4$ K).

When an electron hits an atom or a molecule, it also can be attached to this particle, forming a negative ion. This process is only possible in some gases: in air, molecules of oxygen can attach electrons, while molecules of nitrogen cannot.

Finally, one may conclude that the formation of charge carriers in gases is plain and boring. It is. Liquids are a different case.

Solutions may conduct an electric current—this is an experimental fact which means that these liquids contain free charges inside. How do these charges appear? The old theory, up to the middle of the XIX century, assumed that these charges are born under the influence of this very electrical field applied to the medium. That is, one takes a solution, which has no charge carriers inside it, applies an electrical field, and (a) free charges are born in the liquid, (b) these charges conduct electricity in the liquid. Later it turned out that we cannot observe any additional effects of an electric filed on a medium: both the Ohm law and the Joule–Lenz law are fulfilled. In other words, no additional energy from the electrical field is spent 'sideward'—the liquid behaves in such a manner as if it holds separate charges inside it at all times.

Clausius affirmed this hypothesis. In a main liquid (for example, in water), a solutioned substance (for example, NaCl) is dissociated. Exactly, this substance does two opposite processes: dissociation and association: atoms of sodium and chlorine split and connect once more. The corresponding chemical reaction can be written as

$$NaCl \rightleftarrows Na^+ + Cl^-. \tag{8.2.1}$$

In a liquid, the split ions travel (diffuse), associate again, and so on. Here we may directly note that the probability to associate again must depend on the solution's concentration—the more

---

[8] For the so-called streamers—weak spark discharges—ionizing by photons may possibly play a significant role; generally, this is a debatable issue.

molecules of NaCl are in the liquid, the higher the probability for an ion Na^+ to meet an ion Cl^-. On the contrary, if the concentration of the solution is weak, then the probability for ions to meet each other is very low. In a limiting case, when a single molecule is being dissociated, it is almost impossible for ions of sodium and potassium to meet again in a huge medium which contains $\sim 10^{20}$ particles or more. Thus, we may conclude that weak solutions must be fully dissociated—this is the so-called Ostwald law.

Thus, as we see, a solutioned substance is partially dissociated. In usual conditions—in the absence of an electric field—ions calmly diffuse in the main liquid, meeting each other again, associating, and then dissociating again, and so on. Such liquid solution looks like a non-charged fluid. But if an external electric field is applied, the situation differs significantly. Now ions do not chaotically wander like tourists in Rome, they follow certain directions—positive ions (cations) go to the cathode, negative ions (anions) float to the anode. This army-style behavior diminishes the probability of association; thus, as a whole, the process looks like the amount of charge carriers was increased under the influence of an external electrical field. Actually, as we saw, the number of free charges was not increased there—it rather was not decreased; no energy from the electrical field was spent on ionization of the medium.

The composition of the charged component in a liquid differs from a gas: in a gas, there were electrons and single-charged positive ions, while in a liquid this component is a mixture of ions of both signs, and, moreover, not necessarily single-charged ions. For instance, a solution of zinc sulfate consists of double-charged ions:

$$ZnSO_4 \rightleftarrows Zn^{2+} + SO_4^{2-}. \tag{8.2.2}$$

The fact that liquid solutions contain negative ions instead of electrons leaves an imprint on all the processes that take place in a liquid: ions are heavy, they cannot get significant kinetic energy in an electrical field, and, therefore, they cannot repeat such tricks of electrons that impact ionization, etc. The character of motion of ions in liquids quantitatively differs than in gases.

8.2.2 Dynamics of charge carriers in the bulk of a medium

The question that may liven up the environment in a classroom: in air, under a constant electrical field, how do charged particle move—at certain velocity or at constant acceleration? The trap is in the firmly acquired knowledge that at a constant force a particle moves at constant acceleration; we should note that common answers confirm the greatness of Galileo's discovery of inertia—it was not as obvious as one may assume.

Let us consider a particle that moves in a medium with friction. For our purposes, we will consider an external electrical force; in general case, any force could be taken into account. The second Newton's law is

$$m\vec{a} = q\vec{E} - \vec{F}_f, \tag{8.2.3}$$

where q is the charge of the particle, m is its mass, $\vec{a}$ is acceleration, $\vec{E}$ is the electric field strength, $\vec{F}_f$ is the friction force. The last one can be formulated with the loss of momentum due to friction:

$$\vec{F}_f = \frac{d\vec{p}}{dt} = \frac{m\vec{v}}{\tau} = \nu m\vec{v}, \tag{8.2.4}$$

where τ is the characteristic time of the momentum loss, and $\nu = \tau^{-1}$ is the frequency of collisions—this is, actually, the definition of frequency of collision.

Thus, we see that two forces—the force from the electric field and the friction force—may compensate each other, and the particle will move at constant velocity v which can be determined from that very equation. Therefore, from (8.2.3) at $a = 0$ with (8.2.4) we obtain

$$\vec{\upsilon} = \underbrace{\frac{q}{\nu m}}_{b} E = b\vec{E}. \qquad (8.2.5)$$

Here the coefficient b is the mobility of the particle;[9] it can be calculated if we know the frequency of collisions—the most difficult part of the expression (8.2.5). These collisions are the collisions between the given particle and other particles of the medium: for instance, between an electron and molecules of air or between ion in water solution with molecules of water (and, partially, with other ions of the solution). As we see, a particle in a constant electric filed—in a medium—moves at constant velocity. Note that the mobility of negative ions is negative—in this case $q < 0$, and $\vec{\upsilon}$ directed oppositely to $\vec{E}$, which means an obvious fact that negatively charged particles move against the direction of an electrical field (which was defined, in its turn, as the direction of positively charged particles).

Thus, in a water solution ions move inside the external electric field, providing the current with density j; since the current density is $j = qn\upsilon$, where n is the volume density of the corresponding particles, then we have

$$\vec{j} = \vec{j}_+ + \vec{j}_- = q_+ b_+ n_+ \vec{E} - q_- b_- n_- \vec{E} = \sigma\vec{E}, \qquad (8.2.6)$$

where $\sigma = q_+ b_+ n_+ - q_- b_- n_-$ is the conductivity of the medium; note that usually $n_+ = n_-$ since ions of different signs were born from common parental molecules.

Important notice: in a general case, a charged particle may move not under the influence of only the electrical force. Other forces, of non-electrical nature, may also move charged particles in an organized way; if they exist in the given case, then one has to include the corresponding flow into the relation (8.2.6).

The character of the electrical force influence can be easily understood if we consider the Maxwell equations for electromagnetic field; it is sufficient to take into account only two equations from them:

$$rot\vec{H} = \vec{j} + \varepsilon_0\varepsilon\frac{\partial\vec{E}}{\partial t}, \qquad (8.2.7)$$

$$rot\vec{E} = -\mu_0\mu\frac{\partial\vec{H}}{\partial t}, \qquad (8.2.8)$$

where ε and μ are the dielectric constant and the magnetic permeability of the medium, correspondingly; we use SI here.
Integrating (8.2.8) and substituting the result into (8.2.7), we obtain

$$\frac{\partial\vec{E}}{\partial t} = -\frac{\vec{j}}{\varepsilon_0\varepsilon} - c'^2\int rotrot\vec{E}dt, \qquad (8.2.9)$$

where $c' = \dfrac{1}{\sqrt{\varepsilon_0\varepsilon\mu_0\mu}}$ is the speed of light in the given medium.

Thus, we may see that the electrical field strength may increase only due to the vortex component, i.e., if $rot\vec{E} \neq 0$, or due to external forces that may push charges against the direction of the electrical field. This, by the way, is the main difficulty in the theoretical description of a gas discharge: it cannot be explained with only the potential electrical field $\vec{E} = -\nabla\varphi$.

---

[9] Note that in some references, mobility is defined in another manner: as the proportionality coefficient between the velocity and the external force, not the electrical field strength.

If $rot\vec{E} = 0$ and the electrical current density obeys the correlation (8.2.6), then we see that electric field strength diminishes at the exponent law, since

$$\frac{\partial E}{\partial t} = -\underbrace{\frac{\sigma}{\varepsilon_0 \varepsilon}}_{1/\tau} E = -\frac{E}{\tau},$$
(8.2.10)

$$E = E(0)e^{-t/\tau}.$$
(8.2.11)

Here $\tau = \varepsilon_0 \varepsilon / \sigma$ is the relaxation time—the scale of decrease of the electric field strength; the higher the conductivity, the lower is the value of relaxation time. In simple words, the electrical field causes redistribution of particles which diminishes the value of the electric filed strength.

In a gas plasma, the mobility of heavy positive ions is much lower than the mobility of light fast electrons, so the plasma conductivity is mainly determined by electrons. In a liquid, both kinds of ions have comparable mass and, therefore, play approximately the same role in the current.

Now let us introduce a degree of dissociation α—the ratio of dissociated molecules to the total number of molecules; when all molecules are dissociated, $\alpha = 1$. Thus, if the total number density of molecules is n, then the number of dissociated molecules is αn, that is,

$$n_+ = n_- = \alpha n.$$
(8.2.12)

The rate of dissociation is proportional to the number of non-dissociated molecules $(1 - \alpha)n$, i.e., it can be represented as

$$w_d = C_1 (1 - \alpha)n.$$
(8.2.13)

The rate of association of molecules is proportional to the product of the number of positive ions n_+ and negative ions n_-, i.e., to $n_+ n_-$, or, with (8.2.12)

$$w_a = C_2 \alpha^2 n^2.$$
(8.2.14)

In (8.2.13) and (8.2.14), coefficients C_1 and C_2 represent the rates of the corresponding processes; they depend on the parameters of the medium (density, temperature, etc.) and, form the very number density n, which cannot be ruled out.

In equilibrium, there must be $w_d = w_a$—the rate of dissociation must be equal to the rate of association, and from (8.2.13) and (8.2.14) we have

$$\frac{1-\alpha}{\alpha^2} = \frac{C_2}{C_1} n.$$
(8.2.15)

This relation is a particular case of the equilibrium rule in a chemically reactive system. The equation (8.2.15) also contains the Ostwald law mentioned above: if C_2/C_1 remains limited at $n \to 0$, then, indeed, we see that $\alpha \to 1$ as $n \to 0$; as we considered above, the very weak solution is fully dissociated.

8.2.3 Charge carriers near a metal surface

When ions reach the corresponding electrodes (cations – cathode, anions – anode), they recombine. Positive ions pull out electrons from a cathode and become neutral. Negative ions give away their electrons to an anode and become neutral too. Thus, molecules that correspond to positive and negative parts of initial complex molecules dissolved in the main liquid will be sedimented on different electrodes. Let us calculate the mass foundering on the electrode.

The total mass of ions fixed on the electrode is the product of mass of a single ion m and the number of ions N:

$$M = mN. \tag{8.2.16}$$

The number of ions is proportional to the charge that was brought by ions to this electrode: every ion carries a charge q, which can be represented with the elementary charge e and the multiplicity of charge v: $q = ev$; thus, the common charge that passed through the electrode is

$$Q = Nev. \tag{8.2.17}$$

Combining (8.2.16) and (8.2.17), we get

$$M = \frac{mQ}{ve} = \frac{mN_A}{v}\frac{Q}{eN_A} = \frac{A}{v}\frac{Q}{F}. \tag{8.2.18}$$

Here we multiplied the numerator and denominator by the Avogadro number $N_A \approx 6.022 \cdot 10^{23}$ (the number of particles in 1 mole, where 1 mole is the amount of substance that contains N_A particles—sic), obtaining the molar mass of the substance $A = mN_A$ and the Faraday constant $F = eN_A$. The meaning of (8.2.18) is that the mass M sedimented on the electrode is proportional to the charge Q and to the chemical equivalent A/v.

However, the scheme described at the beginning of this section is too idealistic, so to say. We assumed that an ion approaching a metal surface immediately discharges, that is, only neutral atoms are in vicinity of the surface—no free charges here. Actually, in a real case, ions may be adsorbed at the surface because of many reasons; this charged layer attracts ions of the opposite sign to the surface and repels ions of the same sign. Thus, at the surface two layers of ions can be formed: one layer directly at the metal surface and another one, charged oppositely, in the liquid in vicinity of the surface.

A common scheme of an electrical double layer is presented in Fig. 8.2.1. In accordance with its name, it consists of two kinds of ions: the inner layer is on the solid, the outer layer consists of ions in the bulk of the liquid; meanwhile, the origin of those ions and their distribution in the layer may be very different.

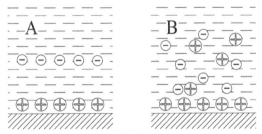

Fig. 8.2.1. Electrical double layer: A—as it may seem to be, B—as it really looks; from the point of view of theoretical models, case B corresponds to the diffusion double layer.

8.2.4 Formation of a double layer

Usually, an electrical double layer is discussed for a solid-liquid interface. We will consider these matters too, but begin with a completely different case—a liquid-gas interface.

A liquid surface may adsorb ions from the surrounding gas; this is especially easy if electrical discharges take place in the gaseous phase. We emphasize: in this case, the surface adsorbs not ions from the very liquid, but gets an electrical charge from the outside; it can be ions from directly the gas or ions that were formed on the liquid interface when, for instance, an electron flow bombards the surface. The outer layer consists of 'regular' ions from the liquid.

A relative to such a way is the formation of a double layer due to processes directly on the solid surface placed in a liquid. Because of the contact with a liquid, the solid may dissociate, producing ions that will detach from the surface and float into the liquid, making the surface charged with the opposite sign. Here, as above, ions that form the double layer are external for the given liquid. An example was considered in (Fridrikhsberg 1986): on the surface of the silicate-type lattice, under the effect of water groups SiOH are formed. Such a group dissociates into SiO^- and H^+, and hydrogen ions go into the liquid, forming the outer layer, while the main parts SiO^- form the inner layer.

A plain, ordinary way to produce a double layer at a solid–liquid interface was mentioned in the previous section: different ions from a liquid have different adsorbing properties, and a solid surface captures only one kind of them. Thereby, this captured kind of ions forms the inner layer, while poor-adsorbing ions stay in a liquid and construct the outer layer.

The most interesting variant of an electrical double layer is the so-called lyoelectrical layer[10] (see Fig. 8.2.2), which is formed by polar molecules of liquid attached for some reasons to the solid surface. This is a fundamentally different case in comparison to the previous one: the liquid molecule is not separated in two ions—positive and negative; in this mechanism of formation, the molecule is 'safe and sound', so the double layer consists of two bounded ions. There may be two reasons why molecules are oriented in such a certain manner:

✓ they were adsorbed by the solid surface

✓ the solid surface is charged, so it attracts the part of a polar molecule with the opposite sign of electrical charge (and repels another tip).

Of course, a mixing case is possible: ions on the surface can be formed in different ways, and the liquid may contain polar molecules.

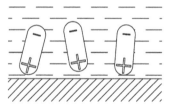

Fig. 8.2.2. The lyoelectrical double layer: liquid contains polar molecules which are specially oriented

8.2.5 Electrostatic potential

Let us consider the distribution of the electrostatic potential φ in a liquid containing ions of both kinds: positive ions with charge q_+ and negative ions with charge n_-, corresponding number densities are n_+ and n_-. In a liquid with a dielectric permeability ε, the Poisson equation for φ reads

$$\frac{d^2\varphi}{dx^2} = -\frac{\rho}{\varepsilon_0\varepsilon}. \tag{8.2.19}$$

Of course, we consider a flat one-dimensional problem. Boundary conditions for (8.2.19) correspond to the value φ_0 on the surface and zero on infinity:

$$\varphi(0) = \varphi0, \ \varphi(\infty) = 0. \tag{8.2.20}$$

---

[10] This is an old and non-common term that was used actively in the middle of the XX century. Here we try to revive it.

The volume charge is

$$\rho = q_+ n_+ + q_- n_-, \tag{8.2.21}$$

as above, we must take into account that $q_- < 0$.

In equilibrium, the number density of ions obeys the Boltzmann distribution: particles of each kind are distributed on their energy β as $\sim e^{-\beta/kT}$, and the energy of an ion in an electrical field with potential φ is $\beta = q\varphi$. Thereby, we have

$$\rho = q_+ n_+^0 e^{-q_+\varphi/kT} + q_- n_-^0 e^{-q_-\varphi/kT}, \tag{8.2.22}$$

where n_+^0 and n_-^0 are number densities of corresponding kinds of ions at infinity, where $\varphi = 0$, in accordance with (8.2.20). If the liquid is neutral at infinity, there must be

$$q_+ n_+^0 + q_- n_-^0 = 0, \tag{8.2.23}$$

Let us analyze the charge density distribution in vicinity of the charged surface, i.e., of the potential φ_0. If $\varphi_0 > 0$—the surface contains positive charges—then near this surface the first (positive) term in (8.2.22) is less than $q_+ n_+^0$, while the second (negative) term is greater than $q_- n_-^0$, since the exponent here is positive. Number densities are shown in Fig. 8.2.3.

Then, let us assume that the potential is comparatively weak, so that

$$\frac{q\varphi}{kT} \ll 1; \tag{8.2.24}$$

so, we may expand exponents in (8.2.22) in series, to get

$$\rho = q_+ n_+^0 \left(1 - \frac{q_+\varphi}{kT} \right) + q_- n_-^0 \left(1 - \frac{q_-\varphi}{kT} \right). \tag{8.2.25}$$

and we have from (8.2.19) with (8.2.23) and (8.2.25) the equation

$$\frac{d^2\varphi}{dx^2} = \underbrace{\frac{q_+^2 n_+^0 + q_-^2 n_-^0}{\varepsilon_0 \varepsilon kT}}_{R_D^{-2}} \varphi. \tag{8.2.26}$$

The factor at φ in (8.2.26) has the dimension of the reverse square length; this length, denoted as R_D, is a characteristic spatial scale for the deviation of the potential. This parameter is called the Debye radius; all the theory presented here is the Debye–Huckel theory.

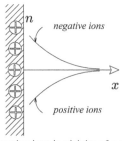

Fig. 8.2.3. Distributions of positive and of negative ions in vicinity of a positively charged surface. Positive charges are repelled from the surface, the negative ones are attracted. In sum, the negative volume charge is concentrated there; this negative charge 'screens' the surface from the outer space of liquid.

The equation (8.2.26) has a simple solution that satisfies boundary conditions (8.2.20):

$$\varphi = \varphi_0 e^{-x/R_D}. \tag{8.2.27}$$

Now let us estimate parameters that we meet during this solution. First, look closer at the condition of linearization made above—in equation (8.2.24). Such linearization is a usual move in plasma physics where, by the way, the Debye–Huckel approach is used in a questionable manner—for a single plasma particle instead of a surface; exactly because of this problem, the length turns into radius (radius around the selected plasma particle). However, in plasma we usually have to deal with a medium at temperature $\sim 10^4$ K; thus, the fulfillment of (8.2.24) looks easier than for a liquid where temperature is lower by two orders of magnitude. For $T \sim 300$ K and for a single-charged ion, i.e., for $q = e$, we see that the potential on the surface must be

$$\varphi_0 \ll \frac{kT}{e} \approx 0.03\,\text{V}, \tag{8.2.28}$$

and we even round this value up. Thus, the potential of the surface must be considerably small, otherwise we cannot use linearization and, therefore, the solution (8.2.27) for the potential.

Some help, however, may come if charges of ions are the same, i.e., $q_+ = q = -q_-$ and, consequently, $n_+^0 = n_-^0 = n$. If so, we have for the charge density, expanding up to quadratic terms

$$\rho = qn\left(1 - \frac{q\varphi}{kT} + \frac{1}{2}\left(\frac{q\varphi}{kT}\right)^2\right) - qn\left(1 + \frac{q\varphi}{kT} + \frac{1}{2}\left(\frac{q\varphi}{kT}\right)^2\right) = -\frac{2q^2 n\varphi}{kT} \tag{8.2.29}$$

again. Thus, to use linearization, we may replace the sign '$\ll$' in (8.2.28) with '$<$'. This is better, of course, but does not solve the problem principally. For more realistic cases, we have to consider the full equation

$$\frac{d^2\varphi}{dx^2} = -\frac{q_+ n_+^0}{\varepsilon_0 \varepsilon}\exp\left(-\frac{q_+\varphi}{kT}\right) - \frac{q_- n_-^0}{\varepsilon_0 \varepsilon}\exp\left(-\frac{q_-\varphi}{kT}\right). \tag{8.2.30}$$

This equation can be rewritten for the electric field strength $E = -d\varphi/dx$ as

$$-\frac{dE}{dx} \equiv -\frac{dE}{dx}\frac{d\varphi}{d\varphi} \equiv E\frac{dE}{d\varphi} = -\frac{q_+ n_+^0}{\varepsilon_0 \varepsilon}\exp\left(-\frac{q_+\varphi}{kT}\right) - \frac{q_- n_-^0}{\varepsilon_0 \varepsilon}\exp\left(-\frac{q_-\varphi}{kT}\right), \tag{8.2.31}$$

and we have the connection between E and φ, integrating from infinity, where $\varphi = 0$ and $E = 0$, to the given point

$$E^2 = \frac{2n_+ kT}{\varepsilon_0 \varepsilon}\left(e^{-\frac{q_+\varphi}{kT}} - 1\right) + \frac{2n_- kT}{\varepsilon_0 \varepsilon}\left(e^{-\frac{q_-\varphi}{kT}} - 1\right). \tag{8.2.32}$$

For instance, from (8.2.32) one may obtain the electric field strength at the surface, calculating the function value at $\varphi = \varphi_0$. Formally, one may continue this way and get a sort of solution for the given problem in the form

$$x = -\int_{\varphi_0}^{\varphi} \frac{d\varphi}{E(\varphi)}, \tag{8.2.33}$$

that is, find the dependence of the coordinate x counted from the surface, from the potential φ, with function $E(\varphi)$ determined by (8.2.32). However, this integral cannot be taken strictly analytically; thus, we stop at this point, especially since it can be taken numerically—we live in the XXI century now.

Then, let us consider the Debye radius R_D. Its physical meaning is the scale on which the disturbance of the potential field fades; in other words, this is the scale of screening of the external charge located in the medium. As it follows from its definition

$$R_D = \sqrt{\frac{\varepsilon_0 \varepsilon k T}{q_+^2 n_+^0 + q_-^2 n_-^0}} \tag{8.2.34}$$

this scale is greater when the temperature is higher (when particles move faster, disordering the screen) and when the volume charge is lower (simply the screen is weak). For single-charged ions of both kinds, i.e., also for $n_+^0 = n_-^0 = n$, and for $\varepsilon = 1$, we have

$$R_D = 69 \sqrt{\frac{T}{n}} [\text{m}] \tag{8.2.35}$$

where temperature must be inserted in kelvins, while the number density in m^{-3}. For instance, if the temperature is 300 K and $n = 10^{20}$ m^{-3}, the Debye length is 0.1 μm.

We will make some comments that, at first glance, may ruin all the consideration given above. As follows from experiments ε and comparison of their results with the theory, the permittivity ε, used in all the equations presented above, does not equal to the 'regular' permittivity of the given substance. Two reasons may be responsible for such a deviation.

The first reason is that in vicinity of the surface, where concentration of charged particles of one kind is high, electrical properties of the medium correspond not to the solution in its normal state (i.e., with number density of charged particles n), but to a charged medium (i.e., with number density $n e^{q\varphi_0/kT}$, and the exponent, as we have seen above, can be large).

The second reason is sad: the spatial scale of a double layer is so small that, actually, the consideration may concern only a few intermolecular distances. In this case, there is no sense in considering Boltzmann-like distributions of charged particles in the form of (8.2.22)—they were derived for macroscopic case, while on such short spatial scales fluctuations (i.e., random deviations from average values) are very strong. Moreover, on short distances many nuances that can be negligible on larger scales play a significant role: at various non-homogeneities of the surface. Of course, the direct application of equations from this section is impossible. On the other hand, if we understand that almost all the potential of the surface (i.e., the potential of the inner layer of ions) drops (i.e., screened by the outer layer) on the short scale, comparable to intermolecular distances in the liquid, then we have a reasonable question: what is the purpose of finding the structure of this short layer? We may represent the problem in two parts: considering the case A from Fig. 8.2.1 as the first, internal part of the double layer, and then supplement it with the outer part in form of B from Fig. 8.2.1. The potential on the border between layers A and B is the shielded potential of the surface.

8.2.6 Double layer and triboelectricity

Double layer may be considered as some reason behind triboelectric effects.

Let us suppose that on the boundary of two contacting bodies, a double layer appeared; as we considered above, that can be caused by many reasons, and friction can be counted as one of them. Then, when bodies suddenly detach from each other, they still carry charges that they gained because of friction.

This mechanism, at first glance, may give the answer to the problem of triboelectricity. On the other hand, such an answer is merely the same question with rearranged words: now one has to explain in detail how a double layer arises in such conditions.

8.3 Electrokinetic effects

8.3.1 The list of effects

Electrokinetic effects, in short, is the complex of phenomena that concerns two matters: (a) electrization, (b) fluid flow, and the casual relationship between them can be of any kind. We may list several effects relative to electrokinetic phenomena (Hunter 1981, Masliyah 2005):

- ✓ electrophoresis;
- ✓ electroosmosis;
- ✓ sedimentation potential;
- ✓ streaming potential.

Note that in some traditions, electrokinetic effects refer only to the first two items, but we will consider the classic set—all of them, especially if we take into account the application of these effects to a cavitating flow, where an external electrical field is absent.

Electrophoresis occurs in two phase systems: the main phase—liquid or gas, and the second suspended phase—solid or liquid particles. The dispersed (colloid) phase can move into the main phase when an external electrical field is applied. Actually, the second part of the word, '-phoresis', always means a similar phenomenon: movement under the applied force, and the first part of the word means exactly the type of that force.

Electroosmosis is the movement of fluid in capillaries or in a porous solid under the influence of an external electrical field. If the liquid contains an excess volume charge, for instance due to separation of charge on the wall of capillary, then the electrical field will affect this charge, and this charge will pull the liquid after it.

The sedimentation potential arises when charged colloidal particles are settled in a gravity field; it is also known as the Dom effect or the migration potential (Masliyah 2005). In a liquid, a colloidal particle is surrounded by a cloud of opposite charge; thereby, the migration potential is established due to a complicated process of interaction between colloidal particles and free charges in a liquid.

Indeed, these three electrokinetic effects do not play a big role in our problems, except for the case when bubbles play the role of colloidal particles: they can be charged and bear that charge on their walls. However, the fourth electrokinetic effect is very important for us.

The streaming potential, in some sense, is the reverse effect in relation to electroosmosis. When a fluid contains excess charges and flows under an external pressure gradient in a capillary[11] in the absence of an external electrical field, then the excess charge will accumulate downstream. This accumulated charge induces an electrical field that prevents the following movement of charge particles (ions) downstream, see Fig. 8.3.1.

The electrization of a fluid due to flow in narrow dielectric channels was observed in many experiments. The fact is undeniable and can be explained by triboelectric effects (adjusted for the unclear nature of triboelectricity itself). However, accompanying effects depend on nuances of the construction of the experimental setup.

If the flanges in Fig. 8.3.1 are connected, that is, the circuit is short and $U_{AB} = 0$, charges produced on the wall may flow free from the working section. Note that this is a normal case based on the aim of that construction: flanges grip the pipe, so they are connected with metal bolts. In this case, no additional effects take place.

---

[11] Actually, not only in a capillary, it can be a pipe of a 'normal' size.

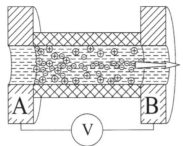

Fig. 8.3.1. Fluid flows from left to right in a plastic pipeclamped between two metal flanges. Due to triboelectric effects, charges separate on the pipe wall: positive charges settle on the wall, negative ones move with a flow downstream and settle on the flange B. The streaming potential U_{AB} can be measured between two flanges; it prevents the electrical current from coming out from the pipe.

However, if the flanges are isolated one from another, then the physical picture is complemented by new important details. The flange B now collects charges escaping from the pipe with a flow, until the field of accumulated charges repels them back from the flange B to the pipe: in equilibrium, the potential (charge) on the flange B establishes from the condition of $dQ/dt = 0$; new charges do not settle on the flange since they are pushed away from it by its own electrical field. Thus, the free charges now cannot leave the pipe, accumulating in the pipe volume and, sooner or later, one may see electrical discharges caused by these charges in gaseous bubbles contained in the liquid. The light emission of this type will be considered in Chapter 9, the plasma parameters of which can be determined with spectroscopic diagnostics, in Section 8.5.

8.3.2 Zeta-potential

Zeta potential, or ζ-potential, plays an important role in the description of electrokinetic effects—effects which take place in the fluid flow with charge separation on the walls. As we know, viscous fluids attach to a wall, or, in other words, the fluid is immobile there. For the problems considered in this chapter, it is important that this layer—the shear plane in corresponding terminology—contains charged particles which are immobile too, see Fig. 8.3.2.

The electrostatic potential on the shear plane is exactly ζ-potential. Note that the surface potential φ_s can be of any value: it can be higher than ζ or lower—that depends on the processes that take place directly on the surface.

The width of the shear layer is about several intermolecular distances in a liquid, that is, of an order of ~ 10 nm. Here, inside this layer, where the potential changes from φ_s to ζ, theoretical

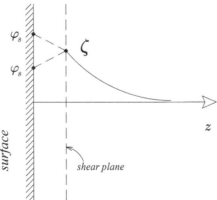

Fig. 8.3.2. The shear plane and ζ-potential. Actually, the charges inside the bulk of the liquid screen zeta-potential ζ, not the potential of the surface φ_s.

description is complicated by the very small spatial scale: usually, the continuum approach cannot be applied here. However, the distribution of charge inside the shear plane is not very interesting, especially if we take into account that ζ potential can be determined in experiments directly.

8.3.3 Interaction of charged macroparticles in a charged medium

Let us imagine a macroscopic charged particle in a medium which contains free charges. It may be a particle of disperse or colloidal phase in an electrolyte, a dust particle in plasma or even a bubble, carrying a charge on its walls, in a charged liquid.

In all these cases, the charged macroparticle is surrounded by free charges from the ambient medium, attracting the charges of opposite sign and repelling the charges of the same sign. In sum, each particle is enveloped by a cloud of predominantly opposite charge (see Fig. 8.3.3).

How do these charged particles interact? The evident part of the coupling are the Coulomb forces directly between two macroparticles: due to electrostatic forces, particles would repel from each other. However, here one may also see another way: each macroparticle attracts the cloud of its neighbor. Actually, we have two macroatoms there: each of them consists of a nucleus (macroparticle) and of a shell (cloud of charged microparticles). As for ordinary atoms, these macroatoms may attract each other due to coupling: in this case, because of coupling of the given macroparticle with the cloud around another macroparticle (Gerasimov and Sinkevich 1999).

Let us consider the potential energy of two macroatoms. It consists of three terms:

$$K(s) = K_0 + K_{pp}(s) + K_{pc}(s), \qquad (8.3.1)$$

where K_0 is the energy of interaction between the given macroparticle and its shell—this term will be omitted further because it does not depend on the distance s between two macroparticles; K_{pp} is the energy of interaction between two macroparticles (between 'nuclei'); K_{pc} is the potential energy for 'crossed' coupling—the given macroparticle and the shell of its neighbor. The shell of microparticles always 'adjusts' for the macroparticle; in a linear case, because of the superposition, the number density of microparticles is determined by the sum of the electrostatic potential from both macroparticles.

Omitting K_0, the equation (8.3.1) can be represented as

$$K(s) = \underbrace{\varphi(s)Q}_{K_{pp}} + \underbrace{\int \varphi(s,\vec{r})\rho(\vec{r})d\vec{r}}_{K_{pc}}. \qquad (8.3.2)$$

Here ρ is the charge density in the shell of the second macroparticle.

To find the potential energy, we have to determine the potential $\varphi(r)$ that is created by the shielded macroparticle: at the point r, this is the potential of the very macroparticle and of the cloud

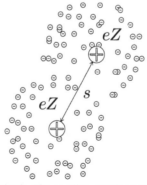

Fig. 8.3.3. Two macroparticles with their clouds. The coupling contains both the repulsive term and the attractive one. Macroparticles have the charge $Q = eZ$ and are displaced on the distance s from each other.

of microparticles that surrounds it. This problem is similar to one considered in the previous section, but now we have a problem in spherical geometry. Considering ions of the same charge e, with number density at infinity n_0, we now write the Poisson equation without excessive comments (for details see Section 8.2):

$$\frac{d^2\varphi}{dr^2} + \frac{2}{r}\frac{d\varphi}{dr} = \frac{en_0}{\varepsilon_0\varepsilon}\left[\exp\left(\frac{e\varphi}{kT}\right) - \exp\left(-\frac{e\varphi}{kT}\right)\right]. \tag{8.3.3}$$

Linearizing the right-hand side of (8.3.3) and using dimensionless variables we get

$$\frac{d^2\Phi}{dx^2} + \frac{2}{x}\frac{d\Phi}{dx} = \Phi, \tag{8.3.4}$$

where $\Phi = \dfrac{e\varphi}{kT}$, $x = \dfrac{r}{R_D} = r\sqrt{\dfrac{2e^2n_0}{\varepsilon_0\varepsilon kT}}$ is the dimensionless coordinate obtained with the Debye radius R_D.

The solution for (8.3.4) is

$$\Phi = \Phi_0\frac{x_0}{x}e^{x_0-x}, \tag{8.3.5}$$

where Φ_0 is the dimensionless potential of the macroparticle, x_0 is its dimensionless radius. Substituting (8.3.5) in (8.3.2), we integrate this and obtain the expression

$$K(s) = K_0\left[2\left(1 + x_0 + x_0^2e^{x_0}\right)\frac{e^{-s}}{s} - e^{x_0}e^{-s}\right], \tag{8.3.6}$$

where $K_0 = 4\pi kTn_0R_D^3\Phi_0^2x_0^2e^{x_0}$. For small particles $x_0 \ll 1$, and we have

$$K(s) = K_0\left(\frac{2e^{-s}}{s} - e^{-s}\right). \tag{8.3.7}$$

Thus, the potential energy has the minimum at $s_{\min} = 1 + \sqrt{3}$. The potential energy of two macroparticles contains the repelling part $\sim e^{-s}/s$ and the attractive one $\sim e^{-s}$. This means that, at some special conditions (Gerasimov and Sinkevich 1999), such particles may form localized condensed structures like liquids or solids, which consist not of atoms or molecules but of colloid particles surrounded by a charged media of positive and negative ions.

8.4 Cavitation accompanied by electrization

8.4.1 Ultrasonic cavitation

Ultrasonic (or, more precisely, acoustic) cavitation is a type of cavitation induced by the impact of a waveguide placed in a liquid and oscillating at high frequency. This phenomenon was discussed in Chapter 6 in general; here we are concerned with a certain special aspect of it—electrization of a fluid.

From the common point of view, the impact of a waveguide on a liquid is intensive friction. Inside the chapter devoted to triboelectricity and related phenomena, it is evident that this kind of influence may cause triboelectricity.

Electrization of liquid during ultrasonic cavitation, partially, was predicted by JakovFrenkel in his monography (Frenkel 1946). To be exact, he discussed a certain scheme of electrization, since, indeed, the very possibility of this phenomenon is obvious. According to Frenkel, the double layer formed near an ultrasonic waveguide may produce additional electrical charges, if its width exceeds the free path of electrons there.

The breeding of electrons can be described, in the simplest case, with the following equation:

$$\frac{dn}{dx} = \alpha n \,, \tag{8.4.1}$$

where n is the electron density and α is the ionization coefficient, which depends on the electric field strength. The physical meaning of α is clear—this is the rate of the electron breeding: an electron hits an atom, produces an additional electron, then two electrons produce two additional electrons impacting two atoms, and so on. For $\alpha = const$, that is, at a constant electric field, the number of electrons after length l becomes

$$n(l) = n(0)e^{\alpha l}. \tag{8.4.2}$$

However, to produce an additional electron in the collision, two conditions must be satisfied:

✓ traveling through the double layer, an electron must meet at least a single atom of the medium, that is, the length of the double layer must be greater than the mean free path of the electron here;

✓ traveling between two consecutive collisions, an electron must gain energy (from the electric field) that is sufficient to ionize the atom of the medium.

In addition to these conditions, we may point out an interesting matter: actually, now we are talking about the breakdown of a liquid. It may be surprising, but today the theory of the breakdown of the liquid—under 'theory' we mean a set of theoretical representations that (a) explains the features of the investigated phenomenon and (b) may answer at least the basic counterarguments of the opponents—does not exist. For the breakdown of a liquid, we deal with a dense media, where the simple breeding of electrons described by the equation (8.4.1) looks questionable. More complicated theories include the formation of the vapor phase in a liquid as a necessary condition; the enemies of this approach point at the breakdown of solids and suggest a common theory for the breakdown of a condensed medium with mechanisms similar to auto-electron emission (see Section 8.1), etc. At least, it is an interesting fact that in experiments (Torshin 2008), there is no huge difference in the breakdown in vicinity of the critical point, despite the fact that below critical parameters the vapor phase forms 'willingly' while above the critical point the vapor phase does not exist, the liquid phase does not exist there (above the critical point) too.

However, we are slightly distracted. For our case, in a liquid in the vicinity of an oscillating ultrasonic waveguide, we may expect the vibrations of the guide to cause the formation of a rarefied medium in this region, since the liquid is not 'glued' to it firmly, and when the guide moves back away from the liquid,[12] the liquid does not get in time to follow it, at least to some extent.

Thus, the presumption about some electric-breakdown-like processes near the waveguide looks reasonable. Moreover, in this case, we may expect, as from any respectable discharge, light emission. This type of light emission is observed in experiments (see Fig. 8.4.1); therefore, we may expect to discover some electrical effects too (Biryukov et al. 2014).

The working area of the experimental setup is shown in Fig. 8.4.2; the whole setup was discussed in Chapter 6. A titanium waveguide placed in glycerol, and we measure the voltage between the waveguide and the copper electrode placed into glycerol. Also, a thermocouple tip is located below the ultrasonic waveguide: we also control the temperature of the fluid at a single point. Of course, such a measurement,at a single point, is not very representative, but the ultrasonic waveguide provokes intensive convection in a liquid, so at the initial approximation one may consider this value as a typical value of temperature in the vessel.

---

[12] We remind that the runtime of the waveguide is very short: no one can detect it with the naked eye.

Fig. 8.4.1. The light emission from the ultrasonic waveguide (in this photo, we look at it from below). Note that this is only one of possible modes of such luminescence. Actually, this is a very dim glow, and not every eye is able to see it; this photo is made with a special cam.

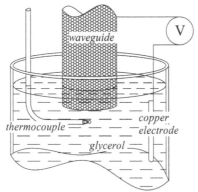

Fig. 8.4.2. The working area of measuring the electricity produced by ultrasonic cavitation. The voltage between the titanium waveguide and the copper electrode placed in a liquid is measured. Temperature is controlled with a thermocouple.

During the experiments, we observe cavitation on the ultrasonic waveguide (see also Chapter 6). We observe light emission (see above).Also, we observe voltage—the difference in the electrical potential of the waveguide and of the copper electrode (see Fig. 8.4.3).

This graph is very interesting from many points of view. At first, hearing about the electrical signal in extraordinary conditions, one may say that this is an induced voltage. This is definitely not the case: as we see from Fig. 8.4.3, voltage is not abruptly connected to the moments of power-on or power-off.

When the power is on, voltage begins to grow smoothly, and when the power is off, we almost do not see any trace of it on the graph. Interestingly, the voltage continues to grow for a few minutes further. Reaching the peak value, if it is possible to call such a weak maximum a 'peak', the voltage begins to decrease very slowly. The maximum values vary between 0.3–0.4 V or lower; the higher the initial temperature of the liquid, the lower the maximum voltage. The 'click' at the moment 800 s is simply a noise: all this setup is sensible to mechanical disturbance.

Then, if it is not an induced voltage, then one may assume that this is the direct effect of the temperature increase. Indeed, the friction of the waveguide in glycerol causes, at first, simply a heating of the liquid. Moreover, with rich imagination, one may find some similarities between the voltage graph and the temperature one. Thus, the origin of voltage might be the change in the physical property of glycerol with temperature.

To verify this hypothesis, we warmed glycerol in that very vessel (with the waveguide placed into glycerol, etc.—the configuration was the same) with an external electrical heater located in the liquid. No voltage appeared on the scheme: this means that the temperature increase is not the cause of the voltage, but, like the voltage,it is a consequence of some processes that occur in the liquid under the intensive ultrasonic impact.

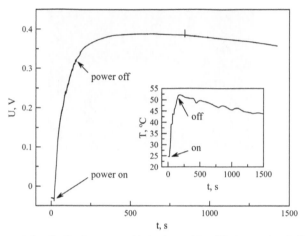

Fig. 8.4.3. The voltage generated on the titanium waveguide (main graph) and the temperature in the liquid with time. When the ultrasonic generator is on, both temperature and voltage begin to grow. When the generator is off, temperature stops rising almost immediately (note that temperature is measured at a distance from the waveguide, so convection takes some time). On the contrary, voltage keeps growing during ~ 2 minutes after that.

As a result of the waveguide impact, weak conductivity appeared in the liquid, of ~ $10^{-1}\mu$S/cm (Biryukov and Gerasimov 2017). Thus, glycerol became an electrolyte, in some sense, and the whole scheme presented in Fig. 8.4.2 is a very weak battery. To check this, we conducted additional experiments.

The pace of the voltage decrease is very slow, as one can see from Fig. 8.4.3. We may accelerate it with a special procedure: remove the waveguide from the vessel, dry it off (with rag) and place it back into glycerol. The result is shown in Fig. 8.4.4: the character of the voltage decrease differs significantly from the case presented in Fig. 8.4.3. Thus, drying the waveguide's surface we remove some important thing—the substance that contained separated charges. Note that, with time, the layer that was mechanically removed begins to repair itself: the character of decay in Fig. 8.4.4 varies—that dependence has an inflection point.

Then, the fact that voltage appears only when both the waveguide and the second electrode are placed into glycerol needs a more detailed proof. As we mentioned above, sometimes we removed the waveguide from the vessel at the stage when the power was off—at the relaxation phase; to be exact, we removed the vessel from under the waveguide: actually, the waveguide and all its equipment (including a cooling system) is a huge heavy construction, so it is much easier to get the vessel away from the waveguide. Corresponding schemes are shown in Fig. 8.4.5.

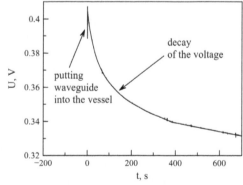

Fig. 8.4.4. Decay of the voltage after the 'mechanical intervention'. At the beginning phase, one may see the growth of voltage when the waveguide was brought into contact with the liquid.

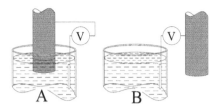

Fig. 8.4.5. Case A: both the waveguide and the second electrode are paced into glycerol. Case B: the waveguide is removed from the vessel. Power is off everywhere.

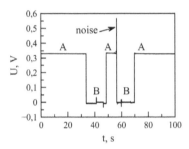

Fig. 8.4.6. Voltage corresponding to schemes presented in Fig. 8.4.6: A—waveguide in, B—waveguide out. In accordance with the classics: 'He was taking the balloon out, and putting it back again, as happy as could be...'[13] Yes, sometimes scientists behave like Eeyore.

As it follows from Fig. 8.4.6, separately, the titanium waveguide and the copper electrode does not produce voltage. Of course, such 'in–out' operation produced some disturbances that may be seen on that figure as a noise on the scheme.

Thus, acoustic cavitation produces some electrical effects in the vessel where it occurs. Note that despite mentioning light emission in this section, we do not state these electrical effects are the cause of that luminescence. This notification seems important: the light that accompanies the cavitation process is a very mysterious thing.

8.4.2 Flow in a narrow channel

The flow in a channel also can be accompanied with electrization, as it was considered in Section 8.3. Indeed, a cavitating flow is a particular case of a common flow which, as we discussed in Section 8.3, may cause electrization because of friction and one type of the so-called electrokinetic effects –the streaming potential, which holds charged particles in the channel and does not let them leak with the flow.

However, a cavitating flow has some important distinction concerning the possibility of electrization: the bubbles inside a liquid are the possible sources of additional charged particles because of electrical discharges in them. In addition, the discharges in bubbles may produce light emission, so the luminescence of the flow in a narrow channel may be the source of information about the medium inside bubbles via spectroscopy analysis.

Let us consider a scheme from Fig. 8.3.1, where the channel has the form of a throttle (see Fig. 8.4.7). The narrow part is the region of low pressure: note that here—slightly non-intuitive, because one may expect the narrow region to be the place of increased stress—pressure is lower than in the wide part: to keep the fluid discharge, the velocity is higher; then, because of the Bernoulli integral, the pressure is lower than at the beginning (wide) part of the throttle. On the other hand, usually—at least, for the problems considered here—the pressure at the entrance to the throttle is very high (of about tens of atmospheres or even higher), so the decrease in pressure in the narrow part does not necessarily lead to cavitation. Meanwhile, one may conclude that cavitation is possible

---

[13] A.A. Miln. 'Winnie-the-Pooh'.

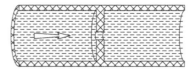

Fig. 8.4.7. Flowing through the throttle, a fluid may begin to glow.

here at least, and sometimes in very astonishing forms: in one of our experiments (Biryukov and Gerasimov 2016), we observed a lighting bubble formed at the end of the narrow part of the throttle.

Another two-phase zone arises at the second wide part of the throttle—at the exit: here the jet sprayed from the narrowest site contains both the liquid phase, as the gaseous one.

The electrical effects in such conditions were observed in several works (Gertsenshtein and Monakhov 2009, Biryukov and Gerasimov 2018). Electricity produced on the flanges in our experiments was definitely strong: of an order of $\sim 10^2$ V. Our team investigated various liquids in the setup pictured in Fig. 8.4.7: oils, water, glycerol; to be exact, we treated these liquids on variations of the setup pictured in this figure, since every liquid demands a certain approach because of the individual physical properties, especially viscosity. The light emission from this channel will be discussed in the next chapter; here we discuss the light emission only as a source of information about the plasma in a bubble.

The spectrum of the I-40A oil (industrial oil used in Russia); its composition is taken from (Ermolaeva and Golubkov 2011) listed in (Biryukov and Gerasimov 2016) is represented in Fig. 8.4.8 (Biryukov et al. 2012). The resolution of the AvaSpec-2048 spectrometer was 2.4 nm; this feature will be important further, during the analysis of the spectrometric data.

At first, we see on spectrum 8.4.8 the continuum, the nature of which will be discussed in Chapter 9. Here we are interested in the peaks.

Actually, the 'peaks' may be of two kinds: lines and bands. Lines are originating from the transition of an electron in the atom, bands correspond to the set of lines caused by vibrational–rotational transitions in molecules. We consider the spectroscopic issues in the next section.

The peaks in Fig. 8.4.8 are the bands of neutral molecules of nitrogen N_2. The I-40A oil contains many components, but not molecular nitrogen. Thus, one may conclude that nitrogen originated from air: from the air bubbles in oil. Indeed, the intensity of light emission depended on the specific conditions of the flow: the barbotage in the liquid tank increased the glow; again, we refer to Chapter 9, where this type of luminescence—the so-called hydroluminescence—is considered. The measured bands of nitrogen correspond to the second positive series of N_2—transition $C^3\Pi\text{-}B^3\Pi$; in Section 8.5, we provide a more detailed information about it.

To excite the molecules, usually, free electrons are required; thus, one may also expect plasma inside the bubbles. We see some confirmation of this fact on the spectrum: we also see a trace of presence of N_2^+ there. But this is all that we have: light emission of other air components is imperceptible. However, even the information obtained from N_2 series is sufficient to calculate some plasma parameters (see further).

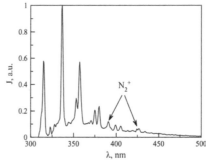

Fig. 8.4.8. The spectrum of light emission of the I-40A oil in a narrow channel. All the lines correspond to the second positive series of N_2 except for the bands of molecular nitrogen ion.

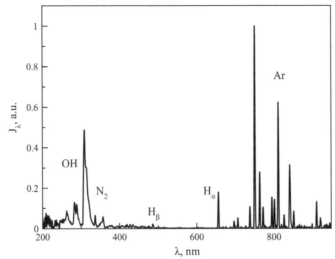

Fig. 8.4.9. The spectrum of light emission from the water flow in a narrow channel.

The luminescence in water gives a much more interesting information (Biryukov and Gerasimov 2018). The corresponding spectrum is presented in Fig. 8.4.9; this is again AvaSpec with 2.4 nm resolution. Note that the streaming potential measured between two flanges was ~ 10^2 V, the spectra presented below are measured in such conditions.

Here we see various components of glow. At first, we again observe the N_2 series (again the 2$^+$-series),which now combine in that spectral range with the OH bands; transition $A^2\Sigma^+$-$X^2\Pi$. This fact means that dissociation of water takes place in the channel:

$$H_2O \rightarrow H^+ + OH^-. \tag{8.4.3}$$

The light emission of H^+ cannot be observed, of course, since this is a naked proton in its nature (except bremsstrahlung, but not at such energies, surely). However, in plasma hydrogen ions recombine, produce neutral atoms of H, and their light emission can be seen on the spectrum—the Balmer series with lines with wavelengths obey the rule

$$\frac{1}{\lambda} = R_H \left(\frac{1}{2^2} - \frac{1}{n^2} \right), \tag{8.4.4}$$

where integers $n > 2$ correspond to the number of upper levels and $R_H \approx 1.097 \cdot 10^7$ m^{-1} is the Rydberg constant. The transition $3 \rightarrow 2$ gives the so-called H_α line with wavelength $\lambda_\alpha = 656$ nm, the transition $4 \rightarrow 2$ corresponds to the H_β line with $\lambda_\beta = 486$ nm and so on, in accordance with the Greek alphabet.

On our spectrum, we may differentiate H_α and H_β lines; indeed, H_γ too, but not on such a scale—with slightly larger magnification. As for another magnification, we also may find oxygen lines on the spectrum, see Fig. 8.4.10.

By the way, two reasons why we present here such details of the spectra: first, to convince the readers; second, to illustrate that spectroscopic investigations are very tangled affairs: sometimes, it feels like decrypting a cipher.

We may also compare the relative intensity of nitrogen bands and hydrogen lines (and hydroxyl bands too, of course) along the narrow part of the throttle, see Fig. 8.4.11. First, we have to note that the exact comparison of intensities in different parts is impossible, because the displacement of the

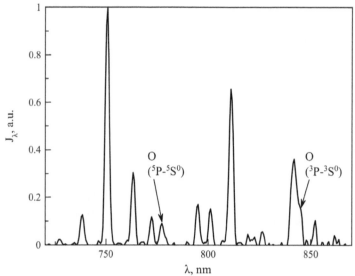

Fig. 8.4.10. Oxygen lines between the set of argon lines. The second oxygen line is overlapped with the argon line—the result of moderate resolution of the spectrometer.

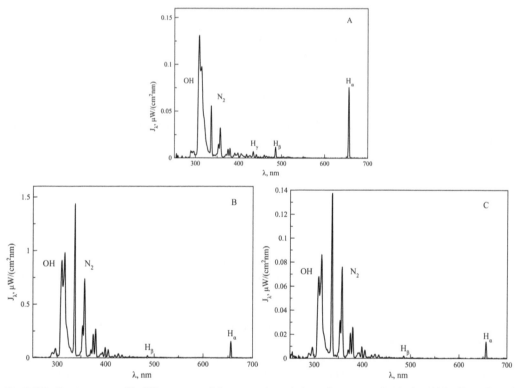

Fig. 8.4.11. Spectra measured in different zones of the narrow channel: A—at the entrance, B—in the middle, C—at the end of the throttle neck. Here we present the results in native dimensions—in absolute units to compare intensity from various regions.

sensitive element of the spectrometer was accompanied by the variation of the angle coefficient:[14] despite the small length of the channel, this factor played a role anyway.

Partially, the result is predictable: the light emission from N_2 series is conditioned by the presence of an air bubble in the liquid, and, as we mentioned above, due to cavitation one may expect the cavitation-induced growth of bubbles in the low-pressure region, i.e., in the narrow part of the throttle. Thus, at the entrance, where the pressure is high, we do not have many gaseous volumes and, consequently, we do not see the intensive light emission from the gaseous phase (of nitrogen from gaseous phase). On the contrary, in the region where bubbles may grow, we see significantly increased intensity of the N_2 component. This explanation, meanwhile, leaves an open question: does the amount of air in the bubble growth in the cavitation region, or the excitation condition differ for a large bubble? This is one of many problems that must be considered if we want to get more information about the gas discharge in bubbles.

Note that at the end of the throttle, the total light emission is dimmer: we should remember that fact in view of consideration of the glycerol luminescence in such conditions, see Chapter 9.

However, we do not see the corresponding increase of intensity of hydrogen lines: in comparison with the gain of brightness of nitrogen bands by an order of magnitude, the enhancement of emission of the H_α line looks modest. It is also interesting that the emission in hydroxyl bands increased more significantly than the emission in hydrogen lines. On the other hand, we see anyway that water dissociation is gained in the cavitation zone too—if the emission is a marker.

Finally, we may note that at the end of the throttle the total light emission is dimmer: we should remember that fact in view of the consideration of the glycerol luminescence in such conditions, see Chapter 9.

Thus, in this section, we presented some qualitative facts that show evidence of plasma in gaseous bubbles. In the following section, we obtain the quantitative information from these spectra.

8.5 The bubble plasma

8.5.1 Types of plasma

In a general case, plasma consists of at least three kinds of particles: neutral particles, electrons, positive ions. In some gases, in oxygen for example, plasma also includes negative ions; in this case, after the collision an electron does not ionize an atom but attaches to it.

Let the number of neutral particles in plasma per the unit of volume be n_0, the number of ions (per the unit of volume) is n_i, the corresponding value for the electrons is n_e. The degree of ionization is the parameter

$$\alpha = \frac{n_i}{n_0 + n_i}. \tag{8.5.1}$$

For a neutral gas $\alpha = 0$, for a fully ionized gas $\alpha = 1$.

Thus, one can try to describe plasma as a multicomponent gas. However, the mixture of ions, electrons and neutral particles should be described in a different manner than the mixture of oxygen, nitrogen and argon, for example. Due to significantly different properties (in mass and charge), various components of plasma have different characteristics of their dynamics. In a general case, different velocity distribution functions correspond to different plasma components. In the simplest situation, we may suppose that all the distribution functions are Maxwellians, but satisfy different temperatures; here and below we define 'temperature' as the measure for the equilibrium energy distribution.

---

[14] Angle coefficient is the fraction of energy from the radiation source to the receiver. When we change the orientation of the spectrometer's sensitive element, we see a different part of the object from a slightly different distance at a slightly different angle; this circumstance should not be forgotten.

Consequently, we may consider all three plasma components mentioned above as three separate sub-systems with, in common, three different temperatures:

$$T_n \neq T_i \neq T_e. \tag{8.5.2}$$

In a case when all the temperatures are equal to the same temperature T, we have an equilibrium plasma. This condition can be fulfilled in a hot dense plasma; for example, in the plasma of an electric arc. In general, this is not the case, and usually we have inequalities (8.5.2).

Ordinarily, the source of energy for a plasma is an external electrical field. In this field, the charged particles—electrons and ions—are accelerated until the friction forces $vm\vec{v}$ (where v is the frequency of collisions) will be balanced out by the electric field forces $q\vec{E}$. Thus, we have the stationary velocity for the simplified scheme when the particle gains an energy from the electric field and loses it in collisions:

$$\vec{v} = \underbrace{\frac{q}{mv}}_{b} \vec{E} , \tag{8.5.3}$$

where b is the mobility coefficient; see also Section 8.2. The corresponding kinetic energy of a particle is

$$\varepsilon = \frac{mv^2}{2} = \frac{q^2 E^2}{2mv^2}. \tag{8.5.4}$$

Thus, electrons obtain much higher kinetic energy from the electric field than ions because they are much lighter: $m_e \ll m_i$. Consequently, electrons tend to have higher temperature than heavy particles: $T_e > T_i$, but the temperature difference may be diminished because of collisions: in every collision an electron gives a part of its energy of an order of $\sim m_e/m_i$ to heavy particles. Since only a small portion of energy is transferred in a single collision, one should not mix the mean free path and the length of energy losses: for electrons, the first parameter is much less than the last one.

Thus, due to collisions, electrons try to transfer their energy to heavy particles, and the final result—the relation between temperatures of different components—depends on the frequency of collisions in the medium, i.e., on the density of plasma. At relatively high pressures (around the atmospheric level; for instance, in many arc discharges in air), the temperature of electrons is not significantly higher than the temperature of ions and neutral particles.

Above we used, as usual, the term 'temperature' as the measure for the kinetic energy of particles that make up the plasma. Defining the corresponding parameter in other distribution functions involves some difficulties.

Both ions and neutral atoms can be found in different electron configurations. For the case of the Boltzmann distribution on energies, the number of particles with energy E (the population of level E) is

$$N \sim \exp\left(-\frac{E}{k\theta}\right) \tag{8.5.5}$$

We also call the parameter θ in the denominator as 'temperature'.

The origin and the meaning of (8.5.5) can be explained by various approaches; particularly, its meaning can be explained as follows. Let us consider an atom that has energy levels E_i, and this atom may jump from one level to another with probability p_i^j—probability to transfer from level i to level j. Thereby, the total number of jumps from level i to level j is $p_i^j N_i$, etc. Considering only transitions to neighboring levels, in equilibrium we must have

$$\underbrace{p_i^{i+1} N_i + p_i^{i-1} N_i}_{\text{rate of outcome}} = \underbrace{p_{i+1}^i N_{i+1} + p_{i-1}^i N_{i-1}}_{\text{rate of income}}. \tag{8.5.6}$$

For the distribution of a kind of (8.5.5) with equal interlevel distance ΔE, we may write

$$N_i \sim \exp\left(-\frac{i\Delta E}{k\theta}\right) \sim \chi^i, \tag{8.5.7}$$

where

$$\chi = e^{-\Delta E / k\theta}. \tag{8.5.8}$$

Thereby, from (8.5.6) we get for probabilities of transitions

$$p_i^{i+1} = p_{i-1}^i = \frac{\chi}{1+\chi}, \ p_{i+1}^i = p_i^{i-1} = \frac{1}{1+\chi}. \tag{8.5.9}$$

Thus, the probability to 'jump up' is lower than the probability to 'jump down' by the value χ.

Note that equations (8.5.9) can also be derived from the detailed equilibrium between two neighboring levels and assuming that $p_i^{i+1} + p_i^{i-1} = 1$ which means, again, only transitions on the neighboring levels. In sum, relations (8.5.7–9) establish correlations between the static parameters (level populations) and the dynamic ones (transition probabilities). Above, we begin from the Boltzmann distribution and obtain correlations for p; one may choose the reverse way and obtain the Boltzmann distribution based on the hypothesis concerning transition probabilities.

The correlation between the parameter θ in the energy distribution like (8.5.5) and the temperature of the corresponding gas T, in general, is a very complicated question.[15] Particularly, for a plasma the equality $\theta = T_e$ is an open question. Old reasons for a positive answer were based on the hypothesis of equilibrium in the processes of 'settlement' of exited levels in 'electron–atom' collisions. However, in modern literature (Sarani 2010, Yanguas-Gil et al. 2006, Garamoon 2007, Biryukov et al. 2021), the equality $\theta = T_e$ is questioned. A kind of a 'compromise' is the opinion that the correlation between the electron temperature and the parameter θ can be stated only for high energy levels.

In molecular gases, we have a much more complicated situation. In addition to the kinetic energy of a molecule and the electron term energy distribution, we may also distinguish the inner energies that correspond to vibration and rotation of atoms in a molecule. In the case of the distribution (8.5.5) for these energies, we may introduce the 'vibrational temperature' and the 'rotational temperature' correspondingly.

Thus, even in the simplest case when each sub-system is in an equilibrium state, we have the set of temperatures: $T_n, T_i, T_e, \theta_e, \theta_r, \theta_v$. Only in the case when all these temperatures are equal to each other, the plasma is in equilibrium. In any other situation, we have to consider a non-equilibrium medium, where the correlation between various temperatures may be different.

The order of temperatures depends on the certain type of plasma. For instance, in a gas discharge plasma where electrons obtain their energy from an external electric field and give it out to other kinds of particles, we have

$$T_e \sim \theta_e > \theta_v > \theta_r \sim T_n \sim T_i. \tag{8.5.10}$$

Put into words: the electron temperature T_e (measure of the kinetic energy of electrons) is roughly equal to the measure θ_e of the distribution on excitation energies of atoms or molecules, the last one is reasonably higher than the measure θ_v of the distribution on vibrational energies of molecules which, in its turn, is larger than the measure θ_r of the distribution on rotational energies of molecules; the temperature of neutral particles T_n and of ions T_i (in the sense of the kinetic energy of their motion) is roughly equal to the parameter θ_r. Simply put, one may replace the term 'measure' with the term 'temperature' here, as it is often done: vibrational temperature, for example.

---

[15] If only we are not restricted with a brief sentence like 'all is equal in equilibrium'.

In many practical cases, however, the distribution on some sort of energy does not satisfy the expression (8.5.5). If so, we cannot use the term 'temperature' entirely: for such non-equilibrium objects,this term makes no sense.

Usually, plasma can be induced by an external electric field: for instance, in gas discharges of various sorts. But above we discuss a more exceptional situation: electricity can be produced by the intrinsic processes in the medium, in the absence of the external voltage. On the one hand, it does not matter what field the electrons are in: in a field of zeta-potential or in a field created by a battery supply. On the other hand, the plasma that originates in such specific conditions is a slightly mysterious object,the properties of which must be determined from the very beginning: now it is easy to consider the cathode layer or the positive column of glow discharges, since we know that these objects do exist. For the bubble plasma, we have no valuable information except our everlasting desire to reduce everything to known subjects (in this particular case, subjects of plasma physics). Meanwhile, the consideration of the properties of a poorly understood object should be started from the initial position: from the methods of obtaining the information about this object.

8.5.2 Spectrometers

Usually, plasma is very hot—the temperature of gas inside it may be 10^4 kelvins or more. Thus, any contact methods to determine the plasma temperature is inappropriate—no device such as a thermocouple, a resistive thermometer, etc., can withstand such heat. Thus, the only choice is spectroscopy—diagnostics of the medium by measuring its spectrum of emission or absorption.

To measure a spectrum, different methods can be used to decompose the initial 'white' light, that is, the light which contains all wavelengths, into a spectrum, obtaining the intensities corresponding to certain wavelengths in the given range. Excluding exotic methods, we must mention two different instruments to obtain the spectrum:

- ✓ prism;
- ✓ diffraction grating.

In their turn, the main parameters of spectrometers are

- ✓ sensitivity: what is the minimal intensity that can be measured with this spectrometer;
- ✓ resolution: what is the minimal spectral interval that can be detected;
- ✓ dispersion characteristic: how the signal depends on the wavelength.

The resolution of a spectrometer is usually described by the parameter FWHM—full width at half maximum (see Fig. 8.5.1). For a spectrometer, regardless of its construction, its signal always 'diffuses' on the wavelengths, more or less. Even a monochromatic signal at a certain wavelength λ_0 is perceived by a spectrometer as the 'hill' shown in Fig. 8.5.1; the measure of width for this hill is FWHM. To draw the form of the line one must know the function $f(\lambda)$ corresponding to the given spectrometer; usually this is a thing-in-itself and is not provided by the manufacturers of the spectrometer who usually present only an estimation of the FWHM parameter. In many cases, the Gaussian can be used for such a function:

$$I(\lambda) = \frac{I_0}{\sqrt{2\pi\sigma^2}} \exp\left(-\frac{(\lambda-\lambda_0)^2}{2\sigma^2}\right).$$ (8.5.11)

This is one more confirmation of the versatility of the Gaussian in physics, so to say. In (8.5.11), I_0 is the total intensity of the line which is independent from the 'diffusion' of the line in the spectrometer; λ_0 is the wavelength of the initial monochromatic light. From this function, we see

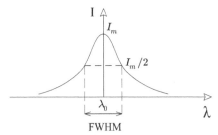

Fig. 8.5.1. Spectrometer turns an infinitely thin line into a hill. The full width of this hill at the half of its maximum is adopted as the measure of resolution of a spectrometer.

that the height of the 'hill' is $I_0 / \sqrt{2\pi\sigma^2}$, and the wavelength corresponding to the half of this height is $\lambda = \lambda_0 \pm \Delta\lambda$, where $\Delta\lambda = \sigma\sqrt{2\ln 2}$. Thus, FWHM is related to σ from (8.5.11) as

$$\text{FWHM} = \sigma 2\sqrt{2\ln 2}. \qquad (8.5.12)$$

In other words, if we know the FWHM from the passport of the spectrometer, we may determine σ for it from (8.5.12) and then calculate form of the lines given by this spectrometer with the expression (8.5.11). If the initial—we may refer to it as 'ideal'—spectrum consists of several lines with wavelengths λ_{0i} and intensities I_{0i}, then the measured spectrum can be represented by the function

$$I(\lambda) = \sum_{i=1}^{N} \frac{I_{0i}}{\sqrt{2\pi\sigma^2}} \exp\left(-\frac{(\lambda - \lambda_{0i})^2}{2\sigma^2}\right) \qquad (8.5.13)$$

which returns the value of intensity at the wavelength λ measured by this spectrometer. Note that a spectrum may not represent only the separate lines with centers at positions λ_{0i}—the wide lines may overlap, composing objects of strange forms (see Fig. 8.4.10 for example, where the oxygen line overlaps the argon line). When there are many overlapped lines, the spectrometer shows a 'hill' of bizarre shape, and the decoding of this 'hill'—decomposition into initial lines to extract the information about I_{0i}—can be done only with accurately adjusted parameter σ. Such a situation can be very complicated when the parameter σ depends on wavelength, and this case is quite possible.

Usually, prism spectrometers have much better resolution, i.e., much lower FWHM, than spectrometers with a diffraction grating. Among other factors, resolution depends on the spectrometer's slit opening width: the wider the slit, the worse the resolution. On the other hand, the wider the slit, the better the sensitivity, since more light enters the device. Therefore, when constructing a spectrometer, we have the dilemma: to choose between the resolution and sensitivity. In a certain case, one of two factors prevails, for instance, our spectrometer (which we use to measure the sonoluminescence spectrum) has the slit of around ~ 1 mm and, correspondingly, very bad resolution up to ~ 10 nm because this glow is very dim.

The next important parameter of a spectrometer is dispersion. Generally, dispersion means the dependence of the phase velocity of light in the medium on the frequency of this light. Or, which is the same, dispersion means the dependence of the refractive index on the frequency. However, for a spectrometer this phenomenon means that after a prism the rays corresponding to different frequencies will be refracted significantly differently (see Fig. 8.5.2). Due to this, the different wavelength interval $\Delta\lambda$ corresponds to the same length Δx in the image plane.

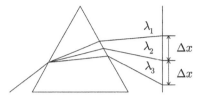

Fig. 8.5.2. Dispersion in a prism.

The dispersion characteristic of a spectrometer shows the dependence on the wavelength of the parameter

$$D(\lambda) = \frac{\Delta\lambda}{\Delta x} \qquad (8.5.14)$$

which is called a dispersion function or simply a dispersion.

For spectrometers with a diffraction grating $D(\lambda)$ is approximately a constant; for prism spectrometers, this function increases significantly with λ, the exact function depends on the kind of glass used in this spectrometer. For instance, for our old ISP-30 spectrometer the dispersion on the 200 nm wavelength is 0.35 nm/mm, while at 600 nm it is 11 nm/mm. Such a great increase in the red end of the spectrum makes such spectrometers not well suited for this part of the spectrum: one of certain inconveniences is the difficulty in calibration.

8.5.3 Plasma spectrometry

The spectroscopic diagnostics of plasma is a developed scientific area. Actually, this is a special scientific discipline, covered by an extensive literature. We may recommend (Ochkin 2009) where many methods are considered.

Spectroscopy can be used to determine all the basic parameters of plasma, such as temperature (exactly, temperatures, see above), electron and ion concentrations. These properties can be determined, in their turns, by various methods. Below we consider only the methods that were used to measure the parameters of the bubble plasma.

Let us begin with temperature. From the common measurement theory, measurement can be direct, when we measure the desired parameter immediately (like the length with a ruler) or indirect, when not the quantity itself is measured, but some characteristics which are somehow connected with this quantity. Any temperature determination is an example of an indirect measurement: the most domestic example is the measurement with a mercury thermometer, when one determines temperature by expanding the volume of mercury in the thermometer.

The measurement of the plasma temperature is very indirect, so to speak. Actually, in experiments we determine some characteristics of energy distributions of the given sub-system, for instance, of the exited atomic levels θ_e or of the rotational sub-system of molecules θ_r. The obtained quantities are directly what we get; for that example—temperature of the excited atomic or of the exited rotational levels. However, applying some theoretical assumptions, one may correspond the parameter θ_e to the electron temperature T_e and quantity θ_r to the temperature of heavy plasma particles T. To apply these theories, one has to know a lot about plasma: for $\theta_e = T_e$, the electron system must be in equilibrium with the atomic system (Biberman et al. 1987), which is rather an exception than a rule—plasma of the most types of gas discharges is non-equilibrium. However, and this is good news, for many cases plasma parameters can be estimated by such methods, albeit with a large margin of error.

The temperature of the electron component can be determined, as we mentioned above twice, through the parameter θ_e—the measure of the atom distribution on the exited states. Because of transitions of electrons from the atomic level i to another level k, the given volume of plasma emits radiation on the wavelength λ

$$J_\lambda = A_{ik} n_i \frac{hc}{\lambda}. \tag{8.5.15}$$

Here A_{ik} is the transition probability factor, $[A_{ik}] = s^{-1}$, n_i is the number density of atoms on the higher level at energy E_i. For the Boltzmann distribution, we have

$$n_i = n_0 \frac{g_i}{Z} \exp\left(-\frac{E_i}{kT_e}\right). \tag{8.5.16}$$

Here we use the electron temperature without equivocal, since, first of all, the equivocal was given above. In (8.5.16), n_0 is the total number density of atoms, g_i is the statistical weight of the level i, and Z is the partition function

$$Z(T) = \sum_i g_i e^{-E_i/kT_e}. \tag{8.5.17}$$

Note that often $T << E_i$, and $Z \approx g_0$—the statistical sum of the ground level where $E_0 = 0$.

Using (8.5.16) with (8.5.15), one may find the electron temperature with the following implementation, which is called the relative intensity method. Let us consider two lines, which are corresponding to two transitions $i1 \rightarrow k1$ and $i2 \rightarrow k2$. For them, we have from (8.5.15) and (8.5.16)

$$J_1 = A_1 \frac{hc}{\lambda_1} n_0 \frac{g_1}{Z} \exp\left(-\frac{E_1}{kT_e}\right), \tag{8.5.18}$$

$$J_2 = A_2 \frac{hc}{\lambda_2} n_0 \frac{g_2}{Z} \exp\left(-\frac{E_2}{kT_e}\right). \tag{8.5.19}$$

Here we simply denote all the quantities corresponding to the given transition by index 1 or 2, that is, $A_1 = A_{i1k1}$, $g_1 = g_{i1}$, and so on. Now we see that the relation of equations (8.5.18) and (8.5.19) eliminates, among other factors, the unknown quantity n_0, and the electron temperature can be found as

$$\frac{1}{T_e} = \frac{k}{E_1 - E_2} \ln \frac{A_1 g_1 J_2 \lambda_2}{A_2 g_2 J_1 \lambda_1}. \tag{8.5.20}$$

From (8.5.20), we may see that the error of estimation of T_e is higher when upper energies of two levels are close: if $E_1 \approx E_2$, then one will obtain from (8.5.20) the combination of an order of $\sim \ln 1/0$ (zero on zero), and because of the inevitable error of experimental data for J_1 and J_2, one would get almost any value of temperature T_e, even negative: this value would be determined mainly by the errors $\Delta J_{1,2}$.

Generally, to enhance the precision of the temperature calculation, another approach is used. If we see many lines in a spectrum, then we may find the electron temperature considering all the possible pairs of lines, i.e., consider the set of equations

$$y_j = \frac{1}{T_e} x_j, \tag{8.5.21}$$

where $x_j = \left(\dfrac{E_1 - E_2}{k}\right)_j$ is the difference of the upper level energies for the given pair of lines (denoted by index j); $y_j = \ln\left(\dfrac{A_1 g_1 J_2 \lambda_2}{A_2 g_2 J_1 \lambda_1}\right)_j$.

From (8.5.21), temperature can be found with the least square method to find the linear approximation of the set of points (x_j, y_j).

Thus, as we see, the electron temperature can be better defined if we have a lot of bright lines. On the contrary, if only one or two lines of the given atom are presented in the spectrum, the determination of T_e is impossible or almost impossible. A good example is the spectrum obtained during the fluid flow in a narrow channel (see the previous section): we may distinct hydrogen lines H_α and H_β there, but cannot normally calculate the electron temperature with them. However, to measure the electron temperature we may apply the method of the 'gaseous thermometer': one may mix up the outer gas—usually, argon is a good choice—and then measure temperature Te analyzing the emission lines of this gas. In Section 8.4, we presented the spectra that contained argon lines; below we calculate temperature with this data.

When the electron temperature is obtained, we may also determine the number density n_0 from equations (8.5.15) and (8.5.16), if we can measure the absolute value of intensity J_λ.

The molecular spectra can give information about rotational and vibrational temperatures, which correspond to distributions of vibrational and of rotational molecular energies in a gas.

The vibrational energy of a molecule can be represented as the energy of a quantum harmonic oscillator with correction on in harmonicity:

$$E_\upsilon = \omega_e\left(\upsilon + 1/2\right) - \omega_e x_e\left(\upsilon + 1/2\right)^2. \qquad (8.5.22)$$

Parameters ω_e and $\omega_e x_e$ are constants; integer υ defines the vibrational level.

The rotational energy can be written like for the rigid rotator at the energy level J with weak correction:

$$E_r = \left(B_e - \alpha_e\left(\upsilon + \frac{1}{2}\right)\right)J\left(J+1\right) + \left(D_e + \beta_e\left(\upsilon + \frac{1}{2}\right)\right)J^2\left(J+1\right)^2. \qquad (8.5.23)$$

Here parameters B_e, α_e, D_e, β_e are constants.

Equations (8.5.12) and (8.5.13) allow one to calculate energies of rotational and vibrational terms separately. This is a presumption, which may be wrong for some types of molecules or in certain conditions. For instance, a molecule of hydroxyl OH shows anomalies of such kind.

Meanwhile, for most cases we may consider electron, vibrational and rotational terms independently. Thus, the probability to have such energies is independent too. Therefore, we may consider the number density of molecules in the state with energy E_e (electron term), E_υ (vibrational term) and E_r (rotational term) as

$$n = n_0 \frac{g_e g_v g_r}{Z_e Z_\upsilon Z_r} e^{-E_e/k\theta_e} e^{-E_\upsilon/k\theta_\upsilon} e^{-E_r/k\theta_r}. \qquad (8.5.24)$$

When a molecule jumps from the level described by v' and J' to the level with v'' and J'', it emits a photon with corresponding wavelength which is determined by the difference in energies at the initial and the final state. Transition probability from the one vibrational level v' to another v'' is described with the so-called Franck–Condon factor $q_{v'v''}$; this is the band strength. Transition probability for rotational levels $J' \rightarrow J''$ is the Honl–London factor $S_{J'J''}$ (line strength). Finally, the intensity of light emission at the wavelength λ is

$$J_{v' \rightarrow v'', J' \rightarrow J''} = C \frac{q_{v'v''} S_{J'J''}}{\lambda^4} \exp\left(-\frac{E_v}{k\theta_v} \right) \exp\left(-\frac{E_r}{k\theta_r} \right), \qquad (8.5.25)$$

where the constant C depends on the number density of molecules.

As a result, the emission spectrum of a molecule consists of many lines which are grouped near certain vibrational transitions. The distance between close rotational lines is very small, and for most types of spectrometers, the resolution of which is insufficient to distinct lines one from another, the rotational lines merge into 'hills'—the top of such a hill corresponds to the given vibrational transition, and the hillside represents those merged lines.

The vibrational temperature describes the relative height of different 'hills' corresponding to different vibrational transitions. In the simplest methods, the value of θ_v can be determined with a kind of a relative intensity method; however, the better way is to calculate the whole spectrum—both vibrational and rotational transitions: this method allows to find out the vibrational and the rotational temperature simultaneously. In their turn, the rotational temperature, when the rotational spectrum is 'unresolved' and represents a 'hill', determines the width of the hill: the higher the θ_r, the wider that hill. However, the width of the hill is also determined simply by the spectrometer's resolution, which leads to the same effect.

Therefore, the determination of the rotational temperature with a spectrometer of a moderate resolution is a delicate work. For the most intriguing phenomena concerning light emission during cavitation, the dim glow can be detected only with very sensitive spectrometers—thereby, with spectrometers of poor resolution. The rotational temperature can be an estimation of temperature of a neutral gas, that is why the measurement of this quantity is very important: possibly, it is preferable to abstain from such estimations at all, but the goal is too seductive (see the next chapter). If so, at least, it is important to consider carefully two impacts: the widening of the vibrational-rotational band due to high rotational temperature or due to bad spectrometer resolution.

Fig. 8.5.3 illustrates this matter. In Fig. 8.5.3, we present two spectra of the N_2: one with a bad resolution and moderate rotational temperature is presented; the other one illustrates the opposite case—moderate resolution and high temperature. As we see from these figures, the result is the same 'by eye'. To determine the rotational temperature more or less correctly, we have to know the spectrometer's resolution well.

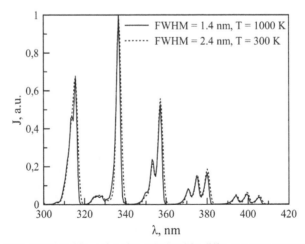

Fig. 8.5.3. Two spectra of the second positive series of N_2 calculated for different spectrometer resolutions and for different rotational temperatures.

In addition, we may emphasize that the rotational temperature—the measure of the distribution of rotational energy—is not necessarily the estimation of the gaseous temperature. An example, again, is a hydroxyl molecule, which originates usually in a reaction of dissociation of water:

$$H_2O \rightarrow H^+ + OH^-. \qquad (8.5.26)$$

As far as we can judge, in this reaction the hydroxyl molecule obtains high rotational energy, which is quite understandable: the water molecule splits in two parts, the total angular momentum must be conserved, so hydroxyl gets the excess. Thus, the high rotational energy of OH is caused by its nascence, not by the high temperature of gas where the dissociation takes place. Usually, in experiments the values of rotational temperature measured by the OH band are much higher than obtained after calculation with other components of the medium. In short, the hydroxyl thermometer is a bad thermometer.

8.5.4 Plasma in bubbles

In this section, we present the results concerning the parameters of plasma observed in experiments described in Section 8.4. These results were obtained in certain conditions, on a certain experimental setup—this is not a generalization of an enormous data set. Thus, they must be considered as an illustration for the common description of bubble plasma, they cannot be regarded as typical values of the plasma parameters, mainly because we don't know ourselves what parameters are typical.

At first, we consider the spectrum of water flowing in a narrow channel (Biryukov and Gerasimov 2018). The pressure at the entrance was 50 – 100 atmospheres; the diameter of the narrowest part of the channel is 1 millimeter. Under these conditions, a light emission was produced, which will be discussed in the next section. The common spectrum given in Fig. 8.4.11 shows that the plasma is formed in the channel. The parameters of this plasma can be determined as it was considered in Section 8.5.3.

In Fig. 8.5.4, we present the simulated spectrum of all molecular components (both positive series of N_2 and A–X transition of *OH*) in comparison with the experimental spectrum. To calculate the theoretical spectrum, constant C in (8.5.15) was chosen to fit the height of the most intensive band of the given molecule. Note that here we omit some special details; more information is presented in (Biryukov and Gerasimov 2018).

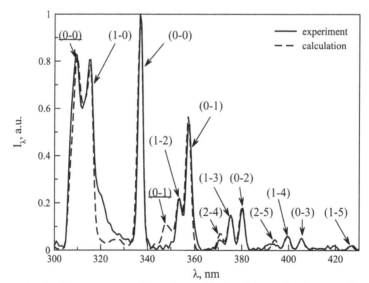

Fig. 8.5.4. Spectra of molecular components: nitrogen N_2 and hydroxyl OH; solid line—experiment, dashed line—calculation. The number of corresponding vibrational transitions is shown (underlined for OH). The part of the hydroxyl spectrum that is not fitted with calculations (on 320 – 330 nm) demands much higher temperature, of around ~ 10^4 K.

Theoretical spectra were calculated for temperatures for both gaseous components

$$\theta_v = 4500 \text{ K}, \; \theta_r = 300 \text{ K}. \tag{8.5.27}$$

The error of the values obtained with our method is up to 10%; thus, we may consider values on (8.5.27) only as estimations; this not extremely bad since above we agreed to look at them as illustrations. The reason behind such a high error, first of all, is the unresolved rotational spectrum. Note that the rotational temperature is much lower than the vibrational one: this may mean that the plasma is out of equilibrium, and the gas has a room temperature.

In Fig. 8.5.4, we see that a part of the experimental spectrum is poorly described by calculations with temperatures (8.5.27). This part corresponds to the hydroxyl level at high excitation energies: to fit the experimental values, we have to put $\theta_r \sim 10^4$ K in calculations. This case is normal for gas discharges in water: two significantly different temperatures are needed to describe the hydroxyl spectrum in such conditions (Sarani et al. 2010, Carrington 1964, Bruggeman et al. 2009, 2010). Thus, despite the fact that in our experiments we have slightly different external conditions—first, we have no external electric field, therefore, we have no gas discharge here, but have to deal with triboelectricity—we may expect that such an issue is quite understandable for us; the explanation was given above (high rotational energy origins from the origin of the hydroxyl molecule). The Franck–Condon factors for $0 \rightarrow 1$ transition differs significantly in various references (Nichols 1956, 1977, Ochkin 2009, Saleh et al. 2005, Felenbok 1963); the result can be seen in Fig. 8.5.4, where the disagreement between calculations and experiment is obvious.

The electron temperature can be estimated with an argon thermometer—by measuring the temperature of excitation states of argon atoms. Using the relative intensity method, we get the results presented in Fig. 8.5.5; as we explained above, in theory, the slope of the line approximating this point set can provide information about temperature.

However, as we can see, this is not the case. The points do not fit on the line, and, therefore, the slope of the curve may give only a rough estimation of the electron temperature:

$$T_e \approx 10^4 \text{ K}. \tag{8.5.28}$$

Actually, that is everything that we can say about this value. On the other hand, such a result is better than nothing, and, at least, we see that this value is much greater than the temperatures in (8.5.27).

The reason behind such a large error may give some useful information about the state of plasma, in its turn. The atoms in plasma do not obey the Boltzmann distribution (8.5.16): this system is out of equilibrium. Using the spectroscopy results, we may measure the population of the upper levels; the corresponding graph is presented in Fig. 8.5.6, where the value N_i is shown:

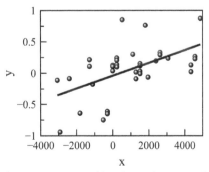

Fig. 8.5.5. An attempt to define the electron temperature with an 'argon thermometer'. Values x and y correspond to (8.5.21). In theory, the dots must fit a line the slope of which determines the electron temperature. Unfortunately, the plasma is out of equilibrium, and this method provides only estimations.

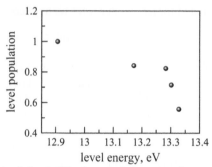

Fig. 8.5.6. The population of argon levels in a bubble plasma; relative units, where unity corresponds to the population of the level with the lowest energy. By the way, if it seems that this graph is 'too empty': we are not able to connect these dots as a matter of principle.

$$N_i = \frac{J_{ik}}{A_{ik} g_i \left(hc / \lambda_{ik} \right)}. \qquad (8.5.29)$$

As it follows from the previous consideration, in equilibrium we expect that this quantity obeys the simple exponent

$$N_i \sim \exp\left(-\frac{E_i}{kT_e} \right), \qquad (8.5.30)$$

and one may find T_e from such an approximation. However, it is easy to see that the dependence $N_i(E_i)$ has nothing in common with the function (8.5.30): the distribution on excited levels is non-equilibrium. Therefore, it is not a surprise that we can only estimate some characteristic value of electron temperature and nothing more.

Now we collect all the information that we obtained with spectrometric investigations:

- ✓ plasma is out of equilibrium: different temperatures are in equal to each other;
- ✓ the gas temperature is low and corresponds to the 'room temperature';[16]
- ✓ the electron temperature is high, nevertheless.

The combination of these facts means that the non-thermal plasma originates in bubbles: this is rather a glow discharge than an arc plasma, despite the fact that glow discharges take place at low pressures.

However, here we discussed only the results of electrization of water; it is possible that electrization of other liquids may give fundamentally different results. To check this, we consider the spectrum of light emission from the oil I-40A in the same conditions: when the fluid flows through a narrow channel.

The common spectrum was shown in Fig. 8.4.8. There we have seen the bands of nitrogen placed over the 'continuum'; the nature of that continuum will be discussed in Chapter 9. The analysis of these lines by comparison to calculations (see Fig. 8.5.7) gives the close values of temperatures for (8.5.27):

$$\theta_v = 4000 \text{ K}, \ \theta_r = 300 \text{ K}. \qquad (8.5.31)$$

---

[16] The room temperature is a beautiful term through the mouths of plasma physicists who usually have a deal with objects of temperature at $\sim 10^{4-5}$ kelvins: this may be a several hundred degrees. In our case, the room temperature means much more moderate values.

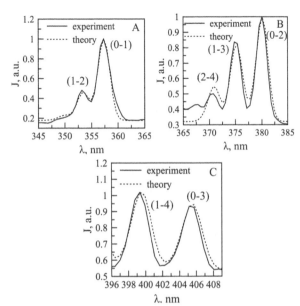

Fig. 8.5.7. Comparison of the measured and the calculated spectrum of nitrogen N_2. Since the experimental spectrum is placed on the continuum, we may compare only local parts of the spectra. The numbers correspond to vibrational transitions.

Taking into account the experimental errors, we may conclude that values (8.5.27) and (8.5.31) are very close; thereby, the plasma conditions do not differ significantly for bubbles in water and in oil.

Thus, we see that the plasma conditions are similar for water and oil. Note that we cannot observe light emission from the channel filled with glycerol, which was mentioned above. To be exact, we observed some glow from glycerol too, but that will be the matter of the next chapter.

The common property of plasma in water and in oil is the non-equilibrium state. For a dense plasma—we may expect that the pressure in bubbles is not less than the atmospheric pressure—we may conclude that this is a non-typical state, since electrons must give their energy to heavy particles. We see that they do not, so, one may assume two explanations of this fact.

Possibly, the bubbles are too small—less than the length of the effective energy transfer (LET). This length should not be mixed with the mean free path (MFP): MFP is the length between two consecutive collisions of the electron, but at a single collision an electron may transfer to a heavy particle only a fraction $\sim 10^{-4}$ of its kinetic energy (of an order of its mass ratio). Thus, LET is much larger than MFP. Note that ionization is a completely different process—it can be described as a collision between two electrons (the impacting electron and the electron from the atom shell); the efficiency of this collision is much higher since the masses of collided particles are equal. Thus, the first explanation is: at short spatial scales, electrons cannot heat a gas.

The second explanation is non-stationarity. If we have a series of microdischarges inside a bubble, then the result will be approximately the same: now, electrons have no sufficient time to transfer their energy to atoms and molecules. A dense gas plasma recombines quickly; thus, the birth and death of charged particles may represent a periodically repeating process. The proximity of each explanation to reality (or may be both, or may be none of them) can be established only after numerous experiments and theoretical treatments.

Here, we have to admit that today we have no sufficient information about the bubble plasma. This is a unique object, the properties of which should be investigated more thoroughly. We are far from the complete description of the bubble plasma now, especially if we take into account that some theories (and results of some experiments) predict the existence of plasma in extreme conditions. This side of the bubble science will be considered in the next chapter.

Conclusion

The electrical effects which arise due to friction are well known and poorly understood. Triboelectricity is an open scientific area, and many discoveries will be made in this field in the future. Today, we understand many static components of the common puzzle, but the whole dynamic picture is still hidden from us.

Restricting ourselves by the static cases, we may consider the formation of the double layer. The solid surface is the place where the charges contained in a liquid tend to separate for many reasons; moreover, the solid surface may produce free charges in a liquid by itself. Thus, the surface takes a charge of a certain sign, while the liquid takes a charge of the opposite sign. The structure of the near-surface charged layer is very complicated but, anyway, it is not as confusing as the theory behind it, since we have to deal here with a spatial scale of an order of the size of an ion. It was once thought that the separation of the double layer is caused by motion, but now we realize that the problem does not have such a simple solution.

The electrokinetic phenomena are a set of various effects that occur in the 'fluid flow—surface charge separation—electrical field' combination, and the casual relationship between the parts of this combination can be any. We are especially interested in the streaming potential—the electrostatic potential which appears in a narrow pipe (or a capillary) and locks the charged particles, formed at the wall, inside that pipe. If a charge cannot leave a pipe, it accumulates there, settles on the bubbles.

When the number of charged particles in a channel reaches some critical level, one may see gas discharges in the air bubbles contained in the liquid; of course, charges must be distributed irregularly on the bubble surface.

The spectroscopy of those discharges proves that we deal with a strongly non-equilibrium plasma, with high energy of electrons $\sim 10^4$ K, high vibrational temperature of molecules $\sim 10^3$ K and moderate rotational temperature of molecules $\sim 10^2$ K. This alignment means the discharge in a cold gas; possibly, its nearest analogue is a spark—a strongly non-stationary and therefore non-equilibrium object. On the other hand, the small spatial scale of the bubble may play a role too.

Another manifestation of electrization can be observed during acoustic cavitation. A metal waveguide placed into a vessel and impacting a liquid at frequency ~ 20 kHz also produces electricity. At first glance, this is a regular triboelectricity; however, it concerns sonoluminescence—one of the most intriguing phenomena in modern physics. Many issues discussed in this chapter will be continued in the following one.

References

Abramson, H. A. 1930. Electrokinetic Phenomena. III: The "Isoelectric point" of normal and sensitized mammalian Erythrocytes. *The Journal of General Physiology*, 20: 163–77. https://doi.org/10.1085/jgp.14.2.163.

Adam, N. K. 1930. *Physics and Chemistry of Surfaces*. Clarendon Press: Oxford.

Asadi, A., B. B. K. Huat and N. Shariatmadari. 2009. Keeping electrokinetic phenomena in tropical peat into perspective. *European Journal of Science and Research*, 29: 281–8.

Asadi, A., H. Moayedi, B. B. Huat and A. Parsaie. 2011. Probable electrokinetic phenomena in tropical peat: A review. *International Journal of the Physical Sciences*, 6: 2184–8. https://doi.org/10.5897/IJPS11.384.

Bailey, A. G. 2001. The charging of insulator surfaces. *Journal of Electrostatics*, 52:82–90.

Balestrin, L. B. S., D. D. Duque, D. S. da Silva and F. Galembeck. 2014. Triboelectricity in insulating polymers: evidence for a mechanochemical mechanism. *Faraday Discussions*, 170: 369–83. https://doi.org/10.1039/C3FD00118K.

Bazant, M. Z. 2008. Nonlinear electrokinetic phenomena. *Encyclopedia of Microfluidics and Nanofluidics*, 14: 1461–70.

Biberman, L.M., Vorob'ev V.S. and I.T. Yakubov. 1987. *Kinetics of Nonequilibrium Low-Temperature Plasmas*. Consultants Bureau: NY.

Biryukov, D. A. and D. N. Gerasimov. 2017. Dynamics of multibubble sonoluminescence intensity. *Technical Physics Letters*, 43: 520–2. https://doi.org/10.1134/S1063785017060049.

Biryukov, D. A. and D. N. Gerasimov. 2018. Spectroscopic diagnostics of hydrodynamic luminescence. *Journal of Molecular Liquids*, 266: 75–81. https://doi.org/10.1016/j.molliq.2018.06.043.

Biryukov, D. A., D. N. Gerasimov, H. T. T. Hoang and E. I. Yurin. 2021. Non-Equilibrium conditions in a luminescent medium during hydrodynamic luminescence of water. *Journal of Luminescence*, 6: 118164. https://doi.org/10.1016/j.jlumin.2021.118164.

Biryukov, D. A., D. N. Gerasimov and O. A. Sinkevich. 2012. Measurement and analysis of hydroluminescence spectrum. *Technical Physics Letters*, 38: 80–1. https://doi.org/10.1134/S1063785012010191.

Biryukov, D. A., Gerasimov, D. N. and O. A. Sinkevich. 2014. Electrization of a liquid during sonoluminescence. *Technical Physics Letters*, 40: 138–40. https://doi.org/10.1134/S1063785014020059.

Biryukov, D. A. and D. N. Gerasimov. 2016. Triboluminescence of liquid dielectrics: On a way to discover the nature of sonoluminescence. *In:* Olawale D., Okoli O., Fontenot R., Hollerman W. (eds.). *Triboluminescence*. Springer, Cham. https://doi.org/10.1007/978-3-319-38842-7_5.

Blakemorem, J. S. 2013. *Solid State Physics*. Cambridge University Press.

Bruggeman, P., D. Schram, M. Á. González, R. Rego, M. G. Kong and C. Leys. 2009. Characterization of a direct dc-excited discharge in water by optical emission spectroscopy. *Plasma Sources Science and Technology*, 18: 025017. https://doi.org/10.1088/0963-0252/18/2/025017.

Bruggeman, P., F. Iza, P. Guns, D. Lauwers, M. G. Kong, Y. A. Gonzalvo, C. Leys and D. C. Schram. 2010. Electronic quenching of OH(A) by water in atmospheric pressure plasmas and its influence on the gas temperature determination by OH(A–X) emission. *Plasma Sources Science and Technology*, 19: 015016. https://doi.org/10.1088/0963-0252/19/1/015016.

Burgo, T. A. L., T. R. D. Ducati, K. R. Francisco, K. J. Clinckspoor, F. Galembeck and S. E. Galembeck. 2012. Triboelectricity: Macroscopic charge patterns formed by self-arraying ions on polymer surfaces. *Langmuir*, 28: 7407–16. https://doi.org/10.1021/la301228j.

Carrington, T. 1964. Angular momentum distribution and emission spectrum of OH in the photo dissociation of H2O. *The Journal of chemical physics*, 41: 2012–8. https://doi.org/10.1063/1.1726197.

Crowley, J. M. 1983. Electrohydrodynamic Droplet Generators. *Journal of Electrostatics*, 14: 121–34. https://doi.org/10.1016/0304-3886(83)90001-3.

Delgado, A. V., F. González-Caballero, R. J. Hunter, L. K. Koopal and J. Lyklema. 2007. Measurement and interpretation of electrokinetic phenomena. *Pure and Applied Chemistry*, 77: 1753–805. https://doi.org/10.1351/pac200577101753.

Ermolaeva, N. V. and Y. V. Golubkov. 2011. Oborudovanie i tekhnologii dlya neftegazovogo kompleksa, 3: 49–53 (in Russian).

Felenbok, P. 1963. Contribution à l'étude du spectre des radicaux OH et OD. *Annales d'Astrophysique*, 26: 393–423.

Franklin, A. 1997. Millikan's Oil-drop experiments. *The Chemical Educator*, 2: 1–14. https://doi.org/10.1007/s00897970102a.

Frenkel, J. 1946. *Kinetic Theory of Liquids*. Oxford University Press: Oxford.

Fridrikhsberg, D. A. 1986. *A course in colloid chemistry*. Mir: Moscow.

Galembeck, F., T. A. L Burgo, L. B. S. Balestrin, R. F. Gouveia, C. A. Silva and A. Galembeck. 2014. Friction, tribochemistry and triboelectricity: recent progress and perspectives. *RSC Advances*, 4: 64280–98. https://doi.org/10.1039/C4RA09604E.

Garamoon, A. A., A. Samir, F. F. Elakshar, A. Nosair and E. F. Kotp. 2007. Spectroscopic study of argon DC glow discharge. *IEEE Transactions on Plasma Science*, 35: 1–6. https://doi.org/10.1109/TPS.2006.889270.

Gerasimov, D. N. and E. I. Yurin. 2018. *Kinetics of Evaporation*. Springer Cham. https://doi.org/10.1007/978-3-319-96304-4.

Gerasimov, D. N. and O. A. Sinkevich. 1999. Formation of ordered structures in thermal dusty plasma. *High Temperature*, 37: 823–7.

Gertsenshtein, S. Ya. and A. A. Monakhov. 2009 Electrization and liquid glow in a coaxial channel with dielectric walls. *Fluid Dynamics*, 44: 430–4. https://doi.org/10.1134/S0015462809030107.

Harper, W. R. 1967. *Contact and Frictional Electrifcation*. Clarendon Press: Oxford.

Hunter, R. J. 1981. *Zeta Potential in Colloid Science*. Academic Press.

Jouniaux, L. 2011. Electrokinetic Techniques for the Determination of Hydraulic Conductivity. inTech, 307–28.

Kittel, C. 2000. *Thermal Physics*. Twenty-first printing.

Klyshko, D. N. 1994. Quantum optics: quantum, classical, and metaphysical aspects. *Physics-Uspekhi*, 37: 1097–122. https://doi.org/10.1070/PU1994v037n11ABEH000054.

Lacks, D. J. and T. Shinbro. 2019. Long-standing and unresolved issues in triboelectric charging. *Nature Reviews Chemistry*, 3: 465–476. https://doi.org/10.1038/s41570-019-0115-1.

Landau, L. D. and E. M. Lifshitz. 1977. *Quantum Mechanics: Non-Relativistic theory*. Pergamon Press: Oxford.

Masliyah, J. H. and S. Bhattacharjee. 2005. *Electrokinetic and colloid transport phenomena*. John Wiley & Sons, Inc.: Hoboken, New Jersey.

Millikan, R. A. On the elementary electrical charge and the avogadro constant. *Electrical Charge and Avogadro Constant*, 2: 109–43. https://doi.org/10.1103/PhysRev.2.109.

Moayedi, H., B. B. K. Huat, T. A. M. Ali, S. A. Moghaddam and P. T. Ghazvinei. 2010. Electrokinetic injection in highly organic soil—A review. *Electronic Journal of Geotechnical Engineering*, 15: 1593–8.

Moore, J. R. and S. D. Glaser. 2007. Self-potential observations during hydraulic fracturing. *Journal of Geophysical Research: Solid Earth*, 112: B02204. https://doi.org/10.1029/2006JB004373.

Moyer, L. S. and H. A. Abramson. 1936. Elektrokinetic Phenomena : XII. Electroosmotic and Electrophoretic mobilities of protein surfaces in dilute salt solutions. *The Journal of General Physiology*, 19: 727–38. https://doi.org/10.1085/jgp.19.5.727.

Nicholls, R. W. 1956. The interpretation of intensity distributions in the CN Violet, C2 Swan, OH Violet and 02 Schumann-Runge Band Systems by use of their r-Centroids and Franck-Condon Factors. *Proceedings of the Physical Society. Section A*, 69: 741–53. https://doi.org/10.1088/0370-1298/69/10/303.

Nicholls, R. W. 1977. Transition probability data for molecules of astrophysical interest. *Annual Review of Astronomy and Astrophysics*, 15: 193–234. https://doi.org/10.1146/annurev.aa.15.090177.001213.

Ochkin, V. N. 2009. *Spectroscopy of Low Temperature Plasma*. Wiley-VCH Verlag GmbH& Co. KGaA, 2009.

Oddy, M. H. and J. G. Santiago. 2003. An electrokinetic mobility measurement technique using AC and DC electrophoresis. *7th International Conference on Miniaturized Chemical and Biochemical Analysts Systems*, 587–590.

Olivier, J. P. and P. Sennett. 1967. Electrokinetic effects in kaolin-water systems. I. The measurement of electrophoretic mobility. *Clays and Clay Minerals*, 15: 345–56. https://doi.org/10.1346/CCMN.1967.0150137.

Pertsov, A.V. and E. A. Zaitseva. 2008. Discovery of electrokinetic phenomena in Moscow University. *ICCCPCM'08*.

Postnikov, V. S. 1978. *Physics and Chemistry of the solid state* (Pfisika I khimiya tverdogo sostoyaniya). Metallurgiya:Moscow. (in Russian).

Rosenberg, B. 1972. Apparatus For fracticing electrophotography. *United State Patent Office*, 3: 649,514. *https://patents. google.com/patent/US3649514A/en.*

Saleh, Z. A., D. H. Al-Ameidy and B. T. Chiad. 2005. Franck-Condon factor and dissociation energy for OH free radical. *ATTI DELLA "FONDAZIONE GIORGIO RONCHI*, 439–42.

Sarani, A., A. Yu. Nikiforov, and C. Leys. 2010. Atmospheric pressure plasma jet in Ar and Ar/H2O mixtures: optical emission spectroscopy and temperature measurements. *Physics of Plasmas*, 17: 063504. https://doi.org/10.1063/1.3439685.

Tamm I. 1932. Über eine mögliche Art der Elektronenbindung an Kristalloberflächen. *Zeitschrift für Physik*, 76: 849–50 (in German) https://doi.org/10.1007/BF01341581.

Tandon, V., S. K. Bhagavatula, W. C. Nelson and B. J. Kirby. 2008. Zeta potential and electroosmotic mobility in microfluidic devices fabricated from hydrophobic polymers: 1. The origins of charge. *Electrophoresis*, 29: 1092–101. https://doi.org/10.1002/elps.200700734.

Tandon, V. and B. J. Kirby. 2008. Zeta potential and electroosmotic mobility in microfluidic devices fabricated from hydrophobic polymers: 2. Slip and interfacial water structure. *Electrophoresis*, 29: 1102–14. https://doi.org/10.1002/elps.200800735.

Torshin, Yu. V. 2008. *Physical Processes of forming the electrical breakdown in condensed dielectrics*. Energoatomizdat: Moscow (in Russian).

Xu, Ch., Y. Zi, A. C. Wang, H. Zou, Y. Dai, X. He, P. Wang, Y. C. Wang, P. Feng, D. Li and Z. L. Wang. 2018. On the electron-transfer mechanism in the contact-electrification effect. *Advanced Materials*, 30: 1706790. https://doi.org/10.1002/adma.201706790.

Yanguas-Gil, A., Cotrino, J. and A. R. Gonzalez-Elipe. 2006. Measuring the electron temperature by optical emission spectroscopy in two temperature plasmas at atmospheric pressure: a critical approach. *Journal of Applied Physics*, 99: 033104. https://doi.org/10.1063/1.2170416.

Zou, H., Y. Zhang, L. Guo, P. Wang, X. He, G. Dai, H. Zheng, C. Chen, A. C. Wang, C. Xu and Z. L. Wang. 2019. Quantifying the triboelectric series. *Nature Communications*, 10: 1427. https://doi.org/10.1038/s41467-019-09461-x.

CHAPTER 9

Cavitation and Light Emission

9.1 Sonoluminescence

9.1.1 The mysterious light

There may be several ways to estimate the weight of a scientific problem. For instance, one can take into account

- ✓ its importance for mankind (atomic energy)[1];
- ✓ its grandeur (black holes);
- ✓ its productivity for the development of science (differential calculus);
- ✓ its citation capacity (this sad story needs no comments);
- ✓ time taken to solve it (sonoluminescence).

For the last nomination, we do not consider the nature of a regular, everyday phenomenon (such as lightning) that took many centuries only because the science in the past was underdeveloped: of course, some physical conceptions must be in our disposal to solve the problem. But if we have those fundamental conceptions, or if we think we do, then we may solve the riddle. If we accept this way, then we meet sonoluminescence as a worthy challenger.

The light emission from a liquid under mechanical influence—sonoluminescence (SL)—was discovered by Frentzel and Schultes (Frenzel and Schultes 1934) in 1934. Since then, this phenomenon was observed many times in different conditions (under different sorts of mechanical effects); thus, the fact that this fluorescence exists is undeniable. The most frequently used method to obtain sonoluminescence is ultrasound cavitation: in conditions described in Chapter 6, bubbles obtained on those experimental setups emit light.

We think that we know all about bubbles (except some special quantitative details, which are connected with the computational complexity of equations that describe the problem), also we suppose that we know a lot about light emission and its origins (in optical range at least). Nevertheless, we may state a fact that we cannot connect these two kinds of knowledge together, to say the least, and sometimes very strange and very specific theories appear to explain the phenomenon of sonoluminescence. Generally, it is a great contrast in physics: today, scientists discuss the existence of the 'fifth' force,[2] but they cannot explain why an oscillating bubble emits light.

The probable main reason why sonoluminescence has no explanation to date is its very low intensity. Usually, it is such a dim glow that not every human can distinguish it by eye; for the same reason, regular measurement devices cannot register it too. Scientists have to use special apparatuses which may catch that weak light, but common parameters of such devices are poor—for instance,

---

[1] We do not discuss the positive or the negative sign of that importance.

[2] If we consider the electroweak interaction as two separate ones: the electromagnetic and the weak one, in addition to the gravitational and nuclear forces.

the spectrometer resolution. It is very sad, since very important information can be obtained with spectrometry: many properties of a luminescent medium can be extracted from its emission spectrum (see also Chapter 8), but sonoluminescence is a difficult object.

Moreover, the obtained information must be properly interpreted. Let us consider, for example, the simplest medium where sonoluminescence can be observed–water. To reach the observer (or the measuring device), the light from a bubble must pass through some layer of water of width x; on its path, the intensity of light I_λ will be reduced as

$$I_\lambda = I_{0\lambda} e^{-\mu_\lambda x}, \tag{9.1.1}$$

where μ_λ is the absorption rate at the given wavelength λ. The main problem is that quantity μ_λ strongly depends on the wavelength for water (Segelstein 1981); in this sense, water is not the simplest but rather the worst medium for experiments.

The absorption rate changes by orders in the quite narrow wavelength range in the infrared part of spectrum; this fact radically influences the measured intensity. For instance, a black body spectrum, which corresponds to the intensity

$$I_{0\lambda} = \frac{2hc^2}{\lambda^5} \frac{1}{e^{hc/\lambda kT} - 1}. \tag{9.1.2}$$

will be turned into a different spectrum shown in Fig. 9.1.1, if the radiation passed through a water layer of a moderate width x.

Now let us imagine that we obtain data from the experimental spectrum in Fig. 9.1.1 and we need to interpret it, retrieving from this curve some certain information about temperature: it is difficult to even recognize that this is a blackbody spectrum at all. Moreover, this is not the single problem with interpretation of a black-body spectrum, see Section 9.4.

Nevertheless, spectroscopy remains the trunk line of sonoluminescence investigations. Some of them give results that look strange at first glance (and at next glances too, probably); the most famous of them concerns very high temperatures of the medium during sonoluminescence (from ~ 10^4 K and higher). The object of investigations is a liquid and a tiny bubble inside it; it is necessary to explain how one of this media can heat up to such enormous temperatures. Moreover, these results were obtained in several ways: from spectroscopy of molecular bands (of hydroxyl), or by determining the continuum temperature. In the latter case, if we assume that the continuum spectrum represents black body radiation, the wavelength corresponding to its maximum would obey the Wien law

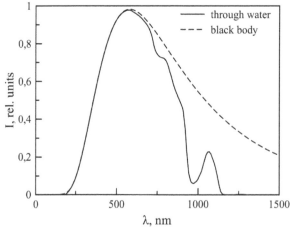

Fig. 9.1.1. The black body spectra after the water layer of 5 cm of width. The temperature of radiation (that is, the parameter T in (9.1.1)) is 5000 K. Data obtained from (Segelstein 1934).

$\lambda \approx 3 \cdot 10^3/T$, where temperature is measured in kelvins and the result in micrometers. Thus, it seems that a solid spectrum is also a thermometer; unfortunately, this thermometer is confused and confusing.

Another strange spectroscopy result is the great difference between the light emission from a single cavitating bubble and from the whole cavitation cloud. Below we will consider the details of this non-coincidence; in short, even the continuum of these spectra differs significantly.

We may conclude that spectroscopy is a very delicate measuring technique. For instance, in Chapter 8 we discussed the matter of hydroxyl-based measuring: it is not a reliable source of information, since the high rotational temperature (see Chapter 8 for details) measured from the OH spectrum is related, as one may judge, to the conditions of its formation (when a hydrogen atom H tears away from the hydroxyl OH in the H_2O molecule), not connected to the real temperature of the medium.

The critical analysis of spectroscopy data concerning the continuum part of SL spectra is given in (Lavrov 2001). In common, some special details usually are absent in works devoted to the subject (Lavrov 2001); in relation to the measurement of the continuum spectra:

✓ intensity calibration should take re-absorption in (possible) plasma into account;

✓ determination of the background level is a particular problem since light is scattered in a vessel;

✓ the continuum intensity should be accurately separated from the light emission or absorption in bands;

✓ by the way, comparison of different spectra measured in different conditions cannot be done with a simple normalization on unity (as we talked above for a single spectrum) because maxima may have different origins.

For measuring the spectral lines or bands, we may add the problems concerning the (poor) resolution of spectrometers, which were discussed in the previous chapter; this matter impedes the determination of plasma parameters with spectroscopic methods.

Thus, experimental research of sonoluminescence meets difficulties at every step. As for the theoretical one, here the situation is even worse. Sonoluminescence accompanies bubble oscillations with strong deviations of the parameters of substance inside it. The common description of bubble oscillations given in Chapter 7 provides a general picture, while to formulate such a point as light emission, we need to analyze the process at a very short time scale on a very short spatial scale. Actually, the main theoretical apparatus—physics of a continuous media (including hydrodynamics, thermodynamics, etc.)—cannot be firmly applied on such scales. All the processes important for sonoluminescence, all the candidates for the role of the possible source of light emission successfully hide behind a tiny pulsating bubble—yes, here it is easier to hide beyond small objects than vice versa.

In this section, we will start from the experimental results. In the following sections, we discuss the phenomena related to sonoluminescence, and then, finally, we will discuss the possible explanations of this phenomenon.

9.1.2 Single-bubble sonoluminescence

As we considered in Chapter 6, we may distinguish single-bubble acoustic cavitation (cavitation in an external sound field) and multi-bubble cavitation. Correspondingly, one may point out single-bubble sonoluminescence (SBSL) for a single luminous bubble, and multi-bubble sonoluminescence (MBSL) for a fluorescent bubble cloud.

Historically, it all began from multi-bubble sonoluminescence. Single-bubble case was described first in (Gaitan et al. 1992); it can be obtained in two ways:

✓ as it was described in Chapter 6—in an external acoustic field;

✓ as a laser-induced cavitating bubble.

The last approach was used, for instance, in (Volgel et al. 1996, Akhatov et al. 2001, Supponen et al. 2017) for the same purpose as acoustic cavitation: to obtain a single bubble—in this case, with the optical breakdown of the medium—the collapse of which generates light emission.

In the case of single-bubble sonoluminescence, it is comparatively easy to establish the moment of light emission and its location. Analyses of a video clearly show that the light originates from a collapsing bubble. To be exact, the flash appears from the point where the collapsing bubble is placed at during its compression stage; optical resolution cannot allow us to determine precisely the certain 'luminous pixels' of the image. Initially, in early works, the coincidence of the flash location with the bubble location was interpreted as justification of the assumption (or even the solid fact) that the vapor in the bubble is the source of light emission itself. Later, another concept was considered: the source of light is not the gaseous phase but the walls of the bubble (the liquid surrounding the bubble).

The spectrum of single-bubble luminescence from (Putterman 2000) is presented in Fig. 9.1.2. We see that this is a continuum without any lines or bands; the detailed discussion concerning the possible nature of sonoluminescence will be given in Section 9.4, but we may note that the absence of atomic lines or molecular bands can be treated as some argument against high temperatures inside a bubble: at high temperatures, atomic lines from the gaseous phase would be observed.

Another important matter is the duration of a sonoluminescent impulse: it was determined in the 1990s that this time is ~ 10–100 picoseconds (Putterman 2000). Above, in Chapter 6, we determined that the time period τ of a bubble oscillation can be estimated as

$$\tau \sim R\sqrt{\rho / p}. \tag{9.1.3}$$

If we only assume that the flash duration can be somehow connected with the oscillation period—the flash occurs at some stage of pulsations, for instance, at the late stage of the collapse—then, we come to conclusion that the time period $\sim 10^{-10}$ s corresponds to the radius of a bubble in water (for density 10^3 kg/m^3 at pressure $\sim 10^5$ Pa) ~ 1 nanometer. Of course, this value is abnormally small; among other reasons, cavity of such a spatial scale contains only several molecules—this object cannot be called a 'bubble' at all, it cannot be described with the continuous medium approach, etc.

Thus, the flash duration is conditioned by some other reasons, not simply by the collapse phase of the cavitating bubble. Some special processes take place during the bubble shrinking, and the time scale of that processes—excitation and deexcitation of the medium—is much shorter than the period of bubble pulsations.

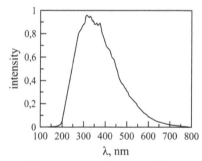

Fig. 9.1.2. Spectrum of SBSL, data from (Putterman 2000). The substance is water.

9.1.3 Multi-bubble sonoluminescence

From a certain point of view, multi-bubble sonoluminescence represents some additional problems: in this case, contrary to SBSL, it is not clear what phase is the source of light emission: the liquid phase or the gaseous one. Of course, we may make the conclusion based on the SBSL results, but this way would leave space for uncertainty. To answer this question, we will explore the morphology of the MBSL itself.

At first, the foam on the waveguide will be very helpful (see Chapter 6): during cavitation, 'fractional structures' are formed on the ultrasonic waveguide. The dynamics of the formation of such a structure are shown in Fig. 9.1.3. As we see, the 'lightning grid' is similar to the foam considered in Chapter 6.

Next, we may point out the 'light streamers'—the thin jets emitted from the ultrasonic waveguide (see Fig. 9.1.4). These luminous formations look like spark discharge, but have no similarities with this object except for the formal resemblance which, we have to admit it, is strong: the first time we saw this lightning pierce the vessel, we just thought that this is really a streamer or something of that sort.

The last confirmation of the fact that the necessary condition of light emission is the vapor phase—the cavern formed on the waveguide. As we mentioned in Chapter 5, cavitation destroys the waveguide surface, and large hollows appear among little caverns. These hollows are the source of especially bright light (see Fig. 9.1.5).

Figure 9.1.5 corresponds to the mode of unadjusted frequency of the ultrasonic generator. As we discussed in Chapter 6, in this mode the cavitation cloud is repelled from the waveguide with aperiodic pushes; light is emitted there from many points inside the liquid (precisely, from the liquid containing the gaseous phase), but not from the waveguide itself, how it takes place (see Fig. 9.1.6).

Fig. 9.1.3. The formation of a lightning structure on the waveguide during MBSL; compare this figure to Fig. 6.4.2. Here and below we present results for light emission in glycerol at acoustic cavitation.

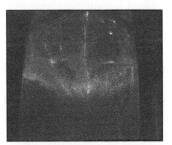

Fig. 9.1.4. MBSL 'streamer'.

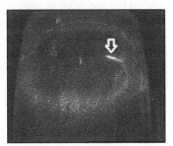

Fig. 9.1.5. The glow from the cavern.

Fig. 9.1.6. Normal mode of MBSL: Glow on the ultrasonic waveguide.

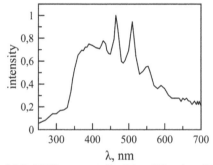

Fig. 9.1.7. MBSL spectrum (Suslick and Flannigan 2008).

The spectrum from multi-bubble sonoluminescence from (Suslick and Flannigan 2008) is presented in Fig. 9.1.7. As we see, this spectrum contains not only the continuum component, but also the bands of C_2. It is a normal fact for MBSL: in many works (Sharipov et al. 2009, 2012a, 2012b, 2014, 2018, 2019), the glow of additional component, mixed with the main medium, was observed.

9.1.4 Comparison between single- and multi-bubble sonoluminescence

At first glance—for a glance of a scientist who is far from the subject—SBSL and MBSL are the same phenomenon 'by definition': indeed, both of them represent the light emission from a cavitating fluid; thus, the physical nature of them must be the same. However, among the specialists in this area there exists an assumption—from the 1990s till now—that SBSL and MBSL are completely different phenomena.

The main reasoning for this assumption is the spectra, see the comparison from (Matula et al. 1995) in Fig. 9.1.8.

The comparison between the SBSL and the MBSL spectra with the same results were presented in other works with the same conclusions: the spectrum of MBSL contains more traces of lines and bands from various elements, while the SBSL spectrum is more 'solid'. Another matter of significant difference can be pointed out for the temperature measured in accordance with the Wien law, or by approximation of the 'continuum tale' of the spectrum at large wavelength with the Planck's curve (9.1.1): these methods give the temperature of about $\sim 10^{4-5}$ K for SBSL, while the temperature of MBSL is usually not higher than $\sim 10^3$ (determined with the same methods).

Thus, one may assume that the difference between SBSL and MBSL is not quantitative but qualitative: it is possible that these phenomena are indeed fundamentally different in their nature. Now we leave the fact that the nature of both SBSL and MBSL is not clear yet: we are comparing two unknown entities, which is funny in itself. Normally, the procedure is of the reverse order: establish the nature of the happenings, then compare them.

Some criticism for such a comparison from (Lavrov 2011) was given above. Continuing these doubts, the author mentioned that SBSL is a refined phenomenon, where MBSL is a poorly

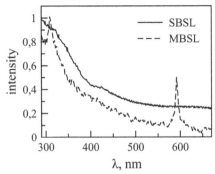

Fig. 9.1.8. Comparison of SBSL and of MBSL spectra from (Matula et al. 1995)

controlled experiment. The difference may be in the gas composition inside bubbles: if 'artificial aeration' is not used to observe single-bubble cavitation, then this bubble consists mainly of vapor, and the jumble of multi-bubble cavitation—especially in vicinity of a free surface—leads to bubbles filled mostly with air. Moreover, the spectra from (Matula et al. 1995) coincide after background subtraction and renormalization (Lavrov 2011).

The common property of SBSL and MBSL is also a sensitivity to temperature: the higher the temperature, the dimmer the glow. Usually, this fact is explained from the mechanical point of view: the higher the temperature, the higher the saturation pressure inside a bubble; therefore, the collapse is obstructed.

This explanation looks suitable, but, first, we need to discuss whether cavitation is an obligatory process to obtain the phenomenon of glow from the 'liquid–gas' system, or this light can be obtained with other methods.

9.2 Hydroluminescence

9.2.1 General line

There may be several ways to determine the nature of such a complex and complicated phenomenon as sonoluminescence. One of them is to reproduce the key feature of the phenomenon in different external conditions—let us call them X; the common issues of the usual sonoluminescence conditions and of X will be, with high probability, crucial elements of sonoluminescence.

Thus, we want to obtain light emission with the same spectrum from the same liquid. Before analyzing the successful attempt, we should briefly mention the unfortunate ones.

As we described above, in Section 9.1, glycerol was the main working liquid in our experiments. Thus, we tried to obtain an emission from liquid glycerol in various ways in the hope to measure the same spectrum as the sonoluminescence one.

First, glycerol is a non-luminescent material,that is, it is impossible to produce a glow from it with external irradiation by light, by light of ultraviolet frequency, by gamma or beta rays. There is no reaction to such external influences, and, commonly, these results are expectable: the nature of sonoluminescence observed in various liquids cannot be reduced to such a simple answer as regular fluorescence.

On the other hand, it is comparatively easy to create an electrical discharge in a liquid. It is not surprising that the liquid, as any other medium, emits light under the influence of an electrical discharge; the problem is that the spectrum of that light emission does not coincide with the spectrum of sonoluminescence.

Glycerol is a comparatively well-conducting liquid; thus, it is impossible to obtain a weak-current discharge in it. Placing electrodes in glycerol and giving the voltage on them, we will observe that electrical current simply closes the circuit. No electrical discharge takes places in such conditions.

Another way is the high-current discharge at a current of ~ 10 A, with a high Joule heating of the medium. However, this electrical discharge will absolutely break the inner structure of molecules of glycerol: at high temperatures, molecules immediately dissociate into atoms, so the emission spectrum will consist only of separate lines of atoms which are consisted in the molecule of glycerol $C_3H_5(OH)_3$ and, of course, of atoms from metallic electrodes.

Thus, simple attempts to reproduce light emission from glycerol failed. We cannot get the same glow from the surface of glycerol, or from its volume without a mechanical impact on it. Then, let us try it.

9.2.2 Hydrodynamic luminescence

We may observe light in a liquid which flows through a narrow dielectric channel—see Chapter 8 where the idea of such an experiment was discussed in relation with observation of triboelectricity in a liquid. This successful way to obtain light emission from a liquid is sometimes also referred to as sonoluminescence, despite the fact that the 'sono-' partin this term in conditions of absence of an external acoustic field means nothing. Therefore, in our opinion, another term is much more convenient—hydrodynamic luminescence or hydroluminescence (HL); this term emphasizes the hydraulic nature of the phenomenon.

The general scheme of the working section of such a setup was given in Chapter 8, the general scheme of the experimental setup is presented in Fig. 9.1.1. The main principle: a liquid flows under significant pressure (up to ~ 100 atm) through a special section—a dielectric throttle—and in this section (or in its vicinity), we may observe the glow.

We used this scheme to investigate the glow of different liquids: of the I-40A oil (Biryukov et al. 2013), water (Biryukov and Gerasimov 2018), and glycerol (Belyaev et al. 2018). It is interesting to compare even the patterns of those glows.

The glow of the I-40A oil is presented in Fig. 9.2.2. This is a very bright light; it can be seen even in daylight. We may see that this glow occurs in the narrowest part of the channel and may also be observed at its end.

The glow of water that was discussed in Chapter 8 is much dimmer. Fortunately, the intensity of that light was above the sensitivity of our spectrometers, so we could analyze the spectra and extract information from them. As we see, the glow has a similar spatial distribution to the luminescence of water; moreover, in Chapter 8 we discussed various forms of the spectrum in dependence on the coordinate in the channel.

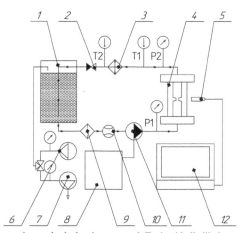

Fig. 9.2.1. The experimental setup to observe hydroluminescence. 1. Tank with distilled water, 2. Non-return valve, 3. Annular tube (pipe-in-pipe) heat exchange unit, 4. Working section for generating luminescence, 5. Spectrometer, 6. Argon tank, 7. Vacuum pump VE 115, 8. E3-9100-010H vector inverter, 9. Filter system, 10. Flowmeter, 11. WS 151 ('InterPumpGroup') high pressure pump, 12. Computer.

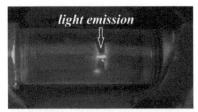

Fig. 9.2.2. Hydroluminescence of the I-40A oil. Fluid flows from left to right.

Fig. 9.2.3. Hydroluminescence of water. Fluid flows from left to right.

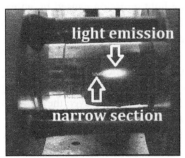

Fig. 9.2.4. Hydroluminescence of glycerol. Fluid flows from left to right.

The glow of glycerol is absolutely different. As we see from Fig. 9.2.4, here we observe light emission not in the narrowest part of the channel, but at its end. This glow is as dim as sonoluminescence: it is very difficult to watch it by the naked eye, and only special, very sensitive camera may fix that.

Thus, even at first glance, one may notice significant differences between various forms of HL in various liquids: light emission arises in different parts of the channel. It may seem to some that the glow in oil and water also contains the same part as in Fig. 9.2.4: some dim torch at the end of the narrowest part of the throttle—the light from the jet lapses from the throttle's neck. This is not the case: a sensitive camera and a special sensitive spectrometer do not register any additional components of the glow.

Thus, we have to discuss hydroluminescence separately in different media, that is, how quickly we come to the strange issues of that type of luminescence.

9.2.3 Glow of water and oil

The light emission of these media looks so similar that we may discuss it simultaneously.

In these liquids, the light emission is produced in the narrowest section of the throttle, that is, in a zone of intense cavitation (see Chapters 2, 8). The jet outflowing to the wide part of the throttle does not emit light, as we see.

The cause of light emission in these conditions was discussed in Chapter 8. Electrization of the liquid leads to electrical discharges in bubbles, and the main part of radiation from both these liquids are bands (of molecules) or lines (of atoms).

However, the radiation of the I-40A oil also contains the 'continuous part' that was presented in Fig. 8.4.8. Actually, this is not a true continuum like black body radiation or something of that sense—this 'solid' part of the spectrum is too narrow for that. Moreover, observing various parts of the channel, one may register significantly different spectra from them: like different points of fluid in the channel are in different conditions; see Fig. 9.2.5. Usually, there is a tendency to call any such spectrum as a solid-body spectrum, the deviations of which from the black-body curve are traditionally explained with strong dependence of the emission coefficient on the wavelength; the wavelength of the maximum of that curve λ_m can be used to determine the temperature of the medium in accordance with the Wien's formula:

$$\lambda_m T \approx 3 \text{ mm·K}. \tag{9.2.1}$$

For instance, if the maximum of the spectrum corresponds to $\lambda_m \sim 400 \div 500$ nm, then the temperature of the medium is ~ 7000 K—too hot for an oil, probably. Then, following this line, we have to conclude that different regions of that channel have diverse temperature: temperature would differ by thousands of kelvins. Finally, this line stops at the point where it is hard to interpret how can regions of fluid that close have such diverse parameters.

It looks much more tenable that these elements of HL spectra look rather like the spectrum of fluorescence of the medium—the light emission caused by an external excitation of the medium. The mechanisms of fluorescence can be different, but commonly, the process looks as follows. An external impact (of any kind: irradiation by photons of sufficient energy, or by electrons, or a chemical excitation, etc.) excites the medium—puts the so-called luminescent center into an excited state. De-excitation of this luminescent center is accompanied by the emission of a photon, the energy of which corresponds to the energy difference from the upper level (of the excited state) to the lower level (possibly, of the ground state).However, the energies of these levels are not constants: in a condensed medium, these values fluctuate; thereby, the energy of the emitted photon fluctuates too. Usually, fluctuations can be described with the Gaussian—not because this is a fundamental law, but in accordance with our expectations—so, the form of the emission spectrum from a single luminescent center is approximately

$$I_\varepsilon \sim I_0 \exp\left(-\frac{(\varepsilon - \varepsilon_0)^2}{2\delta_\varepsilon^2} \right). \tag{9.2.2}$$

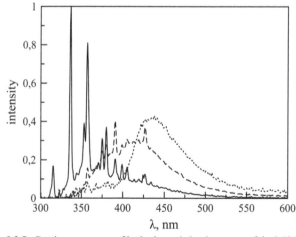

Fig. 9.2.5. Continuous spectra of hydrodynamic luminescence of the I-40A oil.

Here the parameter δ describes the deviation of energy levels from the stationary values, corresponding to the average photon energy ε_0. The expression (9.2.2) gives a peak of the Gaussian form which is much wider than the atomic line broadened due to various processes in the gaseous phase, such as the Doppler effect or collision broadening. On the other hand, usually, the width of the peak (9.2.2) is ~ 100 nm—it is much narrower than the Planck curve.

Continuing these reasonings, we may conclude that, if the 'continuum' is caused by fluorescence of the oil, then the same light emission can be caused by another reason: here we repeat our logical chain from the beginning of this section. The simplest way to excite fluorescence is the external irradiation by light, the wavelength of which is shorter than the wavelength of the emission from the given luminescent center; this type of luminescence is called photoluminescence (PL). We used a UV-lamp, the spectrum of which is represented in Fig. 9.2.6, and observed luminescence of oil, as expected. Note that this is a normal fact for oil: many petroleum-derived products can be luminescent.

Fluorescent spectra in comparison with HL spectra are given in Fig. 9.2.7. We see very well the agreement between two kinds of spectra, that is, we may conclude that the nature of these 'continuous' spectra is regular fluorescence. As for different hydroluminescence spectra from the same oil in the same channel, we may notice that under the UV-irradiation the oil fluorescents similarly: regions which are closer to the source of UV-radiation emit light of a shorter wavelength: the color of oil changes from violet in vicinity of the UV-source to yellow-green far away from it (here 'far' means the distance less than 1 cm).

Thus, we proved that hydroluminescence spectrum contains a part which corresponds to fluorescence; in the throttle, this luminescence is excited by electrons or by another component of radiation, for instance by the light emission of molecular nitrogen considered in Chapter 8.

Hydroluminescence spectra in water and oil are fully explainable, see Chapter 8. All the components of light emission from these media are caused by electrization—a gas discharge is the

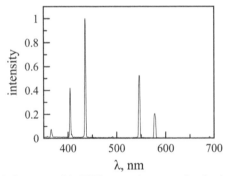

Fig. 9.2.6. Spectrum of the UV-lamp used to cause photoluminescence.

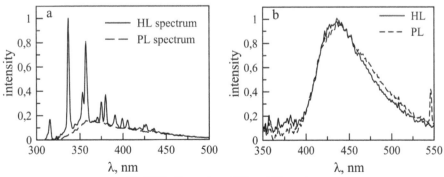

Fig. 9.2.7. Comparison of HL and PL spectra.

clear reason of that light emission. Since two of three liquids show the same radiation properties, we may expect that the third medium—glycerol—would reveal a similar luminescence spectrum, conditioned by electrization of the fluid in the throttle. But we will be surprised.

9.2.4 Glow of glycerol

As we pointed above, glycerol luminates in the wide part of the throttle, after the narrowest section. Actually, in Fig. 9.2.4, we see the luminous jet flowing from the channel's constriction. Avoiding certain description, we may depict the following picture: it looks like in the narrow channel there were no conditions for light emission, but after releasing from the tightness the fluid—to be exact, now it is a two-phase flow, consisting of liquid and gas—obtains radiation properties which, in their turn, vanish very quickly when this jet spreads into the wide section of the throttle.

Note that the glow may also occur at the end of the narrowest part of the throttle (it is difficult to make out the exact starting point of glow), but here we are interested, mainly, in light emission from the wide part of the throttle, where the flow represents, actually, a free jet. For the considered case, the external conditions for such a jet are more complicated in comparison with the pure, refined ones that are usually created to treat such objects (Guha et al. 2011, Or et al. 2011, Lau et al. 1972, Sivakumar et al. 2011), since the outer section of the throttle represents a closed region, but anyway,the shape of the luminous area looks like the potential core region of a free jet.

The potential core region, generally, is a theoretical term which means the region of flow where the potential approach[3] can be applied. For our purposes, the better term is near-field region (Fellouah et al. 2009)—the region, the length of which is approximately ~ $6d$ (d is the diameter of the nozzle; in Fig. 9.2.4, the length of the glowing region is slightly longer, but not significantly); in this region, the parameters of the flow are crucially determined by the conditions at the entrance to this section. In simple words, here the pressure depends on the coordinate inside the emerging jet: on the axis, where the velocity of fluid is higher, the pressure is lower—this fact can be illustrated with a simple trick with a ping-pong ball and a vacuum cleaner (see Fig. 9.2.8). We will not provide any relations, since, really, we have to deal with a complicated turbulent flow, with strong pulsations of all parameters including pressure: in such type of flow, we may operate only with average quantities. Thus,the 'low-pressure zone' in that throttle also contains oscillating pressure.

The glow pattern shown in Fig. 9.2.4, in general, corresponds to the pattern of the low-pressure zone in a jet. It is explainable: as we discussed in the previous section, sonoluminescence arises in cavitation regions, and we may expect that its relative—hydroluminescence—requires similar conditions. Thus, we are one step from concluding that cavitation is absolutely necessary for

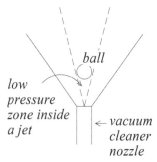

Fig. 9.2.8. A simple experiment to find out that the pressure inside a jet is lower than in the outer region: the ball cannot drop from there.

---

[3] That is, we may restrict the consideration with the scalar potential of velocity; see Chapter 4 and Appendix B.

that type of luminescence, that is, the key role belongs to the dynamical process which leads to a significant change in

✓ the volume of the gaseous phase;

✓ the area of the liquid–vapor interface.

It is difficult to decide which effect dominates: usually, the first circumstance is mentioned (for instance, the theory that implies high temperature inside a gaseous bubble during its collapse phase can be related to this group). Later, we will discuss this matter—which part of the liquid-gas system emits a glow—in detail.

We should emphasize the difference between two types of hydroluminescence: for the 'gas-discharge' type (like in water oroil), the dynamics of the gaseous phase is not necessary. It is enough that such phase exists in the channel, so the gas discharge has a medium where it may occur. For the 'sono-' type of hydroluminescence, the dynamics of caverns is required.

Above, we stated (twice) that hydroluminescence may be related to sonoluminescence, but did not give any evidence to this statement. Indeed, the fact that the glow takes place in the low-pressured region is not an argument. We need a coincidence in quantitative parameters—in emission spectrum, first of all.

9.2.5 Comparison of sonoluminescence and hydroluminescence

It is often repeated that in science a negative result is also a result. It is very fair when we discuss sonoluminescence, where one successful result accounted for several failed experiments.

The intense ultrasound impact on the liquid, theoretically and even experimentally, causes many effects—the liquid gains additional energy in various ways, including electrization, as we have seen in Chapter 8. In this case, one may soundly expect that a fluorescent liquid, for example the I-40A oil, will luminate in these conditions. We do not mean some special 'sonoluminescent' mechanism, but regular fluorescence, i.e., irradiation from luminescent centers, may be reasonably expected.

However, we did not observe any light emission from the I-40A oil during experiments with acoustic cavitation. Our eyes and measuring devices detected absolutely nothing. Of course, it is always possible that the intensity of emission is above the threshold of sensitivity; but such an explanation is beyond the logical area. Fluorescence of that oil was strong enough in close conditions—when the mechanical effect on the liquid was caused by friction in a narrow channel. On the contrary, under the ultrasonic influence, fluorescence of this oil does not appear at all. Physically, it means that luminescent centers do not get excited in an acoustic field. It is a very strange circumstance, the meaning of which may be as important as the results of 'positive experiments'.

As we declared in the beginning of this section, we assume that the emission spectrum is the main criterium for us to make a decision on the coincidence of the physical nature of two phenomena—sonoluminescence and something else. Thus, we represent the spectra of sonoluminescence and hydroluminescence in Fig. 9.2.9.

As we see, these spectra are very similar. They are not identical, but considering the experimental error, the difference between them is almost negligible. They are on a distance of ~ 10 nm from each other, the SL spectrum is slightly wider, but the main character of these spectra looks the same: both of them are similar to the fluorescence spectra discussed in Section 9.2.2.

Let us try to approximate the intensity of SL and HL spectra with the dependence of a kind of (9.2.2); the only difference is that now we need the distribution on wavelength, not on energies. Thus, from the equation

$$I_\varepsilon d\varepsilon = I_\lambda d\lambda, \tag{9.2.3}$$

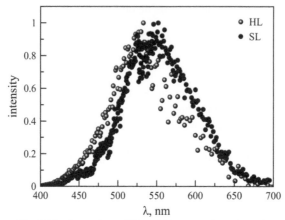

Fig. 9.2.9. Comparison of SL and HL spectra from glycerol.

where we omit the sign '–'which would mean only that the increase in λ corresponds to the decrease in ε, we can find that the distribution of intensity in the fluorescence spectrum is

$$I_\lambda \sim I_0 \lambda^{-2} \exp\left(-\frac{\left(\lambda^{-1} - \lambda_0^{-1}\right)^2}{2\delta_\lambda^2} \right).$$ (9.2.4)

The approximation (9.2.4) suits the observed experimental spectra well; not perfectly, but well; see (Belyaev et al. 2018).

Thus, we come to a strange conclusion: it looks that the hydroluminescence spectrum is similar to the sonoluminescence one, and both of them look like a regular fluorescence spectrum (see, for example, Fig. 9.2.10).

Thus, we finish this section on an intriguing note. In the beginning, we stated that we could not cause luminescence of a sonoluminescent medium (glycerol) with usual methods—like excitation by an external light. Finally, we state that the spectrum of sonoluminescence (and of its sister-in-law—hydroluminescence) looks like a regular fluorescence spectrum.
We need to consider other ways to obtain light emission from liquids now.

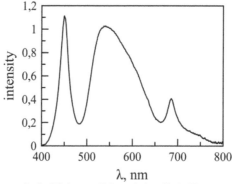

Fig. 9.2.10. Fluorescence: Spectrum of a flashlight on a light-emission diode. Here two characteristic luminescence peaks are distinguished. The third peak (about 700 nm) corresponds to the fluorescence of chlorella vulgaris (green algae), the luminescence of which was caused by that lantern. Yes, independently from nature, these luminescent spectra look similar.

9.3 Other methods to produce light from a liquid

9.3.1 A bullet in a liquid

As we mentioned in the previous section, the best way to understand the nature of sonoluminescence is to obtain the same radiation (radiation with the same spectral characteristics) but in different conditions. If so, one may try various ways to elicit light from liquid–vapor media; the most ordinary of them is hydrodynamic luminescence, while the most exotic, undoubtedly, is the collapse of a cavitation cavern formed in a liquid behind a flying bullet.

Generally, in this case this object is not exactly a liquid: this viscous mass—the ballistic gel (ballistic gelatin)—looks rather like a pudding; it has a parallelepiped shape and does not need a vessel. A bullet in such a medium leaves a specific trace behind it (see Fig. 9.3.1); when collapsing, this trace may produce light.

Unfortunately, in our experiments we did not register such light, but evidences affirm that light emission is possible in such a medium. As we can judge, this phenomenon can be fully explained by analogy of sonoluminescence: the cavern may collapse so intensely, that. ... Actually, we cannot continue from here since we still do not know why the collapsed fluid emits light. Anyway, the intense collapse is an element of the relaxation of the bullet trace in the gel, and the physical conditions there are similar to the conditions in an ultrasound-irradiated fluid, as far as we can imagine. The phenomenon may give additional information about sonoluminescence, and can be explored more carefully, in all senses of the word: these experiments, among other features, are dangerous.

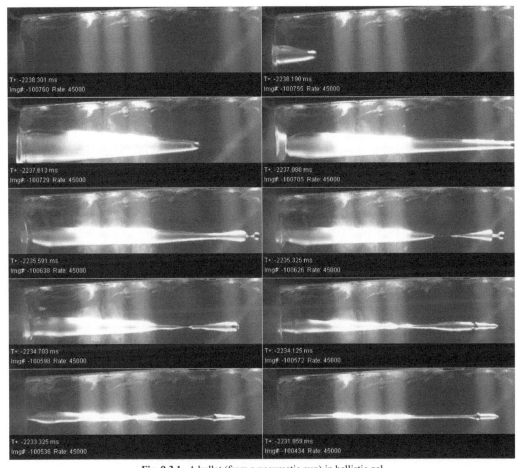

Fig. 9.3.1. A bullet (from a pneumatic gun) in ballistic gel.

Slightly in another setting, experiments devoted to light emission from bullets were conducted in (Fontenot and Hollerman 2011). Special projectiles were designed—bullets with ZnS:Mn (a triboluminescent material); their impact on an aluminum target produced a typical triboluminescent spectrum.

9.3.2 Cavitation mills

A cavitation mill is the device which uses the force of cavitation for a benefit: this apparatus produces suspensions (solid particles inside a liquid), emulsions (liquid droplets inside another liquid), colloid solutions (liquid with added very small particles, the size of which is not much larger than the size of molecules), etc. In simple words, this is a crusher which allows to obtain particles of ~ 1 μm in size or even much smaller. This device was considered in Chapter 2; see the scheme of it there.

For us, this is a very special entity, since the cavitation mill was the first setup, where we observed—occasionally, and we were very surprised by that—a mysterious light emission from a purely mechanical—not electrical, which would be so habitual to us—objects in purely mechanical conditions.

Morphologically, the light emission from cavitation mills is similar to multi-bubble hydroluminescence. The set of the key participants is the same: a liquid with small cavitating bubbles, and the whole system is under an intense mechanical impact (of rotational motion). Intense cavitation leads to intense interfacial processes, and some of those bubbles emitted a blue light, like air bubbles during hydroluminescence.

9.3.3 Radiation emission from an external flow

Another way to organize an intense mechanical effect on a liquid is external flow—the high-speed jet directed to an obstacle or something similar. It is a variant of hydrodynamic luminescence, especially when it is realized as the flow over a hydrofoil in a channel (Leighton et al. 2003, Farhat et al. 2011). In such a flow, the cavitation properties of which were described in Chapter 4, a weak luminescence has been observed, the properties of which were close to sonoluminescence. Generally, the increase in velocity of the flow was accompanied by the increase in luminescence; it is easy to connect that conclusion with the increasing cavitation.

An old work devoted to the same experiments (Konstantinov 1947) is interesting, first of all, because of the author's explanation of the nature of the luminescence phenomenon. Exploring the light emission that arises due to water flowing over cylinders in channels, Konstantinov directly connected the light emission with electrical discharges; he termed the flashes as sparks and finally stated that 'the fact of electrical discharges caused by cavitation was experimentally established'. However, no electrical parameter (voltage, current) was measured, and the conclusion looks slightly astonishing. Nevertheless, we may note that light emission was observed by him at the back side of cylinders, at the boundary of the low-pressure region, in a foam. This is very similar to the hydroluminescence of glycerol in a channel, see the previous section.

The most interesting experimental result concerning light emission during hydrodynamic cavitation is the accompanying X-ray emission. In some experiments, X-ray signals were measured from hydrodynamic cavitation in channels (Kornilov et al. 2007, 2009) or when an intense jet strikes a solid surface (Kornilova et al. 2010). In all those cases, X-ray spectra were registered, with the maximum at energies ~ 1–5 keV. Note that in an earlier work (Koldamasov and Mason 1991), a signal corresponding to ionizing radiation was detected: the dose rate 0.85 μR/s[4] was reported at the energy of gamma-quants at 0.3 MeV. In some works (Kornilova et al. 2009, Monakhov 2013), the attempts at a theoretical explanation of high-energy radiation are presented.

---

[4] To understand what does such a dose rate mean, we may notice that the lethal dose is about 600 R (this is not an exact value since the very term 'lethal dose' has many definitions; the dose which may induce health concerns is about 10 R).

X-ray in such conditions is a very exciting thing because, first, this fact does not represent an unbelievable phenomenon: such quanta may appear due to high-energy electrons, which can be produced in various conditions, including such a famous exercise as tearing off an adhesive tape (during which, by the way, light emission can be observed too). Then, detection of an X-ray may sufficiently help to understand the phenomenon of sonoluminescence (and of associated phenomena): the characteristic energy of X-quants may be connected to the energy of electrons in such a medium, and, therefore, to clarify the physical picture: it is much easier for theorists to explain a well-measured experiment than predict some special details of an uncertain phenomenon. In any case, the appearance of X-ray radiation would mean that the medium contains high-energy electrons,and, knowing this fact, we have a proper starting point for a theory.

We tried to measure the X-ray or the gamma-ray output from our experimental setups, and got different results. In experiments for hydroluminescence of oil (see Section 9.2), in the range of quant energies from 100 keV to 3000 keV,[5] the scintillation spectrometer showed us nothing, see Fig. 9.3.2.

A scintillation spectrometer cannot give results for lower energies of quants (precisely, some of them can be calibrated up to ~ 60 keV, but with a very strong noise). Another device is a semiconductor spectrometer, which can detect quants of energies at ~ 1 – 100 keV: with this equipment, we obtain clear, solid, continuous spectra both from the sonoluminescence setup and from the hydroluminescence one. Characteristic energies of photons in both cases ~ 1eV; they are close to the minimal energy that can be measured by this spectrometer, but these were clearly measured values. Thus, we indeed obtained an X-ray spectra similar to the works (Kornilova et al. 2009, 2010).

At this point, we are ready to include high-energy electrons into the physical picture of the luminescence phenomenon, but first we want to stop for a while and discuss another matter—the correctness of the measured values.

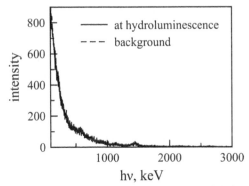

Fig. 9.3.2. Scintillation spectra: the background spectrum and the one measured during hydroluminescence; there is no difference, however.

9.3.4 Some notes about radiation dosimetry

As we all know, and as it was mentioned previously in this book, measurements can be of two kinds: direct or indirect. The measurement of ionizing radiation is of the second kind, and superlatively.

Generally, ionizing radiation consists of photons (gamma-quanta, X-rays) and corpuscular particles (alpha- and beta-particles, neutrons, etc.). Detectors of that radiation must do the following

---

[5] Generally, the difference between X-rays and γ-quants is arguable. Sometimes, it is defined that X-rays are of artificial origin while γ-rays are of the natural one. Another terminology claims that an X-ray represents quants with energies from ~ 10 eV to ~ 10 keV (certain values depend on the taste of a certain author), while γ-quants have higher energies. We use the last variant: X-rays are quanta of lower energy.

thing: convert the microscopic act of interaction between an elementary particle of that radiation and the medium into a macroscopic event.

Detectors of ionizing radiation may be of different constructions; most common types are gas discharge detectors, scintillation detectors, and semiconductor detectors.

Gas discharge detectors (the most famous type of them is the Geiger counter) produce a discharge inside them in response to a single ionization act in a gas. In simple words, a gas discharge detector is a discharge gap, which registers the electrical current appearing due to ionization of the gas by radiation (such a detector is called the ionized camera) or amplifies that current because of the electron avalanche in the detector (the Geiger counter): an electron accelerated in the electrical field ionizes a gas atom, then both electrons—the initial one and the electron from the atom—gain energy from the electrical field, ionize two atoms, producing two additional electrons, and so on.

Scintillation detectors use luminescence[6] of some substances under ionizing radiation; sometimes this type of luminescence is termed as radioluminescence. A high-energy particle (qama-quantum, for example) excites the medium; after that, a scintillation photon can be registered with a photo-electron multiplier or with something else. Scintillation detectors can be used as spectrometers, allowing to define the energy of ionizing particles.

In semiconductor detectors, ionizing radiation produces 'electron–hole' pairs; registering the electron current, before they recombine back with holes, one can measure the energy of a particle of ionizing radiation. These types of detectors have many benefits compared to scintillation spectrometers (can register low-energy gamma-quants, have much better energy resolution), but they are more complicated in use. In addition, their indication needs slightly better understanding than the measurement results from the scintillation concurrent.

There is a problem with all types of particle detectors: they register not only ionizing radiation, or ionizing radiation of the expected kind of particles; for the last problem, to firmly determine the type of a particle, various types of detectors must be used simultaneously. However, the biggest problem is the first case: when a detector reacts not to ionizing radiation, but to something else—for instance, electromagnetic radiation.

For the case discussed in the previous sub-section, we are interested in solid semiconductor detectors. For them, we have to take into account that the electron transfer between zones can be caused not only by an external ionizing radiation, but also due to an intense acoustic impact—because of the electron-phonon[7] coupling. The result will be almost the same—the detector generates some electrical signal which is interpreted by it as some energy spectrum; it is important to distinct this signal from the real spectrum of ionizing radiation.

Let us consider the experimental setup for the sonoluminescence treatment. When the sensitive element of a spectrometer is located in vicinity of the ultrasonic waveguide placed in a liquid, the ultrasound wind affects the entrance window (Fig. 9.3.3); possibly, this fact also influences the spectrometric result—we want to ascertain this matter. Howbeit, in the result of experiments, the detector measured a spectrum shown in Fig. 9.3.4—the continuum as a function of the X-quanta energy. Note that such spectra are always—at least, for all the cases that we saw—located at the left boundary of the spectrometer range, at the minimal threshold 'energy of ionizing radiation'.

Such a spectrum looks 'natural', but actually it is not. First, we can analyze signals that are registered by the detector: primal impulses that are registered by the measurement scheme. The signal incoming from 'sonoluminescence' is presented in Fig. 9.3.5 in comparison with the test signal from a Cs-137 isotope (of gamma-quanta of energy of 32 keV). The difference is clear: Cs-137 produces single flashes corresponding to acts of gamma-quanta emissions after radioactive decays, while 'sonoluminescence' gives continuous wave-like signal. Moreover, the frequency of that wave corresponds to the frequency of the ultrasonic generator: about 20 kHz.

---

[6]   Scintillation is a single luminescent flash.

[7]   The phonon is a quasi-particle corresponding to acoustic vibrations; quasi-particles are a convenient way to describe the energy and momentum transfer in a medium.

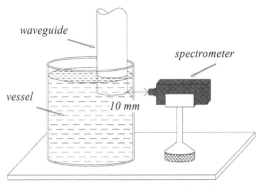

Fig. 9.3.3. Measurement of the X-spectrum from sonoluminescence. A semiconductor spectrometer is located in vicinity of an ultrasonic waveguide and tries to register X-rays from the liquid in the vessel.

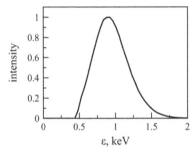

Fig. 9.3.4. X-spectrum measured during sonoluminescence of glycerol.

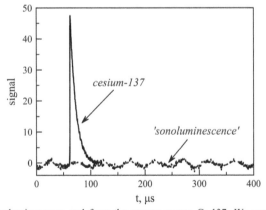

Fig. 9.3.5. Signals from sonoluminescence and from the gamma-source Cs-137. We can clearly see the fundamental difference between them.

Thus, we may assume that the 'sonoluminescence spectrum' is not a real X-ray spectrum indeed: this spectrum is induced by the acoustic effect on the semiconductor detector—that ultrasound wind is a possible cause for such a signal. This assumption can be easily checked: the same spectra were measured from the working setup but when the waveguide was placed in an empty vessel (for the purity of experiment, we left a vessel, however), that is, in this scheme the waveguide pounds air in an empty vessel; see result in Fig. 9.3.6.

Thus, we have to conclude that the X-spectra obtained in our experiments are not X-ray spectra indeed—they are a backward signal caused by the direct effect of the ultrasonic setup, which is very sad.

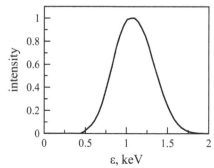

Fig. 9.3.6. 'X-ray spectrum' from the empty vessel: no glycerol, by the signal still present. The problem is solved: this is not an X-spectrum indeed.

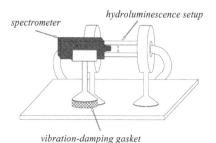

Fig. 9.3.7. Measurement procedure of X-rays from hydroluminescence. When the vibration-damping gasket is off, the signal from the 'X-ray spectrum' is on.

However, we measured the same signal—the same X-ray spectra—from hydroluminescence. In that case (see Fig. 9.3.7), there cannot be any 'acoustic wind'; here we are dealing with a fundamentally different scheme, but the signal—the X-spectrum similar to the one presented in Fig. 9.3.4 from the channel with a cavitating liquid—has been measured. Is this a case of the 'true X-radiation'?

Unfortunately, no again. When the hydroluminescence setup works—driven by a powerful pump, which pushes a fluid through the millimeter-wide channel—all the components in the setup shiver, and all the things around the setup are leaping. This shaking influences the X-ray spectrometer in the same manner, as the acoustic wind: it produces a false signal; to be exact, this signal is true, but it does not correspond to the real ionizing radiation.

It is easy to eliminate an 'X-ray spectrum' measured in such conditions: it is enough to place the stand of the spectrometer on a vibration-damping gasket. In a stable position, the X-ray spectrometer gives zero signal for the X-spectrum.

The moral of the story: when we have to deal with radiation dosimetry, we should be very careful with conclusions. We really may expect high-energy radiation from such an object, especially from hydroluminescence conditions; in the previous chapter, we have seen that various electrical processes accompany the phenomenon. X-rays can be expected in these conditions, i.e., in an intense electrical discharge (we only need to explain the source of such a high electron energy), but the experimental data confirmed that this statement does not stand up to criticism.

Usually, the more special the detail is, the simpler is its explanation. If the presence of ionizing radiation during sono- or hydroluminescence would be a solid experimental fact, then the theory of these phenomena would contain specific signs—'exotic' features intended to explain the appearance of such high-energy radiation. It would be important especially for SL, where the state of the luminescent medium is not as clear as for HL. However, we see nothing of this matter. Thus, considering the theories of luminescence of fluids in the next section, we will have to deal with a humble, cryptic bubble in a liquid.

9.4 Possible physical nature of light emission

9.4.1 Common properties of particular theories

In this section, we expose some theories that may explain light emission from a liquid–vapor mixture under an intense mechanical impact. The common purpose of these theories is obvious—to explain the nature of sonoluminescence; however, certain points of them differ significantly. One may divide all possible theories in two groups:

- ✓ macroscopic theories, which explain how, generally, the conditions for light emission can be achieved in a two-phase fluid;
- ✓ microscopic theories that must determine the certain mechanism of radiation: which exact process causes luminescence—the bound–bound transition of an electron in an atom, bremsstrahlung, etc.

Alas, we have to admit that all the basic theories of sonoluminescence belong to the first group. These explanations describe the picture with 'broad strokes', giving information rather about 'how some light emission can be generated during cavitation' but not concerning 'the exact type of radiation is the nature of sonoluminescence'. Denying to discuss the elementary acts of sonoluminescence, we lose the most important points of the phenomenon. On the other hand, even understanding the macroscopic conditions may give some initial information about the process.

Thus, it is not surprising that macroscopic theories cannot give the final, firm answer to the challenge. Light can be emitted in various conditions, and sometimes, these conditions can be mixed. If so, the casual relationship plays the crucial role: what fact is the cause of luminescence and other observed facts? As we will see below, two main theories (or, precisely, two groups of theories) are the 'thermal hypothesis', considering sonoluminescence as the radiation from a very hot object, and the 'electric theory', implying that the electrization of a liquid–gaseous mixture causes an electrical discharge and the corresponding glow. But these two phenomena—heating and electrization—are connected with each other. Strong electrical current causes heating, while strong heating causes ionization of a substance. Thus, even when macroscopic theories give a certain answer—describe the exact conditions in the radiating medium—some questions can remain unresolved.
However, now we are far from that state.

9.4.2 Thermal radiation

We begin with the most popular hypothesis to date, if it is possible to say so about a scientific theory. According to this, sonoluminescence is the thermal radiation from a bubble: during the collapse stage, the temperature inside a bubble increases significantly in adiabatic conditions, so the heated medium emits light.

In two words, the crucial equation to understand the nature of the temperature growth inside a bubble is the Poisson adiabat:

$$TV^{\gamma-1} = const. \tag{9.4.1}$$

This expression corresponds to a static, the reversible adiabatic process. According to (9.4.1), when a bubble diminishes in its volume V, then its temperature T increases, since the adiabatic exponent $\gamma > 1$.

Let us analyze the thermal scheme of light emission in detail. Mainly, the chain of events can be described as follows:

- ✓ a bubble shrinks;
- ✓ due to non-isothermal conditions—the medium inside a bubble has no sufficient time for heat exchange with the surrounding liquid—the gaseous phase inside a bubble is heated;
- ✓ when the gaseous phase is very heated, it emits light; probably, in some conditions inside a bubble the spectral distribution of this light corresponds to the Planckian;

✓ the heat from the bubble now leaks not only via the conductive or convective heat fluxes, but also with the radiative one; probably, this additional mechanism provides such a strong cooling rate that the whole medium inside a bubble will be chilled fast;

✓ thereby, it is possible that the duration of the flash may be much shorter than the entire collapsing phase.

The last item looks strange, and not in vain: this point is required to explain the short duration of light impulse at acceptance of the hypothesis of thermal nature of sonoluminescence, but it does not look solid. Then let us estimate the cooling rate of a gaseous bubble. Considering a bubble as a continuous object, we can formulate for it

$$c_p \rho \underbrace{\frac{4\pi R^3}{3}}_{V} \frac{dT}{dt} = -\sigma T^4 \underbrace{4\pi R^2}_{S}. \tag{9.4.2}$$

The physical description of the problem, thereby, can be stated in the following manner: the medium inside the bubble represents an optically thick layer, and the radiating heat flux on the bubble surface corresponds to the Stefan–Boltzmann relation.

Assume the air inside a bubble with $c_p = 10^3$ J/kg·K, density ρ is not such a univocal parameter in the shrinking bubble; let it be 1 kg/m3 anyway: in such a case, we diminish the characteristic time of the process (in a denser bubble, its mass is higher, so the time needed for its cooling is greater).

Then, assume that the bubble radius is a constant, i.e., consider the temperature deviation in a quasi-stationary bubble; below we compare the obtained time scale of the heat exchange process with the time scale of bubble dynamics. At all these assumptions, we have for temperature of the bubble

$$T^3 = \frac{T_0^3}{1 + t/\tilde{t}}, \tag{9.4.3}$$

where T_0 is the temperature of the bubble at $t = 0$, and the characteristic time scale

$$\tilde{t} = \frac{c_p R \rho}{9\sigma T_0^3}. \tag{9.4.4}$$

Note that a relation of the form (9.4.4) follows almost directly from the equation (9.4.2) without any solutions; a weak justification of our approach is the nine in the denominator of the expression for $\tilde{t}$: with the usual method which actively uses signs '~', we should obtain the estimation smaller by an order (without the factor '9') in the final expression.

Nevertheless, even the ~ 10 multiplicator does not clarify the picture, since the result strongly depends on the 'initial' temperature of the bubble T_0. Let us consider a bubble of radius 10^{-5} m; for it, with $T_0 = 10^4$ K we get $\tilde{t} = 2 \cdot 10^{-8}$ s, while for $T_0 = 10^3$ K we have, obviously, $\tilde{t} = 2 \cdot 10^{-5}$ s. We must compare it with the time period of bubble pulsations: as usual, we use the simplest expression

$$\tau \sim R\sqrt{\frac{\rho}{p}}, \tag{9.4.5}$$

which gives for a bubble of our favorite ten-micrometer size the period of 10^{-6} s.

Thus, we see that a bubble of 'moderate' temperature, that is, at temperatures 'only' about several thousand kelvins, cannot lose heat very fast. On the contrary, the bubble would keep its energy until the temperature inside it would decrease due to the expansion phase: this process would regulate the temperature, not the radiation losses.

Consequently, we have the first difficulty: the thermal flash is too long. Another problem is the optical length of the radiating medium: thermal radiation can be produced only by an optically thick layer of the medium. A bubble is too thin from this point of view, even if it consists of pure water vapor: water has high absorption coefficient but not at the visible range. Thus, a bubble of diameter less than 1 mm is optically thin, and the light emission from it must represent separate lines (from optical transitions of atoms) and bands (from optical transitions of molecules). Despite all the uncertainties of experimental data, such emission does not correspond to the real picture: whether the sonoluminescence spectrum matches the black body spectrum or not, it still contains a pronounced continuum which cannot be explained by the heated gas.

If so, we have to conclude that another heated substance can be the source of radiation. A suitable candidate for this role is the heated layer of liquid surrounding the bubble: during the collapse, the liquid surrounding the bubble accepts some heat from it. Liquid is much denser than gas; thus, one may expect that the hot liquid walls of the bubble may emit thermal radiation, the spectrum of which will be close to the Planckian curve.

However, this way is problematic too. The reason why the bubble increases its temperature during its collapse is exactly the absence of heat leaks from it, i.e., the adiabatic conditions. In Chapter 7, we discussed this matter in detail, so we do not repeat these reasonings again. Thus, basically, adiabatic conditions mean the absence of hot liquid around the bubble. But we may assume that the heat leaks are negligible: if thermal losses are small, then we still have a very hot bubble and a hot liquid surrounding it.

Let us analyze the last choice. If the collapse is almost adiabatic and we assume a perfect gas inside it, then all the mechanical work is spent to change the internal energy of the gas in a bubble:

$$\Delta U = c_{vg} m_g \Delta T_g, \qquad (9.4.6)$$

where c_{vg} is the specific isochoric thermal capacity of the gas inside a bubble, m_g is the mass of gas and ΔT_g is the deviation of temperature due to heating.

In accordance with the condition stated above, this value must be much greater than the heat that was leaked from the bubble to increase the temperature by ΔT_l of the liquid mass m_l surrounding the bubble:

$$Q = c_{pl} m_l \Delta T_l. \qquad (9.4.7)$$

The inequality

$$\Delta U >> Q \qquad (9.4.8)$$

can be fulfilled only if $\Delta T_l << \Delta T_g$, since c_{vg} is even smaller than c_{pl} (for instance, for an air bubble in water $c_{vg} \approx 700$ J/kg·K and $c_{pl} \approx 4200$ J/kg·K), and we need that $m_l >> m_g$, otherwise the liquid mass will be insufficient to form the thick optical layer. Therefore, to increase the temperature of the liquid by $\sim 10^3$ K in 'almost adiabatic conditions', the bubble must be heated by tens of thousands of kelvins. It is a very questionable case.

Of course, there may be an intermediate case, when the bubble collapses in non-adiabatic and non-isothermal conditions, losing a significant amount of heat to the liquid, but if so, it is hard to explain the great increase in temperature using those quasi-static representations anyway.

An alternative way may include a description with the so-called temperature of the adiabatic drag: when a body moves in a gas (in our case, a liquid wall moves in the bubble gas) with velocity

v, then from the condition of total enthalpy ($h + v^2/2$) conservation[8] we obtain the estimation for temperature at the stagnation point

$$T = T_0 + \frac{v^2}{2c_p}. \tag{9.4.9}$$

The conditions to apply this expression for a tiny gaseous bubble looks questionable; moreover, to obtain temperatures of $\sim 10^3$ K due to the second term in (9.4.9), the velocity must be about ~ 1 km/s (since the heat capacity is about 10^3 J/kg·K).

On the other hand, nevertheless, we may say that theoretically, considering the shrinking as strong as we like, we will obtain temperature as high as we like. Another matter needs to be discussed: how indeed liquid, for instance water,can achieve temperature of $\sim 10^3$ K or more? Is this theory—that liquid can be heated up to thousands of kelvins—nonsense from the very beginning?

Curiously enough, the answer is negative: such theory is more realistic than it looks. The habit to consider—not to deal with, videlicet to consider—only equilibrium state often plays a bad joke with researches. Of course, liquid cannot be found in an equilibrium state at a temperature higher than the critical one (374°C for water). But, first, this number means only the limiting value in the sense of phase transition: at parameters higher than critical, the phase transition is impossible, and the substance represents there a 'supercritical fluid', not a liquid. Then, and this is the most important matter, any non-equilibrium state may exist for some time. Let us consider a particle in a medium—an atom or a molecule—which has some abnormal parameters after an external impact: for instance, it has an enormous kinetic energy or is located in vicinity of another particle (i.e., has a huge potential energy). With time, this abnormal parameter must dissipate, come into balance with the rest of the system. The corresponding (relaxation) time can be calculated if we know the potential of interaction of molecules φ, the interparticle distance l and the mass of the particle m as

$$\tau \sim l\sqrt{\frac{m}{\varphi}}. \tag{9.4.10}$$

For $\varphi \sim 10^{-1}\ldots10^{-2}$ eV, $l \sim 1$ nm and $m \sim 10^{-26}$ kg we have the estimation $\tau \sim 10^{-12}$ s. Thus, the relaxation time—the 'dead time', while the molecules of substance are immobile—for the condensed medium is ~ 1 ps, and during this time, the system, generally, may have any parameters: the system goes to equilibrium only at a time much greater than τ.

Thus, even a strongly non-equilibrium state may exist for some time; how the system came to this state is, of course, a separate problem. Returning to the light emission from a liquid at an enormous temperature—at a temperature where the liquid cannot exist in equilibrium—yes, fundamentally, this process is possible.

9.4.3 The Planck formula

Another problem is the sonoluminescence spectrum. Sometimes, the Planckian spectral distribution was registered, sometimes it was not; considering difficulties with the measurement procedure for such a small, non-stationary object like a collapsing bubble, this fact is not surprising. Generally, some similarity (not necessarily the identity) of a sonoluminescence spectrum to a black-body curve is the fundament of the thermal hypothesis. To realize why the curve of the continuum radiation may deviate from the Planck formula, we consider the structure of this expression.

---

[8]   This condition follows from the first law of thermodynamics in the form $dq = dh + v\,dv + g\,dz$ for $dq = 0$, because of adiabatic conditions, and $dz = 0$, since we cannot seriously discuss the height difference there; meanwhile, in a non-inertial system, for a motion with acceleration a, we will have a similar equation with a instead of g.

This relation occasionally transformed physics, and not only physics, in 1900, bringing the quantum representation into the science. The Planck formula describes the spectral distribution of the internal energy (per volume) for thermal radiation, that is, the radiation in equilibrium with the body that emits it:

$$u_\nu = \underbrace{\frac{8\pi\nu^2}{c^3}}_{Z(\nu)} \underbrace{\frac{h\nu}{e^{h\nu/kT}-1}}_{E(\nu,T)}. \qquad (9.4.11)$$

This relation is the multiplication of two terms.

The first factor $Z(\nu)$ is the mode density—the number of types of oscillations with frequency ν. Generally, the only necessary condition for this term is a sufficiently large spatial scale in comparison with the wavelength, which is normally fulfilled for radiation at the optical range with $\lambda \sim 10^2$ nm. The expression for $Z(\nu)$ was obtained by Planck when he considered the quasi-harmonic motion of an oscillator in an electrical field, assuming that the drag forces (caused by radiation) do not significantly disturb the oscillations at frequency ν.

The second term $E(\nu,T)$ represents the average energy of an oscillator which vibrates at frequency ν and temperature T—and this is a more complicated quantity. The expression for $E(\nu,T)$ used in (9.4.11) has been obtained for thermal equilibrium; actually, this formula led to quantum mechanics, when Planck denied to aspire an elementary portion of energy ε to zero (and get $E(\nu,T) = KT$) but put it as proportional to frequency with the coefficient, which is now named after Planck, i.e., put $\varepsilon = h\nu$. However, in the absence of thermal equilibrium, the relation for $E(\nu,T)$ is different, and we must use another expression instead of (9.4.11) for the spectral energy distribution of,in this case, non-thermal radiation.

Generally, the relation for function $E(\nu,T)$ can be of any kind. However, the generalized Wien law claims that the function u_ν must be of the form

$$u_\nu \sim \nu^3 f(\nu/T). \qquad (9.4.12)$$

This expression has been obtained for the equilibrium process of radiation in an adiabatic medium (in a hollow with absolutely reflective walls); therefore, this expression corresponds to the common property of radiation in the equilibrium process. The problem that we discuss now concerns slightly another case: there is no total equilibrium, but in the intermediate case there may be that function $E(\nu,T)$ as the product of ν and some function of $\xi = \nu/T$:

$$E(\nu,T) \sim \nu f(\nu,T). \qquad (9.4.13)$$

Then, to avoid the ultraviolet catastrophe, one may note that the function f must decrease with frequency at high ν, that is

$$\frac{df}{d\nu} = \frac{1}{T}\frac{df}{d\xi} < 0. \qquad (9.4.14)$$

Note that we meet this problem again: in the absence of equilibrium, we actually have no good recipes for macroscopic parameters; to establish them in a non-equilibrium case, we have to consider the problem on the microscopic level, if we have a clear physical picture of the phenomenon. If we do not have a proper physical model, we may use mathematical tricks, representing this function in a series on ξ. Normally, in equilibrium, there must be

$$f^{-1}(\xi) = g(\xi) = \sum_{i=1}^{\infty}\frac{\xi^i}{i!}. \qquad (9.4.15)$$

The most logical way to generalize such an expression—remember that currently we are discussing only mathematical tricks, so claims like 'Tell us about the physical meaning!' are not valid here—is to substitute the gamma-function instead of the factorial

$$i! = \Gamma(i+1), \tag{9.4.16}$$

and using a more common form of the argument of Γ,[9] we get

$$g(\xi) = \sum_{i=1}^{\infty} \frac{\xi^i}{\Gamma(\alpha i + \beta)}, \tag{9.4.17}$$

where

$$\Gamma(x+1) = x\Gamma(x), \ \Gamma(x) = \int_0^{\infty} t^{x-1} e^{-t} dt. \tag{9.4.18}$$

Such generalization can be reduced to definition of parameters α and β in expansion. For instance, let $\beta = 1$; then, for small frequencies, i.e., for small values of $\xi = v/T$, we should have the Rayleigh–Jeans formula $u_v \sim v^2\theta$ but, probably, with a non-equilibrium parameter θ instead of temperature T, that is, the parameter α can be found from the expression $\Gamma(\alpha + 1) = T/\theta$, where θ can be defined from the experimental data (from the experimental curve).

It should be emphasized that all the methods that were discussed above can be applied for the conditions that are close to equilibrium, when the expression (9.4.11) makes sense with the given function $Z(v)$, but with some other function for $E(v,T)$. This common form of the spectrum is shown in Fig. 9.4.1, where we represent the spectral distribution on wavelengths instead of frequencies: $u_\lambda = cu_v(\lambda)/\lambda^2$.

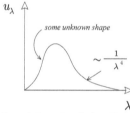

Fig. 9.4.1. The quasi-equilibrium spectrum of quasi-thermal radiation. The right branch corresponds to the Rayleigh–Jeans law, where the mean energy of an oscillator E does not depend on frequency; the left branch corresponds to some unknown dependence $E(\lambda)$ at high oscillation frequencies.

A more 'physics' way, surely, is to find the distribution of the oscillator's energy on frequencies and then to determine the mean value of that energy from this distribution. However, to bring this approach to life, we need to construct the kinetic model of the process, which demands many more details concerning the state of the medium in vicinity of the phase interface under intense cavitation.

9.4.4 The electric hypothesis

As we mentioned above, the 'thermal' theory is very popular today; indeed, many engineers and scientists, who do not have to deal with this phenomenon professionally, know about sonoluminescence that 'this is a light emission that occurs in a collapsing bubble since the temperature inside it increases during the collapse'.

---

[9] Actually, here we use the method to generalize the exponent function, which leads to the Mittag–Leffler function

$$M(x) = \sum_{i=0}^{\infty} \frac{x^i}{\Gamma(\alpha i + \beta)}.$$

However, those who specialize in sonoluminescence know that there exists another way to explain this phenomenon: through electricity which, presumably, may appear during cavitation. Initially, these representations were born in the works by J. Frenkel (Frenkel 1940, 1946).

In the first work (Frenkel 1940), which is only cited, by the way, Frenkel supposed that during cavitation the rupture of the liquid phase leads to free charges on the bubble wall: due to fluctuations, the number of ions of different signs would be never exactly equal on the opposite sides of the bubble. Frenkel assumed that the ions correspond to the ones existing in the liquid initially, but it can be presumed that even neutral molecules can be ruptured in the process of intense cavitation. An intermediate variant described the appearance of charges including reactions of dissociation-recombination of complex molecules and the loss of neutrality due to impossibility of recombination of ions on different sides of the bubble's wall after the rupture.

Later, in his monograph (Frenkel 1946), Frenkel discussed the 'surface cavitation', that is, actually, the multibubble cavitation (see Chapter 6) which takes place on the ultrasound waveguide placed in the liquid. According to Frenkel, the double electrical layer under ultrasonic vibrations , forms a 'gas lens' with electrical charges on the opposite sides (see Fig. 9.4.2).

In both cases, we have the gaseous phase under an external electrical field formed by electrical charges on the boundaries of the phase. Then, it may be possible that an electrical field causes the electrical breakdown in the gaseous phase, if the width of the 'gaseous lens' exceeds the mean free path (MFP) of an electron there. Then, as usual, the electrical breakdown leads to glow; thus, sonoluminescence is represented as a gas discharge glow.

As for the mechanism of electrical breakdown of gas, there are no doubts that it looks possible, except for the thesis that the size of the cavity cavern is larger than the MFP of an electron there; eventually, that depends on the size of the certain cavern. Moreover, in the previous sections, we discussed (and presented an image of the glow) that, indeed, in a regular mode, light emission originates from the near-waveguide zone of the fluid.

But, again, the spectrum—the unique characteristic of the glow—is important. The spectrum of a gas discharge differs significantly from the continuum of a black body radiation. Electrical breakdown emits light, mainly, in atomic lines or molecular bands; these components were discussed in Chapter 8. In addition, this discrete spectrum is accompanied by bremsstrahlung and recombination radiation, the intensity of which in plasma is usually much weaker; both of them give a solid spectrum, but different from the Planckian curve.

Bremsstrahlung corresponds to the energy loss of a free electron: dragging in an electrical field of another charge in plasma, an electron emits an electromagnetic wave. Since the initial value of the electron energy belongs to continuum spectrum as well as the final one, the corresponding bremsstrahlung spectrum is continuum too. The corresponding cross-section[10] of bremsstrahlung for the electron with velocity v (Raizer 1991), which interacts with an ion with a charge Ze, is

$$d\sigma = \frac{2Z^2 e^6}{3\sqrt{3}\varepsilon_0^2 m^2 c^3} \frac{dv}{hv},$$ (9.4.19)

where $\varepsilon_0 = 8.85 \cdot 10^{-12}$ F/m is the dielectric constant, m is the electron mass, c is the speed of light. Thus, the intensity of light emission at frequency v is

$$dI_v = \int\limits_{v_{min}}^{\infty} h v n_i n_e v f(v) \, d\sigma dv,$$ (9.4.20)

---

[10] Cross-section describes the probability of the process.

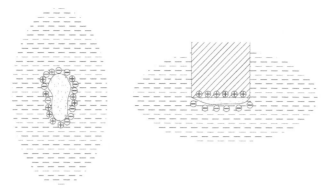

Fig. 9.4.2. Two models of electrical conditions during cavitation by J. Frenkel. A: after the rupture of the liquid, ions of different signs are located on different sides of the gas cavity. B: the double electric layer oscillates since the ultrasonic waveguide moves in the liquid.

where n_i and n_e is the number density of ions and electrons correspondingly; $f(v)$ is the velocity distribution function for electrons; $v_{min} = \sqrt{2h\nu/m}$ is the minimal electron energy: obviously, an electron cannot emit a photon with larger energy than the energy of this electron. For the Maxwellian distribution function, one may obtain from (9.4.20) with (9.4.19)

$$dI_\nu = C\frac{Z^2 n_i n_e}{\sqrt{T}}\exp\left(-\frac{h\nu}{kT}\right)d\nu, \tag{9.4.21}$$

where T is the plasma temperature. Transforming $I_\nu = dI_\nu/d\nu$ into I_λ, we obtain

$$I_\lambda = C_0\frac{Z^2 n_i n_e}{\lambda^2\sqrt{T}}\exp\left(-\frac{hc}{\lambda kT}\right). \tag{9.4.22}$$

In some sense, the function (9.4.22) looks similar to the Planckian curve: the difference is that the long-wavelength tail for the black body curve is proportional to $\sim\lambda^{-4}$ (corresponding to the Rayleigh–Jeans law), while for the bremsstrahlung this is $\sim\lambda^{-2}$ (for the small argument under the exponent in (9.4.22)). Taking into account the experimental error, it is not surprising that sometimes these two functions are confused; one of such examples is mentioned in (Lavrov 2001). One of the possible sources of error of such type is the background subtraction (which was not done properly). For the simplest example, in Fig. 9.4.3 the comparison between two tails—corresponding to bremsstrahlung and black body radiation—is presented. Note that the range of x in Fig. 9.4.3 corresponds to the relative range of wavelengths from the common experimental spectrum: the range 200 nm—800 nm is a normal case, often the range is narrower.

Recombination radiation arises when a free electron is captured by an ion; as a result, the electron is located on some level n with energy E_n. A certain answer can be obtained for a hydrogen-like atom (with charge Z) and for the Maxwellian distribution on energy for electrons:

$$I_\nu = C\frac{Z^2 n_i n_e}{\sqrt{T}}\exp\left(-\frac{h\nu}{kT}\right)\sum_n\frac{2\xi_n e^{\xi_n}}{n}, \tag{9.4.23}$$

where

$$\xi_n = \frac{E_n}{kT}. \tag{9.4.24}$$

We should emphasize that both equations (9.4.22) and (9.2.23) were derived for equilibrium distributions of electrons on their energy. In conditions of a non-equilibrium plasma, the real equations may differ significantly from these expressions.

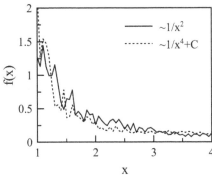

Fig. 9.4.3. Simulation of the experimental spectra: comparison between two tails—corresponding to the dependence $f \sim \lambda^{-2}$ (bremsstrahlung) and $f \sim \lambda^{-4}$ (black body radiation) with background noise added for greater plausibility. As we see, it is very easy to confuse these 'experimental spectra'.

Thus, the gas discharge produces light, but its spectrum is very complicated. In conditions where the sonoluminescence spectrum could be measured with such a high accuracy, that is, the resolution and the error of the measured intensity, that can be easily achieved, for example, for an electric arc, the problem to establish the nature of sonoluminescence would be solved, approximately, in one hour. Instead, we have to guess the character of the continuous spectra—whether it corresponds to bremsstrahlung, thermal radiation, non-equilibrium thermal radiation or something else.

Moreover, in a situation when we cannot point out the exact step-by-step mechanism of sonoluminescence, we may investigate more exotic constructions.

9.4.5 Exotic theories

In this section, we place all other theories; for some of them, the term 'exotic' may seem superfluous.

First of such 'semi-exotic' theories is the chemi-luminescence mechanism applied to sonoluminescence (Suslick 1990). Sonochemistry is a huge and developed scientific area (see a brief introduction to it in Section 6.5); therefore, one may assume that these various chemical reactions that take place under ultrasonic effects lead to light emission. This is a quite reasonable theory, but it has one disadvantage: the exact mechanism of the light emission is still unknown. In the beginning of this section, we divided all the theories into macroscopic and microscopic: this one belongs to the first group.

The main problem that must be explained for sonoluminescence is the solid spectrum—the continuum part of it—of this light emission. As we see above, there are many candidates for its role even at first glance: from the black-body (despite the fact that we could not explain the nature of high temperature clearly) to the complicated plasma spectrum which consists, generally, of many elements. To this assortment, we add one more possible way.

As we discussed in Section 9.4.2, the spectrum of sonoluminescence is similar to the spectrum of hydroluminescence in the same medium (in glycerol). We may point out many differences in these phenomena, but at least, there is one coincidence there: both processes include intense cavitation. In an ultrasound impact, the fact of cavitation is obvious; as for the flow in a narrow channel, the light emission is observed in the dynamic low-pressure zone after the exit from the narrowest part of the channel.

It is hard to expect that the magnitudes of shrinking pressure are very large in conditions of HL: for SL at a strong ultrasonic impact, one may assume something of that sort, but in a free jet this looks questionable. Thus, enormous collapse velocities (of kilometers per second) can be excluded from the consideration, as well as the thermal mechanism of light-emission.

Moreover, the spectra presented in Fig. 9.2.9 for SL and HL differ from the black-body curve and from the bremsstrahlung, *a fortiori*: they are too narrow. These confined solid spectra look rather like the spectra of an LED (light-emission diode) or a similar object. Therefore, it is logical to presume the fluorescent nature of such light emission, despite the fact, of course, that glycerol is a non-fluorescent medium.

Scintillation—the fluorescent flash—can be described as follows. After excitation, the radiant de-excitation process is a quantum transition from the upper energy level to the lower one. In a condensed medium, where atoms are close to each other and, therefore, affect each other strongly, the energy of levels fluctuates—that is why the energy of the emitted photon differs from the single value. If so, then we may establish that the light is emitted from the condensed phase—from the liquid, not from the vapor. Thereby, the source of luminescence is the liquid in vicinity of the gaseous phase because this is the only area of influence of the cavitation process (we do not take into account such matters as deviation of pressure in the bulk of a fluid due to the bubble collapse, etc.). That is, one may say that the emitter is the interface.

It all sounds well, and in Section 9.2 we even present the corresponding expression for the shape of the 'continuum'—actually, following our consideration, for a very broadened line. But the question is: how it can be possible, that is, how does the interface obtain radiation properties? On the microscopic level, this question has a direct answer: due to the special structure of energy zones in the given medium (see Chapter 8). Commonly, the surface is a special object from this point of view: here the zone structure differs from the one in the bulk of the medium. The only problem is that the surface itself also has no luminescent properties in non-luminescent materials (like glycerol), like the bulk of this liquid. The flashes (very short flashes, in comparison with the majority of processes during cavitation) happen only at an intense mechanical impact.

If we accept all the hypotheses above, then we may continue with the following choices as the answer to the last question:

- ✓ an intense mechanical impact changes the structure of energy zones; especially, it must be easy to observe this under ultrasonic irradiation, when a strong ordering force acts on the medium;
- ✓ the filling of zones (by electrons) changes because of electrization during cavitation: for optical transitions, there must be a subject that transits.

Most likely, both processes take place.

Of course, this hypothesis is exotic, as it was promised in the name of this section.

Another exotic hypothesis, which may explain the light emission from the interface in the presence of charged particles, is the transition radiation predicted by Ginzburg and Frank (Ginzburg and Frank 1945). This radiation arises when a charged particle crosses at constant velocity the boundary between two media with different dielectric permittivity. This radiation has a specific spatial (angle) distribution relative to the direction of velocity of the initial electron; this fact does not eliminate the application of this theory to sonoluminescence, since there we have a lot of directions for moving electrons. The spectrum of such radiation depends on the electron energy; again, in general case, because we do not know much about the electron energy in such conditions, we cannot reject this explanation.

The most exotic, but still scientific, theory of sonoluminescence implies that SL is the quantum vacuum radiation ((Eberlein 1996a,b) based on an idea from (Scwinger1992))—the so-called dynamic Casimir effect. This effect is quantum fluctuations of the electromagnetic field, which may cause the attraction of two metal plates placed very close to each other—the most popular manifestation of the Casimir effect. The application of this theory to SL has been strongly criticized (Unnikrasann and Mukhopadhyay 1996) and now it is not even discussed, but, indeed, this hypothesis was very original.

9.4.6 Final thoughts

In the very end, we feel obligated to articulate our personal position, at least for the purpose to be pointed at our own mistakes by our colleagues, when the nature of sonoluminescence will be finally discovered, which will happen sometime.

First of all, we prefer to keep an open mind about this matter. It is very important to admit solid facts when they appear, and under 'solid facts' we mean not only a perfect experiment, but also a consistent theory. But, using the language of probabilistic conclusions, we may proclaim the following statements:

- ✓ it is likely that sonoluminescence has analogues in hydroluminescence—the light emission from a fluid flowing in a throttle; this means that the ultrasonic irradiation itself does not play a crucial role in this phenomenon;
- ✓ considering the continuous spectrum, it is likely that the emission phase is not the gas, but the liquid in vicinity of bubble's walls;
- ✓ nevertheless, it is very likely that some components of a glow—lines and bands—originate from the gas in a bubble; this fact means that the medium inside a bubble is at an excited state;
- ✓ it is unlikely that the emission is caused by the bulk of a liquid;
- ✓ it is likely that electrical effects at least accompany the sonoluminescence phenomenon;
- ✓ it is an open question whether high temperature even accompanies the SL phenomenon necessarily or not; if yes, the second open question is what is the cause, and what is the consequence—high temperature or other effects including light emission? If high temperature accompanies SL, this fact must also be confirmed for HL.

It is very likely that after discovering the true nature of sonoluminescence, many people will say: 'Well, this is what I've been saying all along!'

Conclusion

Sonoluminescence (SL) arises in a fluid under intensive mechanical impacts—under the influence of ultrasound, first of all. The simplest observation shows that this light emission originates from a collapsing bubble.

At this point, the exact data runs out. As usual, for glowing objects—flames, gas discharges, stars, etc.—spectroscopy is the main source of information about the state of the medium. But sonoluminescence is not the case: the luminous bubble is a very

- ✓ small,
- ✓ dim,
- ✓ unstable,
- ✓ moody

object and we cannot explore it carefully, with proper accuracy. Interpretations of SL spectra can be different, especially considering the large experimental error; we cannot extract a lot of useful information from them yet.

Thus, there is no explanation for this effect to date. Being discovered in 1934, sonoluminescence remains a mysterious object in modern physics. It has no proper theoretical explanation or, from some point of view, it has too many possible explanations. Indeed, there may be a lot of external conditions in which a liquid–gas system would produce light, but we cannot distinguish the correct one that explains all the details of SL. Not the least of all, this is due to the fact that the exact details remain hidden from us.

A possible way to explain the nature of sonoluminescence is to compare it with other processes of light emission from fluid systems. Following this way, we discovered that sonoluminescence is somewhat relative to hydroluminescence (HL), the process of light emission from a fluid flowing in a throttle.

The spectra of SL and HL look similar for the same medium (glycerol); this is a good circumstance which helps us to establish the common properties of these two processes—in both cases, the glow arises in the zone of intense cavitation. The ultrasound impact, therefore, is not mandatory to cause the light emission: the irregular cavitating flow in a jet streaming from the narrowest part of a throttle is a quite suitable replacement for it.

On the other hand, we may emphasize one important distinction. During HL of a fluorescent media (to be exact, the I-40A oil), we see the corresponding component in the spectrum. On the contrary, under the ultrasonic influence we see nothing. That is, under the impact of ultrasound we see neither some special components of the glow, nor the ordinary fluorescence of that oil, which can be excited in many ways, including an external light.

The last matter is another contribution the common mystery of sonoluminescence. There are many strange experimental facts there, and a lot of more or less exotic theories too. Even the simplest and the oldest explanation of SL—as the thermal radiation of a medium inside a strongly collapsing bubble—becomes weird when we analyze the details of this conception.

Nevertheless, the nature of the sonoluminescence phenomenon will be established sooner or later—this phraseological unit is questionably applied here since the time for 'sooner' was up several decades ago. However, this will inevitably happen, and after that many theories that were seriously discussed would probably cause a smile (not from us, should we live to see that day). But, anyway, the story of investigation of sonoluminescence deserves to be examined by students of the future: with all the turns, with experiments and their interpretations, with theories and their falsifications—without a doubt, the book 'The Story of Sonoluminescence' will be a scientific bestseller. After everything becomes finally clear, of course.

Anyway, we should remember how much the bubble, not the black hole, resisted the physical science.

References

Abdrachmanov, A. M., G. L. Sharipov, I. V. Rusakov, V. R. Akhmetova and R. G. Bulgakov. 2007. Sonoluminescence of elementary sulfur melt. *Journal of Experimental and Theoretical Physics Letters*, 85: 410–1. https://doi.org/10.1134/S0021364007080139.

Akhatov, I., O. Lindau, A. Topolnikov, R. Mettin, N. Vakhitova and W. Lauterborn. 2001. Collapse and rebound of a laser-induced cavitation bubble. *Physics of Fluids*, 13:2805–19. https://doi.org/10.1063/1.1401810.

Arakeri, V. H. 2003. Sonoluminescence and bubble fusion. *Current Science*, 911–6.

Barber, B. P. and S. J. Putterman. 1991. Observation of synchronous picosecond sonoluminescence. *Nature*, 352: 318–20. https://doi.org/10.1038/352318a0.

Barber, B. P., C. C. Wu, R. Löfstedt, P. H. Roberts and S. J. Putterman. 1994. Sensitivity of sonoluminescence to experimental parameters. *Physical Review Letters*, 72: 1380. https://doi.org/10.1103/PhysRevLett.72.1380.

Bataller, A. W. 2014. *Exploring the Universality of Sonoluminescence*. University of California, Los Angeles.

Belyaev, I. A., D. A. Biryukov and D. N. Gerasimov. 2018. Spectroscopy of a cavitating liquid in the optical and X-ray spectral ranges. *Optics and spectroscopy*, 125: 42–8. https://doi.org/10.1134/S0030400X18070056.

Belyaev, V. B., B. F. Kostenko, M. B. Miller, A. V. Sermyagin and A. S. Topol'nikov. 2003. Super high temperatures and acoustic cavitation. *Joint Institute for Nuclear Research*, JINR-R--3-2003-214.

Biryukov, D. A., M. I. Vlasova, D. N. Gerasimov and O. A. Sinkevich. 2013. Light emitted from a liquid that flows in a narrow channel as triboluminescence. *Optics and Spectroscopy*, 114: 704–8. https://doi.org/10.1134/S0030400X13050032.

Biryukov, D. A. and D. N. Gerasimov. 2018. Spectroscopic diagnostics of hydrodynamic luminescence. *Journal of Molecular Liquids*, 266: 75–81. https://doi.org/10.1016/j.molliq.2018.06.043.

Borisenok, V. A. 2015. Sonoluminescence: Experiments and models. *Acoustical Physics*, 61: 308–32. https://doi.org/10.1134/S1063771015030057.

Brenner, M. P., Hilgenfeldt, S. and D. Lohse. 2002. Single-bubble sonoluminescence. *Reviews of modern physics*, 74: 425. https://doi.org/10.1103/RevModPhys.74.425.

Brodsky, A. M., Burgess, L. W. and A. L. Robinson. 2001. Cooperative effects in sonoluminescence. *Physics Letters A*, 287: 409–14. https://doi.org/10.1016/S0375-9601(01)00511-4.

Chakravarty, A. and A. J. Walton. 2000. Light emission from collapsing superheated steam bubbles in water. *Journal of luminescence*, 92: 27–33. https://doi.org/10.1016/S0022-2313(00)00250-7.

Chen, Q. D. and L. Wang. 2005. Luminescence from transient cavitation bubbles in water. *Physics Letters A*, 339: 110–7. https://doi.org/10.1016/j.physleta.2005.03.029.

Ciuti, P., N. V. Dezhkunov, A. Francescutto, F. Calligaris and F. Sturman. 2003. Study into mechanisms of the enhancement of multibubble sonoluminescence emission in interacting fields of different frequencies. *Ultrasonics sonochemistry*, 10: 337–41. https://doi.org/10.1016/S1350-4177(03)00097-X.

Delgado, A. V., F. González-Caballero, R. J. Hunter, L. K. Koopal and J. Lyklema. 2005. Measurement and interpretation of electrokinetic phenomena (IUPAC technical report). *Pure and Applied Chemistry*, 77: 1753–805. https://doi.org/10.1016/j.jcis.2006.12.075.

Dezhkunov, N. V. 2002. Multibubble sonoluminescence intensity dependence on liquid temperature at different ultrasound intensities. *Ultrasonics sonochemistry*, 9: 103–6. https://doi.org/10.1016/S1350-4177(01)00116-X.

Didenko, Y. T., W. B. McNamara III and K. S. Suslick. 2000a. Effect of noble gases on sonoluminescence temperatures during multibubble cavitation. *Physical review letters*, 84: 777. https://doi.org/10.1103/PhysRevLett.84.777.

Didenko, Y. T., W. B. McNamara III and K. S. Suslick. 2000b. Molecular emission from single-bubble sonoluminescence. *Nature*, 407: 877–9. https://doi.org/10.1038/35038020.

Eberlein, C. 1996a. Sonoluminescence as quantum vacuum radiation. *Physical review letters*, 76: 3842. https://doi.org/10.1103/PhysRevLett.76.3842.

Eberlein, C. 1996b. Theory of quantum radiation observed as sonoluminescence. *Physical Review A*, 53: 2772. https://doi.org/10.1103/PhysRevA.53.2772.

Eddingsaas, N. C. and K. S. Suslick. 2006. Light from sonication of crystal slurries. *Nature*, 444:163. https://doi.org/10.1038/444163a.

Eddingsaas, N. C. and K. S. Suslick. 2007. Intense mechanoluminescence and gas phase reactions from the sonication of an organic slurry. *Journal of the American Chemical Society*, 129: 6718–9. https://doi.org/10.1021/ja0716498.

Farhat, M., A. Chakravarty and J. E. Field. 2011. Luminescence from hydrodynamic cavitation. *Proceedings of the Royal Society A: Mathematical, Physical and Engineering Sciences*, 467: 591–606. https://doi.org/10.1098/rspa.2010.0134.

Fellouah, H., C. G. Ball and A. Pollard. 2009. Reynolds number effects within the development region of a turbulent round free jet. *International Journal of Heat and Mass Transfer*, 52: 3943–54. https://doi.org/10.1016/j.ijheatmasstransfer.2009.03.029.

Ferguson, D. N. 2010. A basic triboelectric series for heavy minerals from inductive electrostatic separation behaviour. *Journal of the Southern African Institute of Mining and Metallurgy*, 110: 75–8.

Fiedler, H. E. 2003. Control of free turbulent shear flows. *In*: Gad-el-Hak, M., A. Pollard and J. P. Bonnet (eds.). *Flow Control: Fundamentals and Practices*. Springer Science & Business Media.

Flannigan, D. J. and K. S. Suslick. 2005. Plasma formation and temperature measurement during single-bubble cavitation. *Nature*, 434: 52–5. https://doi.org/10.1038/nature03361.

Flannigan, D. J., S. D. Hopkins, C. G. Camara, S. J. Putterman and K. S. Suslick. 2006. Measurement of pressure and density inside a single sonoluminescing bubble. *Physical review letters*, 96: 204301. https://doi.org/10.1103/physrevlett.96.204301.

Flint, E. B. 1990. *Spectroscopic studies of sonoluminescence* (Doctoral dissertation). University of Illinois: Urbana-Champaign. http://dx.doi.org/10.1021/acs.jpcb.5b10221.

Flint, E. B. and K. S. Suslick. 1991. The temperature of cavitation. *Science*, 253: 1397–9. https://doi.org/10.1126/science.253.5026.1397.

Fontenot, R. S. and W. A. Hollerman. 2011. Measuring triboluminescence from ZnS: Mn produced by ballistic impacts. *Journal of Instrumentation*, 6: T04001. https://ui.adsabs.harvard.edu/link_gateway/2011JInst...6T4001F/doi:10.1088/1748-0221/6/04/T04001.

Frenkel, J. 1940. On electrical phenomena associated with cavitation due to ultrasonic vibrations in liquids. *Acta Physicochim. USSR*, 12: 317–22.

Frenkel, J. 1946. *Kinetic Theory of Liquids*. Oxford University Press: Oxford.

Frenzel, H. and H. Schultes. 1934. Luminescenz in ultrschallbeschikten Wasser. *Zeitschrift für Physikalische Chemie*, B27: 421–4 (in German).

Gaitan, D. F., Lawrence A. C., Charles C. C. and R. A. Roy. 1992. Sonoluminescence and bubble dynamics for a single, stable, cavitation bubble. *The Journal of the Acoustical Society of America*, 91: 3166–83. https://doi.org/10.1121/1.402855.

Galembeck, F., T. A. L. Burgo, L. B. S. Balestrin, R. F. Gouveia, C. A. Silva and A. Galembeck. 2014. Friction, tribochemistry and triboelectricity: recent progress and perspectives. *Rsc Advances*, 4: 64280-98. https://doi.org/10.1039/C4RA09604E.

Garcia, N., A. P. Levanyuk and V. V. Osipov. 1999. Scenario of the electric breakdown and UV radiation spectra in single-bubble sonoluminescence. *Journal of Experimental and Theoretical Physics Letters*, 70: 428–33. https://doi.org/10.1134/1.568192.

Ginzburg, V. L. and I. M. Frank. 1945. Radiation of a uniformly moving electron due to its transition from one medium into another. *Journal of Physics (USSR)*, 9: 353–62.

Grupen, C., B. A. Shwartz, H. Spieler, S. Eidelman and T. Stroh. 2008. *Particle detectors* (Vol. 11). Cambridge university press: Cambridge.

Guha, A., R. M. Barron and R. Balachandar. 2011. An experimental and numerical study of water jet cleaning process. *Journal of Materials Processing Technology*, 211: 610–8. https://doi.org/10.1016/j.jmatprotec.2010.11.017.

Hale, G. M. and M. R. Querry. 1973. Optical constants of water in the 200-nm to 200-μm wavelength region. *Applied optics*, 12: 555–63. https://doi.org/10.1364/AO.12.000555.

Hammer, D. and L. Frommhold. 2001. Sonoluminescence: how bubbles glow. *Journal of modern optics*, 48: 239–77. http://dx.doi.org/10.1134/S1063771015030057.

Harvey, E. N. 1939. Sonoluminescence and sonic chemiluminescence. *Journal of the American Chemical Society*, 61: 2392–8. https://doi.org/10.1021/ja01878a037.

Hickling, R. 1968. Comments on "Light Emission from Hydrodynamic Cavitation". *Physics of Fluids*, 11: 1586. http://dx.doi.org/10.1063/1.1692152.

Hiller, R., S. J. Putterman and B. P. Barber. 1992. Spectrum of synchronous picosecond sonoluminescence. *Physical Review Letters*, 69: 1182. https://doi.org/10.1103/PhysRevLett.69.1182.

Hiller, R., K. Weninger, S. J. Putterman and B. P. Barber. 1994. Effect of noble gas doping in single-bubble sonoluminescence. *Science*, 266: 248–50. https://doi.org/10.1126/science.266.5183.248.

Holzfuss, J., M. Rüggeberg and R. Mettin. 1998. Boosting sonoluminescence. *Physical Review Letters*, 81: 1961. https://doi.org/10.1103/PhysRevLett.81.1961.

Jarman, P. 1960. Sonoluminescence: A discussion. *The Journal of the Acoustical Society of America*, 32: 1459–62. https://doi.org/10.1121/1.1907940.

Kamath, V., A. Prosperetti and F. N. Egolfopoulos. 1993. A theoretical study of sonoluminescence. *The Journal of the Acoustical Society of America*, 94: 248–60. https://doi.org/10.1121/1.407083.

Koldamasov, A. I. and A. Mason. 1991. Plasma formation in a cavitating dielectric liquid. Soviet physics. *Technical physics*, 36: 234–5.

Konstantinov, V. A. 1947. About electrical discharges during cavitation. *Doklady Akademii Nauk SSSR, URSS*, 56, pp. 259–60 (in Russian).

Kornilova, A. A., V. I. Vysotskii, A. I. Koldamasov, H. I. Yang, D. B. McConnell and A. V. Desyatov. 2007. Generation of intense directional radiation during the fast motion of a liquid jet through a narrow dielectric channel. *Journal of Surface Investigation. X-ray, Synchrotron and Neutron Techniques*, 1: 167–71. https://doi.org/10.1134/S1027451007020103.

Kornilova, A. A., V. I. Vysotskii, N. N. Sysoev and A. V. Desyatov. 2009. Generation of X-rays at bubble cavitation in a fast liquid jet in dielectric channels. *Journal of Surface Investigation. X-ray, Synchrotron and Neutron Techniques*, 3: 275–83. https://doi.org/10.1134/S1027451009020207.

Kornilova, A. A., V. I. Vysotskii, N. N. Sysoev, N. K. Litvin, V. I. Tomak and A. A. Barzov. 2010. Shock-cavitational mechanism of X-ray generation during fast water stream cavitation. *Moscow University Physics Bulletin*, 65: 46–50. https://doi.org/10.3103/S002713491001011X.

Koverda, V. P., V. N. Skokov, A. V. Reshetnikov and A. V. Vinogradov. 2005. Fluctuations with 1/f spectrum under conditions of acoustic cavitation of water. *High Temperature*, 43: 634–9. https://doi.org/10.1007/s10740-005-0106-4.

Kudryashov, V. E., S. S. Mamakin, A. N. Turkin, A. E. Yunovich, A. N. Kovalev and F. I. Manyakhin. 2001. Spectra and quantum efficiency of light-emitting diodes based on GaN heterostructures with quantum wells and their dependence on current and voltage (Spektry i kvantovyj vykhod izlucheniya svetodiodov s kvantovymi yamami na osnove geterostruktur iz GaN-zavisimost'ot toka i napryazheniya). *Fizika i Tekhnika Poluprovodnikov*, 35 (in Russian).

Lau, J. C., M. H. Fisher and V. Fuchs. 1972. The intrinsic structure of turbulent jets. *Journal of Sound and Vibration*, 22: 379–406. https://doi.org/10.1016/0022-460X(72)90451-8.

Lavrov, B. P. 2001. Mysteries and unknows of single bubble sonoluminescence from viewpoint of plasma spectroscopy. *APP spring meeting 'Diagnostics of non-equilibrium high pressure plasmas' Book of papers*, 241–5.

Leighton, T. G., M. Farhat, J. E. Field and F. Avellan. 2001. Luminescence from hydrodynamic cavitation: Method and preliminary analysis.

Leighton, T. G., M. Farhat, J. E. Field and F. Avellan. 2003. Cavitation luminescence from flow over a hydrofoil in a cavitation tunnel. *Journal of Fluid Mechanics*, 480: 43. http://dx.doi.org/10.1017/S0022112003003732.

Lepoint, T., F. Lepoint-Mullie, N. Voglet, S. Labouret, C. Pétrier, R. Avni and J. Luque. 2003. OH/D A2Σ+–X2Πi rovibronic transitions in multibubble sonoluminescence. *Ultrasonics sonochemistry*, 10: 167–74. https://doi.org/10.1016/S1350-4177(03)00082-8.

Liu, Z., J. Luo, C. Shao and L. Yu. 2000. A representation of sonoluminescence spectra. *Journal of Physics B: Atomic, Molecular and Optical Physics*, 33: 4495. https://doi.org/10.1088/0953-4075/33/20/321.

Lockett, R. D. and A. Bonifacio. 2021. Hydrodynamic luminescence in a model diesel injector return valve. *International Journal of Engine Research*, 22: 963–74. https://doi.org/10.1177/1468087419870421.

Margulis, M. A. and I. M. Margulis. 2002. Contemporary review on nature of sonoluminescence and sonochemical reactions. *Ultrasonics sonochemistry*, 9: 1–10. https://doi.org/10.1016/S1350-4177(01)00096-7.

Margulis, M. A. 2000. *Sonoluminescence*. *Physics-Uspekhi*, 43: 259. https://doi.org/10.1070/PU2000v043n03ABEH000455.

Matula, T. J., R. A. Roy, P. D. Mourad, W. B. McNamara III and K. S. Suslick. 1995. Comparison of multibubble and single-bubble sonoluminescence spectra. *Physical review letters*, 75: 2602. https://doi.org/10.1103/PhysRevLett.75.2602.

Matula, T. J., I. M. Hallaj, R. O. Cleveland, L. A. Crum, W. C. Moss and R. A. Roy. 1998. The acoustic emissions from single-bubble sonoluminescence. *The Journal of the Acoustical Society of America*, 103: 1377–82. https://doi.org/10.1121/1.421279.

Monakhov, A. A. 2013. A hydrodynamic X-ray radiation source. *Doklady Physics*, 58: 258–60. https://doi.org/10.1134/S1028335813060116.

Or, C. M., K. M. Lam and P. Liu. 2011. Potential core lengths of round jets in stagnant and moving environments. *Journal of Hydro-environment Research*, 5: 81–91. https://doi.org/10.1016/j.jher.2011.01.002.

Ouerhani, T., R. Pflieger and S. I. Nikitenko. 2015. Spectroscopic studies of sonoluminescence of the N2/Ar/H2O system to probe the non-equilibrium plasma in cavitation bubbles. http://dx.doi.org/10.1021/acs.jpcb.5b10221.

Peterson, F. B. and T. P. Anderson. 1967. Light emission from hydrodynamic cavitation. *The Physics of Fluids*, 10: 874–9. https://doi.org/10.1063/1.1762203.

Putterman, S. 1998. Sonoluminescence: the star in a jar. *Physics world*, 11: 38. https://doi.org/10.1088/2058-7058/11/5/31.

Putterman, S. J. and K. R. Weninger 2000. Sonoluminescence: How bubbles turn sound into light. *Annual Review of Fluid Mechanics*, 32: 445–76. https://doi.org/10.1146/annurev.fluid.32.1.445.

Qi-Dai, C. and W. Long. 2004. Luminescence from tube-arrest bubbles in pure glycerin. *Chinese Physics Letters*, 21: 1822. https://doi.org/10.1088/0256-307X/21/9/041.

Raizer, Y. P. 1991. *Gas Discharge Physics*. Springer-Verlag: Berlin.

Schwinger, J. 1992. Casimir energy for dielectrics. *Proceedings of the National Academy of Sciences of the United States of America*, 89: 4091. https://dx.doi.org/10.1073/pnas.89.9.4091.

Segelstein, D. J. 1981. *The Complex Refractive Index of Water*. Thesis. University of Missouri: Kansas City.

Sharipov, G. L., A. M. Abdrakhmanov and A. A. Tukhbatullin. 2009. Sonotriboluminescence in suspensions of trivalent terbium compounds. *Technical Physics Letters*, 35: 452–5. https://doi.org/10.1134/S1063785009050204.

Sharipov, G. L., A. M. Abdrakhmanov and B. M. Gareev. 2012. Luminescence of sodium atoms in aqueous solution during sonolysis in moving-single-bubble regime. *Technical Physics Letters*, 38(1): 74–76. https://doi.org/10.1134/S1063785012010282.

Sharipov, G. L., A. M. Abdrakhmanov and L. R. Yakshembetova. 2012. Multibubble sonoluminescence of Tb 3+ ion in aqueous solutions of dimethyl sulfoxide. *Russian Chemical Bulletin*, 61: 528–31.

Sharipov, G. L., A. A. Tukhbatullin, M. R. Muftakhutdinov and A. M. Abdrakhmanov. 2014. Luminescence of OD radical as an evidence for water decomposition under destruction of the deuterated terbium sulfate crystal hydrate. *Journal of luminescence*, 148: 79–81. https://doi.org/10.1016/j.jlumin.2013.11.086.

Sharipov, G. L., Gareev, B. M. and A. M. Abdrakhmanov. 2018. Sonoluminescence of suspensions of insoluble chromium carbonyl nanoparticles in water and inorganic acids. *Technical Physics Letters*, 44: 1072–3. https://doi.org/10.1134/S1063785018120350.

Sharipov, G. L., A. M. Abdrakhmanov and B. M. Gareev. 2019. Visualization of luminescence of two types in an acoustic field in a liquid. *Technical Physics Letters*, 45: 1175–7. https://doi.org/10.1134/S1063785019120137.

Sivakumar, S., Sangras, R. and V. Raghavan. 2012. Characteristics of turbulent round jets in its potential-core region. *World Academy of Science, Engineering and Technology*, 61: 526–32. https://doi.org/10.5281/zenodo.1055134.

Skokov, V. N., A. V. Reshetnikov, A. V. Vinogradov and V. P. Koverda. 2007. Fluctuation dynamics and 1/f spectra characterizing the acoustic cavitation of liquids. *Acoustical Physics*, 53: 136–40. https://doi.org/10.1134/S1063771007020042.

Su, C. S. and S. M. Yeh. 1996. UV attenuation coefficient in water determined by thermoluminescence detector. *Radiation measurements*, 26: 83–6. https://doi.org/10.1016/1350-4487(95)00284-7.

Supponen, O., D. Obreschkow, P. Kobel and M. Farhat. 2017. Luminescence from cavitation bubbles deformed in uniform pressure gradients. *Physical Review E*, 96: 033114. https://doi.org/10.1103/PhysRevE.96.033114.

Suslick, K. S., Y. Didenko, M. M. Fang, T. Hyeon, K. J. Kolbeck, W. B. McNamara III, M. M. Mdleleni and M. Wong. 1999a. Acoustic cavitation and its chemical consequences. *Philosophical Transactions of the Royal Society of London. Series A: Mathematical, Physical and Engineering Sciences*, 357: 335–53. https://doi.org/10.1098/rsta.1999.0330.

Suslick, K. S., W. B. McNamara and Y. Didenko. 1999b. *Hot spot conditions during multi-bubble cavitation. In Sonochemistry and sonoluminescence*. Springer:Dordrecht. https://doi.org/10.1007/978-94-015-9215-4_16.

Suslick, K. S. and D. J. Flannigan. 2008. Inside a collapsing bubble: sonoluminescence and the conditions during cavitation. *Annual Review of Physical Chemistry*, 59: 659–83. https://doi.org/10.1146/annurev.physchem.59.032607.093739.

Suslick, K. S., N. C. Eddingsaas, D. J. Flannigan, S. D. Hopkins and H. Xu. 2011. Extreme conditions during multibubble cavitation: Sonoluminescence as a spectroscopic probe. *Ultrasonics sonochemistry*, 18: 842–6. https://doi.org/10.1016/j.ultsonch.2010.12.012.

Suslick, K. S. 1990. Sonochemistry. *Science*, 247: 1439–45. https://doi.org/10.1126/science.247.4949.1439.

Unnikrishnan, C. S. and S. Mukhopadhyay. 1996. Comment on "Sonoluminescence as quantum vacuum radiation". *Physical review letters*, 77: 4690. https://doi.org/10.1103/PhysRevLett.77.4690.

Vogel, A., S. Busch and U. Parlitz. 1996. Shock wave emission and cavitation bubble generation by picosecond and nanosecond optical breakdown in water. *The Journal of the Acoustical Society of America*, 100: 148–65. https://doi.org/10.1121/1.415878.

Webb, S. M. and N. J. Mason. 2003. Single-bubble sonoluminescence: creating a star in a jar. *European journal of physics*, 25: 101. https://doi.org/10.1088/0143-0807/25/1/013.

Williams, M. W. 2015. The increasing role of Chemistry in understanding triboelectric charging of insulating materials–A paradigm shift. *Physical Science International Journal*, 88–92. https://doi.org/10.9734/PSIJ/2015/14033.

Xu, H., N. G. Glumac and K. S. Suslick. 2010. Temperature inhomogeneity during multibubble sonoluminescence. *Angewandte Chemie*, 122: 1097–100. https://doi.org/10.1002/ange.200905754.

Yakobson, M. A., D. K. Nel'son, O. V. Konstantinov and A. V. Matveentsev. 2005. The tail of localized states at the forbidden band of quantum well in system In 0.2 Ga 0.8 N/GaN and its effect on the photoluminescence spectrum at the laser excitation. *Fizika i Tekhnika Poluprovodnikov*, 39: 1459–63 (in Russian).

Yuan, L. 2005. Sonochemical effects on single-bubble sonoluminescence. *Physical Review E*, 72: 046309. https://doi.org/10.1103/physreve.72.046309.

Conclusion

Cavitation is a complex of non-stationary phenomena in a fluid after an abrupt pressure drop in the system. Since pressure decreases below the saturation value at the given temperature, the 'liquid–vapor' phase transition is a part of the common process, but boiling—the common term for a phase transition of such type—is not a significant component of the process. We are interested in cavitation not because some amount of liquid vaporizes during this process, but because of the fast growth of vapor bubbles in a fluid and, partially, because of some side effects related to this phenomenon.

Cavitation can be easily observed in domestic conditions with the help of a regular syringe. Take a syringe, fill it half-way with water, close its nozzle, pull the piston—and you will see the subject of this book. Trying various speeds of pulling, taking water at different temperatures, exploring the reverse process—pushing the piston back into the syringe—and doing many other tests, one may obtain a lot of information to think about, especially if one has the ability to capture the phenomenon with a video camera. This is a very good start for exploring cavitation, and finishing these treatments, one may think that almost everything is clear, and cavitation is a well-understood phenomenon.

Alas, it is not. Cavitation is a well-understood process only in the sense of prediction of its appearance. We know that cavitation begins when pressure decreases below the saturation value, and, in these conditions, the amount of the gaseous phase will increase (*a*) inside the liquid phase, (*b*) outside of the liquid phase. The process *a* is, indeed, the process that we are interested in, and its different sides are described in different chapters of this book. The process *b* is the manifestation of the Le Chatelier – Brown principle: once a stable system has been taken away from equilibrium, processes in that system will tend to return the system back to equilibrium. In simple words, due to the process *b*, the pressure outside the liquid will be tend to get the saturation value; once it is reached, the expansion of bubbles in the liquid (process *a*) will stop.

Nearly at this point, our ability to predict cavitation expires. Cavitation is too irregular of a process for accurate forecasts. Even initial conditions—the parameters of the system right after the sudden pressure drop—depend on the details of that drop. Generally, this is an irreversible adiabatic process, the degree of irreversibility of which is hard to determine. Then, processes in the gaseous phase are non-stationary, and even the qualitative description of them demands very complicated physics.

The main hero of cavitation is a bubble. Its expansion, its collapse and its oscillations give a mental pabulum for scientists of diverse specialties: even such a simple, at first glance, object has not been investigated properly yet, both theoretically and empirically. For instance, most theories imply the evolution of a spherical bubble; it is easy to confirm that the bubble really loses its sphericity during oscillations. Then, the boundary conditions for that bubble are so fuzzy that, indeed, theories usually obtain the results that were initially put into these theories: assuming adiabatic conditions for the bubble instead of isothermal ones, we have approximately the same results for dynamic characteristics, but noticeably different temperature within.

Bubble evolution is an almost inexhaustible theme, but cavitation cannot be reduced to bubble dynamics. Multi-bubble acoustic cavitation is an intermediate problem, where we can observe bubbles interacting and merging. Such experiments teach us about some elementary processes of the inter-bubble physics, but also turn us to the investigation of different types of cavitation.

Acoustic cavitation is the simplest, purified way to observe and explore cavitation. Many results presented in this book were obtained during this process; indeed, this is a more predictable type of cavitation. Another way to observe cavitation is the flow in a throttle. This complicated flow may provoke cavitation in several regions: at the narrowest part or at the exit; the investigation of this type of cavitation helps us measure some parameters that cannot be obtained in refined conditions of acoustic cavitation. Namely, there is a way to establish the composition of the gaseous phase inside a bubble in such conditions: a special effect helps us to conduct these measurements—triboelectricity.

When flowing in a narrow dielectric channel, a fluid obtains some electrical charge, which can be locked in the channel due to the so-called electrokinetic effect (exactly, due to one version of this effect—the stream potential). This locked charge may produce initial charges and then cause gas discharges inside bubbles; in sum, the ionized medium does what it must—emits a glow. That light emission—emission of an excited substance—is called hydrodynamic luminescence. The spectral analysis of that light emission lets us to distinguish separate lines and bands and, therefore, determine the composition of the gaseous phase. Moreover, this spectroscopic analysis may give us some quantitative information about the state of the emitting medium. However, here we are walking on thin ice.

Spectroscopic measurements demand a lot of initial interpretation of the spectrum, so to say, before one may extract some usual information from the measured spectrum. To calculate the quantitative parameters of the emitting medium, one should clearly recognize all the components of the spectrum, including technical details: for instance, noise can be mistaken for a continuum radiation—band broadening caused by poor resolution—for a high rotational temperature, etc.

In our book, we sometimes turn our attention to the interpretation of experimental results—not only of the optical spectroscopy data, but of some other parameters too; we believe that this matter is of interest by itself. This factor is especially important when we analyze side effects of cavitation. Measurement methods are always important, but they are crucial for phenomena of unknown nature: for them, 'measurements' with wrong interpretation distort not only the numerical data collected in experiments, but directly the physical nature of the observed process.

Cavitation is contained within or touches many special issues of unknown or poorly known nature, which may be of interest for general physics; each of these effects deserves special attention. In this book, we consider some of them—the phenomena associated with cavitation—but, indeed, the magistral line of cavitation lies slightly aside—in engineering application.

Scientists find many interesting problems in cavitation, but engineers look at cavitation from a slightly different angle. In their language, cavitation is a synonym of an accident. Often, engineers would not describe the details of an issue; a single word is enough for them—cavitation.

In other words, the main practical meaning of cavitation is destruction. Then, the best way to reduce the impact of cavitation is to prevent it. This is the main 'defensive line' for pumps, for example: it is possible to prognose the parameters of these devices efficiently, so to avoid the cavitation mode of a pump is quite possible. Another case is propeller screws: despite the restriction of their rotational frequency, some cavitation effects are unavoidable.

Generally, the goals of cavitation treatments are to explore the negative outcome of cavitation and to reduce it. The main effect of cavitation on a solid surface is erosion—erosion of a special type, called cavitation erosion. Cavitating bubbles, collapsing near a solid boundary, produce microjets which deform the surface. Moreover, once appeared, a cavern on a surface tends to grow: this spot—a hole—promotes cavitation at that site, which causes further growth of that cavern, and so on. The only good point here is that a huge cavern distorts the cavitation process in such a manner that collapsing bubbles are obstructed there. Thereby, with time, the erosion rate diminishes.

Another component of destructive effects of cavitation is a hydraulic shock. Indeed, this is the main part of impacts on a solid surface caused by the bubble collapse. This process has its own physics; actually, this is one of the points where physics meets engineering—this is a usual thing for cavitation, the problems of which need, first of all, more scientific investigation. One of the illustrations for this matter: are high temperatures achieved in a collapsing bubble? Thermal effects may play a role in the common set of the cavitation influence on a surface. Moreover, if high temperature can be reached in a collapsing bubble, this fact may explain the special sort of luminescence—the most intriguing part of cavitation.

Cavitation is the source of a mysterious light emission. The common name of this phenomenon is sonoluminescence, because early observations of this phenomenon were under conditions of ultrasonic irradiation. However, a similar glow was discovered in other conditions later; as far as we can judge now, cavitation is a common point for these various conditions.

The physical nature of sonoluminescence is unclear. The main set of explanations aims rather at the description of conditions that may cause some light emission and does not account for certain details of the given specific glow. Today, the main cause of sonoluminescence is assumed to be high temperature inside a bubble during its collapse, but this theory cannot point out all the intermediates in the chain from the bubble collapse to the light emission. The main competitor of this theory supposes electrical effects during the bubble collapse; generally, such approach does not explain all the features of that light too.

On the other hand, we should be prepared for an absolutely abnormal nature of sonoluminescence. A scientist is not a human who does not believe in unproven things. Scientist is a human who does not believe in anything unproven, but is ready to believe in anything that satisfies a scientific method.

On this note, we finish our brief excursion into cavitation—the engineering science with a long history and many unsolved problems.

Appendix A
Thermodynamics in brief

1

A simple thermodynamic system can be described by two parameters: internal parameter pressure p and external parameter volume V; we may also use a specific volume $v = V/m$, where m is the mass of the substance. Experience shows that there also exists one additional parameter: when two bodies are in contact, they come with time to the state of thermodynamic equilibrium, which takes place when that additional parameter is equal for both bodies. This additional parameter is temperature T, which is the measure of the thermal state of the system. The relation between p, v, T is called the thermal equation of state (or sometimes the equation of state, without clarification):

$$p = f(v, T). \tag{A.1}$$

The simplest thermodynamic system is a perfect gas—the gas that obeys the Clapeyron equation

$$pV = \frac{m}{\mu} RT \text{ or } pv = \tilde{R}T. \tag{A.2}$$

Here parameter

$$\tilde{R} = \frac{R}{\mu} = \frac{8.314}{\mu}, \tag{A.3}$$

is the thermodynamic constant, dimension is $[\tilde{R}] = \dfrac{J}{kg \cdot K}$ if the molecular mass μ of the substance is expressed in $[\mu] = \dfrac{kg}{mole}$. Gases in room conditions—at atmospheric pressure and at the temperature ~ 300 K—are in good agreement with the Clapeyron equation. Generally, the more rarified the gases are, the better agreement with this equation they have. Dense gases, liquids and solids do not obey the Clapeyron law; to describe such systems, we need special equations of state.

2

The total energy of particles in a body is called the inner energy U; the specific energy per a unit of mass is $u = U/m$. The enthalpy is introduced as

$$H = U + pV \tag{A.4}$$

By analogy, the specific enthalpy is $h = u + pv$.

The first law of thermodynamics states that the heat conducted to the system is spent on increasing the inner energy of the system and on work pdV:

$$\delta Q = dU + pdV \text{ or } dq = du + pdv. \tag{A.5}$$

With enthalpy, this equation can be rewritten as

$$\delta Q = dH - Vdp \text{ or } dq = dh - \upsilon dp. \tag{A.6}$$

This law follows from the experiments and is based on the notion of heat: heat is the way to change the inner energy without changing the volume. In other words, we may change the inner energy in two ways—doing work or conducting heat.

The heat capacity of the system shows how the heat conducted to the system increases the temperature of the system in the given process:

$$C_x = \left(\frac{\delta Q}{dT}\right)_x, \tag{A.7}$$

where index 'x' denotes the given process.[1] For instance, for an isochoric process, the heat capacity is

$$C_V = \left(\frac{\partial U}{\partial T}\right)_V, \tag{A.8}$$

which follows from equations (A.5) and (A.7), and for an isobaric process, by analogy, from (A.6) and (A.7)

$$C_p = \left(\frac{\partial H}{\partial T}\right)_p. \tag{A.9}$$

Heat capacities C_p and C_V are connected. Let us represent (A.5) in the form

$$\delta Q = \underbrace{\left(\frac{\partial U}{\partial T}\right)_V dT + \left(\frac{\partial U}{\partial V}\right)_T dV}_{dU} + pdV. \tag{A.10}$$

Then, with (A.7), dividing by dT and constant p, we have from (A.10) with respect to (A.8)

$$C_p = C_V + \left[\left(\frac{\partial U}{\partial V}\right)_T + p\right]\left(\frac{\partial V}{\partial T}\right)_p. \tag{A.11}$$

In a more common case, we may consider a general process at which the heat capacity remains constant; such a process is called a polytropic process. Thus, if the heat capacity during some process is constant, we may represent it in (A.10) as

$$\delta Q = CdT, \tag{A.12}$$

and also expanding the temperature differential in (A.10) as

$$dT = \left(\frac{\partial T}{\partial V}\right)_p dV + \left(\frac{\partial T}{\partial p}\right)_V dp, \tag{A.13}$$

using the correlation (A.12), from which we have

$$\left(\frac{\partial U}{\partial V}\right)_T + p = \left(C_p - C_V\right)\left(\frac{\partial T}{\partial V}\right)_p \tag{A.14}$$

---

[1] Derivation $(\partial z/\partial x)_y$ means an operation during which the parameter y remains constant.

We finally obtain the correlation between the pressure and the volume during a polytropic process:

$$dp = \frac{C_p - C}{C_V - C} \underbrace{\left[-\left(\frac{\partial T}{\partial V}\right)_p \left(\frac{\partial p}{\partial T}\right)_V \right]}_{(\partial p / \partial V)_T} dV. \tag{A.15}$$

Introducing the polytropic exponent

$$n = \frac{C_p - C}{C_V - C}, \tag{A.16}$$

we may formulate the equation (A.15) in the form (dividing it by dV and obtaining in the left-hand side $(\partial p/\partial V)_n$, where index n corresponds to (A.16)):

$$\left(\frac{\partial p}{\partial V}\right)_n = n\left(\frac{\partial p}{\partial V}\right)_T. \tag{A.17}$$

For a perfect gas, for which

$$\left(\frac{\partial p}{\partial V}\right)_T = -\frac{p}{V}. \tag{A.18}$$

Consequently, from (A.17) with (A.18), we have that during a polytropic process

$$pV^n = const. \tag{A.19}$$

Note that (A.19) was obtained in condition (A.18), which is only for a perfect gas.
An adiabatic process is defined by the condition

$$\delta Q = 0 \tag{A.20}$$

and can be considered with the help of (A.12) as the process with $C = 0$. The corresponding parameter (A.16) is called the adiabatic exponent:

$$\gamma = \frac{C_p}{C_V}. \tag{A.21}$$

Thus, the corresponding equation for the adiabatic process, called the Poisson equation, is

$$pV^\gamma = const. \tag{A.22}$$

The heat capacity during a common polytropic process can be expressed with the polytropic exponent, the adiabatic exponent, and the isochoric heat capacity as

$$C = C_V \frac{n - \gamma}{n - 1}. \tag{A.23}$$

As we see, heat capacity can be negative for $1 < n < \gamma$. In this case, despite the fact that some heat is conducted to the system, its temperature decreases, since the inner energy is spent on the work.

3

For a flow of a fluid, we may use the Bernoulli equation (see Appendix B):

$$\frac{w^2}{2} + gz + \int \upsilon dp = const, \tag{A.24}$$

where we use w for velocity. From here, we get

$$wdw + gdz + \upsilon dp = 0. \tag{A.25}$$

Thus, expressing from (A.25) the term υdp and using it for (A.6), we obtain

$$dq = dh + wdw + gdz . \tag{A.26}$$

Sometimes, this equation is called the 'first law of thermodynamics for a flow'.

4

The heat δQ in the equation (A.5) is not a complete differential. However, it can be reduced to it by the multiplier θ, that is,

$$dS = \frac{\delta Q}{\theta} \tag{A.27}$$

which is a complete differential. The multiplier θ is the integrating denominator for (A.5). If there exists a single denominator, then there exists an infinite number of denominators. Therefore, we may choose a single type of denominator which satisfies conditions for a composite system '1+2', both parts of which are in thermal equilibrium, i.e., at the same temperature T:

$$S_{1+2} = S_1 + S_2. \tag{A.28}$$

In a common case, the integrating denominator may be a function of two thermodynamic parameters – temperature and volume: $\theta(T,V)$. Thus, for a composite system there must be

$$\delta Q_{1+2} = \delta Q_1 + \delta Q_2. \tag{A.29}$$

$$\frac{\delta Q_{1\ 2}}{\theta_{1\ 2}(T,V_1+V_2)} = \frac{\delta Q_1}{\theta_1(T,V_1)} + \frac{\delta Q_2}{\theta_2(T,V_2)} \tag{A.30}$$

The equation (A.30) follows from (A.28). Thus, we see that (A.29) and (A.30) are correct only if

$$\theta_{1+2}(T,V_1+V_2) = \theta_1(T,V_1) = \theta(T,V_2). \tag{A.31}$$

For arbitrary volumes, this can be possible only if θ does not depend on V. Thus, we have

$$\theta = f(T). \tag{A.32}$$

The corresponding function S from (A.27) is termed entropy.

Now consider a thermodynamic cycle which consists of two isotherms and two adiabats, that is, of four processes:

$$1\text{–}2\text{: adiabatic, } S_1 = const; \tag{A.33}$$

$$2\text{–}3\text{: isothermal, } T_2 = const; \tag{A.34}$$

$$3\text{–}4\text{: adiabatic, } S_2 = const > S_1 \tag{A.35}$$

$$4\text{–}1\text{: isothermal, } T_1 = const < T_2. \tag{A.36}$$

For this cycle—the Carnot cycle—we have for the heat conducted in the process '2–3' Q_{2-3} and in the process[2] '1–4' Q_{1-4}:

$$\frac{Q_{2-3}}{Q_{1-4}} = \frac{f(T_2)(S_2 - S_1)}{f(T_1)(S_2 - S_1)} = \frac{f(T_2)}{f(T_1)}. \tag{A.37}$$

Now consider a perfect gas in such cycle. For a perfect gas, there exist the Joule law: its inner energy depends only on temperature:

$$U = U(T), \text{ or } \left(\frac{\partial U}{\partial V}\right)_T = 0. \tag{A.38}$$

Initially, (A.38) was an empirical law which was established by Joule during experiments.

Thereby, we see that because of (A.38) in processes '2–3' and '1–4' $dU = 0$, and, consequently, the heat in these processes is equal to the work of a gas; for example,

$$Q_{1-4} = \int_1^4 pdV = RT_1 \ln \frac{V_4}{V_1}. \tag{A.39}$$

Correspondingly, we have

$$\frac{Q_{2-3}}{Q_{1-4}} = \frac{T_2 \ln(V_3/V_2)}{T_1 \ln(V_4/V_1)}. \tag{A.40}$$

Now let us take a look at processes '1–2' and '4–3'. For these adiabatic processes, the Poisson equation can be written with respect to (A.2) as

$$TV^{\gamma-1} = const, \tag{A.41}$$

that is, we get

for the process '1–2': $T_1 V_1^{\gamma-1} = T_2 V_2^{\gamma-1}$, $\qquad$ (A.42)

for the process '4–3': $T_1 V_4^{\gamma-1} = T_2 V_3^{\gamma-1}$. $\qquad$ (A.43)

Therefore, we see from (A.42) and (A.43) that

$$\frac{V_4}{V_1} = \frac{V_3}{V_2}. \tag{A.44}$$

Consequently, for (A.40) we have

$$\frac{Q_{2-3}}{Q_{1-4}} = \frac{T_2}{T_1}. \tag{A.45}$$

Comparing (A.37) and (A.45), we get that the integrating denominator coincides with the thermodynamic temperature:

$$\theta = T; \tag{A.46}$$

and, thereby,

$$\delta Q = TdS. \tag{A.47}$$

---

[2] Actually, in our cycle 1–2–3–4–1 during the process '4–1' the heat is conducted out from the system. Here we consider the reverse process '1–4' in which the heat is conducted.

Thus, from (A.5 and (A.47), the entropy of the system is

$$dS = \frac{dU + p\,dV}{T}.$$

(A.48)

Expanding dU again as in (A.10), we see from (A.48) that

$$\left(\frac{\partial S}{\partial T}\right)_V = \frac{1}{T}\left(\frac{\partial U}{\partial T}\right)_V,$$

(A.49)

$$\left(\frac{\partial S}{\partial V}\right)_T = \frac{1}{T}\left[\left(\frac{\partial U}{\partial V}\right)_T + p\right].$$

(A.50)

Taking the derivative from (A.49) on V, the derivative from (A.50) on T, and equating these expressions (since a mixed derivative cannot depend on the order of derivations), we have

$$\left(\frac{\partial U}{\partial V}\right)_T + p = T\left(\frac{\partial p}{\partial T}\right)_V.$$

(A.51)

Actually, the Joule law (A.38) follows directly from (A.51): for a perfect gas

$$p = T\left(\frac{\partial p}{\partial T}\right)_V,$$

(A.52)

and, consequently, $(\partial U/\partial V)_T = 0$ from (A.51). Thus, the existence of entropy alone helps us discover some important properties of a substance.

Various formulations about the existence of entropy constitute the second law of thermodynamics. These are different statements of different power. The simplest justification for the affirmation concerning the essence of entropy follows from pure mathematics.

For a simple thermodynamic system, which may perform only a single type of work—the work of expansion $p\,dV$—the differential form

$$\delta Q = X_1 dx_1 + X_2 dx_2,$$

(A.53)

(where $X_1 = 1, x_2 = U, X_2 = p, x_2 = V$) always has an integrating denominator; above, we demonstrated that it is equal to temperature.

In a common case, when many types of work are possible (for instance, of electric or magnetic forces), the integrating denominator may not exist. The existence of such a denominator is defined by the Frobenius rule, which for the form

$$\delta Q = X_1 dx_1 + X_2 dx_2 + X_3 dx_3,$$

(A.54)

(where X_1, X_2, X_3 are defined as functions of x_1, x_2, x_3) demands:

$$X_1\left(\frac{\partial X_3}{\partial x_2} - \frac{\partial X_2}{\partial x_3}\right) + X_2\left(\frac{\partial X_1}{\partial x_3} - \frac{\partial X_3}{\partial x_1}\right) + X_3\left(\frac{\partial X_2}{\partial x_1} - \frac{\partial X_1}{\partial x_2}\right) = 0.$$

(A.55)

5

With entropy, one may construct two additional thermodynamic functions of the Helmholtz free energy F:

$$F = U - TS,$$

(A.56)

and the Gibbs energy Φ:

$$\Phi = H - TS = U + pV - TS = F + pV. \tag{A.57}$$

Combining these relations with (A.5) and (A.47), we have

$$dF = -pdV - SdT, \tag{A.58}$$

$$d\Phi = Vdp - SdT. \tag{A.59}$$

As we see, external conditions $V, T = const$ correspond to

$$dF = 0, F = const, \tag{A.60}$$

while conditions $p, T = const$ give

$$d\Phi = 0, \Phi = const. \tag{A.61}$$

6

Now let us consider a system with a variable amount of substance: $dm \neq 0$. Introducing the specific Gibbs potential $\varphi = \Phi/m$, we get

$$d\varphi = \upsilon dp - sdT, \tag{A.62}$$

$$d\Phi = Vdp - SdT + \varphi dm, \tag{A.63}$$

where $s = S/m$.

The specific Gibbs potential is also referred to as the chemical potential μ:

$$\varphi = \mu = \left(\frac{\partial \Phi}{\partial m}\right)_{p,T}. \tag{A.64}$$

For the specific Helmholtz energy $f = F/m$, we may write

$$df = -pd\upsilon - sdT, \tag{A.65}$$

Multiplying (A.65) by m, we have

$$\underbrace{d(mf) - fdm}_{mdf} = \underbrace{-pd(mv) + pvdm}_{-pmdv} - smdT. \tag{A.66}$$

Then, taking into account that $V = m\upsilon$, $S = sm$, $F = fm$, and $\varphi = f + p\upsilon$, we get

$$dF = -pdV - SdT + \varphi dm. \tag{A.67}$$

Yes, this expression may seem strange: we see that here the specific Gibbs energy is presented, not the specific Helmholtz energy. Similar equations for dU and dH can be obtained by analogy with similar results –the function φ is involved everywhere:

$$dU = -pdV + TdS + \varphi dm, \tag{A.68}$$

$$dH = Vdp + TdS + \varphi dm. \tag{A.69}$$

7

Now let us consider a closed two-phase system at parameters

$$p = const, T = const. \tag{A.70}$$

The mass of one phase is m_1, the corresponding Gibbs potential is φ_1; by analogy, we may introduce m_2 and φ_2. The total mass of the substance is constant:

$$m_1 + m_2 = m = const, \qquad (A.71)$$

that is,

$$dm_1 + dm_2 = 0. \qquad (A.72)$$

Correspondingly, we may write

$$d\Phi = Vdp - SdT + \varphi_1 dm_1 + \varphi_2 dm_2, \qquad (A.73)$$

where $V = V_1 + V_2, S = S_1 + S_2$ are the total volume of two phases and the total entropy correspondingly. Thus, from the condition (A.61) with respect to (A.72), we get the condition of phase equilibrium:

$$\varphi_1(p,T) = \varphi_2(p,T). \qquad (A.74)$$

With the equation (A.74), one may obtain the dependence of parameters at saturation—at phase equilibrium. Considering small deviations of pressure and temperature, we may similarly write

$$\varphi_1(p + dp, T + dT) = \varphi_2(p + dp, T + dT). \qquad (A.75)$$

The equation (A.75) means that Gibbs energies of two phases must be equal anyway, even at the new pressure and temperature: the condition (A.74) must be fulfilled at any parameters. Expanding (A.75), we have

$$\varphi_1(p,T) + \underbrace{\left(\frac{\partial \varphi_1}{\partial p}\right)}_{v_1} dp + \underbrace{\left(\frac{\partial \varphi_1}{\partial T}\right)}_{-s_1} dT = \varphi_2(p,T) + \underbrace{\left(\frac{\partial \varphi_2}{\partial p}\right)}_{v_2} dp + \underbrace{\left(\frac{\partial \varphi_2}{\partial T}\right)}_{-s_2} dT. \qquad (A.76)$$

In view of (A.74), we get from (A.76): integrating equality

$$Tds = dh, \qquad (A.77)$$

we get

$$\frac{dp}{dT} = \frac{s_2 - s_1}{v_2 - v_1}. \qquad (A.78)$$

From (A.6), for $p = const$, we get that during the phase transition, the difference in entropy can be defined as

$$s_2 - s_1 = \frac{h_2 - h_1}{T} = \frac{r}{T}, \qquad (A.79)$$

where $r = h_2 - h_1$ is the latent heat of phase transition—the difference of enthalpy of two phases during the phase transition.

Combining all the relations, we get the Clapeyron–Clausius equation:

$$\frac{dp}{dT} = \frac{r}{T(v_2 - v_1)}. \qquad (A.80)$$

Equation (A.80) shows how the pressure of the phase transition is connected to the temperature of the phase transition.

Appendix B
Hydrodynamics in brief

- -

1

Hydrodynamics is based on the laws of conservation: conservation of mass and conservation of momentum. Each of these equations can be written in the form

$$\frac{\partial A}{\partial t} + div\vec{J} = \dot{S}_A .$$ (B.1)

Here A is the volume density of some quantity, $\vec{J}$ is the flux of the corresponding property, $\dot{S}_A$ is the source of A—the rate of production of this quantity. The equation (B.1) reflects the conservation law of the quantity A: according to it, at the given point of a medium, A may increase or decrease due to fluxes or because of some source.

Note that usually hydrodynamics operates with this approach: we consider a fixed point of space and follow the parameters of the fluid at this point—not follow the moving fluid particles.

The law of conservation of mass follows from (B.1) for $A = \rho$, $\dot{S}_A = 0$, $\vec{J} = \rho\vec{u}$, where $\vec{u}$ is the velocity of the fluid at the given point. We get:

$$\frac{\partial \rho}{\partial t} + div\left(\rho\vec{u}\right) = 0.$$ (B.2)

If density is constant, that means the fluid is incompressible, then we see that

$$div\vec{u} = 0 .$$ (B.3)

The equation (B.2) can be rewritten in a more open form for a 3D-case as

$$\frac{\partial \rho}{\partial t} + \frac{\partial\left(\rho u_1\right)}{\partial x_1} + \frac{\partial\left(\rho u_2\right)}{\partial x_2} + \frac{\partial\left(\rho u_3\right)}{\partial x_3} = 0.$$ (B.4)

In a shortened form, this equation can be expressed as

$$\frac{\partial \rho}{\partial t} + \frac{\partial\left(\rho u_i\right)}{\partial x_i} = 0.$$ (B.5)

Equations (B.4) and (B.5) mean the same; the last equation contains the so-called 'silent summation': we take the sum on the repeating index.

The law of conservation for a component of the momentum $A = \rho u_i$ can be written with a stress momentum flux density tensor Π_{ij} and the source which describes mass forces (for certainty, we will consider only gravity) $\dot{S}_A = \rho g_i$:

$$\frac{\partial\left(\rho u_i\right)}{\partial t} + \frac{\partial \Pi_{ij}}{\partial x_j} = \rho g_i.$$ (B.6)

Tensor Π_{ij} contains two terms: the velocity part $\rho u_i u_j$ and the stress tensor σ_{ij} corresponding to pressure p and viscous forces:

$$\Pi_{ij} = \rho u_i u_j - \sigma_{ij}, \tag{B.7}$$

where the stress tensor is

$$\sigma_{ij} = -p\delta_{ij} + \tau_{ij}. \tag{B.8}$$

Here δ_{ij} is the Kronecker symbol (gives 1 for $i = j$ and 0 otherwise).
The viscous stress tensor in its common form can be represented as

$$\tau_{ij} = \mu\left(\frac{\partial u_i}{\partial x_j} + \frac{\partial u_j}{\partial x_i} - \frac{2}{3}\delta_{ij}\frac{\partial u_k}{\partial x_k}\right) + \zeta\delta_{ij}\frac{\partial u_k}{\partial x_k}. \tag{B.9}$$

Here μ is the dynamic coefficient of viscosity and ζ is the second (or the volume) viscosity. The second viscosity is a subtle parameter; fortunately, for an incompressible fluid it does not matter since for such a fluid $\partial u_k/\partial x_k = 0$ because of (B.3).

Finally, we may represent the equation for the momentum conservation—it is termed as the Navier–Stokes equation—with respect to (B.5) in the form

$$\rho\frac{\partial u_i}{\partial t} + \rho u_j\frac{\partial u_i}{\partial x_j} = -\frac{\partial p}{\partial x_i} + \frac{\partial}{\partial x_j}\left[\mu\left(\frac{\partial u_i}{\partial x_j} + \frac{\partial u_j}{\partial x_i} - \frac{2}{3}\delta_{ij}\frac{\partial u_k}{\partial x_k}\right)\right] + \frac{\partial}{\partial x_i}\left(\zeta\frac{\partial u_k}{\partial x_k}\right) + \rho g_i. \tag{B.10}$$

2

Usually, a hydrodynamic problem is reduced to finding the velocity field in the given configuration. To do this, one has to solve the Navier–Stokes equation (B.10). The obtained velocity field can be represented with streamlines—lines in a space (or on a plane) with tangents corresponding to the velocity of the fluid at the given point.

The equation (B.10) demands boundary conditions. As we see from (B.10), if viscosity is equal to zero, then we have the first-order differential equation for velocity

$$\frac{\partial u_i}{\partial t} + u_j\frac{\partial u_i}{\partial x_j} = -\frac{1}{\rho}\frac{\partial p}{\partial x_i} + g_i. \tag{B.11}$$

For this equation, the only boundary condition for velocity is set: impermeability of a solid boundary: the normal (orthogonal) component of a fluid at a solid boundary is zero, that is

$$u_n = 0. \tag{B.12}$$

For a viscous fluid ($\mu \neq 0$), i.e., for the full equation (B.10), an additional boundary condition appears: the attachment condition;according to it, the tangential projection of velocity at the solid boundary is also zero:

$$u_\tau = 0. \tag{B.13}$$

3

Navier–Stokes equations are nonlinear partial differential equations (nonlinearity appears due to the term $u\partial u/\partial x$). They are very difficult to solve; solutions are possible in rare particular cases.

One of such cases is a laminar flow in a pipe of radius R. Laminar flow is a regular flow at comparatively low velocity, which can be described by the Navier–Stokes equation neglecting the nonlinear term in the left-hand side. Considering a stationary flow, we also set to zero all of the left side of the equation (B.10).

Then, we will have to deal with an incompressible fluid, that is, in condition (B.3) all the divergences in (B.10) vanish. By also neglecting the mass forces, we have

$$-\frac{\partial p}{\partial x_i} + \frac{\partial}{\partial x_j}\left(\mu\frac{\partial u_i}{\partial x_j}\right) = 0. \tag{B.14}$$

Now, we will consider a stable flow in a pipe where the only component of velocity—along a pipe—has meaning; we will denote it simply as u. This velocity may depend only on the radius of the given point; thus, in this case, the Laplacian of velocity can be represented in cylindrical coordinates. Then, if the viscosity of the fluid μ is constant, we have

$$\mu\frac{1}{r}\frac{d}{dr}\left(r\frac{du}{dr}\right) = \frac{dp}{dx}. \tag{B.15}$$

For a stable flow, we may reasonably assume that the pressure gradient is constant, which we denote as $\Delta p/L$; thus, we have, integrating (B.12) twice and demanding the restricted value of velocity at the pipe axis:

$$u = C + \frac{\Delta p r^2}{4\mu L}. \tag{B.16}$$

At the wall of a pipe, at $r = R$, as it follows from (B.13), there must be $u = 0$. This circumstance allows us to define the constant C in (B.16), and to get

$$u(r) = \underbrace{\frac{\Delta p R^2}{4\mu L}}_{u_{max}}\left(1 - \frac{r^2}{R^2}\right). \tag{B.17}$$

This relation is known as the Poiseuille parabola.

The average velocity in the given cross-section can be found through the balance for the mass discharge G

$$G = \rho\bar{u}\pi R^2 = \rho\int u dS = \rho\int_0^R 2\pi r u(r) dr, \tag{B.18}$$

$$\bar{u} = \frac{2\pi\int_0^R u(r) r dr}{\pi R^2} = \frac{u_{max}}{2}.$$

Consequently, we may write for the mass discharge

$$G = \frac{\rho u_{max}\pi R^2}{2} = \frac{\rho\Delta p\pi R^4}{8\mu L}. \tag{B.19}$$

Note that we may use the kinematic coefficient of viscosity $v = \mu/\rho$ there.

Pressure losses in a pipe are caused by friction at the wall of a pipe. We may describe this process with the corresponding forces:

$$\Delta\vec{F}_{pressure} = -\vec{F}_{friction}. \tag{B.20}$$

For the first term in (B.20) we have, obviously, the product of the differential pressure on the area of cross-section:

$$\Delta F_{pressure} = \Delta p \pi R^2. \tag{B.21}$$

The force of friction can be formulated with the corresponding component of the stress tensor

$$F_{friction} = S_w \tau_R = S_w \mu \frac{\partial u}{\partial r}\Big|_{r=R}, \tag{B.22}$$

where the area of contact with the wall at the length L is $S_w = 2\pi RL$; therefore,

$$\Delta p = \frac{2L\tau_R}{R}. \tag{B.23}$$

Using the correlation (B.17) for the laminar distribution of velocity, we have

$$\frac{\partial u}{\partial r}\Big|_{r=R} = \frac{4\bar{u}}{R}, \tag{B.24}$$

(the negative sign is already accounted for in (B.22)), and pressure losses are

$$\Delta p = \frac{8L\mu\bar{u}}{R}. \tag{B.25}$$

Using representation in a form

$$\Delta p = \xi \frac{l}{d} \frac{\rho \bar{u}^2}{2} \tag{B.26}$$

with the diameter $d = 2R$ and the friction factor ξ, comparing with (B.25), we have for ξ

$$\xi = \frac{64}{Re}, \tag{B.27}$$

with the Reynolds number

$$Re = \frac{\bar{u}d}{\nu}. \tag{B.28}$$

Note that this correlation can be obtained with the Poiseuille profile slightly more directly; here we presented the whole circle to explain the physics of the process in more details.

A turbulent flow is an unsteady, chaotic flow of a fluid. In a pipe, a laminar flow transitions into the turbulent one at the critical Reynolds number

$$Re_{cr} \approx 2200. \tag{B.29}$$

The constant in (B.29) is an approximation. Actually, the laminar–turbulent transition can be 'prolonged' for much more values of Re.

The physics of turbulence is more complicated, and, generally, we still do not have a good mathematical/physical description of it. Turbulence is a very special topic not only for hydrodynamics, but for the entire modern physics. Discussion of it may be of any length, so we prefer the shortest one and move forward to the clearer matters.

4

The Euler equation for a non-viscous fluid (B.11) can be written in a vector form as

$$\frac{\partial \vec{u}}{\partial t} + \vec{u}\nabla\vec{u} = -\frac{\nabla p}{\rho} + \vec{g}. \tag{B.30}$$

Here ∇ is the gradient of the corresponding function.
Then, using the known correlation

$$\vec{u}\nabla\vec{u} = \frac{\nabla u^2}{2} - \vec{u} \times rot\vec{u}. \tag{B.31}$$

the equation (B.30) can be represented in the form

$$\frac{\partial \vec{u}}{\partial t} + \frac{\nabla \vec{u}^2}{2} - \vec{u} \times rot\vec{u} = -\frac{\nabla p}{\rho} + \vec{g}. \tag{B.32}$$

Further, we will consider a steady flow $\partial\vec{u}/\partial t = 0$. Let us take the projection of (B.32) on the direction of $\vec{u}$; in other words, we will consider the projection of (B.32) on a streamline l. For this projection, we have $(\vec{u} \times rot\vec{u})_l = 0$. Consequently, for this projection

$$\nabla_l \left(\frac{u^2}{2} + \int \frac{dp}{\rho} + gz \right) = 0, \tag{B.33}$$

where z is the vertical coordinate (opposite to the direction of the acceleration of gravity $\vec{g}$). Here we take into account a possible compressibility of the liquid.
Thus, along a streamline the following equation must be true:

$$\frac{u^2}{2} + \int \frac{dp}{\rho} + gz = const. \tag{B.34}$$

This equation is called the Bernoulli condition; this is one of the most important points in hydrodynamics for cavitation problems.

5

Since Navier–Stokes equations are too complicated for direct solutions, other variables are used to describe a fluid flow.
The first variable that can be used is vorticity—the curl of velocity:

$$\vec{\omega} = rot\vec{u}. \tag{B.35}$$

This function can be useful for clear vortex structures, when rotation plays a crucial role for the flow. In some models, vorticity can be defined for certain types of flow directly, being the starting point of consideration of a flow of such type.
Another common representation of the velocity field is the decomposition of vector $\vec{u}$ into two parts using two types of potentials:

$$\vec{u} = \nabla\varphi + rot\vec{\Psi}. \tag{B.36}$$

Here φ is the scalar potential, and $\vec{\Psi}$ is the vector potential. The joint use of these potentials is possible too, but they are especially convenient in particular cases.

Application of potential $\vec{\Psi}$ is possible for a flat flow (flow on a plane, in two dimensions) of an incompressible fluid. In this case, we see that the condition (B.3) is fulfilled automatically, since

$$div\left(rot\vec{\Psi}\right) = 0 \tag{B.37}$$

for any vector $\vec{\Psi}$.

Then, for a flat flow, only one component of the vector $\vec{\Psi}$ can be considered: for the plane (x, y), it is the component Ψ_z. Thus, components of velocity are

$$u_x = \frac{\partial \Psi_z}{\partial y}, \; u_y = -\frac{\partial \Psi_z}{\partial x}. \tag{B.38}$$

Such a simplified vector potential $\Psi \equiv \Psi_z$ is termed as stream function. A flat vortex is usually described as a pair of variables $\omega - \Psi_z$: vorticity—stream function.

The streamline is defined by conditions

$$\frac{dx}{u_x} = \frac{dy}{u_y}. \tag{B.39}$$

With (B.38), we see that the equation (B.39) gives

$$d\Psi_z = \frac{\partial \Psi_z}{\partial x}dx + \frac{\partial \Psi_z}{\partial y}dy = 0. \tag{B.40}$$

The potential φ can be used solely for an irrotational flow, i.e., for a flow with zero curl of velocity. Indeed, for this flow—also called a potential flow—we have directly

$$rot\left(\nabla\varphi\right) = 0 \tag{B.41}$$

for any function φ.

For a potential flow, we have another form of the Bernoulli equation. If $rot\vec{u} \equiv 0$ for such a flow, then we may consider the non-stationary Euler equation (B.32)

$$\frac{\partial \vec{u}}{\partial t} + \frac{\nabla \vec{u}^2}{2} = -\frac{\nabla p}{\rho} + \vec{g} \tag{B.42}$$

with $\vec{u} = \nabla\varphi$. Thus, from (B.40) we obtain another integral which suits not only the streamline, but the whole space:

$$\frac{\partial \varphi}{\partial t} + \frac{u^2}{2} + \int \frac{dp}{\rho} + gz = const. \tag{B.43}$$

This condition is named after Lagrange–Cauchy.

Index
